Lecture Notes in Computer Science   14766

Founding Editors

Gerhard Goos
Juris Hartmanis

The series Lecture Notes in Computer Science (LNCS), including its subseries Lecture Notes in Artificial Intelligence (LNAI) and Lecture Notes in Bioinformatics (LNBI), has established itself as a medium for the publication of new developments in computer science and information technology research, teaching, and education.

LNCS enjoys close cooperation with the computer science R & D community, the series counts many renowned academics among its volume editors and paper authors, and collaborates with prestigious societies. Its mission is to serve this international community by providing an invaluable service, mainly focused on the publication of conference and workshop proceedings and postproceedings. LNCS commenced publication in 1973.

Anton Eremeev · Michael Khachay ·
Yury Kochetov · Vladimir Mazalov ·
Panos Pardalos
Editors

Mathematical Optimization Theory and Operations Research

23rd International Conference, MOTOR 2024
Omsk, Russia, June 30 – July 6, 2024
Proceedings

 Springer

Editors
Anton Eremeev ⓘD
Dostoevsky Omsk State University
Omsk, Russia

Yury Kochetov ⓘD
Sobolev Institute of Mathematics
Novosibirsk, Russia

Panos Pardalos
University of Florida
Gainesville, FL, USA

Michael Khachay ⓘD
Krasovsky Institute of Mathematics
and Mechanics
Ekaterinburg, Russia

Vladimir Mazalov ⓘD
Karelia Research Center of RAS
Petrozavodsk, Russia

ISSN 0302-9743 ISSN 1611-3349 (electronic)
Lecture Notes in Computer Science
ISBN 978-3-031-62791-0 ISBN 978-3-031-62792-7 (eBook)
https://doi.org/10.1007/978-3-031-62792-7

This Springer imprint is published by the registered company Springer Nature Switzerland AG
The registered company address is: Gewerbestrasse 11, 6330 Cham, Switzerland

If disposing of this product, please recycle the paper.

Preface

This volume contains the refereed proceedings of the 23rd International Conference on Mathematical Optimization Theory and Operations Research (MOTOR 2024)[1] held from June 30 to July 06, 2024, in Omsk, Russia. The MOTOR conference joined several well-known conferences on mathematical programming, optimization, and operations research which had been held in the Urals, Siberia and the Russian Far East for a number of decades. Previous editions of MOTOR were held in Yekaterinburg, Novosibirsk, Irkutsk and Petrozavodsk.

As per tradition, the main conference scope included, but was not limited to, mathematical programming, discrete optimization, computational complexity and approximation algorithms, metaheuristics and local search methods, game theory, optimization in machine learning and data analysis, parallel computations, applications in operations research and mathematical economics. In 2024, the Russian Academy of Sciences celebrates its 300th anniversary. The MOTOR 2024 conference was dedicated to this anniversary as well.

In response to the call for papers, MOTOR 2024 received 170 submissions. Out of 79 full papers considered for review (91 abstracts and short communications were excluded for formal reasons), only 30 papers were selected by the Program Committee (PC) for publication in this volume. Most of the submissions was single-blind reviewed by at least three PC members or invited reviewers, experts in their fields, who evaluated the papers and provided detailed and helpful comments. In addition, the PC recommended 27 more papers for inclusion in the second volume of proceedings in the CCIS series after their presentation and discussion during the conference and subsequent revision with respect to the reviewers' comments.

The conference featured nine invited lectures:

- Xujin Chen (Institute of Applied Mathematics, Academy of Mathematics and Systems Science, Chinese Academy of Sciences, Beijing, China) *Undirected Networks Immune to the Informational Braess's Paradox*
- Enkhbat Rentsen (Business School of National University of Mongolia, Ulaanbaatar, Mongolia) *Recent Advances in Minimax Problem and Bimatrix Games*
- Nikolai N. Guschinsky (United Institute of Informatics Problems, National Academy of Sciences of Belarus, Minsk, Belarus) *Models and Methods for Optimization of Electric Public Transport Systems*
- Roland Hildebrand (Moscow Institute of Physics and Technology, Dolgoprudny, Russia) *Tuning Methods for Minimization of Self-Concordant Functions with Optimal Control*
- Elena V. Konstantinova (China Three Gorges University, Yichang, China; Sobolev Institute of Mathematics, Novosibirsk, Russia) *Recent Progress in Domination Theory*
- Panos M. Pardalos (University of Florida, Gainesville, USA; HSE, Nizhny Novgorod, Russia) *AI and Optimization for a Sustainable Future*

---

[1] https://motor24.oscsbras.ru/.

- Alexander A. Shananin, (Moscow Institute of Physics and Technology, Moscow, Russia) *Mathematical Modeling of the Interest Rates Formation on Consumer Loans in Russia*
- Zaiwen Wen (Peking University, China) *Exploring the Learning-Based Optimization Algorithms*
- Wenwu Yu (Southeast University, China) *"Distributed Optimization +" in Networks: A New Framework.*

We thank the authors for their submissions, and we are grateful to all members of the Program Committee, and all the external reviewers for their efforts in providing exhaustive reviews. We thank our sponsors and partners: The Russian Operational Research Society, Dostoevsky Omsk State University, Sobolev Institute of Mathematics SB RAS, Mathematical Center in Akademgorodok (Novosibirsk), Novosibirsk State University, N. N. Krasovskii Institute of Mathematics and Mechanics UB RAS, the Higher School of Economics (Nizhny Novgorod), Siberian Branch of Russian Academy of Sciences, and BIA Technologies. We are grateful to the colleagues from the Springer LNCS and CCIS editorial boards for their kind and helpful support.

July 2024

Anton Eremeev
Michael Khachay
Yury Kochetov
Vladimir Mazalov
Panos Pardalos

Organization

Program Committee Chairs

Anton Eremeev	Dostoevsky Omsk State University, Russia
Michael Khachay	Krasovsky Institute of Mathematics and Mechanics, Russia
Yury Kochetov	Sobolev Institute of Mathematics, Russia
Vladimir Mazalov	Institute of Applied Mathematical Research, Russia
Panos Pardalos	University of Florida, USA

Program Committee Members

Kamil Aida-zade	Institute of Control Systems, Baku, Azerbaijan
Adil Bagirov	Federation University Australia, Australia
Olga Battaïa	KEDGE Business School, Bordeaux, France
Rene van Bevern	Novosibirsk State University, Russia
Vladimir Beresnev	Sobolev Institute of Mathematics, Russia
George Bolotashvili	Georgian Technical University, Georgia
Vladimir Bushenkov	University of Evora, Portugal
Maxim Buzdalov	Aberystwyth University, UK
Igor Bykadorov	Sobolev Institute of Mathematics, Russia
Xujin Chen	Academy of Mathematics and Systems Science, Chinese Academy of Sciences, China
Yukun Cheng	Jiangnan University, China
Tatiana Davidović	Mathematical Institute of the Serbian Academy of Sciences and Arts, Serbia
Alexander Dolgui	IMT Atlantique, France
Adil Erzin	Sobolev Institute of Mathematics, Russia
Hongwei Gao	Qingdao University, China
Alexander Gasnikov	Innopolis University, Russia
Edward Gimadi	Sobolev Institute of Mathematics, Russia
Aleksander Gornov	Matrosov Institute for System Dynamics and Control Theory, Russia
Evgeni Gurevsky	University of Nantes, France
Milojica Jaćimović	University of Montenegro, Montenegro
Valeriy Kalyagin	HSE University, Nizhny Novgorod, Russia

Vadim Kartak	Ufa State Aviation Technical University, Russia
Alexander Kazakov	Matrosov Institute for System Dynamics and Control Theory, Russia
Oleg Khamisov	Melentiev Energy Systems Institute, Russia
Andrey Kibzun	Moscow Aviation Institute, Russia
Lingchen Kong	Beijing Jiaotong University, China
Igor Konnov	Kazan Federal University, Russia
Alexander Kononov	Sobolev Institute of Mathematics, Russia
Mikhail Kovalyov	United Institute of Informatics Problems, National Academy of Sciences of Belarus, Belarus
Dmitri Kvasov	University of Calabria, Italy
Alexander Lazarev	Institute of Control Sciences, RAS, Russia
Bertrand Lin	National Yang Ming Chiao Tung University, Taiwan
Vittorio Maniezzo	University of Bologna, Italy
Evgeni Nurminski	Far Eastern Federal University, Russia
Leon Petrosyan	Saint Petersburg State University, Russia
Leonid Popov	Krasovskii Institute of Mathematics and Mechanics, Russia
Mikhail Posypkin	Federal Research Center on Computer Science and Control, Russia
Artem Pyatkin	Sobolev Institute of Mathematics, Russia
Soumyendu Raha	Indian Institute of Science, India
Anna Rettieva	Institute of Applied Mathematical Research, Russia
Eugene Semenkin	Siberian State Aerospace University, Russia
Yaroslav Sergeyev	Università della Calabria, Italy
Alexander Shananin	Moscow Institute of Physics and Technology, Russia
Angelo Sifaleras	University of Macedonia, Greece
Andrey Sleptchenko	Khalifa University, UAE
Alexander Strekalovsky	Matrosov Institute for System Dynamics and Control Theory, Russia
Raca Todosijević	Université Polytechnique Hauts-de-France, France
Ider Tseveendorj	Universite de Versailles-Saint Quentin, France
Yury Tsoy	Ailys, Republic of Korea
Igor Vasilyev	Matrosov Institute for System Dynamics and Control Theory, Russia
Alexander Vasin	Lomonosov Moscow State University, Russia
Xin Yao	Southern University of Science and Technology, China

Additional Reviewers

Barkova, Maria
Borisovsky, Pavel
Davydov, Ivan
Deshpande, Neha
Gomoyunov, Mikhail
Il'ev, Victor
Ivanov, Sergey
Kazakovtsev, Lev
Khoroshilova, Elena
Kratica, Josef
Levanova, Tatiana
Neznakhina, Katherine
Novikova, Natalia
Panin, Artem

Plyasunov, Alexander
Pinyagina, Olga
Platonov, Evgenii
Ratushnyi, Alexey
Servakh, Vladimir
Shahrtash, Mohammad
Simanchev, Ruslan
Slastnikov, Alexander
Sorokin, Stepan
Sukhoroslov, Oleg
Takhonov, Ivan
Yong, Xia
Zabudsky, Gennady

Industry Session Chair

Igor Vasilyev

Matrosov Institute for System Dynamics and Control Theory, Russia

Organizing Committee

Anton Eremeev (Chair)	Dostoevsky Omsk State University, Russia
Aleksander Remizov	Pushkin State Scientific Library, Russia
Aleksander Kabanov	Dostoevsky State University, Russia
Ruslan Simanchev	Dostoevsky State University, Russia
Victor Il'ev	Dostoevsky State University, Russia
Anna Romanova	Dostoevsky State University, Russia
Nina Kochetova	Sobolev Institute of Mathematics, Russia
Polina Kononova	Sobolev Institute of Mathematics, Russia
Timur Medvedev	Higher School of Economics, Russia
Gennady Zabudsky	Sobolev Institute of Mathematics, Russia
Pavel Borisovsky	Sobolev Institute of Mathematics, Russia
Svetlana Malakh	Sobolev Institute of Mathematics, Russia
Tatyana Levanova	Sobolev Institute of Mathematics, Russia
Vasily Lobanov	Sobolev Institute of Mathematics, Novosibirsk State University, Russia
Vladimir Servakh	Sobolev Institute of Mathematics, Russia
Lidia Zaozerskaya	Sobolev Institute of Mathematics, Russia

Julia Zakharova   Sobolev Institute of Mathematics, Russia

Vyacheslav Ustyugov   Sobolev Institute of Mathematics, Russia

Organizers

Sobolev Institute of Mathematics, Russia
Dostoevsky Omsk State University, Russia
Krasovsky Institute of Mathematics and Mechanics, Russia
Ural Mathematical Center, Russia
Novosibirsk State University, Russia
Mathematical Center in Akademgorodok, Russia
Higher School of Economics, Russia

Abstracts of Invited Talks

Undirected Networks Immune to the Informational Braess's Paradox

Xujin Chen ⓘD

Institute of Applied Mathematics, Beijing, China
xchen@amss.ac.cn

Abstract. The Informational Braess's paradox exposes a counterintuitive phenomenon that disclosing additional road segments to some selfish travelers results in increased travel times for these individuals. This paradox expands upon the classic Braess's Paradox by relaxing the assumption that all travelers possess identical and complete information about the network. In this presentation, we explore structual conditions that prevent the informational Braess paradox in undirected networks.

Keywords: undirected networks · Braess's Paradox · selfish strategy

Recent Advances in Minimax Problem and Bimatrix Games

Enkhbat Rentsen (iD)

Institute of Mathematics and Digital Technology, Ulaanbaatar, Mongolia
renkhbat46@yahoo.com

Abstract. In this paper, we first survey recent developments in game theory, and then focus on minimax problems and bimatrix games. New optimality conditions for maxmin and minimax problems have been considered. Recently introduced new equilibriums such as Anti-Berge and Anti-Nash equilibriums are discussed with connection to optimization problems. We examine bimatrix games from a view point of global optimization. Recent results in multi player polymatrix game have been given. The existence theorems of all equilibrium are provided. Also, a triple matrix game is introduced and existence of Nash equilibrium in the game has been given.

Keywords: optimization · game theory · optimality condition · maxmin problem · minimax problem · equilibriums · bimatrix game · payoff function · strategy

Mathematical Modeling of the Consumer Loan Market in Russia

Alexander Shananin(iD) and Nikolai Trusov(iD)

Moscow Institute of Physics and Technology, Russia
alexshan@yandex.ru

Abstract. The mathematical modeling of the economic behavior of the households is based on a Ramsey-type model. We introduce an optimal control problem that models the economic behavior of a representative household in an imperfect consumer loan market. The necessary optimality conditions are obtained in the form of Pontryagin's maximum principle in the Clarke form. A synthesis of optimal control can be constructed on an infinite time horizon. The synthesis includes special cases. We present a new model for the formation of interest rates on consumer loans. It is based on the analysis of the interests and logic of behavior of commercial banks that assess the risk of borrowers' default. According to the Feynman–Kac formula, the assessment is reduced to solving a boundary value problem for partial differential equations. It is possible to reduce the solution of the boundary value problem to the Cauchy problem for the heat equation with an external source and obtain a risk assessment in analytical form with a help of the Abel equation. The models of economic behavior of households in the consumer loan market and behavior of commercial banks are identified based on Russian statistics. (joint talk with N.V. Trusov)

Keywords: mathematical modeling · optimal control synthesis · consumer loan · Feynman–Kac formula · Ramsey model

Models and Methods for Optimization of Electric Public Transport Systems

Nikolai Guschinsky (ID)

United Institute of Informatics Problems, Minsk, Belarus
gyshin@newman.bas-net.by

Abstract. The transition towards sustainable cities becomes an increasingly pressing issue because of the growing awareness about the climate change. One of the critical transition directions is to reduce the greenhouse gas emissions generated by transportation. One of efficient answers to this challenge is the implementation of electric public transportation systems. In order to achieve the best impact, it is imperative to design the infrastructure and the network of the public transportation in an optimized way. The following problems concerning strategic, tactical and operational aspects of the electric bus planning process and scheduling are discussed: a) investment of electric bus fleet and charging infrastructure; b) design of charging infrastructure; c) the electric vehicle scheduling; d) the charging scheduling problem. The models and optimization methods for different charging technologies are considered:

- slow plug-in chargers installed at bus depots;
- fast plug-in or pantograph chargers installed at terminals of bus lines or at bus stops;
- overhead contact lines or inductive (wireless) chargers that are used to recharge buses during driving;
- battery swapping.

Keywords: network of the public transportation · electric bus scheduling · charging infrastructure

Tuning Methods for Minimization of Self-concordant Functions with Optimal Control

Roland Hildebrand ⓘ

Moscow Institute of Physics and Technology, Dolgoprudny, Russia
hildebra@mail.ru

Abstract. In the last decade the usage of optimization methods for performance estimation and tuning of other optimization methods has became fashionable. For instance, semi-definite programming allows to analyze the behaviour of accelerated gradient descent methods at minimizing functions of different regularity. Detecting the worst-case performance with respect to the minimized function and maximizing this performance with respect to the parameters of the method allows to obtain the best parameter set. Here we show how to perform a similar analysis of the Newton method when minimizing self- concordant functions. This task arises as a subproblem in more complex structured convex optimization problems such as semi-definite programming inside path-following methods. The appropriate framework for optimizing the parameters, in particular, the step length and the search direction, is optimal control theory. We present a general technique to use the Pontryagin maximum principle in the analysis of the Newton step on a self-concordant function and specialize it in several case studies.

Keywords: convex optimization · self-concordant function · semi-definite programming

Recent Progress in Domination Theory

Elena Konstantinova[1,2] (iD)

[1] China Three Gorges University, Yichang, China
e_konsta@math.nsc.ru
[2] Sobolev Institute of Mathematics, Novosibirsk, Russia

Abstract. In this talk, some recent results in domination theory are presented. In 2020, T. W. Haynes et al. introduced coalitions and coalition partitions based on dominating sets in graphs as a graph-theoretical model to describe political coalitions. The authors have been studied the property of this concept and suggested a list of open problems. In particular, it was suggested to study connected coalitions based on connected dominating sets. In this talk we focus on studying connected coalitions and their partitions in graphs with emphasising to polynomial-time algorithms determining whether the connected coalition number of a graph G of order n is either n or $n - 1$. The talk is based on joint works with S. Alikhani, D. Bakhshesh, and H. Golmohammadi.

Keywords: domination theory · connected coalitions · dominating sets in graph

AI and Optimization for a Sustainable Future

Panos Pardalos[1,2] (iD)

[1] University of Florida, Gainesville, USA
p.m.pardalos@gmail.com
[2] HSE, Nizhny Novgorod, Russia

Abstract. Advances in AI tools are progressing rapidly and demonstrating the potential to transform our lives. The spectacular AI tools rely in part on their sophisticated mathematical underpinnings (e.g. optimization techniques and operations research tools), even though this crucial aspect is often downplayed.

In this lecture, we will discuss progress from our perspective in the field of AI and its applications in Energy systems and Sustainability.

Keywords: energy system · sustainability · artificial intelligence

Exploring the Learning-Based Optimization Algorithms

Zaiwen Wen [ID]

Peking University, China
wenzw@pku.edu.cn

Abstract. The recent revolutionary progress of artificial intelligence has brought significant challenges and opportunities to mathematical optimization. In this talk, we briefly discuss two examples on the integration of data, models, and algorithms for the development of optimization algorithms: ODE-based learning to optimize and learning-based optimization paradigms for solving integer programming. We will also report a few interesting perspectives on formalization and automated theorem proving, highlighting their potential impact and relevance in contemporary mathematical optimization.

Keywords: artificial intelligence · integer programming · automated theorem proving

"Distributed Optimization +" in Networks: A New Framework

Wenwu Yu (iD)

Southeast University, China
wwyu@seu.edu.cn

Abstract. Distributed optimization is solved by the mutual collaboration among a group of agents, which arises in various domains such as machine learning, resource allocation, location in sensor networks and so on. In this talk, we introduce two kinds of distributed optimization problems: (i) the agents share a common decision variable and local constraints; (ii) the agents have their individual decision variables but that are coupled by global constraints. This talk comprehensively introduces the origin of distributed optimization, classical works as well as recent advances. In addition, based on reinforcement learning, shortest path planning, and mixed integer programming, we build the distributed optimization framework of distributed optimization, and also discuss their applications. Finally, we make a summary with future works for distributed optimization.

Keywords: distributed optimization · group of agents · reinforcement learning

Abstracts of Tutorials

Black Box Optimization for Business Applications

Yury Kochetov ⓘ

Sobolev Institute of Mathematics, Novosibirsk, Russia
jkochet@math.nsc.ru

Abstract. The black-box optimization models are characterized by lack of analytical forms for the constraints and the objectives of the problem. In the black-box methods, we need efficiently integrate known analytical part with explicitly unknown correlations obtained from the business simulation models. The direct use of classical global optimization methods is prohibitive due to the lack of exact mathematical expressions. We cannot calculate the derivatives or sub-gradients. Moreover, the computational cost is high due to the simulations. Hence, we have to use the problem specific methods to optimize such black-box systems efficiently. Important applications stem from various disciplines: multi-echelon inventory systems, chemical and mechanical engineering, financial management, network topology design, and others. In this talk, we will discuss some directions in this area and theoretical bounds for global optimization methods. Successful cases for real-world applications will be presented.

Keywords: simulation model · mix-integer programming · multi-objective optimization

Application of GPU Computing to Solving Discrete Optimization Problems

Pavel Borisovsky ⓘD

Sobolev Institute of Mathematics, Omsk, Russia
borisovski@mail.ru

Abstract. Parallel computing on graphic processors (GPUs) is getting more and more popular. Since NVIDIA released the CUDA development tool, it has become convenient to use GPUs for general-purpose computing and not just for graphics display tasks. A feature of a GPU is the presence of a large (hundreds and thousands) number of cores, which allows to significantly speed up the calculation, but requires to design special parallel algorithms. While the development of traditional processors (CPUs) has recently slowed down, the characteristics of the GPU (number of cores, memory size, power consumption, and cost) are improving rapidly. In this tutorial, we will learn the basics of GPU computing in CUDA and OpenCL and briefly review the pros and cons of using GPUs in various discrete optimization algorithms.

Keywords: parallel computation · graphical processing · NP-hard problems

Contents

Game Theory

Operations Research

Invited Papers

Mathematical Modeling of the Interest Rates Formation on Consumer Loans in Russia

Alexander A. Shananin[1,2] and Nikolai V. Trusov[1]($\boxtimes$)

[1] Federal Research Center "Computer Science and Control" of RAS,
Vavilova Street 40, Moscow 119333, Russia
`alexshan@yandex.ru, trunick.10.96@gmail.com`
[2] Moscow Institute of Physics and Technology, National Research University,
9 Institutsky pereulok, Dolgoprudny 141701, Moscow Region, Russia

Abstract. We present the model for the formation of interest rates on consumer loans based on an analysis of commercial interests and the logic of behavior of commercial banks. From the position of the commercial banks, the borrowers' income is a stochastic process described by a geometric Brownian motion. The commercial banks assess the default risk of borrowers. The consumer loan demand is described with the help of representative rational consumer behavior based on Ramsey-type model. The historical data of the interest rates on loans are identified.

Keywords: Mathematical modeling · Ramsey model · Commercial banks · Optimal control synthesis · Identification problem

1 Introduction

In this article we describe the model for the formation of interest rates on consumer loans based on the interests of the commercial banks introduced in [1]. The commercial banks set the interest rate on loans for borrowers that maximizes their profit, taking into account the possible default of the borrowers. The borrower's behavior is described by an optimal control problem of the rational representative economic behavior of the household in an imperfect lending and savings market. The concept of the rational representative economic household arises to F. Ramsey [2]. The household maximizes the discounted consumption with a constant risk aversion, managing the dynamics of its consumptions depending on the current parameters of the economic situation and the behavioral characteristics of the household itself. Analyzing the optimal control problem, we identify four types of borrowers corresponding to various social layers of the population. The Ramsey-type models in the form of optimal control problems have been studied, for example, in [3,4].

The study of the commercial banks behavior is motivated by the circumstances of the Russian economy. For more than ten years, the ratio of interest

---

Supported by RSF (grant No. 24-11-00329).

rates on consumer loans to deposits varied in the range from 2.5 to 3.5. By the end of 2023, the population's consumer loan debt exceeded 15.3 trillion rubles, which is about 10% of GDP. In [5] is shown that about half of borrowers have a high debt burden, and 20% have delays in servicing existing loans. This makes the problem of household debt on consumer loans relevant.

The article is organized as follows. In Sect. 2 we present an optimal control problem of a consumer behavior in an imperfect lending and savings market. Based on the results of [6] we present an optimal control synthesis describing the borrowers behavior in different social layers. The synthesis allows to determine an optimal control depending on the current value of the state variable and the economical conjuncture parameters. In Sect. 3 a model of the commercial banks is presented. From the position of the commercial banks, the borrowers' income is a stochastic process described by a geometric Brownian motion. The problem of maximizing their net percent value has an analytical solution. In Sect. 4 the numerical results of the identification of the volatility parameter of the borrowers' income are presented. Section 5 concludes.

2   The Household Behavior on the Consumer Loan Market

We assume that the dynamics of household liquid assets $M_0(t)$ is described by the ordinary differential equation (ODE)

$$\frac{dM_0(t)}{dt} = S_0(t) - p(t)C(t) + H_L(t) - H_D(t), \tag{1}$$

where $S_0(t)$ is the household income at the time t, which is assumed to grow at the rate $\gamma \in \mathcal{R}$, i.e. $S_0(t) = Se^{\gamma t}$, $S > 0$; $C(t)$ is the household consumption with the consumer price index $p(t)$ that grows at the rate $j \in \mathcal{R}$, i.e. $p(t) = p_0 e^{jt}$; $H_L(t) \in \mathcal{R}$ denotes the borrowings on consumer loans, $H_D(t) \in \mathcal{R}$ denotes the investments in deposits. The positive values of $H_L(t)$ determine that the household borrows the consumer credit at the rate r_L, and the negative values of $H_L(t)$ determine that the household pays off its debt. Same, the positive values of $H_D(t)$ determine that the household saves money in the form of deposits at the rate r_D, and the negative values of $H_D(t)$ determine that the household withdraws money from depository accounts. Denote by $L(t) \geq 0$ the consumer loan debt of the household, and by $D(t) \geq 0$ its depository account. They satisfy the following ODEs

$$\frac{dL}{dt} = H_L(t) + r_L L(t), \tag{2}$$

$$\frac{dD}{dt} = H_D(t) + r_D D(t). \tag{3}$$

The absence of arbitration assumes that the interest rate on loans is higher than the interest rate on deposits, i.e. $r_L > r_D > 0$. The Fisher's equation of exchange relates the liquid assets and demand accounts $M_0(t)$ to the household consumptions $C(t)$ via the coefficient $\theta > 0$: $M_0(t) = \theta p(t)C(t)$, where $\frac{1}{\theta}$ determine the money velocity of the household. Denote by $X(t)$ the financial welfare of the household. It determines as

$$X(t) = M_0(t) + D(t) - L(t). \tag{4}$$

The concept of rational behavior assumes that the household doesn't borrow consumer credits and doesn't save money in the form of deposits at the same time. Thus, we define the consumer loan debt and the deposit account as

$$L(t) = (M(t) - x(t))_+,$$

$$D(t) = (x(t) - M(t))_+,$$

where $(a)_+ = \max\{a, 0\}, \forall a \in \mathcal{R}$. Differentiating the financial welfare (4) by virtue of the Eqs. (1)–(3), we obtain the differential equation on $X(t)$:

$$\frac{dX}{dt} = S_0 - \frac{1}{\theta}M - r_L(M - x)_+ + r_D(x - M)_+, \quad X(0) = X_0. \tag{5}$$

The household seeks to maximize the discounted consumption with a discount factor $\delta_0 > 0$ and a risk aversion parameter $\rho > 0$ on the time interval $[0, T]$, choosing the current consumption

$$\int_0^T \frac{(C(t))^{1-\rho}}{1-\rho} e^{-\delta_0 t} dt \to \max_{C \geq 0}.$$

The motivation to consider the positive real parameter ρ is based on the materials published in [7]. The authors identified that the households in Russia have a risk aversion belonging to the interval $(4, 6)$.

We say that the financial welfare $X(t)$ is liquid if there exists such a control $C(t)$ that ensures $X(T) \geq 0$. In other words, by the final moment of time the household has to pay off its credits. Let us introduce the following substitutions. Denote by $x(t) = X(t)e^{-\gamma t}$, $M(t) = M_0(t)e^{-\gamma t}$, $\delta = \delta_0 + (1 - \rho)j$. Suppose that $\delta > ((1 - \rho)r_L)_+, r_L > r_D > (\gamma)_+$. The optimal control problem at a finite time horizon poses as

$$\int_0^T \frac{M^{1-\rho}}{1-\rho} e^{-(\delta - (1-\rho)\gamma)t} dt \to \max_M, \tag{6}$$

$$\frac{dx}{dt} = S - \gamma x - \frac{1}{\theta}M - r_L(M - x)_+ + r_D(x - M)_+, \tag{7}$$

$$x(0) = x_0, \quad x(T) \geq 0, \tag{8}$$

$$M(t) \geq 0. \tag{9}$$

Remark 1. We suppose that the function $M(t) \in L_1[0, T]$, $x(t) \in AC[0, T]$. To ensure $x(T) \geq 0$, it is necessary that inequality $x > -\frac{S}{r_L - \gamma}$ completes (see [8]). If this inequality incompletes, then the household doesn't have an opportunity to pay off its credit debts, and the financial pyramid arises.

Remark 2. In case $\rho \to 1$ the functional (6) should be considered as

$$\int_0^T \frac{M^{1-\rho} - 1}{1 - \rho} e^{-(\delta - (1-\rho)\gamma)t} dt \to \max_M.$$

Thus, the integrand $\lim_{\rho \to 1} \frac{M^{1-\rho} - 1}{1 - \rho} = \ln M$.

Theorem 1. *Suppose that* $S + (r_L - \gamma)x_0 > 0$, $T > \left(\frac{1}{r_L} \ln \left(\frac{S}{S + (r_L - \gamma)x_0} \right) \right)_+$. *If the optimal control problem has a concave functional, then the optimal control problem (6)–(9) has a solution.*

The proof of Theorem 1 for the case $\rho \in (0, 1)$ can be found in [8]. The more general case $\rho > 0$ does not violate the structure of the proof. Due to the fulfillment of the condition $x(T) \geq 0$, the set of admissible controls is limited in the space L_1. According to Komlos theorem (see [9]), from a norm-bounded sequence of functions $M_i(t)$, $i = 1, 2, \ldots$ in the space L_1, we select such subsequence $M_{i_n}(t)$, $n = 1, 2, \ldots$, whose Cesaro means converge almost everywhere on the interval $[0, T]$ to the function $M_{opt}(t)$. Since the function $\frac{M^{1-\rho}}{1-\rho}$ is concave for $\rho > 0$ (for $\rho = 1$ the integrand of the fractional power function becomes logarithmic, which is also concave) on the set $M \geq 0$, then the limit function $M_{opt}(t)$ will be the maximum of the functional (6).

Denote by $y(\bar{x}, \beta)$ the solution of the equation

$$\left[\frac{1 + \theta r_L}{1 + \theta r_D} \right]^{\frac{\beta}{\rho}} \cdot \frac{y(\bar{x}, \beta)}{\bar{x}} = \left[\left(y(\bar{x}, \beta) - \frac{S\theta}{1 + \gamma\theta} \right) \left(\bar{x} - \frac{S\theta}{1 + \gamma\theta} \right)^{-1} \right]^{\frac{\theta[\delta + \rho\gamma - r_D]}{(1 + \gamma\theta)\rho}}, \tag{10}$$

and by $M_{r_L}(\bar{x}, r_L)$ the solution of the equation

$$x + \frac{S}{r_L - \gamma} = \frac{(1 + \theta r_L)\rho}{\theta(\delta - (1 - \rho)r_L)} M_{r_L}(\bar{x}, r_L) +$$

$$+ \left[\frac{\bar{x}}{M_{r_L}(\bar{x}, r)} \right]^{\frac{\rho(r_L - \gamma)}{\delta + \rho\gamma - r_L}} \cdot \left[\frac{S}{r_L - \gamma} - \frac{r_L - \delta + \frac{\rho}{\theta}}{\delta - (1 - \rho)r_L} \bar{x} \right]. \tag{11}$$

For the optimal control problem (6)–(9) we can construct the synthesis on the infinite time horizon. The synthesis allows us to determine an optimal control depending on the current value of the state variable x and the conjuncture parameters of the mathematical model (6)–(9). The optimal control synthesis depends on the relations between the quantities r_L, r_D, $\delta - \frac{\rho}{\theta}$, $\delta + \rho\gamma$ and defines the social layer of the household. We investigate the borrowers behavior on consumer loan market and present their the optimal control synthesis in the next theorem.

Theorem 2. *The household borrows consumer credits in the following Types:*

Type 1. *If* $r_L < \delta - \frac{\rho}{\theta}$, *then an optimal control synthesis defines as*

$$M(x; S, r_L, \theta, \rho, \gamma, \delta) = \frac{\theta \left(\delta - (1-\rho)r_L\right)}{\rho \left(1 + \theta r_L\right)} \left(x + \frac{S}{r_L - \gamma}\right).$$

Type 2. *If* $\delta - \frac{\rho}{\theta} < r_L < \delta + \rho\gamma$, *and the financial welfare of the household satisfies the inequality* $x < \frac{S(\delta - (1-\rho)r_L)}{(r_L - \gamma)(r_L - \delta + \frac{\rho}{\theta})}$, *then an optimal control synthesis defines as*

$$M(x; S, r_L, \theta, \rho, \gamma, \delta) = \frac{\theta \left(\delta - (1-\rho)r_L\right)}{\rho \left(1 + \theta r_L\right)} \left(x + \frac{S}{r_L - \gamma}\right).$$

Otherwise, if the financial welfare $x \geq \frac{S(\delta - (1-\rho)r_L)}{(r_L - \gamma)(r_L - \delta + \frac{\rho}{\theta})}$, *the household does not borrow consumer credit.*

Type 3. *If* $\max\left\{r_D, \delta - \frac{\rho}{\theta}\right\} < \delta + \rho\gamma < r_L$ *or* $\max\left\{r_D, \delta + \rho\gamma\right\} < \delta - \frac{\rho}{\theta} < r_L$, *and the financial welfare of the household satisfies the inequality* $x < x_{p_1}$, *then an optimal control synthesis defines as*

$$M(x; S, r_L, \theta, \rho, \gamma, \delta) = M_r\left(x_{p_1}, r_L\right),$$

where $x_{p_1} = \frac{\rho S}{r_L - \delta + \frac{\rho}{\theta}}$. *Otherwise, if the financial welfare* $x \geq x_{p_1}$, *the household does not borrow consumer credit.*

Type 4. *If* $\max\left\{\delta + \rho\gamma, \delta - \frac{\rho}{\theta}\right\} < r_D$, *and the financial welfare of the household satisfies the inequality* $x < x_{p_1}$, *then an optimal control synthesis defines as*

$$M(x; S, r_L, \theta, \rho, \gamma, \delta) = M_r\left(x_{p_1}, r_L\right),$$

where $x_{p_1} = \min\left\{\frac{\rho S}{r_L - \delta + \frac{\rho}{\theta}}, y\left(x_{p_2}, 1\right)\right\}$, $x_{p_2} = \frac{S}{r_D - \gamma} \cdot \frac{\delta - (1-\rho)r_D}{r_D - \delta + \frac{\rho}{\theta}}$. *Otherwise, if the financial welfare* $x \geq x_{p_1}$, *the household does not borrow consumer credit.*

The proof of the Theorem 2 can be found in [6]. It is based on the non-smooth analysis for the optimal control problem. Since the righthand side of (7) is non-smooth according to the fact that interest rates on loans and deposits are not equal, the Pontryagin maximum principle in the Clarke form [10,11] should be applied. With its help, three areas of a household economic behavior are identified on the plane of state and conjugate variables: borrowing, non-interaction with the banking system and savings in the form of deposits. The Pontryagin maximum principle in the Clarke form produces an additional control in the area of non-interaction with the banking system, that defines uniquely (see [6]). Tending the time horizon $T \to +\infty$, it is possible to obtain the optimal trajectory on the plane of state and conjugate variables that defines the synthesis of the optimal control problem. In [6] is shown that the Eqs. (10), (11) have unique solutions under assumptions of presented Types of the household.

3   The Commercial Bank Behavior on the Consumer Loan Market

To model the formation of an interest rate in the consumer loan market we use the concept of the Stackelberg equilibrium. Commercial banks set the interest rate on loans assessing the current demand that maximizes their profit. The demand of borrowers depends on the interest rate on loans. Denote by $\hat{H}_L(r_L)$ the demand on consumer loans depending on the current interest rate r_L. Suppose that the household borrow consumer credit for a period $[0, \hat{T}]$. Thus, the loan debt of the household is equal to $\hat{H}_L(r_L) e^{r_L \hat{T}}$ The basic form of payment on a consumer loan is an annuity payment A. It can be found from the equation

$$\hat{H}_L(r_L) e^{r_L \hat{T}} = A \int_0^{\hat{T}} e^{r_L(\hat{T}-t)} dt. \tag{12}$$

From the Eq. (12) it is easy to obtain the formula for the annuity payment of the household

$$A = \frac{r_L}{1 - e^{-r_L \hat{T}}} \hat{H}_L(r_L). \tag{13}$$

The income of the borrower is not stable for the commercial banks. We assume that the commercial banks model the borrower's income as a stochastic differential equation with a volatility parameter $\sigma \geq 0$

$$dS = S\gamma dt + \sigma S dW_t, \quad S(0) = S_0,$$

where γ is the income growth rate as was introduced in Sect. 2, W_t is a Wiener process, dW_t its stochastic differential.

The commercial banks put the risks of the borrower default in their profit. Denote by $\tau \in (0, \hat{T})$ the moment of time when the borrower's income does not cover the annuity payment and the minimal consumption level $\mu > 0$, i.e.

$$\tau = \min \left\{ \inf_{t \in (0, \hat{T})} [S(t) < A + \mu], \hat{T} \right\}.$$

Since the borrower's income is a random process, the quantity τ is a random stopping time.

Commercial banks set the interest rate on loans that maximizes the expectation of the net present value

$$NPV(r_L) = E_\tau \left[A \int_0^\tau e^{-\lambda t} dt - \hat{H}_L(r_L) \right] \to \max_{r_L}.$$

According to formula (13), the profit of the commercial banks describes by maximizing the functional

$$NPV(r_L) = E_\tau \left[\left(\frac{r_L(1 - e^{-\lambda \tau})}{\lambda(1 - e^{-r_L \hat{T}})} - 1 \right) \hat{H}_L(r_L) \right] \to \max_{r_L}, \tag{14}$$

where $\lambda > 0$ is the discount factor for cash flows, which is equal to the cost of funding expenses for commercial banks.

Define by $\alpha_u = \frac{\sigma}{\sqrt{2\gamma - \sigma^2}}$. The following theorem is true.

Theorem 3. *If* $\gamma > \frac{\sigma^2}{2}$, *then*

$$E_\tau e^{-\lambda \tau} = e^{-\lambda \hat{T}} + \frac{\alpha_u}{\pi} \int_0^{\sqrt{\lambda \hat{T}}} e^{-\left(\frac{\lambda}{4\alpha_u^2 \tau^2}\left(\ln\left(\frac{A+\mu}{S_0}\right)\right)^2 + \tau^2\right)} \left(\int_0^{\sqrt{\lambda \hat{T} - \tau^2}} e^{-y^2} dy\right) d\tau. \quad (15)$$

The proof of the Theorem 3 can be found in [1]. According to the Feynman–Kac formula, the risk of borrower default is reduced to a boundary value problem for a partial differential equation. It is possible to reduce the solution of the boundary value problem to the Cauchy problem for the heat equation with an external source, establishing a connection with the Abel equation, and obtain a risk assessment in analytical form.

Remark 3. In case when $\gamma \le \frac{\sigma^2}{2}$, it is unprofitable for a commercial bank to give a consumer loan.

The consumer loan demand $\hat{H}_L$ can be calculated from a model of rational consumer behavior presented in Sect. 2. Suppose that $\hat{H}_L(t) = H_L e^{-\gamma t}$. From the balance equation of liquid asset (1) and the dynamic of the financial welfare (5) and the introduced renormalizations, we obtain the consumer loan demand formula

$$\hat{H}_L = \frac{dM(x)}{dx} \cdot \frac{dx}{dt} - S + \gamma M(x) + \frac{M(x)}{\theta},$$

or,

$$\hat{H}_L = \frac{dM(x)}{dx}\left(S - \gamma x - \frac{M(x)}{\theta} - r_L(M(x) - x)\right) - S + \gamma M(x) + \frac{M(x)}{\theta}, \quad (16)$$

where $M(x) = M(x; r_L, \gamma, \theta, \rho, \delta)$ is the optimal control synthesis presented in Sect. 2.

We next introduce the demand for consumer credit in four types of borrowers presented in Sect. 2.

Consumer Loan Demand in Type 1. If $r_L < \delta - \frac{\rho}{\theta}$, then

$$\frac{dM(x)}{dx} = \frac{\theta\left(\delta - (1-\rho)r_L\right)}{\rho\left(1 + \theta r_L\right)}.$$

The consumer loan demand defines as

$$\hat{H}_L = \frac{\left(\delta - (1-\rho)r_L\right)\left(r_L - \delta + \frac{\rho}{\theta}\right)}{\theta\rho^2\left(1 + \theta r_L\right)} x -$$

$$- \frac{\rho(r_L - \gamma)\left(r_L - \delta + \frac{\rho}{\theta}\right) + \left(\delta - (1-\rho)r_L - \frac{\rho}{\theta}\right)\left(\delta - (1-\rho)r_L\right)}{\theta\rho^2\left(1 + \theta r_L\right)\left(r_L - \gamma\right)} S. \quad (17)$$

Consumer Loan Demand in Type 2. If $\delta - \frac{\rho}{\theta} < r_L < \delta + \rho\gamma$, and the financial welfare of the household satisfies the inequality $x < \frac{S(\delta - (1-\rho)r_L)}{(r_L - \gamma)(r_L - \delta + \frac{\rho}{\theta})}$, then the consumer loan demand defines by formula (17).

Consumer Loan Demand in Type 3. If $\max\{r_D, \delta - \frac{\rho}{\theta}\} < \delta + \rho\gamma < r_L$ or $\max\{r_D, \delta + \rho\gamma\} < \delta - \frac{\rho}{\theta} < r_L$, and the financial welfare of the household satisfies the inequality $x < x_{p_1}$, then

$$
\frac{dM_{r_L}(x_{p_1}, r_L)}{dx} = \frac{\delta - (1-\rho)r_L}{\rho} \cdot \left(\frac{1 + \theta r_L}{\theta} - \left[\frac{x_{p_1}}{M_{r_L}(x_{p_1}, r_L)}\right]^{\frac{\delta - (1-\rho)r_L}{\delta + \rho\gamma - r_L}}\right) \cdot
$$

$$
\cdot \left[S\left(\delta - (1-\rho)r_L\right) + \left(\delta - r_L - \frac{\rho}{\theta}\right) \cdot (r_L - \gamma) x_{p_1}\right]^{-1}.
$$

where $x_{p_1} = \frac{\rho S}{r_L - \delta + \frac{\rho}{\theta}}$. The consumer loan demand defines as

$$
\hat{H}_L = \frac{M_{r_L}(x_{p_1}, r_L)}{\theta} - S + \gamma M_{r_L}(x_{p_1}, r_L) +
$$

$$
+ \frac{\delta - (1-\rho)r_L}{\rho} \cdot \left(\frac{1 + \theta r_L}{\theta} - \left[\frac{x_{p_1}}{M_{r_L}(x_{p_1}, r_L)}\right]^{\frac{\delta - (1-\rho)r_L}{\delta + \rho\gamma - r_L}}\right) \cdot
$$

$$
\cdot \left[S\left(\delta - (1-\rho)r_L\right) + \left(\delta - r_L - \frac{\rho}{\theta}\right) \cdot (r_L - \gamma) x_{p_1}\right]^{-1} \cdot
$$

$$
\cdot \left(S - \gamma x - \frac{M_{r_L}(x_{p_1}, r_L)}{\theta} - r_L\left(M_{r_L}(x_{p_1}, r_L) - x\right)\right). \tag{18}
$$

Consumer Loan Demand in Type 4. If $\max\{\delta + \rho\gamma, \delta - \frac{\rho}{\theta}\} < r_D$, and the financial welfare of the household satisfies the inequality $x < x_{p_1}$, where $x_{p_1} = \min\left\{\frac{\rho S}{r_L - \delta + \frac{\rho}{\theta}}, y(x_{p_2}, 1)\right\}$, $x_{p_2} = \frac{S}{r_D - \gamma} \cdot \frac{\delta - (1-\rho)r_D}{r_D - \delta + \frac{\rho}{\theta}}$, then the consumer loan demand defines by formula (18).

4   Numerical Results

To calculate the net percent value for commercial banks (14) we need to set the parameters λ, σ and the behavioral characteristics of a representative household typical for a given social layer: θ, ρ, δ, γ. The model of household economic behavior was calibrated using Rosstat data. Rosstat provides quarterly statistics of a Household Budget Survey (HBS) [12], questioning about 50000 households living in 82 regions of Russia. Depending on the level of consumption per capita, these regions were divided into three groups: rich, middle, and poor. In each group of regions the low-income and high-income borrowers were identified. They form a social layer. We suppose that the low-income households borrower consumer credit for a 1-year period, while the high-income households borrower consumer credit for a 5-year period. Each social layer can be described by a representative household. With the help of software [13], the behavioral characteristics of a representative household in each social layer were identified. We set parameter λ one percent higher than a key interest rate of the Central Bank [14].

To use the model of commercial banks behavior we need to identify the income volatility parameter $\sigma \geq 0$. In this section we present the results of identification the income volatility parameter σ for the high-income borrowers from the middle group of regions. To identify the parameter σ we consider the historical data of the interest rates on loans for the high-income borrowers on the time period 2015–2022, and solve the inverse problem, searching such parameter σ that gives maximum of NPV (14) at interest rate r_L that coincide with the historical data at each moment of time. The result of the identification of the interest rate on loans is presented in Fig. 1, and the identified volatility parameter σ is presented in Fig. 2.

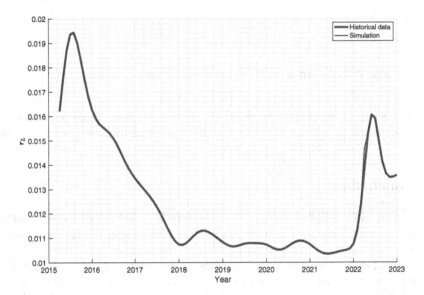

Fig. 1. The results of the identification of the historical interest rates on loans. Monthly data.

Note that the simulated data of the interest rates on loans coincides with historical. The dynamic of the identified volatility parameter σ has the following explanation. The volatility parameter is high during the crisis period 2015, and the COVID-19 pandemic 2020–2021. The crisis time period motivates the households to change their professional spheres in the labor market. The pandemic time period influences the working conditions. These factors influence on the income of the households, thus, the volatility parameter grows up. We note that during February-March 2022 the volatility parameter is close to zero. At this time period sanctions were imposed. Sanctions were aimed at lowering the standard of households' living due to the crisis of the banking system and inflation.

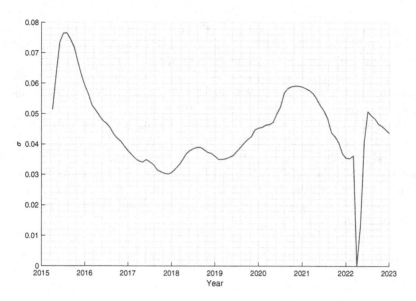

Fig. 2. The identified income volatility parameter. Monthly data.

This was a period of uncertainty. After this period the households have adapted to new economic realities, and the volatility parameter returned to a standard level.

5   Conclusion

In this work we presented the model of the economic behavior of a representative rational household in an imperfect lending and savings market. The model is formalized as an optimal control problem of the Ramsey-type. With the help of statistical data presented by Rosstat, the representative households from different groups of regions were identified. The formation of a credit interest rate for high-income borrowers was modeled by maximizing the net present value for commercial banks, choosing the interest rate on loans is based in the concept of the Stackelberg equilibrium. The choice of the interest rate on loans influence the consumer loan demand that is defined from the Ramsey-type model of the representative rational household. The commercial banks assess the possible default of the borrowers' income and consider their income as a stochastic process described by a geometric Brownian motion. The volatility parameter of the borrowers' income is identified by solving the inverse problem, reproducing the historical data of the interest rate on loans for a high-income borrowers.

The mathematical model of household economic behavior can be used to identify behavioral characteristics of both representative and group behavior of households from different social layers. The group behavior models allow us to examine the economic behavior of a large number of households. From a mathematical point of view, such models are based on the concept of mean field games and represent a system of partial differential equations: the Hamilton–Jacobi–Bellman equation, which describes the choice of strategy for household behavior, and the Kolmogorov–Fokker–Planck equation, which describes the dynamics of the state of the system. This model was earlier presented in [15]. The identification of group behavior of households is based on the identification of a representative household characteristic of the considered social layer. The topic for the further research could be the study of the group behavior model of the borrowers from one social layer with identified volatility parameter of the borrowers' income. This model would help to study the structural changes in the economic behavior of households and commercial banks under the changes in external economic and social factors.

References

1. Shananin, A., Trusov, N.: Mathematical modeling of the consumer loan market in Russia under sanctions. Dokl. Math. **507**(6), 71–80 (2022)
2. Ramsey, F.: A mathematical theory of savings. Econ. J. **152**(38), 543–559 (1928)
3. Rudeva, A.V., Shananin, A.A.: Control synthesis in a modified Ramsey model with a liquidity constraint. Differ. Equations **45**(12), 1835–1839 (2009)
4. Gimaltdinov, I.: Research of the demand for consumer loans and money. Math. Modeling **24**(2), 84–98 (2012)
5. Kuzina, O.E., Krupenskiy, N.A.: Over-indebtedness of Russians: Myth or reality? Vopr. Ekon. **11**, 85–104 (2018). [In Russian.]
6. Shananin, A., Trusov, N.: Optimal control synthesis in a Ramsey-type model. Comput. Math. Math, Phys. (2024). [In print]
7. Koval, P.K., Polbin, A.V.: Estimates of households consumer behavior in Russia. Vopr. Ekon. **3**, 98–117 (2022). (In Russian)
8. Shananin, A., Tarasenko, M., Trusov, N.: Consumer loan demand modeling of households in Russia. Comput. Math. Math. Phys. **61**, 1030–1051 (2021)
9. Komlos, J.: A generalization of a problem of Steinhaus. Acta Mathematica Academiae Scientiarum Hungaricae. **18**, 217–229 (1967)
10. Clarke, F.: Optimization and Nonsmooth Analysis. Wiley, New York (1983)
11. Clarke, F.H., Ledyaev, Yu.S., Stern, R.J., Wolenski, P.R.: Nonsmooth Analysis and Control Theory. Springer, New York (1998). 276 p. https://doi.org/10.1007/b97650
12. Federal State Statistics Service of the Russian Federation. Russian Federation Household Budget Survey (HBS) 2015-2022. Dataset downloaded from https://obdx.gks.ru. Accessed Mar 2024
13. Certificate of state registration of the computer program No.2022619524. "Analysis of demand for consumer credit in the Russian Federation". Copyright holder: Trusov Nikolai Vsevolodovich. Application No. 2022618580. Date of state registration in the Register of computer programs May 23, 2022
14. Interest rates on monetary policy operations. Website of the Bank of Russia. https://www.cbr.ru/hd_base/ProcStav/IR_MPO/. Accessed Mar 2024
15. Trusov, N.V.: Identification of the household behavior modeling based on modified Ramsey model. Lobachevskii J. Math. **44**(1), 455–469 (2023)

Recent Advances in Minimax Problem and Bimatrix Games

Enkhbat Rentsen$^{(\boxtimes)}$ (iD)

Institute of Mathematical and Digital Technology, Mongolian Academy of Sciences,
Ulaanbaatar 13330, Mongolia
renkhbat46@yahoo.com

Abstract. In this paper, we first survey recent developments in game theory, and then focus on minimax problems and bimatrix games. New optimality conditions for maxmin and minimax problems have been considered. Recently introduced new equilibriums such as Anti-Berge and Anti-Nash equilibriums are discussed with connection to optimization problems. We examine bimatrix games. Recent results in multi player polymatrix game have been given. The existence theorems of all equilibrium are provided. Also, a triple matrix game is introduced and existence of Nash equilibrium in the game has been established.

Keywords: optimization · game theory · optimality condition · maxmin problem · minimax problem · equilibriums · bimatrix game · payoff function · strategy

Mathematics Subject Classification: 49J35 · 90C26 · 90C27 · 49K35

1 Introduction

Game theory is an important branch of optimization, operations research and decision making theory. Many research works have been devoted to game theory. Classical minimax theory due to Von Neumann plays an important role in optimization and game theory. Minimax problems and techniques appear in different fields of research including game theory, optimization, and control theory. Many engineering and economics problems such as combinatorial optimization problems of scheduling, location, allocation, and packing as well as inventory problems are formulated as minimax applications. There are many papers [7,11,12,14,15,19,24,25] devoted to theory and algorithms of minimax and maxmin problems.

In recent years, many optimality conditions have been developed for minimax problems. For instance, Truong et al. [2] proposed necessary optimality conditions in terms of upper or lower subdifferentials of both cost and constraint functions for minimax optimization problems. Zhong and Jin [18] derived optimality conditions for minimax optimization problems with an infinite number

A. Eremeev et al. (Eds.): MOTOR 2024, LNCS 14766, pp. 14–36, 2024.
https://doi.org/10.1007/978-3-031-62792-7_2

of constraints. Dhara and Mehrs [1] are concerned with the second-order optimality conditions for minimax problems. Dai and Zhang [30] provided both necessary optimality conditions and sufficient optimality conditions for the local minimax points of constrained minimax optimization problems. Also, more and more attention has been paid recently to the numerical methods of game theory [37–44]. A hidden nonconvexity of the bimatrix games was revealed in [43]. The polymatrix games for the first time was introduced by E.B. Yanovskaya in [45]. Among the numerical methods to seek equilibria, the well-known Lemke-Howson method [38], methods of reducing the polymatrix game to the problem of linear complementarity [42], and the methods of renumbering all simplex vertices defining the Nash equilibria [42] deserve special mentioning. Reduction to some nonconvex optimization problem [46–48] with subsequent application of the global search theory [43, 44, 49] seems to be sufficiently efficient, for example, from the standpoint of the high-dimensionality bimatrix games.

Most of literature on game theory examines a Nash equilibrium [54, 55, 63, 64, 67–69, 75, 79, 80, 82]. Computational Nash equilibriums have been discussed in [59–61, 63, 65, 66, 70, 79, 81].

In recent years, Berge equilibrium is getting attention and finding applications in economics and social sciences [52, 53, 56, 57, 72, 76, 77, 85–88].

Berge equilibrium is supposed to be a model of cooperation in social dilemmas, including the Prisoner's Dilemma games [84].

Connection of Nash and Berge equilibriums has been shown in [86]. Most recently, the Berge equilibrium was examined in Enkhbat and Batbileg [78] for bimatrix game. Also, a notion of Anti-Berge equilibrium was introduced in [62]. Based on Nash and Berge as well as Anti-Berge equilibriums, a new concept of equilibrium so-called Anti-Nash equilibrium was introduced in [97]. Such equilibrium may arise in a game where mutual trust and obligation of two players to help earch other breaks down. Anti-Nash equilibrium for bimatrix game was examined in [97].

2   Maxmin and Minimax Problems

Minimax problem has an important role in optimization, global optimization, game theory, and operations research. In [6], optimality conditions have been formulated for the maxmin problem. In a general case, since the maxmin and minimax values are not always equal, the optimality conditions for the both problems might be different. The classical minimax theorem of von Neuman [20] deals with the equality conditions of maxmin and minimax values. In this paper, we derive new optimality conditions for the minimax problem based on Duvobizkii-Milyution theory [5]. Also, minimax problems find applications in areas such as machine learning, and signal processing, and it has been extensively studied in recent years in [16, 17, 23]. There are many generalizations of minimax theorems. Assume that X and Y are nonempty sets and $f : X \times Y \to \mathbb{R}$.

A *minimax theorem* is a theorem which asserts that, under certain conditions,

$$\min_Y \max_X f = \max_X \min_Y f.$$

Theorem 1 [19]. *Let X and Y be nonempty compact, convex subsets of Euclidean space, and f be continuous.*
Suppose that f is quasiconcave on X, that is to say,

for all $y \in Y$ and $\lambda \in \mathbb{R}, L_f(\lambda, y)$ is a nonempty and convex

and f is quasiconvex on Y, that is to say,

for all $x \in X$ and $\lambda \in \mathbb{R}, L_f(x, \lambda)$ is a nonempty and convex.

Then

$$\min_Y \max_X f = \max_X \min_Y f,$$

where
$L_f(\lambda, y) = \{x : x \in X, f(x, y) \geq \lambda\}$ and $L_f(x, \lambda) = \{y : y \in Y, f(x, y) \leq \lambda\}$.

In 1941, Kakutani [8] analyzed von Neumann's proof from a viewpoint of the fixed point theorem. In 1952, Fan [9] generalized Theorem 1 to the case when X and Y are compact, convex subsets of (infinite dimensional) locally convex spaces and the quasiconcave and quasiconvex conditions are somewhat relaxed, while Nikaido [21], using Brouwer's fixed-point theorem directly, generalized the same result to the case when X and Y are nonempty compact, convex subsets of (not necessary locally convex) topological vector spaces and f is only required to be separately continuous. Nikaido also showed in [22] that, if the quasiconcave and quasiconvex functions are replaced by concave and convex, then the minimax theorem can be proven by elementary calculus.

Theorem 2 [28]. *Let X be a topological space, Y be a compact separable topological space, and $f : X \times Y \to \mathbb{R}$ be separately continuous.*
Suppose that, for all $x_0, x_1 \in X$, there exits a continuous map $h : [0, 1] \to X$ such that $h(0) = x_0$, $h(1) = x_1$ and, for all $y \in Y$ and $\lambda \in \mathbb{R}$,

$$\{t : t \in [0, 1], f(h(t), y) \geq \lambda\}$$

is connected in [0,1].
Suppose also that, for all nonempty finite subsets W of X and $\lambda \in \mathbb{R}$,

$$L_f(W, \lambda) = \{y : y \in Y, f(x, y) < \lambda\}$$

is connected in Y.
Then

$$\min_Y \sup_X f = \sup_X \min_Y f.$$

In 1953, Fan took the theory of minimax theorems out of the context of convex subsets of vector spaces when he established the following results.

Theorem 3 [29]. *Let X be a nonempty set and Y be a nonempty compact topological space. Let $f : X \times Y \to \mathbb{R}$ be lower semicontinuous on Y.*
Suppose that f is concave like on X and convex like on Y, that is to say:

for all $x_1, x_2 \in X$ and $\alpha \in [0,1]$, there exists $x_3 \in X$ such that
$$f(x_3, \cdot) \geq \alpha f(x_1, \cdot) + (1 - \alpha) f(x_2, \cdot) \text{ on } Y,$$

and

for all $y_1, y_2 \in Y$ and $\beta \in [0,1]$, there exists $y_3 \in Y$ such that
$$f(\cdot, y_3) \geq \beta f(\cdot, y_1) + (1 - \beta) f(\cdot, y_2) \text{ on } X.$$

Then
$$\min_{Y} \sup_{X} f = \sup_{X} \min_{Y} f.$$

In 1972, Terkelsen proved first mixed minimax theorem.

Theorem 4 [26]. *Let X be a nonempty set and Y be a nonempty compact topological space. Let $f : X \times Y \to \mathbb{R}$ be lower semicontinuous on Y.*
Suppose that for all $x_1, x_2 \in X$, there exists $x_3 \in X$ such that

$$f(x_3, \cdot) \geq [f(x_1, \cdot) + f(x_2, \cdot)]/2 \text{ on } Y.$$

Suppose also that,
for all nonempty finite subsets W of X and $\lambda \in \mathbb{R}$, $L_f(W, \lambda)$ is connected in Y.
Then
$$\min_{Y} \sup_{X} f = \sup_{X} \min_{Y} f.$$

In [4], the following discrete minimax problem has been considered.

$$\min_{x \in D} \psi(x) \tag{1}$$

where, $\psi(x) = \max_{1 \leq i \leq m} h_i(x)$ and $h_i(x)$ are continuously differentiable functions on a convex closed set $D \subset \mathbb{R}^n$.

The optimality conditions for this problem are given by the following assertion.

Theorem 5 [4]. *If x^* is a solution to problem (1), then*

$$\inf_{x \in D} \max_{i \in I(x^*)} \langle \frac{\partial h_i(x^*)}{\partial x}, x - x^* \rangle = 0$$

where, $I(x) = \{i \in \{1, 2, \cdots, m\} \mid h_i(x) = \psi(x)\}$.

The optimality conditions for saddle points of maxmin problems in Banach space has been derived in 1981 [51].

Let X, Y, Z_1, Z_2 be Banach spaces and $f : X \times Y \to \mathbb{R}$, $g_k : X \to \mathbb{R}$, $k = 1, 2, \ldots, m$, $h_i : Y \to \mathbb{R}$, $i = 1, 2, \ldots, s$, $F : X \to Z_1, G : Y \to Z_2$, and sets $A \subset X$, $B \subset Y$ are convex, and $int A \neq \varnothing$, $int B \neq \varnothing$.

Consider the following maxmin problem

$$\max_{x \in D_1} \min_{x \in D_2} f(x, y), \tag{2}$$

where

$$D_1 = \{x \in A \mid g_k(x) \geq 0, \ k = 1, 2, \ldots, m, \ F(x) = 0\}$$
$$D_2 = \{x \in B \mid h_i(y) \leq 0, \ i = 1, 2, \ldots, s, \ G(y) = 0\}$$

Assume that a point $(x^0, y^0) \in D_1 \times D_2$ is a solution to the maxmin problem, and $f(x, y)$ is *Fréchet* differentiable at the points x^0 and y^0, respectively.

Definition 1 [51]. *A point (x^0, y^0) is called a local-saddle point of maxmin problem if there exist neighborhoods $O_1(x^0)$ and $O_2(y^0)$ of x^0 and y^0 such that*

$$f(x, y^0) \leq f(x^0, y^0) \leq f(x^0, y), \ \forall x \in O_1(x^0), \forall y \in O_2(y^0).$$

Theorem 6 [51]. *Let (x^0, y^0) be a local-saddle point for maxmin problem. Then there exists Lagrange multipliers*

$$(\lambda_0, \lambda_1 \ldots \lambda_m, z_1^*) \neq 0 \in \mathbb{R}^{m+1} \times Z_1^*$$
$$(\mu_0, \mu_1 \ldots \mu_m, z_2^*) \neq 0 \in \mathbb{R}^{s+1} \times Z_2^*$$

such that

$$\begin{cases} \lambda_k \geq 0, \ k = 0, 1, \ldots, m, \ \mu_k \geq 0, \ i = 0, 1, \ldots, s \\ \lambda_k g_k(x^0) = 0, \ k = 1, 2, \ldots, m \\ \mu_i h_i(y^0) = 0, \ i = 1, 2, \ldots, s \\ \langle \lambda_0 f_x'(x^0, y^0) + \sum_{k=1}^m \lambda_k g_k(x^0) + F'^*(x^0) z_1^*, x - x^0 \rangle_X \geq 0, \ \forall x \in A \\ \langle \mu_0 f_y'(x^0, y^0) + \sum_{i=1}^s \lambda_k h_i(y^0) + G'^*(y^0) z_2^*, y - y^0 \rangle_Y \geq 0, \ \forall y \in B \end{cases}$$

In [6], the following maxmin problem has been examined from a viewpoint of optimality conditions.

$$\max_{x \in D} \tilde{\psi}(x) \tag{3}$$

$$D = \{x \in Q \mid f_i(x) \leq 0, \ i = 1, 2, ..., m\} \tag{4}$$

where, $\tilde{\psi}(x) = \min_{y \in A} f(x, y)$, Q is a convex in $\mathbb{R}^n$, $int Q \neq \emptyset$, $f_i : \mathbb{R}^n \to \mathbb{R}$, $i = 1, ..., m$ are differentiable functions, $f(x, y)$ and $\frac{\partial f(x,y)}{\partial x}$ are continuous on $D \times A$, A is compact in $\mathbb{R}^s$.

Theorem 7 [6]. *If x^0 is a solution to problem (3)–(4), then there exist scalars $\alpha_1, \alpha_2, ..., \alpha_{n+1}, \lambda_1, \lambda_2, ...\lambda_m$ and points $y^i \in Y(x^0)$, $i = 1, 2, ...(n+1)$ such that*

$$\begin{cases} \lambda_i \geq 0, \ \lambda_i f_i(x^0) = 0, \ i = 1, 2, ..., m \\ \alpha_i \geq 0, \ i = 1, 2, ..., n+1 \\ \langle \sum_{i=1}^{n+1} \alpha_i \frac{\partial f(x^0, y^i)}{\partial x} + \sum_{i=1}^{m} \lambda_i \frac{\partial f_i(x^0)}{\partial x}, x - x^0 \rangle \geq 0, \forall x \in Q \end{cases} \quad (5)$$

where, $Y(x^0) = \{y \in A \mid f(x^0, y) = \tilde{\psi}(x^0)\}$.

In 2001, I. Tseveendorj in [25] considered the following type of discrete maxmin problems called piecewise convex maximization problem.

Definition 2 [25]. *A function $\varphi : \mathbb{R}^n \rightarrow \mathbb{R}$ is called a piecewise convex function if it can be decomposed into:*

$$\varphi(x) = \min_{1 \leq j \leq m} f_j(x),$$

where $f_j : \mathbb{R}^n \rightarrow \mathbb{R}$ is convex for all $j = 1, 2, \cdots, m$.

Definition 3 [25]. *A problem*

$$\max_{x \in D} \varphi(x) \qquad (PCMP)$$

is called a piecewise convex maximization problem, if $\varphi(x)$ is a piecewise convex function.

Piecewise convex maximization problem (PCMP) has been studied also in [3] and has many applications [19].

Proposition 1 [25]. *If $z \in D$ is global maximizer of (PCMP) then for all $k \in I(z) = \{i \in \{1, 2, \cdots, m\} \mid f_i(z) = \varphi(z)\}$*

$$\partial f_k(y) \bigcap N(D_k(z), y) \neq 0$$

$$\forall \ y \ s.t. \ f_k(y) = \varphi(z),$$

where

$$D_k(z) = D \cap \{x \mid f_j(x) > \varphi(z), \ \forall j = 1, 2, \cdots, m \setminus k\},$$
$$\partial f(y) = \{y^* \in \mathbb{R} \mid f(x) - f(y) \geq \langle y^*, x - y \rangle, \forall x \in \mathbb{R}^n\},$$

and

$$N(D, c) = \{c \in \mathbb{R}^n \mid \langle c, x - c \rangle \leq 0, \ \forall x \in D\}.$$

The following minimax problem has been considered in [89]:

$$\min_{x \in D} \max_{y \in A} f(x, y), \qquad (6)$$

where $f(x, y)$ and $\frac{\partial f(x,y)}{\partial x}$ are continuous on $D \times A$, A is compact in $\mathbb{R}^s$, D is convex and compact in $\mathbb{R}^n$.

According to the minimax Theorem [5] there exists a value v:

$$v = \min_{x \in D} \max_{y \in A} f(x, y).$$

Introduce the function

$$\varphi(x) = \max_{y \in A} f(x, y).$$

Definition 4. *A point $x^0 \in D$ is called a local minimax solution if it is a local minimizer of the function $\varphi(x)$.*

Now we consider the following minimax problem and formulate optimality conditions.
Problem (6) can be written as:

$$\min_{x \in D} \varphi(x), \tag{7}$$

$$D = \{x \in Q \mid g_i(x) \le 0, \ i = 1, 2, ..., m\} \tag{8}$$

where, $\varphi(x) = \max_{y \in A} f(x, y)$, Q is a convex set in $\mathbb{R}^n$, $int Q \ne \emptyset$, $g_i : \mathbb{R}^n \to \mathbb{R}$, $i = 1, 2 ..., m$ are differentiable functions.
The optimality conditions for problem (7)–(8) are given by the following theorem.

Theorem 8 [89]. *If x^0 is a minimax solution to problem (7)–(8), then there exist scalars $\alpha_1, \ \alpha_2 ..., \alpha_{n+1}, \ \lambda_1, \lambda_2, ... \lambda_m$ with $\sum_{i=1}^{n+1} \alpha_i^2 + \sum_{i=1}^{m} \lambda_i^2 \ne 0$ and points $y^i \in Y(x^0)$, $i = 1, 2, ... n+1$ such that*

$$\begin{cases} \langle \sum_{i=1}^{n+1} \alpha_i \frac{\partial f(x^0, y^i)}{\partial x} + \sum_{i=1}^{m} \lambda_i \frac{\partial g_i(x^0)}{\partial x}, x - x^0 \rangle \le 0, \forall x \in Q \\ \lambda_i \ge 0, \ \lambda_i g_i(x^0) = 0, \ i = 1, 2 ..., m \\ \alpha_i \ge 0, \ i = 1, 2, ..., n+1. \end{cases}$$

3   Bimatrix Game and Equilibriums

Consider the bimatrix game in mixed strategies with matrices (A, B) for players 1 and 2.

$$A = (a_{ij}), \ i = 1, \ldots, m,$$
$$B = (b_{ij}), \ j = 1, \ldots, n.$$

Denote by X and Y the sets

$$X = \{x \in R^m \mid \sum_{i=1}^{m} x_i = 1, \ x_i \ge 0, \ i = 1, \ldots, m\},$$

$$Y = \{y \in R^n \mid \sum_{j=1}^{n} y_j = 1, \ y_j \geq 0, \ j = 1, \ldots, n\}.$$

A mixed strategy for player 1 is a vector $x = (x_1, x_2, \ldots, x_m)^T \in X$ representing the probability that player 1 uses a strategy i. Similarly, the mixed strategies for player 2 is $y = (y_1, y_2, \ldots, y_n)^T \in Y$. The expected payoffs of players are given by:

$$f_1(x, y) = x^T A y, \qquad f_2(x, y) = x^T B y.$$

The following definition is given by J.Nash in 1951:

Definition 5. *A pair strategy* $(x^1, y^1) \in X \times Y$ *is a Nash equilibrium if*

$$\begin{cases} f_1(x^1, y^1) \geq f_1(x, y^1), \ \forall x \in X, \\ f_2(x^1, y^1) \geq f_2(x^1, y), \ \forall y \in Y. \end{cases}$$

$$f_1(x^1, y^1) = \max_{x \in X} f_1(x, y^1),$$

$$f_2(x^1, y^1) = \max_{y \in Y} f_2(x^1, y).$$

The following definition is given by C. Berge in 1957:

Definition 6. *A pair strategy* $(x^2, y^2) \in X \times Y$ *is a Berge equilibrium if*

$$\begin{cases} f_1(x^2, y^2) \geq f_1(x^2, y), \ \forall y \in Y, \\ f_2(x^2, y^2) \geq f_2(x, y^2), \ \forall x \in X. \end{cases}$$

$$f_1(x^2, y^2) = \max_{y \in Y} f_1(x^2, y),$$

$$f_2(x^2, y^2) = \max_{x \in X} f_2(x, y^2).$$

Definition of Anti-Berge equilibrium was introduced by Enkhbat in [62]:

Definition 7. *A pair strategy* $(x^3, y^3) \in X \times Y$ *is an Anti-Berge equilibrium(with respect to player 2) if*

$$\begin{cases} f_1(x^3, y^3) \geq f_1(x^3, y), \ \forall y \in Y, \\ f_2(x^3, y^3) \leq f_2(x, y^3), \ \forall x \in X. \end{cases}$$

$$f_1(x^3, y^3) = \max_{y \in Y} f_1(x^3, y),$$

$$f_2(x^3, y^3) = \min_{x \in X} f_2(x, y^3).$$

Definition 8. *A pair strategy* $(x^4, y^4) \in X \times Y$ *is an Anti-Berge equilibrium(with respect to player 1) if*

$$\begin{cases} f_1(x^4, y^4) \leq f_1(x^4, y), \ \forall y \in Y, \\ f_2(x^4, y^4) \geq f_2(x, y^4), \ \forall x \in X. \end{cases}$$

$$f_1(x^4, y^4) = \min_{y \in Y} f_1(x^4, y),$$

$$f_2(x^4, y^4) = \max_{x \in X} f_2(x, y^4).$$

Introduce Berge and Anti-Berge equilibriums for 3-person game.

Definition 9. *A triple strategy* $(x^*, y^*, z^*) \in X \times Y \times Z$ *is a Berge equilibrium if*

$$\begin{cases} \hat{f}_1(x^*, y^*, z^*) \geq \hat{f}_1(x^*, y, z), & \forall (y, z) \in Y \times Z, \\ \hat{f}_2(x^*, y^*, z^*) \geq \hat{f}_2(x, y^*, z), & \forall (x, z) \in X \times Z, \\ \hat{f}_3(x^*, y^*, z^*) \geq \hat{f}_1(x, y, z^*), & \forall (x, y) \in X \times Y, \end{cases}$$

where the functions $\hat{f}_i(x, y, z), i = 1, 2, 3$ defined on a set $X \times Y \times Z$ of strategies are payoff functions of the players.

Now we introduce Anti-Berge equilibrium in the following.

Definition 10. *A triple strategy* $(x^*, y^*, z^*) \in X \times Y \times Z$ *is an Anti-Berge equilibrium (with respect to player 3) if*

$$\begin{cases} \hat{f}_1(x^*, y^*, z^*) \geq \hat{f}_1(x^*, y, z), & \forall (y, z) \in Y \times Z, \\ \hat{f}_2(x^*, y^*, z^*) \geq \hat{f}_2(x, y^*, z), & \forall (x, z) \in X \times Z, \\ \hat{f}_3(x^*, y^*, z^*) \leq \hat{f}_3(x, y, z^*), & \forall (x, y) \in X \times Y. \end{cases}$$

Definition of Anti-Nash equilibrium was introduced in 2022 [97]:

Definition 11. *A pair strategy* $(x^*, y^*) \in X \times Y$ *is an Anti-Nash equilibrium if*

$$\begin{cases} f_1(x^*, y^*) \leq f_1(x^*, y), & \forall y \in Y, \\ f_2(x^*, y^*) \leq f_2(x, y^*), & \forall x \in X. \end{cases}$$

It is clear that

$$f_1(x^*, y^*) = \min_{y \in Y} f_1(x^*, y),$$

$$f_2(x^*, y^*) = \min_{x \in X} f_2(x, y^*).$$

In Nash equilibrium both players maximizes their payoff functions simultaneously. In Berge equilibrium both players mutually supports each other to maximize their payoffs while in Anti-Berge equilibrium one of them betrays other. In Anti-Nash equilibrium each player minimizes payoff function. In other words, each behaves extremely badly and is antagonistic to other.

Theorem 9 [97]. *There exists Anti-Nash equilibrium in a bimatrix game for mixed strategies.*

Theorem 10. *A pair strategy* (x^*, y^*) *is an Anti-Nash equilibrium if and only if*

$$f_1(x^*, y^*) \leq \left[x^{*^T} A \right]_j, \ j = 1, 2, \ldots, n,$$

$$f_2(x^*, y^*) \leq [By^*]_i, \ i = 1, 2, \ldots, m.$$

Theorem 11 [97]. *A pair strategy* (x^*, y^*) *is an Anti-Nash equilibrium for the bimatrix game if and only if there exist scalars* (p^*, q^*) *such that* (x^*, y^*, p^*, q^*) *is a solution to the following quadratic programming problem*:

$$\min_{(x,y,p,q)} F(x, y, p, q) = x^T (A + B)y - p - q$$

subject to:

$$\left[x^T A\right]_j \geq p, \ j = 1, \dots, n,$$

$$[By]_i \geq q, \ i = 1, \dots, m,$$

$$\sum_{i=1}^{m} x_i = 1, \ x_i \geq 0, \ i = 1, \dots, m,$$

$$\sum_{j=1}^{n} y_j = 1, \ y_j \geq 0, \ j = 1, \dots, n.$$

Theorem 12 [62]. *A pair strategy* $(\hat{x}^*, \hat{y}^*)$ *is an Anti-Berge equilibrium (with respect to player 1) for the bimatrix game if and only if there exist scalars* $(\hat{p}^*, \hat{q}^*)$ *such that* $(\hat{x}^*, \hat{y}^*, \hat{p}^*, \hat{q}^*)$ *is a solution to the following quadratic programming problem*:

$$\max_{(x,y,p,q)} F(x, y, p, q) = x^T (B - A)y + p - q$$

subject to:

$$\left[x^T A\right]_j \geq p, \ j = 1, \dots, n,$$

$$[By]_i \leq q, \ i = 1, \dots, m,$$

$$\sum_{i=1}^{m} x_i = 1, \ x_i \geq 0, \ i = 1, \dots, m; \ \sum_{j=1}^{n} y_j = 1, \ y_j \geq 0, \ j = 1, \dots, n.$$

Theorem 13 [62]. *A pair strategy* (x^*, y^*) *is an Anti-Berge equilibrium (with respect to player 2) for the bimatrix game if and only if there exist scalars* (p^*, q^*) *such that* (x^*, y^*, p^*, q^*) *is a solution to the following quadratic programming problem*:

$$\max_{(x,y,p,q)} F(x, y, p, q) = x^T (A - B)y - p + q$$

subject to:

$$\left[x^T A\right]_j \leq p, \ j = 1, \dots, n,$$

$$[By]_i \geq q, \ i = 1, \dots, m,$$

$$\sum_{i=1}^{m} x_i = 1, \ x_i \geq 0, \ i = 1, \dots, m,$$

$$\sum_{j=1}^{n} y_j = 1, \ y_j \geq 0, \ j = 1, \dots, n.$$

4   Nonzero Sum n-Person Game

Consider the n-person game in mixed strategies with matrices $(A_q, \quad q = 1, 2, \ldots, n)$ for players $1, 2, \ldots, n$.

$$A_q = \left(a^q_{i_1 i_2 \ldots i_n} \right), q = 1, 2, \ldots, n$$

$$i_1 = 1, 2, \ldots, k_1, \ \ldots, i_n = 1, 2, \ldots, k_n,$$

Denote by D_q the set

$$D_p = \{ u \in R^p \mid \sum_{i=1}^{p} u_i = 1, \ u_i \geq 0, \ i = 1, \ldots, p \}$$

$$p = k_1, k_2, \ldots, k_n$$

A mixed strategy for player 1 is a vector $x^1 = (x^1_{i_1}, x^1_{i_2}, \ldots, x^1_{i_{k_1}}) \in D_{k_1}$ representing the probability that player 1 uses a strategy i. Similarly, the mixed strategies for q-th player is $x^q = (x^q_{i_1}, x^q_{i_2}, \ldots, x^q_{i_{k_q}}) \in D_{k_q}$, $q = 1, 2, \ldots, n$.

The expected payoffs are given by for 1-th person:

$$f_1(x^1, x^2, \ldots, x^n) = \sum_{i_1=1}^{k_1} \sum_{i_2=1}^{k_2} \cdots \sum_{i_n=1}^{k_n} a^1_{i_1 i_2 \ldots i_n} x^1_{i_1} x^2_{i_2} \ldots x^n_{i_n}.$$

and for q-th person

$$f_q(x^1, x^2, \ldots, x^n) = \sum_{i_1=1}^{k_1} \sum_{i_2=1}^{k_2} \cdots \sum_{i_n=1}^{k_n} a^q_{i_1 i_2 \ldots i_n} x^1_{i_1} x^2_{i_2} \ldots x^n_{i_n},$$

$$q = 1, 2, \ldots, n$$

Definition 12. *A vector of mixed strategies $\tilde{x}^q \in D_{k_q}$, $q = 1, 2, \ldots, n$ is a Nash equilibrium if*

$$\begin{cases} f_1(\tilde{x}^1, \tilde{x}^2, \ldots, \tilde{x}^n) \geq f_1(x^1, \tilde{x}^2, \ldots, \tilde{x}^n), \ \forall x^1 \in D_{k_1} \\ \cdots \quad \cdots \quad \cdots \quad \cdots \quad \cdots \quad \cdots \\ f_q(\tilde{x}^1, \tilde{x}^2, \ldots, \tilde{x}^n) \geq f_q(x^1, \tilde{x}^2, \ldots, \tilde{x}^n), \ \forall x^q \in D_{k_q} \\ \cdots \quad \cdots \quad \cdots \quad \cdots \quad \cdots \quad \cdots \\ f_n(\tilde{x}^1, \tilde{x}^2, \ldots, \tilde{x}^n) \geq f_n(x^1, \tilde{x}^2, \ldots, \tilde{x}^n), \ \forall x^n \in D_{k_n}. \end{cases}$$

It is clear that

$$f_1(\tilde{x}^1, \tilde{x}^2, \ldots, \tilde{x}^n) = \max_{x^1 \in D_{k_1}} f_1(x^1, \tilde{x}^2, \ldots \ \ldots, \tilde{x}^n),$$

$$\cdots \quad \cdots \quad \cdots \quad \cdots$$

$$f_q(\tilde{x}^1, \tilde{x}^2, \ldots, \tilde{x}^n) = \max_{x^q \in D_{k_q}} f_q(\tilde{x}^1, \ldots, \ldots, \tilde{x}^{q-1}, x^q, \ldots, \tilde{x}^{q+1}, \ldots, \tilde{x}^n),$$

$$\cdots \quad \cdots \quad \cdots \quad \cdots$$

$$f_n(\tilde{x}^1, \tilde{x}^2, \ldots, \tilde{x}^n) = \max_{x^n \in D_{k_n}} f_n(\tilde{x}^1, \tilde{x}^2, \ldots, \tilde{x}^{n-1}, x^n).$$

Denote by

$$\sum_{i_1=1}^{k_1}\sum_{i_2=1}^{k_2}\cdots\sum_{i_n=1}^{k_n} a^q_{i_1 i_2 \ldots i_n} x^1_{i_1} x^2_{i_2} \ldots x^n_{i_n} \triangleq \sum_{i_1 i_2 \ldots i_n=1}^{k_1,k_2,\ldots,k_n} a^q x^1 x^2 \ldots x^n \triangleq$$

$$\triangleq \sum_{i_j=1}^{k_j} a^q \left(\prod x^j\right) \triangleq \sum_{i_1 i_2 \ldots i_n=1}^{k_1,k_2,\ldots,k_n} a^q x \triangleq f_q(x^1, x^2, \ldots, x^n) \triangleq f_q(x),$$

$$q = 1, 2, \ldots, n.$$

and

$$\sum_{i_1=1}^{k_1}\sum_{i_2=1}^{k_2}\cdots\sum_{i_{q-1}=1}^{k_{q-1}}\sum_{i_{q+1}=1}^{k_{q+1}}\cdots\sum_{i_n=1}^{k_n} a^q_{i_1 i_2 \ldots i_n} x^1_{i_1} x^2_{i_2} \ldots x^{q-1}_{i_{q-1}} x^{q+1}_{i_{q+1}} \ldots x^n_{i_n} \triangleq$$

$$\sum_{i_1,\ldots i_{q-1},i_{q-1},\ldots i_n=1}^{k_1,\ldots,k_{q-1},k_{q+1},\ldots,k_n} a^q x^1 \ldots x^{q-1} x^{q+1} \ldots x^n \triangleq$$

$$\sum_{\substack{i_j=1 \\ j\neq q}}^{k_j} a^q x^1 \ldots x^{q-1} x^{q+1} \ldots x^n \triangleq$$

$$\triangleq \sum_{i_j=1}^{k_j} a^q \left(\prod_{j=1,\ j\neq q}^{n} x^j\right) \triangleq f_q(x^1, x^2, \ldots, x^{q-1}, x^{q+1}, \ldots x^n) \triangleq f_q(x \setminus x^j),$$

$$j, q = 1, 2, \ldots, n.$$

For further purpose, it is useful to formulate the following statement.

Theorem 14. *A vector strategy $(\tilde{x}^1, \tilde{x}^2, \ldots, \tilde{x}^n)$ is a Nash equilibrium if and only if*

$$\sum_{i_j=1}^{k_j} a^q \left(\prod \tilde{x}^j\right) \geq \sum_{i_j=1}^{k_j} a^q \left(\prod_{j=1,\ j\neq q}^{n} \tilde{x}^j\right) \tag{9}$$

for

$$i_j = 1, 2, \ldots, k_j,$$
$$j = 1, 2, \ldots, n,$$
$$q = 1, 2, \ldots, n.$$

Theorem 15 [94]. *A mixed strategy $\tilde{x}$ is a Nash equilibrium for the nonzero sum n-person game if and only if there exists vector $\tilde{p}$ such that vector $(\tilde{x}, \tilde{p})$ is a solution to the following nonlinear programming problem:*

$$\max_{(x,p)} F(x,p) = \sum_{i_1=1}^{k_1}\sum_{i_2=1}^{k_2}\cdots\sum_{i_n=1}^{k_n} \left(\sum_{q=1}^{n} a^q_{i_1 i_2 \ldots i_n}\right) x^1_{i_1} x^2_{i_2} \ldots x^n_{i_n} - \sum_{q=1}^{n} p_q \tag{10}$$

subject to:

$$\sum_{i_j=1}^{k_j} a^q \left(\prod_{j=1,\ j\neq q}^{n} x^j \right) \leq p_q, \quad \forall i_q = 1, 2, \ldots, k_q, \tag{11}$$

$$\sum_{i=1}^{k_q} x_i^q = 1, \quad q = 1, 2, \ldots, n. \tag{12}$$

In order to solve problem (10)–(12) numerically, special global search methods and algorithms have been developed in [90,94]. Numerical results of the problem for $n = 3, 4, 5, 6$ have been obtained in [92,93,95,96].

5   Three Players Polymatrix Game

The three-player polymatrix game known in [45] is given by $\Gamma_3 = \{I, J, K, h_1, h_2, h_3\}$, where $I = \{1, \ldots, m\}$, $J = \{1, \ldots, n\}$, $K = \{1, \ldots, l\}$ are the sets of strategies of players 1, 2, and 3, respectively. Payoffs of the players are defined on a strategy triple $(i, j, k) \in I \times J \times K$ in the following.

$$\left. \begin{array}{c} h_1(i, j, k) = a_{ij}^1 + a_{ik}^2, \quad h_2(i, j, k) = b_{ji}^1 + b_{jk}^2, \\ h_3(i, j, k) = c_{ki}^1 + c_{kj}^2, \quad i = \overline{1, m}, \ j = \overline{1, n}, \ k = \overline{1, l}. \end{array} \right\}$$

Thus, given are six matrices A_1, A_2, B_1, B_2, C_1, and C_2 whose elements describe the player's payoffs. In this game each player maximizes his payoff. Now we can introduce again Nash equilibrium in game, as in the bimatrix games [34,35], which may not exist in a general. We introduce sets of mixed strategies.

$$x \in S_m, \quad y \in S_n, \quad z \in S_l,$$

where $S_p = \left\{ u = (u_1, \ldots, u_p)^T \in \mathbb{R}^p \ \middle| \ u_i \geq 0, \ \sum_{i=1}^{p} u_i = 1 \right\}, \quad p = m, n, l.$

As usual, x_i may be treated as the probability that player 1 chooses its pure strategy $i \in I = \{1, \ldots, m\}$, and so on. Then the payoff functions are constructed as follows [37–42, 45–48]:

$$\left. \begin{array}{c} F_1(x, y, z) = \langle x, A_1 y \rangle + \langle x, A_2 z \rangle, \quad F_2(x, y, z) = \langle y, B_1 x \rangle + \langle y, B_2 z \rangle, \\ F_3(x, y, z) = \langle z, C_1 x \rangle + \langle z, C_2 y \rangle, \end{array} \right\}$$

The payoff functions have to be maximized:

$$\max_x F_1(x, y, z), \ x \in S_m, \ \max_y F_2(x, y, z), \ y \in S_n, \ \max_z F_3(x, y, z), \ z \in S_l.$$

Definition 13 [33–36,45]. *(a) In three-player game $\Gamma_3 = \Gamma(A, B, C)$ $(A = (A_1, A_2), B = (B_1, B_2), C = (C_1, C_2))$ a triple strategy $(x^*, y^*, z^*) \in S_m \times S_n \times S_l$ is a Nash equilibrium if*

$$F_1(x^*, y^*, z^*) \geq F_1(x, y^*, z^*) \; \forall x \in S_m, \; F_2(x^*, y^*, z^*) \geq F_2(x^*, y, z^*) \Bigg\}$$
$$\forall y \in S_n, F_3(x^*, y^*, z^*) \geq F_3(x^*, y^*, z) \;\; \forall z \in S_l.$$

(b) The payoffs of the players at Nash equilibrium (x^*, y^*, z^*) are computed as:

$$v_1^* = v_1(x^*, y^*, z^*) := \langle x^*, A_1 y^* \rangle + \langle x^*, A_2 z^* \rangle, v_2^* = v_2(x^*, y^*, z^*) := \langle y^*, B_1 x^* \rangle +$$
$$+ \langle y^*, B_2 z^* \rangle, v_3^* = v_3(x^*, y^*, z^*) := \langle z^*, C_1 x^* \rangle + \langle z^*, C_2 y^* \rangle.$$

The set of all Nash equilibriums in game $\Gamma_3 = \Gamma(A, B, C)$ is denoted by $NE(\Gamma_3)$.

We consider the following optimization problem:

$$\Phi(\sigma) \triangleq \langle x, (A_1 + B_1^T) y \rangle + \langle x, (A_2 + C_1^T) z \rangle + \langle y, (B_2 + C_2^T) z \rangle - \atop - \alpha - \beta - \gamma \uparrow \max_{\sigma}, \tag{13}$$

$$\sigma := (x, y, z, \alpha, \beta, \gamma), \;\; (x, y, z) \in S_m \times S_n \times S_l =: S \Bigg\}$$
$$(y, z, \alpha) \in X, \;\; (x, z, \beta) \in Y, \;\; (x, y, \gamma) \in Z, \tag{14}$$

where, the sets X, Y, and Z are defined as

$$X = \{(y, z, \alpha) \in \mathbb{R}^{n+l+1} \mid A_1 y + A_2 z \leq \alpha e_m\} \Bigg\}$$
$$Y = \{(x, z, \beta) \in \mathbb{R}^{m+l+1} \mid B_1 x + B_2 z \leq \beta e_n\} \tag{15}$$
$$Z = \{(x, y, \gamma) \in \mathbb{R}^{m+n+1} \mid C_1 x + C_2 y \leq \gamma e_l\}.$$

Lemma 1. *The following inequality always holds at feasible points $\sigma = (x, y, z, \alpha, \beta, \gamma)$ of the problem, that is*

$$\Phi(\sigma) = \Phi(x, y, z, \alpha, \beta, \gamma) \leq 0 \qquad \forall \sigma \in D, \tag{16}$$

here the set D is defined by conditions (14) and (15).

Proof. If a point $\sigma = (x, y, z, \alpha, \beta, \gamma)$ is feasible, then $x \in S_m$, $y \in S_n$, $z \in S_l$. Therefore, multiplying the inequalities $x \geq 0$, $y \geq 0$, $z \geq 0$ by the corresponding inequalities from (15) in sets X, Y and Z, respectively, we get

$$\langle x, A_1 y \rangle + \langle x, A_2 z \rangle \leq \alpha, \;\; \langle y, B_1 x \rangle + \langle y, B_2 z \rangle \leq \beta, \;\; \langle z, C_1 x \rangle + \langle z, C_2 z \rangle \leq \gamma.$$

Now, summing up these inequalities, we arrive at (16) which proves the lemma.

Theorem 16 [91].

A triple strategy (x^*, y^*, z^*) is a Nash equilibrium in game Γ_3 if and only if there exist scalars $\alpha_*, \beta_*, \gamma_*$ such that $\sigma_* = (x^*, y^*, z^*, \alpha_*, \beta_*, \gamma_*) \in \mathbb{R}^{m+n+l+3}$ is a solution to problem (13)–(15).

As consequence of the proof we have:

$$
\left.
\begin{aligned}
\alpha_* &= v_1(x^*, y^*, z^*) \triangleq \langle x^*, A_1 y^* \rangle + \langle x^*, A_2 z^* \rangle \\
\beta_* &= v_2(x^*, y^*, z^*) \triangleq \langle y^*, B_1 x^* \rangle + \langle y^*, B_2 z^* \rangle \\
\gamma_* &= v_3(x^*, y^*, z^*) \triangleq \langle z^*, C_1 x^* \rangle + \langle z^*, C_2 y^* \rangle.
\end{aligned}
\right\}
$$

Additionally,

$$
\Phi(x^*, y^*, z^*, \alpha_*, \beta_*, \gamma_*) = 0.
$$

6   Multi Players Polymatrix Game

Now we consider a polymatrix game with a finite number of N players. Let $I = \{1, 2, \ldots, N\}$ and S_i defined by

$$
S_i = \left\{ x_i = (x_{i1}, \ldots, x_{in_i})^T \in \mathbb{R}_+^{n_i} \ \Big|\ \sum_{j=1}^{n_i} x_{ij} = 1 \right\}, \quad i = 1, \ldots, N
$$

be the set of mixed strategies of player $i \in I$.

It is clear that $x_i \in S_i$ and $x = (x_1, \ldots, x_N) \in \mathbb{R}^n$, $n = \sum_{i=1}^{N} n_i$.

For a feasible point $x \in S := S_1 \times S_2 \times \ldots \times S_N \subset \mathbb{R}^n$, the payoff functions of the players are given by

$$
F_i(x) = \sum_{j \neq i} \langle x_i, A_{ij} x_j \rangle = \langle x_i, \sum_{j \neq i} A_{ij} x_j \rangle, \quad i = 1, \ldots, N,
$$

where A_{ij} —are the $(n_i \times n_j)$ matrices, $i, j = 1, \ldots, N$.

Therefore, in this game $\Gamma = (\{I\}, \{S_i, i \in I\}, \{F_i, i \in I\})$, each player has to maximize his payoff function

$$
\max_{x_i} F_i(x), \quad x_i \in S_i, \quad i = 1, \ldots, N.
$$

Introduce the following notation:

$$
(x \| z_i) = (x_1, \ldots, x_{i-1}, z_i, x_{i+1}, \ldots, x_N).
$$

Definition 14 [91]. *A strategy* $x^* = (x_1^*, \ldots, x_N^*)$ *is called Nash equilibrium, if*

$$F_i(x^* \parallel x_i) \leq F_i(x^*), \quad \forall x_i \in S_i, \quad i = 1, \ldots, N.$$

Denote by v_i^* values of the payoffs of players. That is

$$v_i^* = v_i(x^*) = F_i(x^*), \quad i = 1, \ldots, N.$$

Now we can formulate the following assertion.

Theorem 17. *A strategy* $x^* = (x_1^*, \ldots, x_N^*)$ *is a Nash equilibrium if and only if the following inequalities hold:*

$$\left(\sum_{j \neq i} A_{ij} x_j^* \right)_p \leq v_i^* \ \forall p = 1, \ldots, n_i, \ \forall i \in I.$$

Theorem 18 [91]. *The strategy* $x^* = (x_1^*, \ldots, x_N^*) \in S$ *is the Nash equilibrium if and only if there exists a vector* $\alpha^* = (\alpha_1^*, \ldots, \alpha_N^*) \in \mathbb{R}^N$ *satisfying*

$$\sum_{j \neq i} A_{ij} x_j^* \leq \alpha_i^* e_{n_i}, \quad i \in I;$$

$$\sum_{i=1}^N \alpha_i^* = \sum_{i=1}^N F_i(x^*) \triangleq \sum_{i=1}^N \langle x_i^*, \sum_{j \neq i} A_{ij} x_j^* \rangle,$$

where $e_{n_i} = (1, \ldots, 1)^T \in \mathbb{R}^{n_i}$.

Now we consider the following optimization problem:

$$\min_{\sigma} \Phi_N(\sigma) := \sum_{i=1}^N [\alpha_i - F_i(x)] = \sum_{i=1}^N (\alpha_i - \langle x_i, \sum_{j \neq i} A_{ij} x_j \rangle), \qquad (17)$$

$$\sigma = (x, \alpha) \in \mathbb{R}^n \times \mathbb{R}^N, \quad x = (x_1, \ldots, x_N) \in S := S_1 \times S_2 \times \ldots \times S_N, \\ \alpha = (\alpha_1, \ldots, \alpha_N) \in \mathbb{R}^N, \qquad (18)$$

$$\sum_{j \neq i} A_{ij} x_j \leq \alpha_i e_{n_i}, \quad i = 1, \ldots, N. \qquad (19)$$

The following results can be proved.

Lemma 2. *For the objective function* $\Phi_N(\sigma)$, $\sigma = (x, \alpha) \in \mathbb{R}^{n+N}$ *the following condition is satisfied*

$$\Phi(\sigma) \geq 0, \qquad \forall \sigma \in D,$$

where the feasible set D *consists of (18) and (19).*

Theorem 19 [91]. *The strategy* $x^* = (x_1^*, \ldots, x_N^*) \in S \subset \mathbb{R}^n$ *is a Nash equilibrium in the polymatrix game* $\Gamma = (\{I\}, \{S_i, i \in I\}, \{F_i, i \in I\})$ *if and only if there exists a vector* $\alpha^* = (\alpha_1^*, \ldots, \alpha_N^*)$ *such that* $\sigma_* = (x^*, \alpha^*) \in \mathbb{R}^{n+N}$ *is a solution to optimization problem (17)–(19).*

It can be shown that

$$\alpha_i^* = v_i(x^*) \triangleq v_i^* = F_i(x^*), \quad i = 1, \ldots, N.$$

Additionally,

$$\Phi_N(\sigma_*) = \Phi_N(x^*, \alpha^*) = 0.$$

7   The Four-Players Triple Polymatrix Game

The four-players triple game is given by $\Gamma_4 = \{i, j, k, t, a, b, c, d\}$, where $i = 1, \ldots, m$, $j = 1, \ldots, n$, $k = 1, \ldots, s$, $\ell = 1, \ldots, p$ are the sets of pure strategies of the respective players 1, 2, 3 and 4.

Payoffs of players defined on a strategy $(i, j, k, t) \in I \times J \times K \times T$ are given by:

$$a(i, j, k, \ell) = a_{ijk}^1 + a_{ij\ell}^2 + a_{ik\ell}^3, \quad b(i, j, k, \ell) = b_{ijk}^1 + b_{ij\ell}^2 + b_{jk\ell}^3,$$

$$c(i, j, k, \ell) = c_{ijk}^1 + c_{ik\ell}^2 + c_{jk\ell}^3, \quad d(i, j, k, \ell) = d_{ij\ell}^1 + d_{jk\ell}^2 + c_{ik\ell}^3,$$

$$i = 1, \ldots, m, \quad j = 1, \ldots, n, \quad k = 1, \ldots, s, \quad \ell = 1, \ldots, p.$$

Thus, twelve matrices are given for players A, B, C, D, where $A = (A_1, A_2, A_3), B = (B_1, B_2, B_3), C = (C_1, C_2, C_3)$ and $D = (D_1, D_2, D_3)$. The payoff functions of the first and second players are defined as:

$$F_1(x, y, z, t) = \sum_{i=1}^m x_i \left(\sum_{j=1}^n \sum_{k=1}^s a_{ijk}^1 y_j z_k + \sum_{j=1}^n \sum_{\ell=1}^p a_{ij\ell}^2 y_j t_\ell + \sum_{k=1}^s \sum_{\ell=1}^p a_{ik\ell}^3 z_k t_\ell \right),$$

$$F_2(x, y, z, t) = \sum_{j=1}^n y_j \left(\sum_{i=1}^m \sum_{k=1}^s b_{ijk}^1 x_i z_k + \sum_{i=1}^m \sum_{\ell=1}^p b_{ij\ell}^2 x_i t_\ell + \sum_{k=1}^s \sum_{\ell=1}^p b_{ik\ell}^3 z_k t_\ell \right),$$

where (x, y, z, t) vector of mixed strategies of four players. Similarly, we can define the payoff functions F_3 and F_4 of other players.

Denote by S_q the set $S_q = \{u \in \mathbb{R}^q | \sum_{\tau=1}^q u_i = 1, u_i \geq 0, \tau = 1, \ldots, q\}$, $q = m, n, s, p$.

Definition 15 [95]. *A strategy* $(x^*, y^*, z^*, t^*) \in S_m \times S_n \times S_s \times S_p$ *is called a Nash equilibrium of the four-person matrix game if the following conditions are satisfied:*

$$F_1(x^*, y^*, z^*, t^*) \geq F_1(x, y^*, z^*, t^*), \quad \forall x \in S_m,$$

$$F_2(x^*, y^*, z^*, t^*) \geq F_2(x^*, y, z^*, t^*), \quad \forall y \in S_n,$$

$$F_3(x^*, y^*, z^*, t^*) \geq F_3(x^*, y^*, z, t^*), \quad \forall z \in S_s,$$

$$F_4(x^*, y^*, z^*, t^*) \geq F_4(x^*, y^*, z^*, t), \quad \forall t \in S_p.$$

We consider the following optimization problem:

$$F(u) = F_1(x,y,z,t) + F_2(x,y,z,t) + F_3(x,y,z,t) + F_4(x,y,z,t) - \alpha - \beta - \gamma - \delta \uparrow \max_u \tag{20}$$

$$u = (x,y,z,t,\alpha,\beta,\gamma,\delta) \in S_m \times S_n \times S_s \times S_p \times \mathbb{R}^4, \tag{21}$$

$$(y,z,t,\alpha) \in X, \ (x,z,t,\beta) \in Y, \ (x,y,z,\gamma) \in Z, \ (x,y,z,\delta) \in T, \tag{22}$$

where the sets X, Y, Z and T obey the conditions:

$$X = \{(y,z,t,\alpha) \in \mathbb{R}^{n+\ell+p+1} \mid \sum_{j=1}^{n}\sum_{k=1}^{s} a_{ijk}^1 y_j z_k + \sum_{j=1}^{n}\sum_{\ell=1}^{p} a_{ij\ell}^2 y_j t_\ell +$$

$$+ \sum_{k=1}^{s}\sum_{\ell=1}^{p} a_{ik\ell}^3 z_k t_\ell \le \alpha e_m\}$$

$$Y = \{(x,z,t,\beta) \in \mathbb{R}^{m+\ell+p+1} \mid \sum_{i=1}^{m}\sum_{k=1}^{s} b_{ijk}^1 x_i z_k + \sum_{i=1}^{m}\sum_{\ell=1}^{p} b_{ij\ell}^2 x_i t_\ell +$$

$$+ \sum_{k=1}^{s}\sum_{\ell=1}^{p} b_{jk\ell}^3 z_k t_\ell \le \beta e_n\},$$

$$Z = \{(x,y,t,\gamma) \in \mathbb{R}^{m+n+p+1} \mid \sum_{i=1}^{m}\sum_{j=1}^{n} c_{ijk}^1 x_i y_j + \sum_{i=1}^{m}\sum_{\ell=1}^{p} c_{ik\ell}^2 x_i t_\ell +$$

$$+ \sum_{j=1}^{n}\sum_{\ell=1}^{p} c_{jk\ell}^3 y_j t_\ell \le \gamma e_s\},$$

$$T = \{(x,y,z,\delta) \in \mathbb{R}^{m+n+\ell+1} \mid \sum_{i=1}^{m}\sum_{j=1}^{n} d_{ij\ell}^1 x_i y_j + \sum_{i=1}^{m}\sum_{k=1}^{s} d_{ik\ell}^2 x_i z_k +$$

$$+ \sum_{j=1}^{n}\sum_{k=1}^{s} d_{jk\ell}^3 y_j z_k \le \delta e_p\}$$

and $e_q = (1,1,\ldots,1)^T, \quad q = m,n,s,p.$

Theorem 20 [95]. *The strategy (x^*, y^*, z^*, t^*) is the Nash equilibrium in the game $\Gamma(A,B,C,D)$ if and only if there exist scalars $(\alpha_*, \beta_*, \gamma_*, \delta_*)$ such that $u^* = (x^*, y^*, z^*, t^*, \alpha_*, \beta_*, \gamma_*, \delta_*) \in \mathbb{R}^{m+n+s+p+4}$ is a global solution to problem (20)–(22).*

Thus, finding a Nash equilibrium reduces to solving problem (20)–(22). In order to solve problem (20)–(22), we apply a global search method proposed in [31].

8   Conclusion

We have considered recent developments in maxmin and minimax problems, and bimatrix games. We have shown how to apply newly introduced equilibriums such as Anti-Berge and Anti-Nash to bimatrix games with connection of global optimization problems. Further research should focus on multi players polymatrix game and its connection with global optimization.

References

1. Anulekha, D., Aparna, M.: Second-order optimality conditions in minimax optimization problems. J. Optim. Theory Appl. **156**, 567–590 (2013)
2. Bao, Q.T., Khahn, P., Gupta, P.: Necessary optimality conditions for minimax programming problems with mathematical constraints. Optim. J. Math. Program. Oper. Res. **66**(11) (2017)
3. Dem'yanov, V.F., Vasil'ev, L.V.: Nondifferentiable optimization, Translation Series in Mathematics and Engineering. Optimization Software Inc., New York (1985)
4. Dem'yanov, V.F., Malozemov, V.N.: Introduction to Minimax. Dover Publications Inc., New York (1974)
5. Duvobitzkii, A.Y., Milyuton, A.A.: Extremum Problems in presence of resrictions. USSR Comput. Math. Math. Phys. **5**, 1–80 (1965)
6. Enkhbat, R., Enkhbayar, J.: A note on maxmin problem. Optim. Lett. **13**, 475–483 (2019)
7. Floudas, C.A., Pardalos, P.M.: Encyclopedia of Optimization. Springer, Berlin (2008). https://doi.org/10.1007/978-0-387-74759-0
8. Kakutani, S.: A generalization of Brouwer's fixed-point theorem. Duka. Math. J. **8**, 457–459 (1941)
9. Fan, K.: Fixed-point and minimax theorems in locally convex topological linear space. Proc. Nat. Acad. Sci. USA **8**, 412–416 (1956)
10. Girsanov, I.V: Lectures on Mathematical Theory of Extremum Problems, Lecture Notes in Economics and Mathematical Systems, vol. 67 . Springer, Heidelberg (1972). 10.1007/978-3-642-80684-1
11. Horst,R., Pardalos, P.M., Thoai, N.V.: Introduction to global optimization, volume 3 of Nonconvex Optimization and its Applications. Kluwer Academic Publishers, Dordrecht (1995)
12. Horst, R., Tuy, H.: Global Optimization, 2nd edn. Springer, Berlin (1993)
13. Ferrera, J.: An Introduction to Nonsmooth Analysis. Elsevier (2014)
14. Kim, D., Pardalos, P.M.: A dynamic domain contraction algorithm for nonconvex piecewise linear network flow problems. J. Global Optim. **17**(1–4), 225–234 (2000). (Dedicated to the memory of Professor P.D. Panagiotopoulos)
15. Kneser, H.: Sur un théorème fondamental de la théorie des jeux. C. R. Acad. Sci. Paris, 234, 2418–2420 (1952)
16. Lin, T., Jin, C., Jordan, M.: On gradient descent ascent for nonconvex-concave minimax problems. In: International Conference on Machine Learning, PMLR, pp. 6083–6093 (2020)
17. Lu, S., Tsaknakis, I., Hong, M., Chen, Y.: Hybrid block successive approximation for one- sided non-convex min-max problems: algorithms and applications. IEEE Trans. Signal Process. **68**, 3676–3691 (2020)

18. Zhong, L., Jin, Y.: Optimality conditions for minimax optimization problems with an infinite number of constraints and related applications. Acta Math. Appl. Sin. Engl. Ser. **37**(2), 251–263 (2021)
19. von Neumann, J.: Über ein ökonomisches Gleichungssystem und eine Verallgemeinerung des Brouwerschen Fixpunktsatzes. Ergebn. MathKolloq. Wein **8**, 73–83 (1937)
20. von Neumann, J.: Zur Theorie der Gessellschaftspiele. Math. Ann. **100**, 295–320 (1928)
21. Nikaido, L.: On von Neumann's minimax theorem. Pac. J. Math. **4**, 65–72 (1954)
22. Nikaido, L.: On method of proof for the minimax theorem. Proc. Amer. Math. Soc. **10**, 205–212 (1959)
23. Nouiehed, M., Sanjabi, M., Huang, T., Lee, J.D., Razaviyayn, M.: Solving a class of non-convex min-max games using iterative first order methods. In: Advances in Neural Information Processing Systems, pp. 14934–14942 (2019)
24. Sion, M.: On general minimax theorems. Pac. J. Math. **8**, 171–176 (1958)
25. Tseveendorj, I.: Piecewise-convex maximization problems: global optimality conditions. J. Global Optim. **21**(1), 1–14 (2001)
26. Terkelsen, F.: Some minimax theorems. Math. Scand. **31**, 405–413 (1972)
27. Ville, J.: Sur la théorie générale des jeux où interviennent l'habilité des joueurs. Traité des probablités et de ses applications **2**, 13–42 (1959)
28. Wu Wen-Tsun, W.: A remark on the fundamental theorem in the theory of games. Sci. Rec. New. Ser. **3**, 229–233 (1959)
29. Fan, K.: Minimax theorems. Proc. Nat. Acad. Sci. USA **39**, 42–47 (1953)
30. Dai, Y.-H., Zhang, L.: Optimality conditions for constrained minimax optimization. CSIAM Trans. Appl. Math. 296–315 (2020)
31. Gornov, A., Zarodnyuk, T.: Computing technology for estimation of convexity degree of the multiextremal function. Mach. Learn. Data Anal. **10**(1), 1345–1353 (2014). http://jmlda.org
32. Neumann, J., Morgenshtern, O.: Theory of Games and Economic Behavior. Princeton University Press, Princeton (1953)
33. Nash, J.F.: Equilibrium points in n-person games. Proc. Nat. Acad. Sci. USA **36**, 48–49 (1950)
34. Vasin, A.A., Morozov, V.V.: Game Theory and Models of Mathematical Economics. MAKS Press, Moscow (2005)
35. Vorobiev, N.N.: Game Theory for Economists and Cyberneticians. Nauka, Moscow (1985)
36. Germier, Y.B.: Games with Noncontradictory Interests. Nauka, Moscow (1976)
37. Kuhn, H.W.: An algorithm for equilibrium points in bimatrix games. Proc. Natl. Acad. Sci. **47**, 1657–1662 (1961)
38. Lemke, C.E., Howson, T.T.: Equilibrium points of bimatrix games SIAM. J. Appl. Math. **12**, 413–423 (1961)
39. Dickhaut J., Kaplan, T.: A program for finding Nash equilibria. Math. J. (1), 87–93 (1991)
40. McKelvey R.D., McLennan, A.: Computation of equilibria in finite games. In: Amman, H.M., Kendrick, D.A., Rust, J. (eds.) Handbook of Computational Economics, vol. 1, pp. 87–142 . Elsevier, Amsterda, (1996)
41. Howson, J.T.: Equilibria of polymatrix games. Manage. Sci. **18**, 312–318 (1972)
42. Audet, C., Belhaiza, S., Hansen, P.: Enumeration of all the extreme equilibria in game theory: bimatrix and polymatrix games. J. Opt. Theory Appl. **29**(3), 349–372 (2006)

43. Strekalovskii, A.S., Orlov, A.V.: Bimatrix Games and Bilinear Programming. Fiz-matlit, Moscow (2007)
44. Vasiliev, I.L., Klimentova, K.B., Orlov, A.V.: Parallel Global Search of Equilibrium Situations in the Bimatrix Games. Vychisl. Metod. Progr. **8**, 233–243 (2007)
45. Yanovskaya, E.B.: Equilibrium points in polymatrix games. Latvian Math. Coll. **8**, 381–384 (1968)
46. Mills, H.: Equilibrium points in finite games. J. Soc. Ind. Appl. Math. **8**(2), 397–402 (1960)
47. Mangasarian, O.L.: Equilibrium points in bimatrix games. J. Soc. Ind. Appl. Math. **12**, 778–780 (1964)
48. Mangasarian, O.L., Stone H.: Two-person nonzero games and quadratic program-ming. J. Math. Anal. Appl. (9), 348–355 (1964)
49. Strekalovskii, A.S.: Elements of Nonconvex Optimization. Nauka, Novosibirsk (2003)
50. Pang, J.-S.: Three modelling paradigms in mathematical programming. Math. Pro-gram. Ser. B. **125**, 297–323 (2010)
51. Strekalovski, A.S., Enkhbat, R.: Optimality conditions for a problem of operations research. J. Discr. Distrib. Syst. Irkutsk, (1), 151–159 (1981). (in Russian)
52. Abalo, K.Y., Kostreva, M.M.: Some existence theorems of Nash and Berge equi-libria. Appl. Math. Lett. **17**, 569–573 (2004)
53. Abalo, K.Y., Kostreva, M.M.: Berge equilibrium: Some recent results from fixed-point theorems. Appl. Math. Comput. **169**, 624–638 (2005)
54. Ben, A., Andrea, C., Marco, S., Ziwen, Z.: Pure Nash equilibria and best-response dynamics in random games. Math. Oper. Res. **173** (2021)
55. Chinchuluun, A., Pardalos,P.M., Migdalas. A., Pitsoulis, L.: Pareto Optimality, Game Theory and Equilibria. Springer Optimization and Its Applications, vol. 17. Springer, New York (2008). https://doi.org/10.1007/978-0-387-77247-9
56. Berge, C.: Theorie generale des jeux n-personnes. Gauthier Villars, Paris (1957)
57. Colman, A.M., Korner, T.W., Musy, O., Tazdait, T.: Mutual support in games: some properties of Berge equilibria. J. Math. Psychol. **55**(2), 166–175 (2011)
58. Crettez, B.: On Sugdens mutually beneficial practice and Berge equilibrium. Int. Rev. Econ. **64**(4), 357–366 (2017)
59. Daskalakis, C., Goldberg, P.W., Papadimitriou, C.H.: The complexity of computing a Nash equilibrium. SIAM J. Comput. **39**(1), 195–259 (2009)
60. ElenaSaiz, M., Hendrix, E.M.T.: Methods for computing Nash equilibria of a loca-tion quantity game. Comput. Oper. Res. **35**, 3311–3330 (2008)
61. Enkhbat, R., Batbileg, S., Tungalag, N., Anikin, A., Gornov, A.A.: Computational method for solving n-person game, Izv. Irkutsk. Gos. Univ. Ser. Mat. (Series Math-ematics http://isu.ru/izvestia), **20**, 109–121 (2017)
62. Enkhbat, R.: A note on anti-berge equilibrium, Izv. Irkutsk. Gos. Univ. Ser. Mat. (Series Mathematics, http://isu.ru/izvestia) **36**, 3–13 (2021)
63. Gilboa, I., Zemel, E.: Nash and correlated equilibria: some complexity considera-tions. Games Econom. Behav. **1**, 80–93 (1989)
64. Harsanyi, J.C., Selten, R.: A General Theory of Equilibrium Selection in Games. MIT Press, Cambridge (1988)
65. Kontogiannis, S., Spirakis, P.: Well supported approximate equilibria in bimatrix games. Algorithmica **57**, 653–667 (2010)
66. Kuhn, H.W.: An algorithm for equilibrium points in bimatrix games. Proc. Natl. Acad. Sci. U.S., **47**, 1657–1662 (1961)
67. Kukushkin, N.S.: Nash equilibrium in compact-continuous games with a potential. Int. J. Game Theory **40**, 387–392 (2011)

68. Lemke, C., Howson, J.: Equilibrium points of bimatrix games. J. SIAM 413–423 (1964)
69. Margarida, C., Andrea, L., Pedroso, J.P.: Existence of Nash equilibria on integer programming games. Oper. Res. 11–23 (2017)
70. Mihalis, Y.: Computational aspects of equilibria. Algorithmic Game Theory 2–13 (2009)
71. Harlan, M.: Equilibrium point in finite game. J. Soc. Indust. Appl. Math **8**(2), 397–402 (1960)
72. Salukvadze, M.E., Zhukovskiy, V.I.: The Berge equilibrium: A Game-Theoretic Framework for the Golden Rule of Ethics, Birkhauser (2020)
73. Nash, J.: Non-cooperative games. Ann. Math. **54**, 289–295 (1951)
74. Nau, R., Gomez-Canovas, S., Hansen, P.: On the geometry of Nash equilibria and correlated equilibria INT. J. Game Theory **32**, 443–453 (2003)
75. Neyman, A., Sorin, S. (eds.): Stochastic Games and Applications. Kluwer, Dordrecht (2003)
76. Nessah, R.: Non cooperative games. Ann. Math. **54**, 286–295 (1951)
77. Nessah, R., Larbani, M., Tazdait, T.: A note on Berge equilibrium. Appl. Math. Lett. **20**(8), 926–932 (2007)
78. Enkhbat, R., Sukhee, B.: Optimization approach to Berge equilibrium for bimatrix game. Optim. Lett. **15**(2), 711–718 (2021)
79. Robinson, J.: An iterative method of solving a game. Ann. Math. **54**, 296–301 (1951)
80. Rosen, J.: Existence and uniqueness of equilibrium points for concave n-person games. Econometrica **33**(3), 520–534 (1965)
81. Savani, R., von Stengel, B.: Hard to solve bimatrix games. Econometrica **74**, 397–429 (2006)
82. Spirakis, P.: Approximate equilibria for strategic two-person games. In: Proceedings 1st Symposium on Algorithm Game Theory, pp. 5–21 (2008)
83. On mutual concavity and strategically-zero-sum bimatrix games: Spyros Kontogiannis., Paul Spirakis. Theoret. Comput. Sci. **432**, 64–76 (2012)
84. Tucker, A.: A two-person dilemma, Stanford University. In: Rassmussen, E. (ed.) Readings in Games and Information, pp. 7–8 (1950)
85. Zhukovskiy, V.I.: Some Problems of Non-antagonistic Differential Games. Mathematical Methods in Operation Research, pp. 103–195. Bulgarian Academy of Sciences, Sofia (1985)
86. Zhukovskiy, V.I., Kudryavtsev, K.N.: Mathematical foundations of the Golden Rule. I. Static case. Autom. Remote Control **78**, 1920–1940 (2017)
87. Vaisman, K.S.: Berge equilibrium, Ph.D thesis, St. Petersburg: St. Petersburg. Gos. Univ. (1995)
88. Vaisman, K.S.: Berge equilibrium, in Linear quadratic differential games. In: Zhukovskiy, V.I., Chikrii, A.A. (eds.) Naukova Dumka, Kiev, pp. 119–143 (1994)
89. Enkhbat, R., Battur, G.: Some optimality conditions for minimax problem. Pacific J. Optimization (2024, accepted and to appear)
90. Batbileg, S., Enkhbat, R.: Global optimization approach to Nonzero sum N person game. Int. J. Adv. Modeling Optim. **13**(1), 59–66 (2011)
91. Strekalovskii, A.S., Enkhbat, R.: Polymatrix games and optimization problems. Autom. Remote. Control. **75**(4), 632–644 (2014)
92. Enkhbat, R., Tungalag, N., Gornov, A., Anikin, A.: The curvilinear search algorithm for solving three-person game, pp. 574–583. Supplementary Proceedings of the 9th International Conference on Discrete Optimization and Operations Research (DOOR2016), Vladivostok, Russia (2016)

93. Enkhbat, R., Batbileg, S., Tungalag, N., Anikin, A., Gornov, A.: a note on solving 5-person game. Adv. Modeling Optim. **19**(2), 227–232 (2017)
94. Enkhbat, R., Batbileg, S.,Tungalag, N., Anikin, A., Gornov, A.: A computational method for solving N-person game. J. Irkutsk State Univ. (Series Mathematics, http://isu.ru/izvestia) **20**, 109–121 (2017)
95. Enkhbat, R., Batbileg, S., Tungalag, N., Gornov, A., Anikin, A.: A Note on Four-Players Triple Game, Contributions to Game Theory Management, vol. XII, pp. 100–113, Saint Petersburg State University, Russia (2019)
96. Enkhbat, R., Batbileg, S., Tungalag, N., Anikin, G., Gornov, G.: A global optimization approach to nonzero sum six-person game. In: Yeung, D., Lucraz, S., Leong, C.K. (eds.) Frontiers in Games and Dynamic Games: Theory, Applications, and Numerical Methods, pp. 219–227, Birkhauser (2020)
97. Enkhbat, R.: A note on Anti-Nash equilibrium. Optim. Lett. **16**(6), 1927–1933 (2022)
98. Enkhbat, R., Sorokovikov, J., Zarodnyuk, T., Gornov, A.: The globalized modification of rosenbrock algorithm for finding ant-Nash equilibrium in bimatrix game. In: Enkhbat, R., Chinchuluun, A., Pardalos, P.M. (eds.) Optimization. Simulation and Control, pp. 153–165. Springer, Cham (2023). https://doi.org/10.1007/978-3-031-41229-5_12

Mathematical Programming

Assessing the Perron-Frobenius Root of Symmetric Positive Semidefinite Matrices by the Adaptive Steepest Descent Method

Zulfiya R. Gabidullina[✉][iD]

Kazan Federal University, Kazan, Russia
zulfiya.gabidullina@kpfu.ru
https://kpfu.ru/Zulfiya.Gabidullina

Abstract. We discuss the maximum eigenvalue problem which is fundamental in many cutting-edge research fields. We provide the necessary theoretical background required for applying the fully adaptive steepest descent method (or ASDM) to estimate the Perron-Frobenius root of symmetric positive semidefinite matrices. We reduce the problem of assessing the Perron-Frobenius root of a certain matrix to the problem of unconstrained optimization of the quadratic function associated with this matrix. We experimentally investigated the ability of ASDM to approximate the Perron-Frobenius root and carry out a comparative analysis of the obtained computational results with some others presented earlier in the literature. This study also provides some insight into the choice of parameters, which are computationally important, for ASDM. The study revealed that ASDM is suitable for estimating the Perron-Frobenius root of matrices regardless of whether or not their elements are positive and regardless of the dimension of these matrices.

Keywords: adaptive steepest descent method · matrix norm · Perron-Frobenius root · spectral radius · dominant eigenvalue

1 Introduction

The maximum eigenvalue problem is fundamental in many active research areas. There is a lot of literature that is most relevant and helpful for our research. We mention for only a few of those parts of the literature, and refer to [1–4] and the references therein for detailed information.

This paper presents new applications of the fully adaptive steepest descent method (ASDM) recently developed for solving unconstrained optimization problems (see [5]). To the best of our knowledge, the application of adaptive optimization methods for approximating the Perron-Frobenius root (spectral radius) of positive semidefinite symmetric matrices are truly new. The present study has been motivated by the existing various applications of the spectral

A. Eremeev et al. (Eds.): MOTOR 2024, LNCS 14766, pp. 39–54, 2024.
https://doi.org/10.1007/978-3-031-62792-7_3

radius of matrices in linear algebra, mathematical physics, statistics, combinatorics, spectral graph theory, chemistry, digital filters design, digital signal processing, network systems, deep learning and etc. (see [6–16]). Here, we explore the approximations of the spectral radius for the so-called Laplacian and Pascal matrices. The considered examples show that ASDM allows one to find the upper bound of the Perron root (which is very close to its real value) of the matrices enumerated above. In addition, visualization of the obtained experimental results and comparative analysis are carried out.

In the context of deep learning, it is common to optimize the so-called loss function or error function using its quadratic approximation. The eigenvalues of the Hessian determine the scale of the learning rate [16]. Moreover, the maximal eigenvalue of the Hessian matrix is used there for calculating the step-size during the process of minimizing the loss function.

The main contributions of this paper are summarized as follows:

The study revealed a new optimization tool (ASDM) that is suitable for estimating the Perron-Frobenius root of matrices regardless of whether or not their elements are positive and regardless of the dimension of these matrices. Namely, we reduce the problem of assessing the Perron-Frobenius root of a certain matrix to the problem of unconstrained optimization of the quadratic function associated with this matrix. As far as we know, there are no similar results for adaptive optimization methods. For the maximum eigenvalue problem, this study also provides some insight into the choice of parameters, which are computationally important, for ASDM.

Paper Organization: Section 2 presents the problem setting of this research and theoretical background for iterative approximating the spectral radius of symmetric positive definite matrices (or positive semidefinite matrices having at least one nonzero eigenvalue). Section 3 discusses the results of computational experiments with ASDM and compares them with the results of other studies. Section 4 concludes the paper. Appendices contain computer printouts corresponding to some of numerical experiments.

2   Estimating the Spectral Radius of Symmetric Positive Semidefinite Matrices

This section is devoted to providing new theoretical tools for estimating, using ASDM, the spectral radius (or leading eigenvalue, Perron root, norm) of symmetric positive definite matrices (or positive semidefinite matrices having at least one nonzero eigenvalue).

Let some real matrix $A \in \mathbb{R}^{n \times n}$ be symmetric (or Hermitian, for definitions see, for instance, [9]). Let us also assume that A is positive semidefinite. Recall that a square matrix A has this type of definiteness if and only if every principal minor of it is nonnegative. Recall also that a matrix is positive semidefinite if and only if all its eigenvalues are nonnegative. For a symmetric real matrix, all its eigenvalues are real.

Suppose that A has at least one nonzero eigenvalue. This assumption is necessary in order to exclude from consideration such a situation as in the following simple example.

Example 1. The matrix $A \in \mathbb{R}^{1\times 1}, A = (0)$ is symmetric positive semidefinite, since its principal minor is nonnegative: $det(A) = 0$. This matrix does not have a nonzero eigenvalue.

For some given matrix A, having solved the so-called characteristic equation of A ($det(A - \bar{\lambda}I) = 0$, where $I \in \mathbb{R}^{n\times n}$ is an identity matrix) one can obtain its eigenvalues. A matrix can have multiple (or repeating) eigenvalues. This means that the characteristic equation may have several roots with a certain multiplicity. Some eigenvalue of A is called simple if its algebraic multiplicity is equal to one.

It is well known that the spectral radius of a symmetric positive semidefinite matrix A is $\varrho(A) = \bar{\lambda}_{max}$, where $\bar{\lambda}_{max}$ is the lagest eigenvalue of A which is usually refered to as the leading (or dominant (majorant)) eigenvalue.

The property of monotonicity of the spectral radius is characterized as follows (see, for instance, [17]): $A \geq B \Rightarrow \varrho(A) \geq \varrho(B)$, where $A, B \in \mathbb{R}_+^{n\times n}$ — the set of all real nonnegative matrices. For matrices, the sign "$\geq$" means that $a_{ij} \geq b_{ij}$, $\forall i, j = 1, 2, \ldots$ We denote the (i, j)-th component of the matrix A and B by a_{ij} and b_{ij}, respectively.

The next lemma helps to establish conditions providing strict monotonicity.

Lemma 1. (monotonicity conditions) [17]. If the elements of $A, B \in \mathbb{R}_+^{n\times n}$ satisfy the following conditions with some positive constant γ:

$$a_{ij} \geq \gamma b_{ij}, \forall i, j = 1, 2, \ldots, \text{ then } \varrho(A) \geq \gamma\varrho(B).$$

The assumptions of Lemma 1 provide strict monotonicity of spectral radii only under the condition that $A + B$ is an irreducible matrix (for details, see [17]).

A nonnegative matrix is called reducible if, after some renumbering of indices, it acquires a block upper-triangular form [9,18]. A matrix is irreducible precisely when the directed graph associated with this matrix is strongly connected.

As is widely known from matrix algebra, the leading eigenvalue of A can be obtained, for example, by solving the following optimization problem:

$$\max_{\|x\|=1} \{f(x) = x^T A x\}. \tag{1}$$

Namely, $\bar{\lambda}_{max} = f(x^*)$, where x^* is the optimizer of (1).

O. Perron ([19]) proved that a real positive square matrix has a unique dominant real eigenvalue.

Theorem 1 (Perron Theorem for positive matrices) [9]. A matrix with positive elements always has a simple real positive eigenvalue that is greater than or equal to the absolute values of all other eigenvalues.

In the real world, nonnegative matrices are applied more frequently than positive ones. F.G. Frobenius ([20]) generalized the Perron Theorem to the case of nonnegative matrices. By a nonnegative matrix one means a matrix which is element-wise nonnegative.

Theorem 2 (Frobenius Theorem for nonnegative matrices) [9]. An irreducible matrix with nonnegative entries always has a simple real positive eigenvalue that is greater than or equal to absolute values of all other eigenvalues.

Thanks to Frobenius estimates (see, for instance, [2]) for nonnegative matrices, the spectral radius of some matrix $A \in \mathbb{R}_+^{n \times n}$ satisfies the inequalities:

$$\min_{i=\overline{1,n}} \sum_{j=1}^{n} a_{ij} \leq \varrho(A) \leq \max_{i=\overline{1,n}} \sum_{j=1}^{n} a_{ij}. \tag{2}$$

Theorem 3 (Lipschitz continuous gradient of a quadratic function). The gradient of a function $f(x) = x^T A x$, associated with a real symmetric matrix A having at least one nonzero eigenvalue, satisfies a Lipschitz condition on the whole space $\mathbb{R}^n$.

Proof. For a symmetric real matrix A, one can always find an orthogonal matrix $C \in \mathbb{R}^{n \times n}$ such that $A' = C^T A C$ is a diagonal matrix in which all elements off the main diagonal are zero [9] (see p. 279). The elements of the main diagonal are determined by the eigenvalues of A (in descending order (including eigenvalues, repeated in accordance with their multiplicity)) as follows

$$A' = \begin{pmatrix} \bar{\lambda}_1 & 0 & \cdots & 0 \\ 0 & \bar{\lambda}_2 & \cdots & 0 \\ \cdots & \cdots & \cdots & \cdots \\ 0 & 0 & \cdots & \bar{\lambda}_n \end{pmatrix}.$$

Recall that some matrix C is called orthogonal if the following conditions hold:

1) C is a nonsingular matrix (this ensures that this matrix is invertible),
2) $C^T = C^{-1}$.

By substituting the variables, namely $x = Cy$, we can arrive at $y = C^{-1}x$. On the one hand, we therefore obtain

$$y^T(C^T A C)y = (C^{-1}x)^T(C^T A C)(C^{-1}x) = x^T(C^T)^T C^T A C C^{-1}x$$
$$= x^T(CC^{-1})A(CC^{-1})x = x^T(IAI)x = x^T A x. \tag{3}$$

On the other hand, we have

$$\bar{f}(y) = y^T A' y = \bar{\lambda}_1 \eta_1^2 + \bar{\lambda}_2 \eta_2^2 + \ldots + \bar{\lambda}_n \eta_n^2, \text{ where } y = (\eta_1, \eta_2, \ldots, \eta_n).$$

Because the function $\bar{f}(y)$ is differentiable, its gradient can be calculated as follows $\nabla \bar{f}(y) = (2\bar{\lambda}_1 \eta_1, 2\bar{\lambda}_2 \eta_2, \ldots, 2\bar{\lambda}_n \eta_n)$. Moreover, we have

$$\|\nabla \bar{f}(y_1) - \nabla \bar{f}(y_2)\| \leq 2\tilde{\lambda}\|y_1 - y_2\|, \forall y_1, y_2 \in \mathbb{R}^n,$$

where $\tilde{\lambda} = \max\limits_{i=\overline{1,n}} |\bar{\lambda}_i| > 0$. This means that $\nabla \bar{f}(y)$ has a Lipschitzian property on the whole space $\mathbb{R}^n$ with the Lipschitz constant $L = 2\tilde{\lambda}$. Taking into account (3), we can assert that the gradient of $f(x) = x^T A x$ also satisfies a Lipschitz condition on the whole space $\mathbb{R}^n$ with the same coefficient L. This is what we wanted to prove. □

Next we will need the following special case of Condition A with $\tau(x,y) = |x-y|^2$ (see [21], p.1081).

Definition 1 (a particular case of Condition A). For a continuous function $f : \mathbb{R}^n \to \mathbb{R}^1$, Condition A is said to be satisfied on the whole space $\mathbb{R}^n$ if and only if there exists a constant $\mu > 0$ such that

$$f(\alpha x + (1-\alpha)y) \geq \alpha f(x) + (1-\alpha)f(y) - \alpha(1-\alpha)\mu\|x-y\|^2,$$
$$\forall x, y \in \mathbb{R}^n, \ \alpha \in [0,1].$$

Theorem 4 (a quadratic function satisfying Condition A). Let $A \in \mathbb{R}^{n \times n}$ be a real symmetric positive semidefinite matrix having the dominant eigenvalue $\bar{\lambda}_{max} > 0$, then a function $f(x) = x^T A x$ satisfies Condition A with the constant $\mu = \bar{\lambda}_{max}$ and function $\tau(x,y) = \|x-y\|^2$.

Proof. Indeed, for all $\alpha \in [0, 1]$ and all $x, y \in \mathbb{R}^n$ we observe

$$f(\alpha x + (1-\alpha)y) = \alpha^2\langle x, Ax \rangle + 2\alpha(1-\alpha)\langle x, Ay \rangle + (1-\alpha)^2\langle y, Ay \rangle$$
$$= \alpha f(x) + (1-\alpha)f(y) - \alpha(1-\alpha)\langle x-y, A(x-y)\rangle.$$

According to the known Rayleigh Theorem (see, for instance, [22]), we have $0 \leq \langle x-y, A(x-y)\rangle \leq \bar{\lambda}_{max}\|x-y\|^2$. Consequently, the assertion of the theorem is true. □

Corollary 1 (interdependence of the Lipschitz constant and the coefficient μ from Condition A). Let $A \in \mathbb{R}^{n \times n}$ be a real symmetric positive semidefinite matrix having the dominant eigenvalue $\bar{\lambda}_{max} > 0$, then $L = 2\mu = 2\bar{\lambda}_{max}$.

Proof. Due to Theorem 3, the gradient of $f(x) = x^T A x$ is Lipschitz continuous with the constant $L = 2\bar{\lambda}_{max}$. The assertion of the corollary then follows from Theorem 4. □

Next, we focus our attention on the application of ASDM to estimate the leading eigenvalue (or spectral radius) of some symmetric positive definite matrix (or positive semidefinite matrix having at least one nonzero eigenvalue). Namely, we propose to use ASDM for solving the following problem

$$\min_{x \in \mathbb{R}^n} \{f(x) = x^T A x\}. \tag{4}$$

Due to the specific property (adaptivity) of ASDM, at the end of the process of solving the problem, we will obtain the adapted value of the coefficient of

ε-normalizing the steepest descent direction (for more details, see [5]). From Theorem 4 and the theoretical aspects of constructing ASDM, it follows that this value allows us to evaluate from above the spectral radius of the matrix A. Computational confirmation of this theoretical fact can be found in the next section (see Examples 2–7). The presented applications of ASDM for approximating the Perron root of positive semidefinite symmetric matrices are novel.

Note that according to the first-order characterization of generalized convexity (see [23]), the positive semidefiniteness of A ensures quasiconvexity of the quadratic function $f(x)$. Note also that for symmetric matrices there is fulfilled the following equality [24]: $\varrho(A) = \|A\|$, where $\|A\| = \max\limits_{\|x\|=1} \|Ax\|$. This means that by solving the problem (4) we can also estimate the matrix norm from above.

3   Numerical Experiments

Computational experiments were carried out on an $Intel^R$ $Core^{TM}$ $I5 - 8400$ CPU @ 2.80 GHz 2.81 GHz computer. For coding of the adaptive steepest descent algorithm, the programming language C^{++} was chosen. The results of numerical experiments are presented in figures constructed using Matlab. In these figures, the first and final values of the $\varepsilon-$parameter are indicated by ε_{start} and $\varepsilon_{end,,}$ respectively. In Appendices, the notations $f(x^*)$ and x^* stand for the computationally obtained optimal value of the objective function and its optimizer. The gradient norm and runtime values are presented in corresponding computer printouts. See also appendices for more information. During testing, we checked the time required to execute a computer program on the processor.

3.1   Practical Applications of ASDM to Estimating the Perron Root of Symmetric Matrices

We have implemented our computational tests with many Laplacian and Pascal matrices (some of them are presented here). Optimization problems, corresponding to these matrices, are characterized as ill-conditioned ones. However, ASDM has been shown to be efficient in estimating the Perron root of these matrices.

Example 2. (estimation of the spectral radius of the Laplacian matrix)

$$f(x) = x^T Ax = 3\xi_1^2 + 3\xi_2^2 + 2\xi_3^2 + 2\xi_4^2 + 2\xi_1\xi_2 + 2\xi_1\xi_3 + 2\xi_1\xi_4 + 2\xi_2\xi_3 + 2\xi_2\xi_4, \text{ where}$$

$$A = \begin{pmatrix} 3 & 1 & 1 & 1 \\ 1 & 3 & 1 & 1 \\ 1 & 1 & 2 & 0 \\ 1 & 1 & 0 & 2 \end{pmatrix}$$

is the Laplacian matrix (or Kirchhoff matrix) associated with some undirected graph (see [6]). This matrix is not only symmetric, but also positive definite.

Indeed, all its principal minors are positive: $\Delta_1 = 3$, $\Delta_2 = 8$, $\Delta_3 = 12$, $\Delta_4 = 16$. The following set of eigenvalues of A can be checked by a simple calculation:

$$\bar{\lambda}_{max} = \bar{\lambda}_1 = \sqrt{5} + 3 \approx 5.2361, \bar{\lambda}_2 = 2, \bar{\lambda}_3 = 2, \bar{\lambda}_4 = -\sqrt{5} + 3 \approx 0.7629.$$

All elements of the considered Laplacian matrix are nonnegative. Therefore, by applying (2), we can also obtain the following lower and upper bounds of the spectral radius of A:

$$4 \leq \varrho(A) \leq 8. \tag{5}$$

Below we use the same notations as in [5]. Next, we will demonstrate how $\varrho(A)$ can be estimated from above by using ASDM. The computer printout in Appendix A illustrates that

$$f(x_k) - f(x^*) = 4.647289e - 20, \ 1/k^6 = 1/1184^6 = 3.629859e - 19.$$

This means that the convergence rate of ASDM (in the sense, $f(x_k) \to f(x^*)$, $k \to +\infty$) for this problem is faster than $O(1/k^6)$. Therefore, this convergence rate is much faster than the sublinear rate $O(1/k)$ which has been theoretically established for ASDM. In addition, we observe that the values of the steplength and the ε-normalization parameter of the descent direction are recalculated only at some initial iterations (namely, $i_k = 193$ when $k = 0$ and $i_k = 2$ when $k = 7$).

Here, i_k denotes the number of diminishing the initial steplength at the current iteration. This value is also used to calculate the parameter ε_k. For the remaining iterations, these parameters are constant. Namely, we have

$$\lambda_k = 0.995, \ \varepsilon_k = 5.235986e + 00, \ i_k = 1, \forall k = \overline{1, 6}.$$
$$\lambda_k = 0.995, \ \varepsilon_k = 5.262298e + 00, \ i_k = 1, \forall k = \overline{8, 1184}.$$

Having chosen the value of parameter ε_{end} (obtained from the previous experiment) as the value of the parameter ε_{start} and preserving the values of the remaining parameters, we implemented the next experiment (i.e. $\varepsilon_{start} = 5.262298e+00$). Since we obtained that $(i_k = 1)$ & $(\varepsilon_k = \varepsilon_{start})$ for all iterations, then, due to Lemma 3.2 from [21], this is characteristic of the case when $\varepsilon_k \geq \mu$. Then, this inequality yields $\varepsilon_{end} \geq \bar{\lambda}_{max}$, where $\varepsilon_{end} = 5.262298e+00$. This computational trick turned out to be quite efficient. It is used to be convinced that the obtained value ε_{end} really exceeds $\mu = \bar{\lambda}_{max}$.

As shown in Fig. 9, the following parameter values correspond to our experiments for Example 2 (the remaining initial parameters are the same as in the computer printout (see Appendix A)):

$$\varepsilon_{start} = (0.1, 1, 2, 3, 4, 5, 5.2361, 6, 7),$$
$$\varepsilon_{end} = (5.245250, 5.254849, 5.262298,$$
$$5.259400, 5.243408, 5.257015, 5.2361, 6, 7).$$

For the last three values of ε_{start} in this list, we observe that

$$(\varepsilon_{end} = \varepsilon_{start}) \ \& \ (i_k = 1, \forall k = 0, 1 \ldots).$$

According to Lemma 3.2 from [21], for these cases, the above equalities mean that

$$\varepsilon_{start} = \varepsilon_{end} \geq \mu = \bar{\lambda}_{max}.$$

For the first six tests (when $\varepsilon_{start} < \bar{\lambda}_{max}$), we get much better upper bounds of $\bar{\lambda}_{max}$ as compared with the theoretical upper bound from (5). Also note that our results are better or about the same as those presented in [6].

Example 3 (estimation of the dominant eigenvalue of the Laplacian matrix corresponding to the Naphthalene graph G).

Consider the problem (4) with the following matrices:

$$A = \begin{pmatrix}
2 & -1 & 0 & 0 & 0 & 0 & 0 & 0 & 0 & -1 \\
-1 & 2 & -1 & 0 & 0 & 0 & 0 & 0 & 0 & 0 \\
0 & -1 & 2 & -1 & 0 & 0 & 0 & 0 & 0 & 0 \\
0 & 0 & -1 & 2 & -1 & 0 & 0 & 0 & 0 & 0 \\
0 & 0 & 0 & -1 & 3 & -1 & 0 & 0 & 0 & -1 \\
0 & 0 & 0 & 0 & -1 & 2 & -1 & 0 & 0 & 0 \\
0 & 0 & 0 & 0 & 0 & -1 & 2 & -1 & 0 & 0 \\
0 & 0 & 0 & 0 & 0 & 0 & -1 & 2 & -1 & 0 \\
0 & 0 & 0 & 0 & 0 & 0 & 0 & -1 & 2 & -1 \\
-1 & 0 & 0 & 0 & -1 & 0 & 0 & 0 & -1 & 3
\end{pmatrix}, B = \begin{pmatrix}
0 & 1 & 0 & 0 & 0 & 0 & 0 & 0 & 0 & 1 \\
1 & 0 & 1 & 0 & 0 & 0 & 0 & 0 & 0 & 0 \\
0 & 1 & 0 & 1 & 0 & 0 & 0 & 0 & 0 & 0 \\
0 & 0 & 1 & 0 & 1 & 0 & 0 & 0 & 0 & 0 \\
0 & 0 & 0 & 1 & 0 & 1 & 0 & 0 & 0 & 1 \\
0 & 0 & 0 & 0 & 1 & 0 & 1 & 0 & 0 & 0 \\
0 & 0 & 0 & 0 & 0 & 1 & 0 & 1 & 0 & 0 \\
0 & 0 & 0 & 0 & 0 & 0 & 1 & 0 & 1 & 0 \\
0 & 0 & 0 & 0 & 0 & 0 & 0 & 1 & 0 & 1 \\
1 & 0 & 0 & 0 & 1 & 0 & 0 & 0 & 1 & 0
\end{pmatrix}.$$

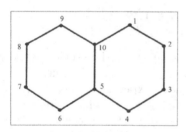

Fig. 1. The molecular graph of Naphthalene.

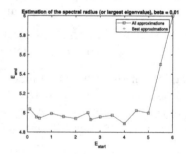

Fig. 2. Results of experiments using ASDM with Rule 1 for the Laplacian matrix corresponding to the Naphthalene graph.

Here, A is the Laplacian matrix of a simple graph G not having loops and multiple edges (see, for instance, [7]). It is a real symmetric $n \times n$ matrix, $n = 10$. The matrix B represents a sparse (square) adjacency matrix of the Naphthalene graph G (see Fig. 1). For $i, j = \overline{1, n}$, we have

$$b_{ij} = \begin{cases} 1, & \text{if } (i, j) \text{ is an edge in } G, \\ 0, & \text{otherwise.} \end{cases}$$

If $B = B^T$,, then the graph is said to be undirected (all edges are bidirectional). Let $D \in \mathbb{R}^{n \times n}$ be a diagonal matrix. Each diagonal element $d_{ii} = \sum_{j=1}^{n} b_{ij}$ represents the degree (or valency) of the vertex (or node) i in G. Now, the Laplacian matrix associated with G can be expressed as the following difference of matrices: $A = D - B$. We calculated all the principal minors of the matrix A using the Matlab R2021b software package: $\Delta_i \geq 0$, $i = \overline{1, 10}$. We utilized Matlab to calculate the leading eigenvalue of A with the purpose of checking the results obtained using ASDM: $\bar{\lambda}_{max} = \bar{\lambda}_1 \approx 4.8608058531$. The smallest eigenvalue of the Laplacian matrix is always equal to zero. In this case, the Perron-Frobenius Theorem can not be applied to estimate the spectral radius of A (since there are some negative entries in this matrix). Owing to the obtained results, the matrix A can be considered as being positive semidefinite. In a certain sense, the function $f(x)$ can be considered as being singular. Indeed, one can easily see that the gradient of $f(x)$ is equal to the zero vector always when $\xi_1 = \xi_2 = \ldots = \xi_{10}$.

The following parameter values of our experiments with ASDM are illustrated in Fig. 2:

$$\varepsilon_{start} = (0.1, 0.382, 0.5, 1, 1.5, 2, 2.5, 2.618, 3, 3.5, 4, 4.5, 5, 5.5, 6),$$
$$\varepsilon_{end} = (5.038185, 4.955514, 4.944765, 4.993084, 4.960245, 4.941542,$$
$$5.001630, 4.931197, 4.958628, 4.975494, 4.890532, 5.026033, 5, 5.5, 6).$$

For the first twelve tests (when $\varepsilon_{start} < \bar{\lambda}_{max}$), we get the upper estimates of $\bar{\lambda}_{max}$ which are sufficiently close to its real value. For the remaining tests (when we have $\varepsilon_{start} \geq \bar{\lambda}_{max}$), we refer the interested reader to Example 2, where we explained in detail how (using some computational trick which is based on theoretical results) one can be convinced that ε_{end} really exceeds $\bar{\lambda}_{max}$. The computer printout in Appendix B shows the work of ASDM (for instance, the rate of convergence, the number of iterations and etc) for Example 3.

Example 4 (estimation of the Perron root of the Laplacian matrix corresponding to the Carbon Skeleton of 2,4-Dimethylhexane).

Let us have matrices

$$A = \begin{pmatrix} 1 & -1 & 0 & 0 & 0 & 0 & 0 & 0 \\ -1 & 3 & -1 & 0 & 0 & 0 & -1 & 0 \\ 0 & -1 & 2 & -1 & 0 & 0 & 0 & 0 \\ 0 & 0 & -1 & 3 & -1 & 0 & 0 & -1 \\ 0 & 0 & 0 & -1 & 2 & -1 & 0 & 0 \\ 0 & 0 & 0 & 0 & -1 & 1 & 0 & 0 \\ 0 & -1 & 0 & 0 & 0 & 0 & 1 & 0 \\ 0 & 0 & 0 & -1 & 0 & 0 & 0 & 1 \end{pmatrix}, B = \begin{pmatrix} 0 & 1 & 0 & 0 & 0 & 0 & 0 & 0 \\ 1 & 0 & 1 & 0 & 0 & 0 & 1 & 0 \\ 0 & 1 & 0 & 1 & 0 & 0 & 0 & 0 \\ 0 & 0 & 1 & 0 & 1 & 0 & 0 & 1 \\ 0 & 0 & 0 & 1 & 0 & 1 & 0 & 0 \\ 0 & 0 & 0 & 0 & 1 & 0 & 0 & 0 \\ 0 & 1 & 0 & 0 & 0 & 0 & 0 & 0 \\ 0 & 0 & 0 & 1 & 0 & 0 & 0 & 0 \end{pmatrix},$$

which are the Laplacian matrix corresponding to the Carbon Skeleton of 2,4-Dimethylhexane and its adjacency matrix, respectively (see Fig. 3). We calculated all the principal minors for the matrix A by the Matlab R2021b software package: $\Delta_i \geq 0$, $i = \overline{1, 8}$. Our following calculation of the leading eigenvalue of

A in Matlab serves the purpose of verifying the results obtained using ASDM: $\bar{\lambda}_{max} \approx 4.4763465754$. In this case, the Perron-Frobenius Theorem can not be used to estimate the spectral radius of A (since this matrix has some negative elements). From the above, it follows that the matrix A can be considered as being positive semidefinite. In a certain sense, the function $f(x) = x^T A x$ can be considered as being singular. Indeed, we can easily observe that the gradient of $f(x)$ is equal to the zero vector whenever $\xi_1 = \xi_2 = \ldots = \xi_8$.

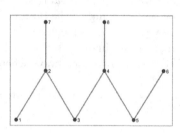

Fig. 3. The Carbon Skeleton of 2,4-Dimethylhexane.

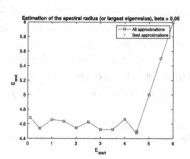

Fig. 4. Results of experiments using ASDM with Rule 1 for the Laplacian matrix corresponding to the Carbon Skeleton of 2,4-Dimethylhexane.

The next parameter values corresponding to our experiments by making use of ASDM are reported in Fig. 4:

$$\varepsilon_{start} = (0.1,\ 0.5,\ 1,\ 1.5,\ 2,\ 2.5,\ 3,\ 3.5,\ 4,\ 4.476347,\ 4.5,\ 5,\ 5.5,\ 6),$$
$$\varepsilon_{end} = (4.685216,\ 4.537901,\ 4.658991,\ 4.636304,\ 4.544146,\ 4.626545,$$
$$4.522019,\ 4.523244,\ 4.665403,\ 4.476347,\ 4.5,\ 5,\ 5.5,\ 6).$$

For the first nine tests (in the case of $\varepsilon_{start} < \bar{\lambda}_{max}$), we obtain the majorants of $\bar{\lambda}_{max}$ which are close enough to its real value. For the rest of the experiments (when $\varepsilon_{start} \geq \bar{\lambda}_{max}$), we refer the interested reader to Example 2, where we described in detail how (making use of some computational trick which is based on the theory) we can be convinced that ε_{end} really majorises $\bar{\lambda}_{max}$ (Fig. 6).

The computer printout in Appendix C corresponds to Example 4.

Example 5 (estimation of the spectral radius of the Laplacian matrix associated with the carbon skeleton of 1,4-Divinylbenzene)

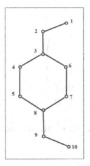

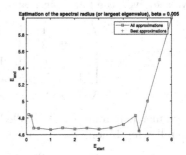

Fig. 5. The Carbon Skeleton of 1,4-Divinylbenzene.

Fig. 6. Results of experiments using ASDM with Rule 1 for the Laplacian matrix associated with the Carbon Skeleton of 1,4-Divinylbenzene.

Suppose that we are given the following matrices:

$$A = \begin{pmatrix} 1 & -1 & 0 & 0 & 0 & 0 & 0 & 0 & 0 & 0 \\ -1 & 2 & -1 & 0 & 0 & 0 & 0 & 0 & 0 & 0 \\ 0 & -1 & 3 & -1 & 0 & 0 & 0 & -1 & 0 & 0 \\ 0 & 0 & -1 & 2 & -1 & 0 & 0 & 0 & 0 & 0 \\ 0 & 0 & 0 & -1 & 2 & -1 & 0 & 0 & 0 & 0 \\ 0 & 0 & 0 & 0 & -1 & 3 & -1 & 0 & -1 & 0 \\ 0 & 0 & 0 & 0 & 0 & -1 & 2 & -1 & 0 & 0 \\ 0 & 0 & -1 & 0 & 0 & 0 & -1 & 2 & 0 & 0 \\ 0 & 0 & 0 & 0 & 0 & -1 & 0 & 0 & 2 & -1 \\ 0 & 0 & 0 & 0 & 0 & 0 & 0 & 0 & -1 & 1 \end{pmatrix}, B = \begin{pmatrix} 0 & 1 & 0 & 0 & 0 & 0 & 0 & 0 & 0 & 0 \\ 1 & 0 & 1 & 0 & 0 & 0 & 0 & 0 & 0 & 0 \\ 0 & 1 & 0 & 1 & 0 & 0 & 0 & 1 & 0 & 0 \\ 0 & 0 & 1 & 0 & 1 & 0 & 0 & 0 & 0 & 0 \\ 0 & 0 & 0 & 1 & 0 & 1 & 0 & 0 & 0 & 0 \\ 0 & 0 & 0 & 0 & 1 & 0 & 1 & 0 & 1 & 0 \\ 0 & 0 & 0 & 0 & 0 & 1 & 0 & 1 & 0 & 0 \\ 0 & 0 & 1 & 0 & 0 & 0 & 1 & 0 & 0 & 0 \\ 0 & 0 & 0 & 0 & 0 & 1 & 0 & 0 & 0 & 1 \\ 0 & 0 & 0 & 0 & 0 & 0 & 0 & 0 & 1 & 0 \end{pmatrix}.$$

The Laplacian matrix A is positive semidefinite, since all of its principal minors are nonnegative: $\Delta_i > 0$, $i = \overline{1, 9}$, $\Delta_{10} \approx 0$. The Perron root of A is equal to $\bar{\lambda}_{max} = \varrho(A) \approx 4.6411864762$. According to calculations in Matlab, the smallest eigenvalue of A equals $2.5234e{-}16$. Consequently, the condition number of A is very large. This allows us to conclude that the problem of minimizing the function $f(x) = x^T A x$ is very ill-conditioned. The following parameter values corresponding to our computational experiments with ASDM are demonstrated in Fig. 6:

$$\varepsilon_{start} = (0.1, 0.2, 0.3, 0.5, 1, 1.5, 2, 2.5, 3, 3.5, 4, 4.5, 4.6412, 5, 5.5, 6),$$
$$\varepsilon_{end} = (4.841005, 4.823627, 4.678138, 4.676016, 4.659231, 4.680065; 4.665835,$$
$$4.677943, 4.663266, 4.680900, 4.719525, 4.827134, 4.6412, 5, 5.5, 6).$$

Example 6 (estimation of the spectral radius of the Pascal matrix)

We are given the following matrices:

$$A = \begin{pmatrix} 1 & 1 & 1 & 1 & 1 \\ 1 & 2 & 3 & 4 & 5 \\ 1 & 3 & 6 & 10 & 15 \\ 1 & 4 & 10 & 20 & 35 \\ 1 & 5 & 15 & 35 & 70 \end{pmatrix}, \quad B = \begin{pmatrix} 0 & 1 & 1 & 1 & 1 \\ 1 & 0 & 0 & 1 & 1 \\ 1 & 0 & 0 & 1 & 1 \\ 1 & 1 & 1 & 0 & 1 \\ 1 & 1 & 1 & 1 & 0 \end{pmatrix}.$$

Any symmetric real matrix A is called a Pascal matrix if its elements are defined as the following binomial coefficients:

$$a_{ij} = \binom{i+j-2}{j-1} = \frac{(i+j-2)!}{(j-1)!}, \quad i, j = 1, 2, \ldots n.$$

Note that the Pascal matrices have many applications in linear algebra, combinatorics, filter design, graph theory and etc. (see, for instance, [8–12]). The matrix B is the adjacency matrix for the graph G corresponding to this Pascal matrix (see Fig. 7). By analysing its principal minors, one can easily verify that the matrix A is positive definite: $\Delta_1 = \Delta_2 = \Delta_3 = \Delta_4 = \Delta_5 = 1$. Owing to the small dimension of A, one can quite easily calculate its eigenvalues (which are positive):

$$\bar{\lambda}_{max} = \bar{\lambda}_1 \approx 92.290435, \ \bar{\lambda}_2 \approx 5.517487, \ \bar{\lambda}_3 = 1, \ \bar{\lambda}_4 \approx 0.181242, \ \bar{\lambda}_5 \approx 0.010835.$$

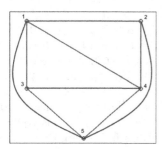

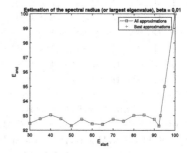

Fig. 7. The graph G associated with the Pascal matrix.

Fig. 8. Results of experiments using ASDM with Rule 1 for the Pascal matrix.

Consequently, the matrix A has the following spectral radius: $\varrho(A) = 92.290435$. The condition number of the matrix A equals $\dfrac{92.290435}{0.010835} \approx 8517.8066$, i.e. it is large enough. This characterizes the problem of minimizing $f(x)$ as very ill-conditioned.

All entries of the Pascal matrix are positive. Therefore making use of (2), one can get the lower and upper bounds of the spectral radius of A as follows

$$5 \le \varrho(A) \le 126. \tag{6}$$

The following parameters correspond to our numerical experiments with the purpose of the estimation of $\varrho(A)$ for the given Pascal matrix (for the illustration of the results, see Fig. 8):

$$\varepsilon_{start} = (30, 35, 40, 45, 50, 55, 60, 65, 70, 75,$$
$$80, 85, 90, 92.290435, 93, 95, 100),$$
$$\varepsilon_{end} = (92.46544, 92.77996, 93.04727, 92.78516, 92.30456, 92.75398,$$
$$92.43531, 92.40204, 92.74972, 92.62371, 93.01694, 93.04696,$$
$$92.75491, 92.290435, 93, 95, 100).$$

For the first thirteen tests (when $\varepsilon_{start} < \bar{\lambda}_{max}$), we get the much better upper estimates of $\bar{\lambda}_{max}$ as compared to the theoretical upper bound from (6). In Example 2, we described in detail how (using some computational trick) one can be completely convinced that ε_{end} exceeds $\bar{\lambda}_{max}$. The computer printout in Appendix E demonstrates that

$$f(x_k) - f(x^*) = 2.307013e - 19, \quad \frac{1}{k^4} = 1.0572041e - 19.$$

Consequently, the convergence rate of ASDM (in the sense, $f(x_k) \to f(x^*)$, $k \to +\infty$) for this problem is nearly the same as $O\left(\dfrac{1}{k^4}\right)$.

Example 7 (comparative analysis of approximations of the Perron root for the positive definite symmetric matrix from [15])

Let us assume that the goal function in (4) is specified by

$$A = \begin{pmatrix} 1 & 1 & 1 & 1 & 1 & 1 \\ 1 & 2 & 2 & 2 & 2 & 2 \\ 1 & 2 & 3 & 3 & 3 & 3 \\ 1 & 2 & 3 & 4 & 4 & 4 \\ 1 & 2 & 3 & 4 & 5 & 5 \\ 1 & 2 & 3 & 4 & 5 & 6 \end{pmatrix}.$$

This matrix is positive definite, since all of its principal minors are positive: $\Delta_1 = \Delta_2 = \Delta_3 = \Delta_4 = \Delta_5 = \Delta_6 = 1$. Due to the Perron-Frobenius formula for positive matrices, the spectral radius of A satisfies the inequalities:

$$6 \leq \varrho(A) \leq 21. \tag{7}$$

According to Brauer's and Lu's formulas (see [14]), we have, respectively:

$$7.899 \leq \varrho(A) \leq 20.2596, \tag{8}$$
$$15.6944 \leq \varrho(A) \leq 18.0498. \tag{9}$$

These upper bounds for $\varrho(A)$ can be evaluated as good enough, since the calculated in Matlab value $\varrho(A) \approx 17.2068572674$.

The following parameter values for our tests with ASDM are shown in Fig. 10:

$$\varepsilon_{start} = (1, 4, 5, 6, 7, \mathit{9}, \mathit{10}, 13, \mathit{15}, 16, 17, \mathit{17.35}, 18, 19),$$

$$\varepsilon_{end} = (17.67983, 17.70416, 18.02519, 17.61796, 17.62273, \mathit{17.53217},$$

$$\mathit{17.58087}, 17.68486, \mathit{17.49526}, 17.72853, 17.89474, \mathit{17.350}, 18, 19).$$

For the first eleven experiments (when $\varepsilon_{start} < \bar{\lambda}_{max}$), we obtain the better upper bounds for $\varrho(A)$ than in (7)-(9). For the rest of the cases (when $\varepsilon_{start} \geq \bar{\lambda}_{max}$), the obtained parameter values $(i_k = 1, \forall k = 0, 1, \ldots) \& (\varepsilon_{end} = \varepsilon_{start})$ clearly demonstrate that ε_{end} exceeds $\bar{\lambda}_{max}$.

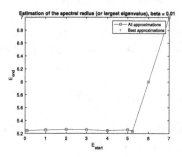

Fig. 9. Results of experiments using ASDM with Rule 1 for the Laplacian matrix given in Example 2.

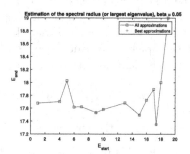

Fig. 10. Results of experiments using ASDM with Rule 1 for the positive definite matrix from [15].

4   Concluding Discussion

We proposed a novel application of the fully adaptive steepest descent method (ASDM). Namely, we theoretically and experimentally investigated the ability of ASDM to approximate the Perron-Frobenius root of symmetric positive definite matrices (or positive semidefinite matrices having at least one nonzero eigenvalue (including Laplacian and Pascal matrices)). This type of matrices have many applications in deep learning, digital filter design, network analysis, graph theory, chemistry, linear algebra, statistics, and mathematical physics. For the convenience of readers, the computational results are visualized in the form of computer-drawn figures. We analyzed the results of our calculations in comparison with some others previously presented in the literature. We have shown that in practice the speed of convergence of ASDM is much faster than the sublinear rate of convergence confirmed in theory. This is not only very exciting but useful from both the theoretical and practical points of view for various cutting-edge research fields.

Appendix A.   (The computer printout corresponding to Example 2)

```
dimension of space = 4, Eps_start =  2.000000e+00,  beta = 0.005,  Eta = 0.995, Rule 1
x_0 = (1.000000000, 1.000000000, 1.000000000, 1.000000000),  f(x_0) = 2.000000e+01
i_k = 193,  k= 0, Eps_k = 5.235986e+00
i_k = 2,   k= 7, Eps_k = 5.262298e+00
i_k = 1,   k= 1184, Eps_k = 5.262298e+00
k= 1184, 1/(k^6)= 3.629859e-19
x_k = (5.666743e-11, 5.666743e-11, 3.502240e-11, 3.502240e-11)
gradient norm = 9.865804e-10, f(x_k)= 4.647289e-20
Eps_end = 5.262298e+00, Time taken: 0.0030s
```

Appendix B.   (The computer printout associated with Example 3)

```
dimension of space = 10, Eps_start =  4.000000e+00, beta =  0.010, Eta =  0.990, Rule 1
x_0 = (1, 0, 0, 0, 0, 0, 0, 0, 0, 0), f(x_0) = 2.000000e+00
i_k = 9, k = 1, Eps_k = 4.334894e+00
i_k = 9, k = 2, Eps_k = 4.697825e+00
i_k = 4, k = 3, Eps_k = 4.841627e+00
i_k = 2, k = 4, Eps_k = 4.890532e+00
i_k = 1, k = 474, Eps_k = 4.890532e+00
k = 474, 1/(k^5) = 4.179346e-14
x_k = (1.000000e-01, 9.999999e-02, 1.000000e-01, 9.999997e-02, 1.000001e-01,
9.999987e-02, 1.000000e-01, 9.999999e-02, 1.000000e-01, 9.999994e-02)
gradient norm = 9.775916e-07, f(x_k)= 4.915512e-14
Eps_end = 4.890532e+00, Time taken: 0.0050s
```

Appendix C.   (The computer printout corresponding to Example 4)

```
dimension of space = 8, Eps_start =  3.000000e+00, beta =  0.050, Eta =  0.950, Rule 1
x_0 = (1, 1, 0, 0, 0, 0, 0, 0), f(x_0) = 2.000000e+00
i_k = 6, k= 0, Eps_k = 3.877066e+00
i_k = 3, k= 1, Eps_k = 4.295918e+00
i_k = 2, k= 3, Eps_k = 4.522019e+00
k= 160, 1/(k^6)= 5.960464e-14
x_k = (2.500001e-01, 2.500001e-01, 2.500000e-01,
2.500000e-01, 2.499999e-01, 2.499999e-01, 2.500001e-01, 2.499999e-01)
gradient norm = 9.565216e-08, f(x_k)= 1.067202e-14
Eps_end = 4.522019e+00, Time taken: 0.0020s
```

Appendix D.   (The computer printout corresponding to Example 5)

```
dimension of space = 10, Eps_start = 0.500000000, beta = 0.005, Eta = 0.995, Rule 1
x_0 =(0,  0,  0,  0,  0,  0,  0,  -1,  1,  0), f(x_0)= 4.000000e+00
i_k = 293, k= 0, Eps_k = 2.160879e+00
i_k = 155, k= 1, Eps_k = 4.676016e+00
k= 802, 1/(k^6)= 3.757974e-18
x_k = (6.344928e-11, -2.310309e-10, 5.467462e-10, -3.331409e-10, 3.331408e-10,
-5.467463e-10, 3.331408e-10, -3.331409e-10, 2.310308e-10, -6.344939e-11)
gradient norm = 9.982743e-09, f(x_k)= 5.367979e-18
Eps_end = 4.676016e+00, Time taken: 0.0040s
```

Appendix E.   (The computer printout associated with Example 6)

```
dimension of space = 5, Eps_start =  5.000000e+01, beta =  0.010, Eta =  0.990, Rule 1
x_0 = (1.0000000,1.0000000,1.000000000,1.0000000,1.0000000), f(x_0) = 2.510000e+02
i_k = 62, k= 0, Eps_k = 9.230456e+01
k = 55433, 1/(k^4)= 1.0572041e-19
x_k = (7.75359e-10,-2.54562e-09,3.24174e-09,-1.87849e-09,4.15402e-10)
gradient norm = 9.999463e-11, f(x_k)= 2.307013e-19
Eps_end = 9.230456e+01, Time taken: 0.0360s
```

References

1. Liu, B.: On an upper bound of the spectral radius of graphs. Discrete Math. **308**, 5317–5324 (2008)
2. Horn, R.-R., Johnson, C.-R.: Matrix Analysis, 2nd edn. Univ Press, Cambridge (2019)
3. Liu, S.-L.: Bounds for the greatest characteristic root of a nonnegative matrix. Linear Algebra Appl. **239**, 151–160 (1996)
4. Xing, R., Zhou, B.: Sharp bounds on the spectral radius of a nonnegative matrices. Linear Algebra Appl. **449**, 194–209 (2014)
5. Gabidullina, Z.-R.: A Fully Adaptive Steepest Descent Method (2021). https://arxiv.org/abs/2108.05027
6. Adam, M., Aggeli, D., Aretaki, A.: Some new bounds on the spectral radius of nonnegative matrices. AIMS Mathematics **5**(1), 701–716 (2020)
7. Trinajstic, N., Babic, D., Nikolic, S., Plavsic, D., Amic, D., Mihalic, Z.: The Laplacian matrix in chemistry. J. Chem. Inf. Comput. Sci. **34**(2), 368–376 (1994)
8. Meyer, C.: Matrix Analysis and Applied Linear Algebra. SIAM, Philadelphia (2000)
9. Gantmacher, F.-R.: The Theory of Matrices. AMS Chelsea Publishing, New York (1960)
10. Cheon, G.-S., Kim, J.S., Mojallal, S.A.: Spectral properties of Pascal graphs. Linear Multilinear Algebra **66**(7), 1–15 (2017)
11. Psenicka, B., Garcia-Ugalde, F., Herrera-Camacho, A.: The bilinear Z transform by Pascal matrix and its application in the design of digital filters. IEEE Signal Process. Lett. **9**(11), 368–370 (2002)
12. Konopacki, J.: The Frequency transformation by matrix operation and its application in IIR filters design. IEEE Signal Process. Lett. **12**(1), 5–8 (2005)
13. O'Leary, D.-P., Stewart, G.-W., Vandergraft, J.-S.: Estimating the largest eigenvalue of a positive definite matrix. Math. Comput. **33**(148), 1289–1292 (1979)
14. Lin-Zhang, L.: Perron complement and Perron root. Linear Algebra Appl. **341**(1), 239–248 (2002)
15. Neumann, M.: Inverse of Perron complements of inverse M-matrices. Linear Algebra Appl. **313**, 163–171 (2000)
16. Goodfellow, I., Bengio, Y., Courville, A.: Deep Learning. The MIT Press, Cambridge (2016)
17. Nesterov, Y., Protasov, V.-Y.: Computing closest stable nonnegative matrix. SIAM J. Matrix Anal. Appl. **41**(1), 1–28 (2020)
18. Pinkus, A.: Totally Positive Matrices. University Press, Cambridge (2010)
19. Perron, O.: Zur Theorie der Matrices. Mathematiche Annalen **64**(2), 248–263 (1907)
20. Frobenius, F.-G.: Uber matrizen aus nicht negativen elementen. S.-B. Preuss Akad. Wiss. 456–477 (1912)
21. Gabidullina, Z.-R.: Adaptive conditional gradient method. J. Optim. Theory Appl. **183**(3), 1077–1098 (2019)
22. Lancaster, P., Tismenetsky, M.: The Theory of Matrices: With Applications. Academic Press Inc, London (1985)
23. Crouzeix, J.-P.: First and Second Order Characterization of generalized Convexity. Summer school on Generalized Convexity, Samos (1999)
24. Polyak, B.: Introduction to Optimization. Optimization Software. Inc., Publications Division, , New York (1987)

How to Use Barriers and Symmetric Regularization of Lagrange Function in Analysis of Improper Nonlinear Programming Problems

Leonid D. Popov[1,2]([✉])[iD]

[1] Krasovskii Institute of Mathematics and Mechanics UB RAS, Ekaterinburg, Russia
popld@imm.uran.ru, l.d.popov@urfu.ru
[2] Ural Federal University, Ekaterinburg, Russia

Abstract. In the note, we present a new approach to numerical analysis of improper mathematical programming problems based on ideas of symmetrically regularization of their Lagrange functions, additionally equipped with barrier terms for both groups of variables, primal and dual. It makes it possibly not to specify in advance the type of incorrectness of the problem being solved as well as apply second-order optimization methods for them. The description of the approach, convergence theorems and meaningful interpretation of the obtained generalized solutions are given.

Keywords: nonlinear programming · improper (ill-posed) problems · generalized solutions · penalty and barrier methods · regularization

1 Introduction

In the theory of mathematical programming, the reducing of the original optimization problem to the problem of finding the saddle point of its Lagrange function plays a central role. If such reduction can be successfully carried out for some problem, then the latter is called regular or proper one. Otherwise, the problem is called improper [1,2].

In linear programming, a problem may be improper only if its constraints and/or the constraints in the dual problem are incompatible. This creates a simple classification of improper linear programs [2]. In the non-linear case, the causes of the impropriety are more diverse.

In practice, the inconsistency of constraint systems in optimization problems can be caused by errors in setting their initial data and even by the presence of real contradictions in economic or technical objects (subjects) modeled by these systems [2–5]. Therefore it would be better if numerical methods applied to improper optimization model not only stated its insolubility, but also gave a

The work was performed as part of research conducted in the Ural Mathematical Center with the financial support of the Ministry of Science and Higher Education of the Russian Federation (Agreement number 075-02-2024-1377).

A. Eremeev et al. (Eds.): MOTOR 2024, LNCS 14766, pp. 55–68, 2024.
https://doi.org/10.1007/978-3-031-62792-7_4

complete list of its ≪bottlenecks≫ and estimated the depth of its contradictions. It would be even better if reasonable options for adjusting of improper model were automatically proposed and a solution to the adjusted problem was given (such a solution is often called a generalized one).

In the scientific literature, the development and analysis of such methods has been going on for a long time. The very various tools to analysis and correction of improper problems of mathematical programming have been proposed (see, for example, [2, 5–17]). Among them one can find original duality schemes for such problems, their regularization according to Tikhonov, the use of various penalty functions, including barrier functions, the inclusion of multicriteria optimization methods, and many others.

Below, we present a new approach to solve discussed topics based on a simultaneous and symmetrical inclusion of both regularizing and barrier terms in the classical Lagrange function of the initial improper problem. This approach makes it possible not to specify in advance the type of impropriety of this problem and effectively apply second-order optimization methods for its numerical analysis. Convergence theorems are proved for the method and a meaningful interpretation of the obtained generalized solution is discussed. Similar results for linear programming problems were given earlier in [18].

2   Problem Statement, Definitions and Assumptions

Let's consider a convex program, may be improper,

$$\inf\{f_0(x)\colon f_j(x) \le 0 \ (j = 1, ..., m), \ x \ge 0\}, \tag{1}$$

where $x \in \mathbf{R}^n$ is a vector of primal variables, and the functions $f_j(x) : \mathbf{R}^n \to \mathbf{R}$ are convex and twice continuously differentiable $(j = 0, 1, ..., m)$. The Lagrange function of this problem is

$$L(x, y) = f_0(x) + \sum_{j=1}^{m} y_j f_j(x), \quad x \ge 0, \quad y \ge 0,$$

and it generates a standard dual problem

$$\sup_{y \ge 0} \inf_{x \ge 0} L(x, y). \tag{2}$$

In a proper case, the optimal values of the problems (1) and (2) are finite, attainable, and coincide with each other. Besides, their optimal vectors form a saddle point of the Lagrange function $L(x, y)$ with respect to the domain $\mathbf{R}^n_+ \times \mathbf{R}^m_+$, where $\mathbf{R}^s_+ = \{z \in \mathbf{R}^s : z \ge 0\}$. In the improper case, these relations are violated.

Obviously, being improper, the problem (1) must be corrected (adjusted).

We restrict ourselves to a linear correction scheme of (1) as the most simple and natural one (see also [1, 2]). According to this scheme, the original problem is immersed in a parametric family of the convex programs

$$\inf\{f_0(x) + \langle \Delta c, x \rangle \colon f_j(x) \le \Delta b_j \ (j = 1, ..., m), \ x \ge 0\}, \tag{3}$$

where $\Delta c \in \mathbf{R}^n$ and $\Delta b = (\Delta b, ..., \Delta b_m) \in \mathbf{R}^m$ are the vectors of correction parameters, $\langle \cdot , \cdot \rangle$ denotes the scalar product of vectors. Accordingly, the dual problem (2) will be placed in a parametric family of the problems

$$\sup_{y \geq 0} \inf_{x \geq 0} \left[L(x,y) + \langle \Delta c, x \rangle - \langle y, \Delta b \rangle \right]. \tag{4}$$

Let discuss a structure of the set

$$\Omega(L) = \left\{ (\Delta c, \Delta b) \in \mathbf{R}^n \times \mathbf{R}^m : L(x,y) + \langle \Delta c, x \rangle - \langle y, \Delta b \rangle \text{ has a saddle point} \right\}.$$

To do that, introduce the sets $\Omega_1 = \{ \Delta c : \exists y \geq 0 \inf_{x \geq 0} \left[L(x,y) + \langle \Delta c, x \rangle \right] > -\infty \}$ and $\Omega_2 = \{ \Delta b : \exists x \geq 0 \sup_{y \geq 0} \left[L(x,y) - \langle \Delta b, y \rangle \right] < +\infty \}$. Due to the conjugate saddle functions theory by Rockafellar R.T. [19]

$$\mathsf{ri}(\Omega_1 \times \Omega_2) \subseteq \Omega(L) \subseteq \Omega_1 \times \Omega_2, \quad \mathsf{cl}\Omega(L) = \mathsf{cl}(\Omega_1 \times \Omega_2), \quad \mathsf{ri}(\mathsf{cl}\Omega(L)) = \mathsf{ri}(\Omega_1 \times \Omega_2),$$

where cl denotes the closure of the corresponding set, and ri is its relative interior. Moreover, the sets Ω_1 and Ω_2 are always non-empty, while the sets $\mathsf{cl}\Omega_1$ and $\mathsf{cl}\Omega_2$ are convex.

The statements and properties above allow us to define the unique vectors of optimal linear correction of the original improper problems (1), (2) as

$$\overline{\Delta c} = \arg \min_{\Delta c \in \mathsf{cl}\Omega_1} \frac{1}{2} \|\Delta c\|^2 \quad \text{and} \quad \overline{\Delta b} = \arg \min_{\Delta b \in \mathsf{cl}\Omega_2} \frac{1}{2} \|\Delta b\|^2,$$

where $\|x\|$ and $\|y\|$ are Euclidean norms in the corresponding spaces. Obviously, if the problem (1) is proper, then $\overline{\Delta c} = 0$, $\overline{\Delta b} = 0$.

Using the vectors $\overline{\Delta c}$ and $\overline{\Delta b}$ introduced above, let's write out the optimally corrected primal and dual problems as

$$\inf \{ f_0(x) + \langle \overline{\Delta c}, x \rangle : f_j(x) \leq \overline{\Delta b}_j \ (j = 1, ..., m), \ x \geq 0 \}, \tag{5}$$

$$\sup_{y \geq 0} \inf_{x \geq 0} \left[L(x,y) + \langle \overline{\Delta c}, x \rangle - \langle y, \overline{\Delta b} \rangle \right]. \tag{6}$$

Everywhere below, we will assume that these problems are proper, and will denote their common optimal value by $\overline{\mathbf{opt}}$. Also we will denote their optimal vectors of minimal norms by $\bar{x} \geq 0$ and $\bar{y} \geq 0$ respectively. Remind, that these vectors satisfy typical KKT-conditions [4]

$$\nabla_x L(\bar{x}, \bar{y}) + \overline{\Delta c} \geq 0, \quad \langle \bar{x}, \nabla_x L(\bar{x}, \bar{y}) + \overline{\Delta c} \rangle = 0, \tag{7}$$

$$F(\bar{x}) \leq \overline{\Delta b}, \quad \langle \bar{y}, F(\bar{x}) - \overline{\Delta b} \rangle = 0. \tag{8}$$

Here

$$\nabla_x L(x,y) = \nabla f_0(x) + J(x)^T y, \quad F(x) = \begin{pmatrix} f(x) \\ \vdots \\ f_m(x) \end{pmatrix}, \quad J(x) = \begin{pmatrix} \nabla f(x)^T \\ \vdots \\ \nabla f_m(x)^T \end{pmatrix}.$$

Concluding the section, we note that the vectors $\bar{\Delta c}$ and $\bar{\Delta b}$ introduced above satisfy the standard optimality conditions

$$\langle \bar{\Delta c}, \bar{\Delta c} - \Delta c \rangle \leq 0 \quad (\forall \Delta c \in \Omega_1) \quad \text{and} \quad \langle \bar{\Delta b}, \bar{\Delta b} - \Delta b \rangle \leq 0 \quad (\forall \Delta b \in \Omega_2),$$

or, what is the same, the inequalities hold

$$\|\bar{\Delta c} - \Delta c\|^2 \leq \langle \Delta c, \Delta c - \bar{\Delta c} \rangle \quad (\forall \Delta c \in \Omega_1), \tag{9}$$

$$\|\bar{\Delta b} - \Delta b\|^2 \leq \langle \Delta b, \Delta b - \bar{\Delta b} \rangle \quad (\forall \Delta b \in \Omega_2). \tag{10}$$

We will use these inequalities some later.

3   Proposed Extension of the Lagrange Function

Following to [18], let us form the following extension of the classical Lagrange function

$$L^\nu(x,y) = L(x,y) + \frac{\alpha}{2}\|x\|^2 - \frac{\beta}{2}\|y\|^2 - \epsilon_1 \sum_{i=1}^n \ln x_i + \epsilon_2 \sum_{j=1}^m \ln y_j, \tag{11}$$

where $\nu = (\alpha, \beta, \epsilon_1, \epsilon_2) > 0$ is a set of small numerical parameters. Here, to the classical Lagrange function, we add the regularizing terms (the squares of the Euclidean norms of x and y) and the sum of logarithmic terms, which play the role of barriers, preventing both the variables from leaving the nonnegative orthants of the corresponding spaces (more precisely, their interiors).

Obviously, the function (11) is strong convex in primal variables and strong concave in dual ones. In addition, it is proper and closed in terminology of [19], and therefore, for any set of its parameters $\nu > 0$, there exists a unique saddle point (x^ν, y^ν) inside its domain of definition, i.e., a point which satisfy the inequalities

$$L^\nu(x^\nu, y) \leq L^\nu(x^\nu, y^\nu) \leq L^\nu(x, y^\nu) \quad \forall x > 0, \ \forall y > 0.$$

From a computational point of view, the saddle point of a function (11), which is an interior point of its domain of definition, can be found (and this is very convenient) by solving a simple system of nonlinear equations

$$\nabla_x L^\nu(x,y) = \underbrace{\nabla f_0(x) + J(x)^T y}_{=\nabla_x L(x,y)} + \alpha x - \epsilon_1 X^{-1} e = 0, \qquad x > 0, \tag{12}$$

$$\nabla_y L^\nu(x,y) = F(x) - \beta y + \epsilon_2 Y^{-1} e = 0, \qquad y > 0; \tag{13}$$

where $X = \text{diag}(x, ..., x_n)$ and $Y = \text{diag}(y, ..., y_m)$ are diagonal matrices with entries x_i and y_j on the diagonal, $e = (1, 1, ..., 1)$ is a vector of appropriate dimension composed of 1.

The written system (12), (13) is very similar to the equations describing the classical central path for a pair of initial settings (1), (2) when they are linear [20]. Solutions of this system for $\nu = (\alpha, \beta, \epsilon_1, \epsilon_2) \rightarrow +0$ will allow us to obtain the ordinary solutions of the original primal and dual problems in the proper case, or the vectors of their optimal linear correction $\overline{\Delta c}$ and $\overline{\Delta b}$ and the solutions of the corrected primal and dual problems in the improper case.

Let's justify the announced convergence.

Theorem 1. *Let (x^ν, y^ν) be a saddle point of the new Lagrange function (11). Then x^ν is an admissible vector for the primal corrected problem*

$$\inf\{f_0(x) + \alpha\langle x^\nu, x\rangle : \ F(x) \le \beta y^\nu, \ x \ge 0\}, \tag{14}$$

and y^ν is an admissible vector for its dual one

$$\sup_{y \ge 0} \phi(y), \quad \phi(y) = \inf_{x \ge 0}\left[L(x,y) + \langle \alpha x^\nu, x\rangle - \langle \beta y^\nu, y\rangle\right]. \tag{15}$$

Proof. Admissibility of the vector x^ν for the problem (14) follows from its positivity and the fact that, by virtue of the relation (13),

$$F(x^\nu) = \beta y^\nu - \epsilon_2 Y_\nu^{-1} e < \beta y^\nu.$$

The vector y^ν is admissible for the problem (15), since it is positive too, and, due to the relations (12) and the properties of convex functions and their gradients,[1] for all $x \ge 0$ the following chain of inequalities and equations is true

$$L(x, y^\nu) + \langle \alpha x^\nu, x\rangle - \langle \beta y^\nu, y^\nu\rangle$$

$$\ge L(x^\nu, y^\nu) + \langle \nabla_x L(x^\nu, y^\nu), x - x^\nu\rangle + \langle \alpha x^\nu, x\rangle - \langle \beta y^\nu, y^\nu\rangle$$

$$= L(x^\nu, y^\nu) + \langle \epsilon_1 X_\nu^{-1} e - \alpha x^\nu, x - x^\nu\rangle + \langle \alpha x^\nu, x\rangle - \langle \beta y^\nu, y^\nu\rangle$$

$$= L(x^\nu, y^\nu) + \epsilon_1 \langle X_\nu^{-1} e, x\rangle - \epsilon_1 n - \langle \alpha x^\nu, x\rangle + \alpha\|x^\nu\|^2 + \langle \alpha x^\nu, x\rangle - \beta\|y^\nu\|^2$$

$$\ge L(x^\nu, y^\nu) - \epsilon_1 n - \beta\|y^\nu\|^2.$$

Consequently,

$$\phi(y^\nu) = \inf_{x \ge 0}\left[L(x, y^\nu) + \alpha\langle x^\nu, x\rangle - \beta\langle y^\nu, y^\nu\rangle\right] \ge L(x^\nu, y^\nu) - \epsilon_1 n - \beta\|y^\nu\|^2 > -\infty.$$

The proof is complete.

Corollary 1. *If (x^ν, y^ν) is a saddle point of the new Lagrange function (11), then*

$$\Delta c = \alpha x^\nu \in \Omega_1, \qquad \Delta b = \beta y^\nu \in \Omega_2.$$

---

[1] Thus, for a convex differentiable function $g(x)$, the inequality $g(a) - g(b) \le \langle \nabla g(a), a - b\rangle$ is true for all a and b.

Theorem 2. *Let (x^ν, y^ν) be a saddle point of the new Lagrange function (11). Then*

$$\lambda = \max\left\{\|\alpha x^\nu - \overline{\Delta c}\,\|, \|\beta y^\nu - \overline{\Delta b}\,\|\right\} \le 2\gamma N(\nu) + \sqrt{\gamma N(\nu)},$$

where $N(\nu) = \max\{\|\bar{x}\|, \|\bar{y}\|\}$, $N(\nu) = \epsilon_1 n + \epsilon_2 m$, $\gamma = \max\{\alpha, \beta\}$.

Proof. Since the inclusion $\Delta c = \alpha x^\nu \in \Omega_1$ have already been proved, we can substitute it into (9) and obtain the first inequality

$$\|\overline{\Delta c} - \alpha x^\nu\|^2 \le \langle \alpha x^\nu, \alpha x^\nu - \overline{\Delta c}\,\rangle$$

$$= \langle \alpha x^\nu, \alpha x^\nu - \overline{\Delta c}\,\rangle = \alpha \langle x^\nu - \bar{x}, \alpha x^\nu - \overline{\Delta c}\,\rangle + \alpha \langle \bar{x}, \alpha x^\nu - \overline{\Delta c}\,\rangle$$

$$= \alpha \langle x^\nu - \bar{x}, \alpha x^\nu \rangle - \alpha \langle x^\nu, \overline{\Delta c}\,\rangle + \alpha \langle \bar{x}, \overline{\Delta c}\,\rangle + \alpha \langle \bar{x}, \alpha x^\nu - \overline{\Delta c}\,\rangle.$$

Taking into account the relations (7) and (12), as well as the properties of convex functions and their gradients, the first term here can be estimated

$$\alpha \langle x^\nu - \bar{x}, \alpha x^\nu \rangle = \alpha \langle x^\nu - \bar{x}, \epsilon_1 X_\nu^{-1} e - \nabla_x L(x^\nu, y^\nu)\rangle$$

$$= \alpha \langle x^\nu, \epsilon_1 X_\nu^{-1} e \rangle - \alpha \langle \bar{x}, \epsilon_1 X_\nu^{-1} e \rangle + \alpha \langle \bar{x} - x^\nu, \nabla_x L(x^\nu, y^\nu)\rangle$$

$$\le \alpha \epsilon_1 n + \alpha \left(L(\bar{x}, y^\nu) - L(x^\nu, y^\nu)\right).$$

On the same reasons, for the second and the third terms, it is valid $-\alpha \langle x^\nu, \overline{\Delta c}\,\rangle \le \alpha \langle x^\nu, \nabla_x L(\bar{x}, \bar{y})\rangle$ and $\alpha \langle \bar{x}, \overline{\Delta c}\,\rangle = -\alpha \langle \bar{x}, \nabla_x L(\bar{x}, \bar{y})\rangle$, i.e. their sum does not exceed

$$-\alpha \langle x^\nu, \overline{\Delta c}\,\rangle + \alpha \langle \bar{x}, \overline{\Delta c}\,\rangle \le \alpha \langle x^\nu - \bar{x}, \nabla_x L(\bar{x}, \bar{y})\rangle \le \alpha \left(L(x^\nu, \bar{y}) - L(\bar{x}, \bar{y})\right).$$

The fourth term is estimated through the product of norms

$$\alpha \langle \bar{x}, \alpha x^\nu - \overline{\Delta c}\,\rangle \le \alpha \|\bar{x}\| \cdot \|\alpha x^\nu - \overline{\Delta c}\,\|.$$

Combining all these estimates into a single whole, we get

$$\|\overline{\Delta c} - \alpha x^\nu\|^2$$

$$\le \epsilon_1 \alpha n + \alpha \left(L(\bar{x}, y^\nu) - L(x^\nu, y^\nu) + L(x^\nu, \bar{y}) - L(\bar{x}, \bar{y})\right) + \alpha \|\bar{x}\| \cdot \|\alpha x^\nu - \overline{\Delta c}\,\|$$

$$= \epsilon_1 \alpha n + \alpha \langle F(\bar{x}) - F(x^\nu), y^\nu - \bar{y}\rangle + \alpha \|\bar{x}\| \cdot \|\alpha x^\nu - \overline{\Delta c}\,\|. \qquad (16)$$

Next, turn to the vector $\Delta b = \beta y^\nu \in \Omega_2$. Taking into account the relations (8), (10), and (13), we elementary get the symmetric inequality

$$\|\overline{\Delta b} - \beta y^\nu\|^2 \le \langle \beta y^\nu, \beta y^\nu - \overline{\Delta b}\,\rangle$$

$$\le \langle \beta y^\nu, \beta y^\nu - F(\bar{x})\rangle = \langle \beta y^\nu, \beta y^\nu - F(x^\nu)\rangle + \langle \beta y^\nu, F(x^\nu) - F(\bar{x})\rangle$$

$$= \epsilon_2 \beta m + \langle \beta y^\nu, F(x^\nu) - F(\bar{x})\rangle. \qquad (17)$$

Now we can add the inequalities (16) and (17) term by term, previously dividing the first of them by $\alpha > 0$ and the second one by $\beta > 0$. We get

$$\alpha^{-1}\|\overline{\Delta c} - \alpha x^\nu\|^2 + \beta^{-1}\|\overline{\Delta b} - \beta y^\nu\|^2$$

$$\leq \epsilon_1 n + \epsilon_2 m + \langle y^\nu, F(x^\nu) - F(\bar{x})\rangle - \langle F(x^\nu) - F(\bar{x}), y^\nu - \bar{y}\rangle + \|\bar{x}\| \cdot \|\alpha x^\nu - \overline{\Delta c}\|$$

$$= \epsilon_1 n + \epsilon_2 m + \langle F(x^\nu), \bar{y}\rangle - \langle F(\bar{x}), \bar{y}\rangle + \|\bar{x}\| \cdot \|\alpha x^\nu - \overline{\Delta c}\|$$

$$= \epsilon_1 n + \epsilon_2 m + \langle F(x^\nu), \bar{y}\rangle - \langle \overline{\Delta b}, \bar{y}\rangle + \|\bar{x}\| \cdot \|\alpha x^\nu - \overline{\Delta c}\|$$

$$\leq \epsilon_1 n + \epsilon_2 m + \langle \bar{y}, \beta y^\nu - \overline{\Delta b}\rangle + \|\bar{x}\| \cdot \|\alpha x^\nu - \overline{\Delta c}\|$$

$$\leq \epsilon_1 n + \epsilon_2 m + \|\bar{y}\| \cdot \|\beta y^\nu - \overline{\Delta b}\| + \|\bar{x}\| \cdot \|\alpha x^\nu - \overline{\Delta c}\|$$

$$\leq \epsilon_1 n + \epsilon_2 m + 2\max\{\|\bar{x}\|, \|\bar{y}\|\} \cdot \max\{\|\alpha x^\nu - \overline{\Delta c}\|, \|\beta y^\nu - \overline{\Delta b}\|\}.$$

This implies that $\lambda = \max\{\|\alpha x^\nu - \overline{\Delta c}\|, \|\beta y^\nu - \overline{\Delta b}\|\}$ satisfies the relation[2]

$$\lambda^2 = \max\{\|\alpha x^\nu - \overline{\Delta c}\|^2, \|\beta y^\nu - \overline{\Delta b}\|^2\}$$

$$\leq \|\alpha x^\nu - \overline{\Delta c}\|^2 + \|\beta y^\nu - \overline{\Delta b}\|^2 \leq \gamma(\alpha^{-1}\|\overline{\Delta c} - \alpha x^\nu\|^2 + \beta^{-1}\|\overline{\Delta b} - \beta y^\nu\|^2)$$

$$\leq \gamma\left[\epsilon_1 n + \epsilon_2 m + 2\max\{\|\bar{x}\|, \|\bar{y}\|\} \cdot \max\{\|\alpha x^\nu - \overline{\Delta c}\|, \|\beta y^\nu - \overline{\Delta b}\|\}\right]$$

$$= 2\gamma\lambda\max\{\|\bar{x}\|, \|\bar{y}\|\} + \gamma(\epsilon_1 n + \epsilon_2 m),$$

where $\gamma = \max\{\alpha, \beta\} > 0$.

So, we have the simplest quadratic inequality with respect to the variable λ, from which we conclude[3]

$$\lambda \leq \gamma\max\{\|\bar{x}\|, \|\bar{y}\|\} + \sqrt{\left(\gamma\max\{\|\bar{x}\|, \|\bar{y}\|\}\right)^2 + \gamma(\epsilon_1 n + \epsilon_2 m)}$$

$$\leq 2\gamma\max\{\|\bar{x}\|, \|\bar{y}\|\} + \sqrt{\gamma(\epsilon_1 n + \epsilon_2 m)}.$$

The proof is complete.

Theorem 3. *Let* (x^ν, y^ν) *be a saddle point of the new Lagrange function* (11) *and* $\nu = (\alpha, \beta, \epsilon_1, \epsilon_2) \to +0$. *Then*[4]

$$\|(F(x^\nu) - \overline{\Delta b})^+\| \to 0, \qquad |f_0(x^\nu) - \langle \overline{\Delta c}, x^\nu\rangle - \overline{\text{opt}}| \to 0,$$

The rates of convergence to zero are $O(\gamma + \sqrt{\gamma(\epsilon_1 + \epsilon_2)})$ *for the first sequence, and* $O(\gamma + \sqrt{\gamma(\epsilon_1 + \epsilon_2)} + \epsilon_1 + \epsilon_2)$ *for the second one, where* $\gamma = \max(\alpha, \beta)$.

---

[2] We use the fact that $\gamma\alpha^{-1} = \max\{\alpha, \beta\}\alpha^{-1} \geq 1$ and $\gamma\beta^{-1} = \max\{\alpha, \beta\}\beta^{-1} \geq 1$, as well as the simple relation $\max(|a|, |b|) \leq |a| + |b| \leq 2\max(|a|, |b|)$..

[3] Here we use a simple relation $\sqrt{|a| + |b|} \leq \sqrt{|a|} + \sqrt{|b|}$..

[4] Here $a^+ = \max\{a, 0\}$ for $a \in \mathbf{R}$, and $p^+ = (p_1^+, ..., p_r^+)$ for $p = (p, ..., p_r) \in \mathbf{R}^r$.

Proof. Indeed, under the assumptions made about the parameters of the method, as well as due to the relations (12), (13) and the previous statement, we have

$$F(x^\nu) - \overline{\Delta b} = \beta y^\nu - \epsilon_2 Y_\nu^{-1} e - \overline{\Delta b} < \beta y^\nu - \overline{\Delta b} \to 0.$$

From the same statements, as well as from the relations (9), (10), (12), (13) and properties of convex functions and their gradients, it follows that

$$\overline{\text{opt}} - f_0(x^\nu) - \langle \overline{\Delta c}, x^\nu \rangle = f_0(\bar{x}) + \langle \overline{\Delta c}, \bar{x} \rangle - f_0(x^\nu) - \langle \overline{\Delta c}, x^\nu \rangle$$

$$= \langle \overline{\Delta c}, \bar{x} - x^\nu \rangle + f_0(\bar{x}) - f_0(x^\nu) \geq \langle \overline{\Delta c}, \bar{x} - x^\nu \rangle + \langle \nabla f_0(x^\nu), \bar{x} - x^\nu \rangle$$

$$= \langle \overline{\Delta c}, \bar{x} - x^\nu \rangle - \langle \alpha x^\nu + J(x^\nu)^T y^\nu - X_\nu^{-1} e, \bar{x} - x^\nu \rangle$$

$$\geq \langle \overline{\Delta c} - \alpha x^\nu, \bar{x} - x^\nu \rangle + \langle y^\nu, F(x^\nu) - F(\bar{x}) \rangle - \epsilon_1 n$$

$$= \langle \overline{\Delta c} - \alpha x^\nu, \bar{x} - x^\nu \rangle + \langle y^\nu, F(x^\nu) \rangle - \langle y^\nu, F(\bar{x}) \rangle - \epsilon_1 n$$

$$= \langle \overline{\Delta c} - \alpha x^\nu, \bar{x} - x^\nu \rangle + \langle y^\nu, \beta y^\nu - \epsilon_2 Y_\nu^{-1} e \rangle - \langle y^\nu, F(\bar{x}) \rangle - \epsilon_1 n$$

$$\geq \langle \overline{\Delta c} - \alpha x^\nu, \bar{x} - x^\nu \rangle + \langle y^\nu, \beta y^\nu \rangle - \langle y^\nu, \overline{\Delta b} \rangle - \epsilon_1 n - \epsilon_2 m$$

$$= \langle \overline{\Delta c} - \alpha x^\nu, \bar{x} \rangle + \alpha^{-1} \langle \alpha x^\nu - \overline{\Delta c}, \alpha x^\nu \rangle + \beta^{-1} \langle \beta y^\nu, \beta y^\nu - \overline{\Delta b} \rangle - \epsilon_1 n - \epsilon_2 m$$

$$\geq -\|\alpha x^\nu - \overline{\Delta c}\| \cdot \|\bar{x}\| - \epsilon_1 n - \epsilon_2 m \to 0.$$

This implies a simple lower estimate

$$\liminf_{\nu \to 0} \left(\overline{\text{opt}} - f_0(x^\nu) - \langle \overline{\Delta c}, x^\nu \rangle \right) \geq 0.$$

To obtain an upper estimate, we use the fact that the vector x^ν is admissible for a problem

$$\min \{ f_0(x) + \langle \overline{\Delta c}, x \rangle : \ F(x) \leq \overline{\Delta b} + \Delta b, \ x \geq 0 \},$$

where $\Delta b = (F(x^\nu) - \overline{\Delta b})^+$, and therefore by virtue of marginal value theorems,

$$f_0(x^\nu) + \langle \overline{\Delta c}, x^\nu \rangle \geq \text{opt}(\Delta b) \geq \overline{\text{opt}} + \langle \bar{y}, (F(x^\nu) - \overline{\Delta b})^+ \rangle$$

$$\geq \overline{\text{opt}} - \|\bar{y}\| \cdot \|(F(x^\nu) - \overline{\Delta b})^+\|,$$

that is

$$\overline{\text{opt}} - f_0(x^\nu) - \langle \overline{\Delta c}, x^\nu \rangle \leq \|\bar{y}\| \cdot \|(F(x^\nu) - \overline{\Delta b})^+\| \to 0.$$

From here,

$$\limsup_{\nu \to 0} \left(\overline{\text{opt}} - f_0(x^\nu) - \langle \overline{\Delta c}, x^\nu \rangle \right) \leq 0.$$

Therefore for $\nu = (\alpha, \beta, \epsilon_1, \epsilon_2) \to +0$ the vectors x^ν form generalized solution of the problem (5) and even converge in distance to its optimal set if the latter is bounded. The proof is complete.

4 Algorithmic Aspects of the New Method

Let's represent the system of equations (12), (13) as

$$\Phi_\nu(z) = \begin{pmatrix} \alpha x + \nabla_x L(x,y) - \epsilon_1 X^{-1} e \\ \beta y - F(x) - \epsilon_2 Y^{-1} e \end{pmatrix} = 0, \quad z = (x,y) > 0. \quad (18)$$

To solve this system, one can apply Newton's method, which generates a sequence of approximations according to the rules

$$z^{(0)} > 0, \quad z^{(k+1)} = z^{(k)} - \tau_k \left(\nabla \Phi_\nu(z^{(k)}) \right)^{-1} \Phi_\nu(z^{(k)}), \quad k = 0,1,.... \quad (19)$$

The step parameter τ_k is usually taken equal to 1. The Jacobian of the system is positive defined and calculated by a simple formula

$$\nabla \Phi_\nu(z) = \begin{pmatrix} D_{1,\nu} & J(x)^T \\ -J(x) & D_{2,\nu} \end{pmatrix},$$

where $D_{1,\nu} = \alpha E + \nabla_{xx}^2 L(x,y) + \epsilon_1 X^{-2}$, $D_{2,\nu} = \beta E + \epsilon_2 Y^{-2}$, $E = \mathrm{diag}(1,1,...,1)$ is identity matrices of the desired order. Respectively,

$$\left(\nabla \Phi_\nu(z) \right)^{-1} = \begin{pmatrix} D_{1,\nu}^{-1}(E - J(x)^T Q_\nu J(x) D_{1,\nu}^{-1}) & -D_{1,\nu}^{-1} J(x)^T Q_\nu \\ -Q_\nu J(x) D_{1,\nu}^{-1} & Q_\nu \end{pmatrix},$$

where $Q_\nu = P_\nu^{-1}$, $P_\nu = D_{2,\nu} + J(x) D_{1,\nu}^{-1} J(x)^T$.

Thus, the most time-consuming part of the algorithm consists of inverting a matrix of size $n \times n$.

Another way to solve the system (18) cosists of reduction its dimension by eliminating the variables y from it. Let multiply the lower equation of the system on the left by the matrix Y. We obtain a series of simple quadratic equations with respect to dual variables

$$\beta y_i^2 - f_i(x) y_i - \epsilon_2 = 0, \quad i = 1,2,...,m.$$

The positive roots of these equations are

$$y_i(x) = \frac{1}{2\beta} \left(f_i(x) + \sqrt{f_i(x)^2 + 4\epsilon_2 \beta} \right).$$

Substituting these relations into the upper subsystem (18), we obtain equations

$$\bar\Phi_k^\nu(x) := \frac{\partial f_0(x)}{\partial x_k} + \frac{1}{2\beta} \sum_{j=1}^m \left(f_j(x) + \sqrt{f_j(x)^2 + 4\epsilon_2\beta} \right) \frac{\partial f_j(x)}{\partial x_k} + \alpha x_k - \frac{\epsilon_1}{x_k} = 0,$$

that can be put together

$$\bar\Phi_1^\nu(x) = 0, \quad \bar\Phi_2^\nu(x) = 0, \quad ..., \quad \bar\Phi_{n-1}^\nu(x) = 0, \quad \bar\Phi_n^\nu(x) = 0. \quad (20)$$

Newton's method can also be applied to solve this system. Let's rewrite its recursion

$$x^{(0)} > 0, \quad x^{(k+1)} = x^{(k)} - \tau_k \left(\nabla \bar{\Phi}^\nu \left(x^{(k)} \right) \right)^{-1} \bar{\Phi}^\nu \left(x^{(k)} \right), \quad k = 0, 1, 2\ldots$$

Using the Kronecker symbol δ_i^k, the Jacobian elements here can be written as

$$\frac{\partial \bar{\Phi}_k^\nu(x)}{\partial x_i} = \frac{\partial^2 f_0(x)}{\partial x_k \partial x_i} + \frac{1}{2\beta} \sum_{j=1}^m \left(\frac{\partial f_j(x)}{\partial x_k} \cdot \left(1 + \frac{f_j(x)}{\sqrt{f_j(x)^2 + 4\epsilon_2 \beta}} \right) \cdot \frac{\partial f_j(x)}{\partial x_k} \right)$$

$$+ \frac{1}{2\beta} \sum_{j=1}^m \left(\frac{\partial^2 f_j(x)}{\partial x_k \partial x_i} \cdot \left(f_j(x) + \sqrt{f_j(x)^2 + 4\epsilon_2 \beta} \right) \right) + \delta_i^k \left(\alpha + \frac{\epsilon_1}{x_k^2} \right).$$

It is interesting, that substantially (and it is not difficult to verify this), the system (20) describes the necessary and sufficient conditions for the vector x^ν to be optimal in the optimization problem

$$\min_{x>0} \left(f_0(x) + \frac{\alpha}{2} \|x\|^2 + \frac{1}{4\beta} \sum_{j=1}^m \mathcal{R}\left(f_j(x), 4\epsilon_2 \beta \right) - \epsilon_1 \sum_{i=1}^n \ln x_i \right),$$

where $\mathcal{R}(t, \delta) = \delta \ln \left(\sqrt{t^2 + \delta} + t \right) + t \left(\sqrt{t^2 + \delta} + t \right)$ is a generalized external penalty function of the problem (1). Unlike the usual external penalties, the function $\mathcal{R}(t, \delta)$ is not only convex, but also smooth, and for it we have standard penalty properties: $\lim_{\delta \to 0} \mathcal{R}(t, \delta) = 0$ for $t < 0$ and $\lim_{\delta \to 0} \mathcal{R}(t, \delta) = +\infty$ for $t > 0$.

Concluding the section, let's consider an alternative form of the system (18):

$$\Psi_\nu(w) = \begin{pmatrix} \alpha x + \nabla_x L(x, y) - u \\ \beta y - F(x) - v \\ Ux - \epsilon_1 e \\ Vy - \epsilon_2 e \end{pmatrix} = 0, \quad w = (x, y, u, v) > 0. \quad (21)$$

Here we introduce additional variables $u = \epsilon_1 X^{-1} e$ and $v = \epsilon_2 Y^{-1} e$ and the abbreviations $U = \text{diag}(u, \ldots u_n)$, $V = \text{diag}(v, \ldots, v_m)$.

For system (21), Newton's method generates a sequence of approximations according to the rule

$$w^{(0)} > 0, \quad w^{(k+1)} = w^{(k)} - \tau_k p^{(k)}, \quad k = 0, 1, 2\ldots,$$

where τ_k is the step parameter and the direction $p^{(k)} = \left(\nabla \Psi_\nu(w^{(k)}) \right)^{-1} \Psi_\nu(w^{(k)})$ is determined from the solution of the following sparse block system of linear equations

$$\begin{pmatrix} \alpha + \nabla_{xx}^2 L(x, y) & J(x)^T & -E & 0 \\ -J(x) & \beta & 0 & -E \\ U_k & 0 & X_k & 0 \\ 0 & V_k & 0 & Y_k \end{pmatrix} \begin{pmatrix} p_x \\ p_y \\ p_u \\ p_v \end{pmatrix} = \begin{pmatrix} h \\ h \\ h_3 \\ h_4 \end{pmatrix}. \quad (22)$$

This solution, despite its increased dimension, can be obtained no more difficult then finding the direction of descent in the previous algorithm scheme. Indeed, by the Gauss elimination method, this system can be reduced to the form

$$
\begin{pmatrix}
\alpha + \nabla^2_{xx}L(x,y) & J(x)^T & -E & 0 \\
-J(x) & \beta & 0 & -E \\
E & H_k X_k J(x)^T & 0 & 0 \\
0 & E & 0 & 0
\end{pmatrix}
\begin{pmatrix}
p_x \\
p_y \\
p_u \\
p_v
\end{pmatrix}
=
\begin{pmatrix}
h \\
h \\
H_k(h_3 + X_k h) \\
G_k Q_k
\end{pmatrix},
$$

where $H_k = \left(U_k + X_k(\alpha + \nabla^2_{xx}L(x,y))\right)^{-1}$, $Q_k = h_4 + Y_k h + Y_k J(x)H_k(h_3 + X_k h)$, $G_k = \left(V_k + \beta Y_k + Y_k J(x)H_k X_k J(x)^T\right)^{-1}$.

This leads to

Theorem 4. *The solution of the system* (22) *can be found sequentially by the formulas*

(1) $p_y = G_k\left(h_4 + Y_k h + Y_k J(x)H_k(h_3 + X_k h)\right)$,

(2) $p_x = H_k(h_3 + X_k h) - H_k X_k J(x)^T p_y$,

(3) $p_u = (\alpha + \nabla^2_{xx}L(x,y))p_x + J(x)^T p_y - h$,

(4) $p_v = \beta p_y - J(x)p_x - h$.

Thus, here, at each iteration, the only non-trivial matrix of the same size as in the scheme (19), but, as shown computational experiments, the method is now faster due to extension the region of its effective convergence. Also its error tolerance increases, as well as rounding and revealed the ability to work with smaller values of the parameters of regularization.

5   About Effective Controlling the Parameters of the Algorithm

Newton's method effectively converges only in a sufficiently small neighborhood of the desired solution, which in our case depends on the value of the penalty parameters and the regularization parameters. Since the starting points of calculations are usually chosen arbitrarily and may be far from this solution, in practice it is advisable to start calculations with moderate values of these parameters, and only gradually and smoothly decrease them from iteration to iteration as the discrepancy (mismatch) of the system being solved at the current point of the iterative process decreases to zero.

Therefore we must suggest some criterion of accuracy for relations (21), which we will verify before the next decreasing of penalty and regularization parameters.

For example, let choose some positive constants $\delta_1 > 0$, $\delta_2 > 0$, $\delta_3 > 0$, $\delta_4 > 0$ and write the following requirements:

$$0 \leq \nabla_x L(x, y) + \alpha x \leq (1 + \delta_1) u, \quad 0 < u_i x_i \leq (1 + \delta_3)\epsilon_1 \quad (i = 1, ..., n), \quad (23)$$

$$0 \leq \beta y - F(x) \leq (1 + \delta_2) v, \quad 0 < v_j y_j \leq (1 + \delta_4)\epsilon_2 \quad (j = 1, ..., m). \quad (24)$$

Here, as before, all variables are strictly positive.

It is obvious, that the exact solutions both of the systems (18) and (21) satisfy the requirements (23), (24) with some «margin». So, for any starting points and values of penalty and regularization parameters, after a finite number of iterations of Newton's method applied to these systems, the requirements (23), (24) will be fulfilled. After that, the regularization and penalty parameters may be a bit decreased, and Newton's method be continued with a new values of them from the last iteration point.

Theorem 5. *Let a set of the vectors $(x^\nu, u^\nu, y^\nu, v^\nu)$ satisfy the relations (23), (24). Then x^ν is an admissible vector for the problem*

$$\inf\{f_0(x) + \alpha\langle x^\nu, x\rangle : f_j(x) \leq \beta y_j^\nu \ (j = 1, ..., m), \ x \geq 0\}, \quad (25)$$

and vector y^ν is an admissible vector for the dual problem

$$\sup_{y \geq 0} \phi(y), \quad where \quad \phi(y) = \inf_{x \geq 0}\left[L(x, y) + \langle \alpha x^\nu, x\rangle - \langle \beta y^\nu, y\rangle\right]. \quad (26)$$

Therefore the inclusions are valid

$$\Delta c = \alpha x^\nu \in \Omega_1, \quad \Delta b = \beta y^\nu \in \Omega_2.$$

Theorem 6. *Let a set of the vectors $(x^\nu, u^\nu, y^\nu, v^\nu)$ satisfy the relations (23), (24). Then*

$$0 < \lambda = \max\{\|\overline{\Delta c} - \alpha x^\nu\|, \|\overline{\Delta b} - \beta y^\nu\|\} \leq \gamma N_5(\nu) + \sqrt{\gamma N_6(\nu, \delta)},$$

where

$$N_5 = 2\max(\|\bar{x}\|, \|\bar{y}\|),$$

$$N_6(\nu, \delta) = n(1 + \delta_1)(1 + \delta_3)\epsilon_1 + m(1 + \delta_2)(1 + \delta_4)\epsilon_2,$$

$$\gamma = \max(\alpha, \beta).$$

Corollary 2. *Let a set of the vectors $(x^\nu, u^\nu, y^\nu, v^\nu)$ satisfy the relations (23), (24) and let δ_{1-4} be constants, and $\nu = (\alpha, \beta, \epsilon_1, \epsilon_2) \to +0$. Then*

$$\|(F(x^\nu) - \overline{\Delta b})^+\| \to 0, \quad |f_0(x^\nu) - \langle \overline{\Delta c}, x^\nu\rangle - \overline{opt}| \to 0.$$

The prooving of these propositions is very similar to the proofs of Theorems 1, 2 and 3 and we omit them here.

Thus we see that for $\nu = (\alpha, \beta, \epsilon_1, \epsilon_2) \to +0$ the sequences of vectors x^ν form a generalized solution to problem (5) and even converge in distance to its optimal set when the latter is bounded.

6   Conclusion

In this note, we transfer to the nonlinear case our previous results on the use of a symmetrically regularized Lagrange function, additionally equipped with barrier terms, for a numerical analysis of improper linear programming problems. The main distinguishing feature of the new approach is the ability to apply second-order optimization methods in internal calculations and not specify in advance the type of incorrectness of the original problem. The article presents proofs of convergence of new computational schemes, interpretation of the got quasi-solution (the usual solution in proper case), as well as accuracy estimates and some rules for controlling the parameters of the method.

References

1. Eremin, I.I.: Duality for improper problems of linear and convex programming. Dokl. Akad. Nauk SSSR **256**(2), 272–76 (1981). In Russian
2. Eremin, I.I., Mazurov, Vl.D., Astaf'ev, N.N.: Improper Problems of Linear and Convex Programming. Nauka, Moscow (1983). In Russian
3. Tikhonov, A.N., Arsenin, V.Ya.: Methods for Solutions of Ill-Posed Problems. Nauka, Moscow (1979). Wiley, New York (1981)
4. Evtushenko, Y.G.: Methods for solving extremal problems and their application in systems of optimization. Optimization and Operations Research. Nauka, Moscow (1982)
5. Kochikov, I.V., Matvienko, A.N., Yagola, A.G.: A generalized discrepancy principle for solving incompatible equations. USSR Comput. Math. Math. Phys. **24**(4), 78–80 (1984)
6. Parametric optimization and methods for approximation of improper mathematical programming problems. [Digest of articles] Edited by I. I. Eremin and L. D. Popov. Akad. Nauk SSSR, Ural. Nauchn. Tsentr, Sverdlovsk (1985). In Russian
7. Skarin, V.D.: On a regularization method for inconsistent convex programming problems. Russian Math. (Iz. VUZ) **39**(12), 78–85 (1995)
8. Ivanitskii, A.Y., Morozov, V.A., Karmazin, V.N.: The pointwise residual method for inconsistent systems of equations and inequalities with approximate data. Fundam. Prikl. Mat. **4**(3), 937–945 (1998). (Russian)
9. Mazurov, Vl.D., Khachai, M.Yu.: Committee constructions. Izv. Ural. Gos. Univ. Mat. Mekh. **172**(2), 77–108 (1999)
10. Dax, A.: The smallest correction of an inconsistent system of linear inequalities. Optim. Eng. **2**, 349–359 (2001)
11. Eremin, I.I.: Theory of Linear Optimization. Inverse and Ill-Posed Problems Series. VSP. Utrecht, Boston, Koln, Tokyo (2002)
12. Erokhin, V.I., Krasnikov, A.S., Khvostov, M.N.: Matrix corrections minimal with respect to the Euclidean norm for linear programming problems. Autom. Remote. Control. **73**, 219–231 (2012)
13. Barkalova, O.S.: Correction of improper linear programming problems in canonical form by applying minmax criterion. Comput. Math. Math. Phys. **52**(1), 1624–1634 (2012)
14. Artem'eva, L.A., Vasil'ev, F.P., Potapov, M.M.: Extragradient method for correction of inconsistent linear programming problems. Comput. Math. Math. Phys. **58**(12), 1919–1925 (2018)

15. Popov, L.D.: Search of generalized solutions to improper linear and convex programming problems using barrier functions. Bull. Irkutsk State Univ. Series Math. **4**(2), 134–146 (2011)
16. Popov, L.D.: Use of barrier functions for optimal correction of improper problems of linear programming of the 1st kind. Autom. Remote. Control. **73**(3), 417–424 (2012)
17. Skarin, V.D.: On the application of the residual method for the correction of inconsistent problems of convex programming. In: Proceedings of the Steklov Institute of Mathematics, vol. 289, pp. 182–191 (2015)
18. Popov, L.D.: Barriers and symmetric regularization of Lagrange function in analysis of improper problems of linear programming. Proc. Steklov Inst. Math. **15**(3), 115–128 (2023)
19. Rockafellar, R.T.: Convex Analysis. Princeton Mathematical Series, Princeton University Press, Princeton (1970)
20. Roos, C., Terlaky, T., Vial, J.-P.: Theory and Algorithms for Linear Optimization. Wiley, Chichester (1997)

Accelerated Stochastic Gradient Method with Applications to Consensus Problem in Markov-Varying Networks

Vladimir Solodkin[1,2], Savelii Chezhegov[1,2], Ruslan Nazikov[1,2],
Aleksandr Beznosikov[1,2,3(✉)], and Alexander Gasnikov[1,2,3]

[1] Moscow Institute of Physics and Technology, Moscow, Russian Federation
anbeznosikov@gmail.com
[2] Institute for Information Transmission Problems RAS, Moscow, Russian Federation
[3] Innopolis University, Innopolis, Russian Federation

Abstract. Stochastic optimization is a vital field in the realm of mathematical optimization, finding applications in diverse domains ranging from operations research to machine learning. In this paper, we introduce a novel first-order optimization algorithm designed for scenarios where Markovian noise is present, incorporating Nesterov acceleration for enhanced efficiency. The convergence analysis is performed by using an assumption on noise depending on the distance to the solution. We also delve into the consensus problem over Markov-varying networks, exploring how this algorithm can be applied to achieve agreement among multiple agents with differing objectives during the changes into communication system. To show the performance of our method on the problem above, we conduct experiments to demonstrate the superiority over the classic approach.

Keywords: convex optimization · stochastic optimization · Markovian noise · accelerated methods · decentralized communications

1 Introduction

Stochastic optimization encompasses a suite of methodologies aimed at minimizing or maximizing an objective function in the presence of randomness. These methods have evolved into indispensable tools across a spectrum of disciplines including science, engineering, business, computer science, and statistics. Applications are diverse, ranging from refining the placement of acoustic sensors on a beam through simulations, to determining optimal release times for reservoir water to maximize hydroelectric power generation, to fine-tuning the parameters of statistical models based on given datasets. The introduction of randomness typically occurs through the cost function or the constraint set. While the term

The research was supported by Russian Science Foundation (project No. 23-11-00229).

A. Eremeev et al. (Eds.): MOTOR 2024, LNCS 14766, pp. 69–86, 2024.
https://doi.org/10.1007/978-3-031-62792-7_5

"stochastic optimization" may encompass any optimization approach that incorporates randomness within certain communities, our focus here is on scenarios where the objective function is stochastic.

As with deterministic optimization, no universal solution method generally excels across all problems. Structural assumptions play a pivotal role in making problems tractable. Given that solution methodologies are intricately linked to problem structures, our analysis heavily relies on problem type, with a detailed exposition of associated solution approaches.

Related Work. A considerable body of research has documented substantial advancements achieved by accelerating gradient descent in a Nesterov way [37]. Building upon this foundation, Nesterov accelerated stochastic gradient descent [2,5] emerged as a powerful tool for optimizing different objectives in stochastic settings. In the earlier works [39,43], the proof of convergence was done using an assumption on bounded variance, which narrows down the application perspective significantly. Later, [40] succeeded in relaxing this assumption to strong growth condition, which partially solved the aforementioned problem. At the same time, several papers delved into applying acceleration to specific stochastic cases, e.g., coordinate descent [38], heavy tailed noise [41], distributed learning [42]. However, all of these works investigate i.i.d. noise setup, while a more general case could be considered.

As of late, there has been an emergence of scholarly works aimed at addressing the existing gap in the analysis of the Markovian noise configuration. Nonetheless, it is noteworthy that this domain continues to be a dynamically evolving field of study. Specifically, [14] examined a variant of the Ergodic Mirror Descent algorithm yielding optimal convergence rates for smooth and nonconvex problems. More recently, [18] proposed a random batch size algorithm tailored for nonconvex optimization within a compact domain. In the realm of Markovian noise, the finite-time analysis of non-accelerated SGD-type algorithms has been investigated in [19,21]. However, [19] heavily relies on the assumption of a bounded domain and uniformly bounded stochastic gradient oracles and [21] achieves only suboptimal dependence on initial conditions for strongly convex problems when employing SGD. In the exploration of accelerated SGD in the presence of Markovian noise, [22] achieved an optimal rate of forgetting the initial condition, but suboptimal variance terms. Recently, [1] proposed the accelerated version of SGD achieving linear dependence on the mixing time.

The aforementioned studies predominantly address general Markovian noise optimization. Recently, a surge of papers has emerged, focusing on the specialized scenario of distributed optimization [24,25]. [26] investigates the generalization and stability of Markov SGD with specific emphasis on excess variance guarantees. Simultaneously, specific results such as those from [27] offer lower bounds for particular finite-sum problems within the Markovian setting.

Our Contributions. We present the analysis of an accelerated version of SGD in the Markovian noise setting under the assumption of a gradient estimator bounded by the distance to the optimum. We obtain sharp convergence rate and optimal dependence in terms of the mixing time of the underlying Markov chain. Moreover, for $k = 1$ Markovian scheme reduces to a classical *i.i.d.* noise setup. To the best of our knowledge, analysis in this case (even for the *i.i.d.* stochasticity) under suggested assumptions has not been presented in the literature before. To show the practicality of our method, we perform numerical experiments on the consensus search problem on time-varying networks, showing a better convergence rate compared to classical approaches for solving this problem.

1.1 Technical Preliminaries

Let $(\mathsf{Z}, \mathsf{d}_\mathsf{Z})$ be a complete separable metric space endowed with its Borel σ-field $\mathcal{Z}$. Let $(\mathsf{Z}^\mathbb{N}, \mathcal{Z}^{\otimes \mathbb{N}})$ be the corresponding canonical process. Consider the Markov kernel Q defined on $\mathsf{Z} \times \mathcal{Z}$, and denote by $\mathbb{P}_\xi$ and $\mathbb{E}_\xi$ the corresponding probability distribution and the expected value with initial distribution ξ. Without loss of generality, we assume that $(Z_k)_{k \in \mathbb{N}}$ is the corresponding canonical process. By construction, for any $A \in \mathcal{Z}$, it holds that $\mathbb{P}_\xi(Z_k \in A | Z_{k-1}) = \mathrm{Q}(Z_{k-1}, A)$, $\mathbb{P}_\xi$-a.s. If $\xi = \delta_z$, $z \in \mathsf{Z}$, we write $\mathbb{P}_z$ and $\mathbb{E}_z$ instead of $\mathbb{P}_{\delta_z}$ and $\mathbb{E}_{\delta_z}$, respectively. For $x^1, \ldots, x^k$ being the iterates of any stochastic first-order method, we denote $\mathcal{F}_k = \sigma(x^j, j \le k)$ and write $\mathbb{E}_k$ as an alias for $\mathbb{E}[\cdot | \mathcal{F}_k]$.

Lemma 1 (*Cauchy Schwartz inequality*). *For any $a, b, x_1, \ldots, x_n \in \mathbb{R}^d$ and $c > 0$ the following inequalities hold:*

$$2\langle a, b \rangle \le \frac{\|a\|^2}{c} + c\|b\|^2, \tag{1}$$

$$\|a + b\|^2 \le \left(1 + \frac{1}{c}\right)\|a\|^2 + (1 + c)\|b\|^2. \tag{2}$$

2 Problem and Assumptions

In this paper, we study the minimization problem

$$\min_{x \in \mathbb{R}^d} \left[f(x) := \mathbb{E}_{Z \sim \pi}[F(x, Z)] \right], \tag{3}$$

where the access to the function f and its gradient are available only through the noisy oracle $F(x, Z)$ and $\nabla F(x, Z)$ respectively. We start by presenting two classical regularity constraints on the target function f:

Assumption 1. *The function f is L-smooth on $\mathbb{R}^d$ with $L > 0$, i.e., it is continuously differentiable and there is a constant $L > 0$ such that the following inequality holds for all $x, y \in \mathbb{R}^d$:*

$$\|\nabla f(x) - \nabla f(y)\| \le L\|x - y\|.$$

Assumption 2. *The function f is μ-strongly convex on $\mathbb{R}^d$, i.e., it is continuously differentiable and there is a constant $\mu > 0$ such that the following inequality holds for all $x, y \in \mathbb{R}^d$:*

$$\frac{\mu}{2}\|x - y\|^2 \le f(x) - f(y) - \langle \nabla f(y), x - y \rangle.$$

Next we specialize our assumption on the sequence of noise variables $\{Z_i\}_{i=0}^{\infty}$. Assumption 3 is also considered to be classical in case of stochastic optimization with the Markovian noise [18,19,22]. It allows us to deal with finite state-space Markov chains with irreducible and aperiodic transition matrix.

Assumption 3. $\{Z_i\}_{i=0}^{\infty}$ *is a stationary Markov chain on* $(\mathsf{Z}, \mathcal{Z})$ *with Markov kernel* Q *and unique invariant distribution* π. *Moreover,* Q *is uniformly geometrically ergodic with mixing time* $\tau \in \mathbb{N}$, *i.e., for every* $k \in \mathbb{N}$,

$$\Delta(\mathsf{Q}^k) = \sup_{z,z' \in \mathsf{Z}} (1/2)\|\mathsf{Q}^k(z, \cdot) - \mathsf{Q}^k(z', \cdot)\|_{\mathsf{TV}} \le (1/4)^{\lfloor k/\tau \rfloor}.$$

Now we specify our assumption on the stochastic gradient estimator. Most of the existing literature on stochastic first order methods for solving (3) utilizes *strong growth condition* [40] or *uniformly bounded variance* [39] as they allow to prove the convergence quite straightforwardly. However, these assumptions narrow down the set of target functions that can be considered rather strongly and there are several kinds of relaxation of it [9], where gradient differences are bounded by the norm of the true gradient and a certain bias. Instead of this, we propose to use the following assumption:

Assumption 4. *For all $x \in \mathbb{R}^d$ it holds that $\mathbb{E}_\pi[\nabla F(x, Z)] = \nabla f(x)$. Moreover, for all $z \in \mathsf{Z}$ and $x \in \mathbb{R}^d$ it holds that*

$$\|\nabla F(x, z) - \nabla f(x)\|^2 \le \sigma^2 + \delta^2 \|x - x^*\|^2. \tag{4}$$

It is one way or another much weaker, then strong grows condition and uniformly bounded variance and, to the best of our knowledge, seem to be new for analyzing accelerated methods for solving stochastic optimization problems. One can notice that unlike the i.i.d. case, we are forced to require the almost sure bound in (4) rather than in expectation. This issue inevitably arises when dealing with Markovian stochasticity due to the impossibility of using the expectation trick [20], and has not yet been solved by any authors dealing with such type of stochasticity [18,19,21]. Either way, there are advantages to it as well: if we additionally require our noisy oracle $F(x, Z)$ to be $\tilde{L}$-Lipschiz, Assumption 4 is satisfied by itself consequently. Formally, if for any $x, y \in \mathbb{R}^d$,

$$\|\nabla F(x, z) - \nabla F(y, z)\| \le \tilde{L}(z)\|x - y\|,$$

for $\tilde{L} : \mathsf{Z} \to \mathbb{R}^+$ with $\sup |\tilde{L}| < \infty$, then

$$\begin{aligned}
\|\nabla F(x, z) - \nabla f(x)\|^2 &\le 3\|\nabla F(x, z) - \nabla F(x^*, z)\|^2 + 3\|\nabla F(x^*, z) - \nabla f(x^*)\|^2 \\
&\quad + 3\|\nabla f(x) - \nabla f(x^*)\|^2 \\
&\le 3(\|\tilde{L}\|^2 + L^2)\|x - x^*\|^2 + 3\|\nabla F(x^*, z) - \nabla f(x^*)\|^2,
\end{aligned}$$

taking $\sigma = \sqrt{3}\|\nabla F(x^*, z) - \nabla f(x^*)\|$ and $\delta = \sqrt{6\max(L, \|\tilde{L}\|)}$ gives Assumption 4.

3 Main Results

We start by introducing our version of Nesterov accelerated SGD. It utilizes the idea from [1] of using exactly the number of samples that comes from the truncated geometric distribution with truncation parameter to be specified later (see Theorem 1) in order to obtain optimal computational complexity of the algorithm.

Algorithm 1. Markov Accelerated GD

1: **Parameters:** stepsize $\gamma > 0$, momentums θ, η, number of iterations N, batchsize limit M
2: **Initialization:** choose $x^0 = x_f^0$
3: **for** $k = 0, 1, 2, \ldots, N-1$ **do**
4: $\quad x_g^k = \theta x_f^k + (1-\theta)x^k$
5: $\quad$ Sample $J_k \sim \text{Geom}(1/2)$
6: $\quad g^k = g_0^k + \begin{cases} 2^{J_k}(g_{J_k}^k - g_{J_k-1}^k), & \text{if } 2^{J_k} \leq M \\ 0, & \text{otherwise} \end{cases}$ with $g_j^k = \frac{1}{2^j}\sum_{i=1}^{2^j} \nabla F(x_g^k, Z_{T^k+i})$
7: $\quad x_f^{k+1} = x_g^k - \gamma g^k$
8: $\quad x^{k+1} = \eta x_f^{k+1} + (1-\eta)x_f^k$
9: $\quad T^{k+1} = T^k + 2^{J_k}$
10: **end for**

The key idea behind randomized batch size is to reduce the bias of the stochastic gradient estimator. Motivation for this is irrefutably natural as under the Markovian stochastic gradients oracles this bias appears by itself. Indeed, one can easily show the fact that:

$$\mathbb{E}_k[\nabla F(x^k, Z_{T^k+i})] \neq \nabla f(x^k).$$

In a subsequent part, we show how the bias of the gradient estimator introduced in line 6 of Algorithm 1 scales with the truncation parameter M. To obtain proper dependence, we first need to introduce auxiliary Lemma 2, which is to constrain the gradient estimator with a simpler structure. In particular, we bound MSE for sample average approximation computed over batch size n under arbitrary initial distribution. We emphasise that it is extremely essential to have the bound for MSE under arbitrary initial distribution ξ, because in the proof of our Theorem 1 we will unavoidably manage the conditional expectations w.r.t. the previous iterate.

Lemma 2. *Consider Assumptions 3 and 4. Then, for any $n \geq 1$ and $x \in \mathbb{R}^d$, it holds that*

$$\mathbb{E}_\pi\left[\|\frac{1}{n}\sum_{i=1}^n \nabla F(x, Z_i) - \nabla f(x)\|^2\right] \leq \frac{8\tau}{n}\left(\sigma^2 + \delta^2\|x - x^*\|^2\right). \tag{5}$$

Moreover, for any initial distribution ξ on $(\mathsf{Z}, \mathcal{Z})$, that

$$\mathbb{E}_\xi\left[\|\frac{1}{n}\sum_{i=1}^n \nabla F(x, Z_i) - \nabla f(x)\|^2\right] \leq \frac{C_1\tau}{n}\left(\sigma^2 + \delta^2\|x - x^*\|^2\right), \tag{6}$$

where $C_1 = 16(1 + \frac{1}{\ln^2 4})$.

Proof. By [31, Lemma 19.3.6 and Theorem 19.3.9], for any two probabilities ξ, ξ' on $(\mathsf{Z}, \mathcal{Z})$ there is a *maximal exact coupling* $(\Omega, \mathcal{F}, \tilde{\mathbb{P}}_{\xi,\xi'}, Z, Z', T)$ of $\mathbb{P}_\xi^Q$ and $\mathbb{P}_{\xi'}^Q$, that is,

$$\|\xi Q^n - \xi' Q^n\|_{TV} = 2\tilde{\mathbb{P}}_{\xi,\xi'}(T > n). \tag{7}$$

We write $\tilde{\mathbb{E}}_{\xi,\xi'}$ for the expectation with respect to $\tilde{\mathbb{P}}_{\xi,\xi'}$. Using the coupling construction (7),

$$\mathbb{E}_\xi^{1/2}\left[\|\sum_{i=1}^n \{\nabla F(x, Z_i) - \nabla f(x)\}\|^2\right] \leq \mathbb{E}_\pi^{1/2}\left[\|\sum_{i=0}^{n-1}\nabla F(x, Z_i) - \nabla f(x)\|^2\right] +$$

$$\tilde{\mathbb{E}}_{\xi,\pi}^{1/2}\left[\|\sum_{i=0}^{n-1}\{\nabla F(x, Z_i) - \nabla F(x, Z_i')\}\|^2\right].$$

The first term is bounded with (5). Moreover, with (7) and Assumption 4, we get

$$\|\sum_{i=0}^{n-1}\{\nabla F(x, Z_i) - \nabla F(x, Z_i')\}\|^2 \leq 8\left(\sigma^2 + \delta^2\|x - x^*\|^2\right)\left(\sum_{i=0}^{n-1}\mathbb{1}_{\{Z_i \neq Z_i'\}}\right)^2$$

$$= 8\left(\sigma^2 + \delta^2\|x - x^*\|^2\right)\left(\sum_{i=0}^{n-1}\mathbb{1}_{\{T>i\}}\right)^2$$

$$\leq 16\left(\sigma^2 + \delta^2\|x - x^*\|^2\right)\sum_{i=1}^{\infty} i\,\mathbb{1}_{\{T>i\}}.$$

Thus, using the Assumption 3, we bound

$$\tilde{\mathbb{E}}_{\xi,\pi}\left[\sum_{i=1}^{\infty} i\,\mathbb{1}_{\{T>i\}}\right] = \sum_{i=1}^{\infty} i\tilde{\mathbb{P}}_{\xi,\xi'}(T > i) = \sum_{i=1}^{\infty} i(1/4)^{\lfloor i/\tau \rfloor} \leq 4\sum_{i=1}^{\infty} i(1/4)^{i/\tau}.$$

Now we set $\rho = (1/4)^{1/\tau}$ and use an upper bound

$$\sum_{k=1}^{\infty} k\rho^k \leq \rho^{-1}\int_0^{+\infty} x^p \rho^x\,dx \leq \rho^{-1}\left(\ln \rho^{-1}\right)^{-2}\Gamma(2)$$

$$= \rho^{-1}\left(\ln \rho^{-1}\right)^{-2} = \frac{\tau^2}{(1/4)^{1/\tau}\ln^2 4}.$$

Combining the bounds above yields

$$\mathbb{E}_\xi\left[\|\frac{1}{n}\sum_{i=1}^n \nabla F(x, Z_i) - \nabla f(x)\|^2\right] \le \left(\frac{c_1\tau}{n} + \frac{c_2\tau^2}{n^2}\right)(\sigma^2 + \delta^2\|x - x^*\|^2),$$

where $c_1 = 16$, $c_2 = \frac{128(1/4)^{-1/\tau}}{\ln^2 4}$. Now we consider the two cases. If $n < c_1\tau$, we get from Minkowski's inequality that

$$\mathbb{E}_\xi\left[\|\frac{1}{n}\sum_{i=1}^n \nabla F(x, Z_i) - \nabla f(x)\|^2\right] \le 2\sigma^2 + 2\delta^2\|x - x^*\|^2,$$

and (6) holds. If $n > c_1\tau$, it holds that

$$\frac{c_2\tau^2}{n^2}\left(\sigma^2 + \delta^2\|\nabla f(x)\|^2\right) \le \frac{c_2\tau^2}{nc_1\tau}\left(\sigma^2 + \delta^2\|x - x^*\|^2\right),$$

and we gain (6) too. $\qquad\qquad\qquad\qquad\qquad\qquad\qquad\qquad\qquad\qquad\qquad\square$

We are now ready to bound the MSE for the gradient estimator introduced in line 6 of Algorithm 1. From Lemma 3, we obtain a desired linear dependence of the error reduction on the parameter M.

Lemma 3. *Consider Assumptions 3 and 4. Then for the gradient estimates g^k from line 6 Algorithm 1 it holds that $\mathbb{E}_k[g^k] = \mathbb{E}_k[g^k_{\lfloor\log_2 M\rfloor}]$. Moreover,*

$$\mathbb{E}_k[\|\nabla f(x_g^k) - g^k\|^2] \le 13C_1\tau\log_2 M(\sigma^2 + \delta^2\|x_g^k - x^*\|^2), \qquad (8)$$
$$\|\nabla f(x_g^k) - \mathbb{E}_k[g^k]\|^2 \le 2C_1\tau M^{-1}(\sigma^2 + \delta^2\|x_g^k - x^*\|^2),$$

where C_1 is defined in (6).

Proof. To show that $\mathbb{E}_k[g^k] = \mathbb{E}_k[g^k_{\lfloor\log_2 M\rfloor}]$ we simply compute conditional expectation w.r.t. J_k:

$$\mathbb{E}_k[g^k] = \mathbb{E}_k\left[\mathbb{E}_{J_k}[g^k]\right] = \mathbb{E}_k[g_0^k] + \sum_{i=1}^{\lfloor\log_2 M\rfloor}\mathbb{P}\{J_k = i\}\cdot 2^i\mathbb{E}_k[g_i^k - g_{i-1}^k]$$

$$= \mathbb{E}_k[g_0^k] + \sum_{i=1}^{\lfloor\log_2 M\rfloor}\mathbb{E}_k[g_i^k - g_{i-1}^k] = \mathbb{E}_k[g^k_{\lfloor\log_2 M\rfloor}]. \qquad (9)$$

We start with the proof of the first statement of (8) by taking the conditional expectation for J_k:

$$\mathbb{E}_k[\|\nabla f(x_g^k) - g^k\|^2] \le 2\mathbb{E}_k[\|\nabla f(x_g^k) - g_0^k\|^2] + 2\mathbb{E}_k[\|g^k - g_0^k\|^2]$$

$$= 2\mathbb{E}_k[\|\nabla f(x_g^k) - g_0^k\|^2] + 2\sum_{i=1}^{\lfloor\log_2 M\rfloor}\mathbb{P}\{J_k = i\}\cdot 4^i\mathbb{E}_k[\|g_i^k - g_{i-1}^k\|^2]$$

$$= 2\mathbb{E}_k[\|\nabla f(x_g^k) - g_0^k\|^2] + 2\sum_{i=1}^{\lfloor\log_2 M\rfloor}2^i\mathbb{E}_k[\|g_i^k - g_{i-1}^k\|^2]$$

$$\le 2\mathbb{E}_k[\|\nabla f(x_g^k) - g_0^k\|^2] +$$

$$+ 4\sum_{i=1}^{\lfloor\log_2 M\rfloor}2^i\left(\mathbb{E}_k[\|\nabla f(x_g^k) - g_{i-1}^k\|^2 + \mathbb{E}_k[\|g_i^k - \nabla f(x_g^k)\|^2]\right).$$

To bound $\mathbb{E}_k[\|\nabla f(x_g^k) - g_0^k\|^2]$, $\mathbb{E}_k[\|\nabla f(x_g^k) - g_{i-1}^k\|^2$, $\mathbb{E}_k[\|g_i^k - \nabla f(x_g^k)\|^2]$, we apply Lemma 2 and get

$$\mathbb{E}_k[\|\nabla f(x_g^k) - g^k\|^2]$$
$$\leq 2\sigma^2 + 2\delta^2\|x_g^k - x^*\|^2 + 12\sum_{i=1}^{\lfloor \log_2 M \rfloor} 2^i \cdot \frac{C_1\tau}{2^i}(\sigma^2 + \delta^2\|x_g^k - x^*\|^2)$$
$$\leq 13C_1\tau\log_2 M(\sigma^2 + \delta^2\|x_g^k - x^*\|^2).$$

To show the second part of the statement, we use (9) and get

$$\|\nabla f(x_g^k) - \mathbb{E}_k[g^k]\|^2 = \|\nabla f(x^k) - \mathbb{E}_k[g_{\lfloor \log_2 M \rfloor}^k]\|^2.$$

Using Lemma 2 and $2^{\lfloor \log_2 M \rfloor} \geq M/2$ finishes the proof. $\qquad\square$

We also note that our proofs of Lemma 2 and Lemma 3 rely on the proofs of Lemmas 1 and 2 of [1], but for the sake of clarity of the narrative we give them in full.

Now, before we move on to the proof of our major result, we first need to introduce two descent lemmas:

Lemma 4. *Consider Assumptions 1 and 2 be satisfied. Then for the iterates of Algorithm 1 with* $\theta = (1 - \eta)/(\beta - \eta)$, $\theta > 0$, $\eta \geq 1$, *it holds that*

$$\mathbb{E}_k[\|x^{k+1} - x^*\|^2] \leq (1 + \alpha\gamma\eta)(1 - \beta)\|x^k - x^*\|^2 + (1 + \alpha\gamma\eta)\beta\|x_g^k - x^*\|^2$$
$$+ (1 + \alpha\gamma\eta)(\beta^2 - \beta)\|x^k - x_g^k\|^2 + \eta^2\gamma^2\mathbb{E}_k[\|g^k\|^2]$$
$$- 2\eta\gamma\langle\nabla f(x_g^k), \eta x_g^k + (1 - \eta)x_f^k - x^*\rangle$$
$$+ \frac{\eta\gamma}{\alpha}\|\mathbb{E}_k[g^k] - \nabla f(x_g^k)\|^2, \tag{10}$$

where $\alpha > 0$ *is any positive constant.*

Proof. We start with lines 8 and 7 of Algorithm 1:

$$\|x^{k+1} - x^*\|^2 = \|\eta x_f^{k+1} + (1 - \eta)x_f^k - x^*\|^2 = \|\eta x_g^k - \eta\gamma g^k + (1 - \eta)x_f^k - x^*\|^2$$
$$= \|\eta x_g^k + (1 - \eta)x_f^k - x^*\|^2 + \gamma^2\eta^2\|g^k\|^2 - 2\gamma\eta\langle g^k, \eta x_g^k + (1 - \eta)x_f^k - x^*\rangle.$$

Using straightforward algebra, we get

$$\|x^{k+1} - x^*\|^2 = \|\eta x_g^k + (1 - \eta)x_f^k - x^*\|^2 - 2\gamma\eta\langle\nabla f(x_g^k), \eta x_g^k + (1 - \eta)x_f^k - x^*\rangle$$
$$- 2\gamma\eta\langle\mathbb{E}_k[g^k] - \nabla f(x_g^k), \eta x_g^k + (1 - \eta)x_f^k - x^*\rangle + \gamma^2\eta^2\|g^k\|^2$$
$$- 2\gamma\eta\langle g^k - \mathbb{E}_k[g^k], \eta x_g^k + (1 - \eta)x_f^k - x^*\rangle$$
$$\leq (1 + \alpha\eta\gamma)\|\eta x_g^k + (1 - \eta)x_f^k - x^*\|^2 + \frac{\gamma\eta}{\alpha}\|\mathbb{E}_k[g^k] - \nabla f(x_g^k)\|^2.$$
$$- 2\gamma\eta\langle\nabla f(x_g^k), \eta x_g^k + (1 - \eta)x_f^k - x^*\rangle + \gamma^2\eta^2\|g^k\|^2$$
$$- 2\gamma\eta\langle g^k - \mathbb{E}_k[g^k], \eta x_g^k + (1 - \eta)x_f^k - x^*\rangle$$

In the last step we also applied Cauchy-Schwartz inequality in the form (1) with $c > 0$. Taking the conditional expectation, we get

$$\mathbb{E}_k[\|x^{k+1} - x^*\|^2] \leq (1 + \alpha\eta\gamma)\|\eta x_g^k + (1 - \eta)x_f^k - x^*\|^2$$
$$- 2\gamma\eta\langle\nabla f(x_g^k), \eta x_g^k + (1 - \eta)x_f^k - x^*\rangle$$
$$+ \gamma^2\eta^2\mathbb{E}_k[\|g^k\|^2] + \frac{\gamma\eta}{\alpha}\|\mathbb{E}_k[g^k] - \nabla f(x_g^k)\|^2. \qquad (11)$$

Now let us handle expression $\|\eta x_g^k + (1 - \eta)x_f^k - x^*\|^2$ for a while. Taking into account line 4 and the choice of θ such that $\theta = (1 - \eta)/(\beta - \eta)$ (in particular, $\beta = \eta + (1 - \eta)/\theta$ and $(1 - \eta)(\theta - 1)/\theta = 1 - \beta$), we get

$$\eta x_g^k + (1 - \eta)x_f^k = \eta x_g^k + \frac{(1 - \eta)}{\theta}x_g^k - \frac{(1 - \eta)(1 - \theta)}{\theta}x^k = \beta x_g^k + (1 - \beta)x^k$$

Substituting into $\|\eta x_g^k + (1 - \eta)x_f^k - x^*\|^2$, we get

$$\|\eta x_g^k + (1 - \eta)x_f^k - x^*\|^2 = \|\beta x_g^k + (1 - \beta)x^k - x^*\|^2$$
$$= \|x^k - x^* + \beta(x_g^k - x^k)\|^2$$
$$= \|x^k - x^*\|^2 + 2\beta\langle x^k - x^*, x_g^k - x^k\rangle + \beta^2\|x^k - x_g^k\|^2$$
$$= \|x^k - x^*\|^2 + \beta\left(\|x_g^k - x^*\|^2 - \|x^k - x^*\|^2 - \|x_g^k - x^k\|^2\right) + \beta^2\|x^k - x_g^k\|^2$$
$$= (1 - \beta)\|x^k - x^*\|^2 + \beta\|x_g^k - x^*\|^2 + (\beta^2 - \beta)\|x^k - x_g^k\|^2. \qquad (12)$$

Combining (12) with (11), we finish the proof. $\qquad\square$

Lemma 5. *Let Assumptions 1 and 2 be satisfied. Let problem (3) be solved by Algorithm 1. Then for any $u \in \mathbb{R}^d$, we get*

$$\mathbb{E}_k[f(x_f^{k+1})] \leq f(u) - \langle\nabla f(x_g^k), u - x_g^k\rangle - \frac{\mu}{2}\|u - x_g^k\|^2 - \frac{\gamma}{2}\|\nabla f(x_g^k)\|^2$$
$$+ \frac{\gamma}{2}\|\mathbb{E}_k[g^k] - \nabla f(x_g^k)\|^2 + \frac{L\gamma^2}{2}\mathbb{E}_k[\|g^k\|^2].$$

Proof. Using Assumption 1 and line 7 of Algorithm 1, we get

$$f(x_f^{k+1}) \leq f(x_g^k) + \langle\nabla f(x_g^k), x_f^{k+1} - x_g^k\rangle + \frac{L}{2}\|x_f^{k+1} - x_g^k\|^2$$
$$= f(x_g^k) - \gamma\langle\nabla f(x_g^k), g^k\rangle + \frac{L\gamma^2}{2}\|g^k\|^2$$
$$= f(x_g^k) - \gamma\langle\nabla f(x_g^k), \nabla f(x_g^k)\rangle - \gamma\langle\nabla f(x_g^k), \mathbb{E}_k[g^k] - \nabla f(x_g^k)\rangle$$
$$- \gamma\langle\nabla f(x_g^k), g^k - \mathbb{E}_k[g^k]\rangle + \frac{L\gamma^2}{2}\|g^k\|^2$$
$$\leq f(x_g^k) - \gamma\|\nabla f(x_g^k)\|^2 + \frac{\gamma}{2}\|\nabla f(x_g^k)\|^2 + \frac{\gamma}{2}\|\mathbb{E}_k[g^k] - \nabla f(x_g^k)\|^2$$
$$- \gamma\langle\nabla f(x_g^k), g^k - \mathbb{E}_k[g^k]\rangle + \frac{L\gamma^2}{2}\|g^k\|^2.$$

Here we also used Cauchy-Schwartz inequality (1) with $a = \nabla f(x_g^k)$, $b = \nabla f(x_g^k) - \mathbb{E}_k[g^k]$ and $c = 1$. Taking the conditional expectation, we get

$$\mathbb{E}_k[f(x_f^{k+1})] \leq f(x_g^k) - \frac{\gamma}{2}\|\nabla f(x_g^k)\|^2 + \frac{\gamma}{2}\|\mathbb{E}_k[g^k] - \nabla f(x_g^k)\|^2 + \frac{L\gamma^2}{2}\mathbb{E}_k[\|g^k\|^2].$$

Using Assumption 2 with $x = u$ and $y = x_g^k$, one can conclude that for any $u \in \mathbb{R}^d$ it holds

$$\mathbb{E}_k[f(x_f^{k+1})] \leq f(u) - \langle \nabla f(x_g^k), u - x_g^k \rangle - \frac{\mu}{2}\|u - x_g^k\|^2 - \frac{\gamma}{2}\|\nabla f(x_g^k)\|^2$$
$$+ \frac{\gamma}{2}\|\mathbb{E}_k[g^k] - \nabla f(x_g^k)\|^2 + \frac{L\gamma^2}{2}\mathbb{E}_k[\|g^k\|^2]. \qquad \square$$

Taking into account all of the considerations above, we can prove the following result:

Theorem 1. *Consider Assumptions 1 – 4. Let the problem (3) be solved by Algorithm 1. Then for $\beta, \theta, \eta, \gamma, M$ satisfying*

$$M = (1 + 2/\beta), \quad \beta = \sqrt{\frac{4\mu\gamma}{9}}, \quad \eta = \frac{9\beta}{2\mu\gamma} = \sqrt{\frac{9}{\mu\gamma}},$$

$$\gamma \lesssim \min\left\{\frac{\mu^3}{\delta^4\tau^2}; \frac{1}{L}\right\}, \quad \theta = \frac{1-\eta}{\beta-\eta},$$

it holds that

$$\mathbb{E}\left[\|x^N - x^*\|^2 + \frac{18}{\mu}(f(x_f^N) - f(x^*))\right]$$

$$\lesssim \exp\left(-N\sqrt{\frac{\mu\gamma}{9}}\right)\left[\|x^0 - x^*\|^2 + \frac{18}{\mu}(f(x^0) - f(x^*))\right] + \frac{\sqrt{\gamma}}{\mu^{3/2}}C_1\tau\log_2 M\sigma^2.$$

Proof. Using Lemma 5 with $u = x^*$ and $u = x_f^k$, we get

$$\mathbb{E}_k[f(x_f^{k+1})] \leq f(x^*) - \langle \nabla f(x_g^k), x^* - x_g^k \rangle - \frac{\mu}{2}\|x^* - x_g^k\|^2 - \frac{\gamma}{2}\|\nabla f(x_g^k)\|^2$$
$$+ \frac{\gamma}{2}\|\mathbb{E}_k[g^k] - \nabla f(x_g^k)\|^2 + \frac{L\gamma^2}{2}\mathbb{E}_k[\|g^k\|^2],$$

$$\mathbb{E}_k[f(x_f^{k+1})] \leq f(x_f^k) - \langle \nabla f(x_g^k), x_f^k - x_g^k \rangle - \frac{\mu}{2}\|x_f^k - x_g^k\|^2 - \frac{\gamma}{2}\|\nabla f(x_g^k)\|^2$$
$$+ \frac{\gamma}{2}\|\mathbb{E}_k[g^k] - \nabla f(x_g^k)\|^2 + \frac{L\gamma^2}{2}\mathbb{E}_k[\|g^k\|^2].$$

Summing the first inequality with coefficient $2\gamma\eta$, the second with coefficient $2\gamma\eta(\eta-1)$ and (10), we obtain

$$
\mathbb{E}_k[\|x^{k+1} - x^*\|^2 + 2\gamma\eta^2 f(x_f^{k+1})]
$$

$$
\leq (1+\alpha\gamma\eta)(1-\beta)\|x^k - x^*\|^2 + (1+\alpha\gamma\eta)\beta\|x_g^k - x^*\|^2
$$

$$
+ (1+\alpha\gamma\eta)(\beta^2 - \beta)\|x^k - x_g^k\|^2 - 2\eta\gamma\langle\nabla f(x_g^k), \eta x_g^k + (1-\eta)x_f^k - x^*\rangle
$$

$$
+ \eta^2\gamma^2\mathbb{E}_k[\|g^k\|^2] + \frac{\eta\gamma}{\alpha}\|\mathbb{E}_k[g^k] - \nabla f(x_g^k)\|^2
$$

$$
+ 2\gamma\eta\Big(f(x^*) - \langle\nabla f(x_g^k), x^* - x_g^k\rangle - \frac{\mu}{2}\|x^* - x_g^k\|^2 - \frac{\gamma}{2}\|\nabla f(x_g^k)\|^2
$$

$$
+ \frac{\gamma}{2}\|\mathbb{E}_k[g^k] - \nabla f(x_g^k)\|^2 + \frac{L\gamma^2}{2}\mathbb{E}_k[\|g^k\|^2]\Big)
$$

$$
+ 2\gamma\eta(\eta-1)\Big(f(x_f^k) - \langle\nabla f(x_g^k), x_f^k - x_g^k\rangle - \frac{\mu}{2}\|x_f^k - x_g^k\|^2 - \frac{\gamma}{2}\|\nabla f(x_g^k)\|^2
$$

$$
+ \frac{\gamma}{2}\|\mathbb{E}_k[g^k] - \nabla f(x_g^k)\|^2 + \frac{L\gamma^2}{2}\mathbb{E}_k[\|g^k\|^2]\Big)
$$

$$
= (1+\alpha\gamma\eta)(1-\beta)\|x^k - x^*\|^2 + 2\gamma\eta\,(\eta-1)\,(f(x_f^k) - 2\gamma\eta f(x^*))
$$

$$
+ ((1+\alpha\gamma\eta)\beta - \gamma\eta\mu)\,\|x_g^k - x^*\|^2
$$

$$
+ (1+\alpha\gamma\eta)(\beta^2 - \beta)\|x^k - x_g^k\|^2 - \gamma^2\eta^2\|\nabla f(x_g^k)\|^2
$$

$$
+ \left(\frac{\eta\gamma}{\alpha} + \gamma^2\eta^2\right)\|\mathbb{E}_k[g^k] - \nabla f(x_g^k)\|^2 + (\eta^2\gamma^2 + \gamma^3\eta^2\,L)\,\mathbb{E}_k[\|g^k\|^2]
$$

$$
\leq (1+\alpha\gamma\eta)(1-\beta)\|x^k - x^*\|^2 + 2\gamma\eta\,(\eta-1)\,(f(x_f^k) - 2\gamma\eta f(x^*))
$$

$$
+ ((1+\alpha\gamma\eta)\beta - \gamma\eta\mu)\,\|x_g^k - x^*\|^2
$$

$$
+ (1+\alpha\gamma\eta)(\beta^2 - \beta)\|x^k - x_g^k\|^2 - \gamma^2\eta^2\|\nabla f(x_g^k)\|^2
$$

$$
+ \eta\gamma\left(\frac{1}{\alpha} + \gamma\eta\right)\|\mathbb{E}_k[g^k] - \nabla f(x_g^k)\|^2 + 8\eta^2\gamma^2\,(1+\gamma L)\,\mathbb{E}_k[\|g^k - \nabla f(x_g^k)\|^2]
$$

$$
+ \frac{1}{2}\eta^2\gamma^2\,(1+\gamma L)\,\mathbb{E}_k[\|\nabla f(x_g^k)\|^2]\,.
$$

In the last step, we also used (2) with $c = 4$. Since $\gamma \leq \frac{9}{16L}$, the choice of $\alpha = \frac{\beta}{2\eta\gamma}$, $\beta = \sqrt{16\mu\gamma/9}$ gives

$$
\beta = \sqrt{16\mu\gamma/9} \leq \sqrt{\mu/L} \leq 1,
$$

$$
(1+\alpha\eta\gamma)(1-\beta) = \left(1 + \frac{\beta}{2}\right)(1-\beta) \leq \left(1 - \frac{\beta}{2}\right),
$$

and, therefore,

$$\mathbb{E}_k\left[\|x^{k+1} - x^*\|^2 + 2\gamma\eta^2 f(x_f^{k+1})\right]$$
$$\leq (1 - \beta/2)\|x^k - x^*\|^2 + 2\gamma\eta\,(\eta - 1)\,(f(x_f^k) - 2\gamma\eta f(x^*))$$
$$+ \eta^2\gamma^2\,(1 + 2/\beta)\,\|\mathbb{E}_k[g^k] - \nabla f(x_g^k)\|^2$$
$$+ 8\eta^2\gamma^2\,(1 + \gamma L)\,\mathbb{E}_k[\|g^k - \nabla f(x_g^k)\|^2]$$
$$+ ((1 + \alpha\gamma\eta)\beta - \gamma\eta\mu)\,\|x_g^k - x^*\|^2$$

Subtracting $2\gamma\eta^2 f(x^*)$ from both sides, we get

$$\mathbb{E}_k\left[\|x^{k+1} - x^*\|^2 + 2\gamma\eta^2(f(x_f^{k+1}) - f(x^*))\right]$$
$$\leq (1 - \beta/2)\,\|x^k - x^*\|^2 + (1 - 1/\eta)\cdot 2\gamma\eta^2(f(x_f^k) - f(x^*))$$
$$+ \eta^2\gamma^2\,(1 + 2/\beta)\,\|\mathbb{E}_k[g^k] - \nabla f(x_g^k)\|^2$$
$$+ 8\eta^2\gamma^2\,(1 + \gamma L)\,\mathbb{E}_k[\|g^k - \nabla f(x_g^k)\|^2]$$
$$+ ((1 + \alpha\gamma\eta)\beta - \gamma\eta\mu)\,\|x_g^k - x^*\|^2$$

Applying Lemma 3 and $\gamma L \leq 1$, one can obtain

$$\mathbb{E}_k\left[\|x^{k+1} - x^*\|^2 + 2\gamma\eta^2(f(x_f^{k+1}) - f(x^*))\right]$$
$$\leq (1 - \beta/2)\,\|x^k - x^*\|^2 + (1 - 1/\eta)\cdot 2\gamma\eta^2(f(x_f^k) - f(x^*))$$
$$+ \eta^2\gamma^2\,(1 + 2/\beta)\cdot 2C_1\tau M^{-1}(\sigma^2 + \delta^2\|x_g^k - x^*\|^2)$$
$$+ 16\eta^2\gamma^2\cdot 13C_1\tau\log_2 M(\sigma^2 + \delta^2\|x_g^k - x^*\|^2)$$
$$+ ((1 + \alpha\gamma\eta)\beta - \gamma\eta\mu)\,\|x_g^k - x^*\|^2$$

With $M \geq (1 + 2/\beta)$, $\sqrt{\gamma} \leq \frac{\mu^{\frac{3}{2}}}{1872 C_1\tau\delta^2\log_2 M}$, $\alpha = \frac{\beta}{2\gamma\eta}$, $\beta = \frac{2}{3}\sqrt{\mu\gamma}$ and $\eta = \sqrt{\frac{9}{\mu\gamma}}$, we have:

$$(1 + \alpha\gamma\eta)\beta - \gamma\eta\mu + \eta^2\gamma^2\delta^2\big((1 + 2/\beta)\cdot 2C_1\tau M^{-1} + 208C_1\tau\log_2 M\big)$$
$$\leq (1 + \alpha\gamma\eta)\beta - 3\sqrt{\mu\gamma} + \frac{C_1\tau\delta^2\gamma}{\mu}(18 + 1872\log_2 M)$$
$$\leq \sqrt{\mu\gamma} - 3\sqrt{\mu\gamma} + 2\sqrt{\mu\gamma} \leq 0,$$

and then,

$$\mathbb{E}_k\left[\|x^{k+1} - x^*\|^2 + 2\gamma\eta^2(f(x_f^{k+1}) - f(x^*))\right]$$
$$\leq (1 - \beta/2)\|x^k - x^*\|^2 + (1 - 1/\eta)\cdot 2\gamma\eta^2(f(x_f^k) - f(x^*))$$
$$+ \left(\eta^2\gamma^2\,(1 + 2/\beta)\cdot 2C_1\tau M^{-1} + 16\eta^2\gamma^2\cdot 13C_1\tau\log_2 M\right)\sigma^2$$
$$\leq \max\left\{(1 - \beta/2),(1 - 1/\eta)\right\}\left[\|x^k - x^*\|^2 + 2\gamma\eta^2(f(x_f^k) - f(x^*))\right]$$
$$+ \left(\eta^2\gamma^2\,(1 + 2/\beta)\cdot 2C_1\tau M^{-1} + 16\eta^2\gamma^2\cdot 13C_1\tau\log_2 M\right)\sigma^2.$$

Using that $\eta\gamma = 9\beta/(2\mu)$, $\beta/2 = 1/\eta$ and $\gamma \le L^{-1}$, we have

$$\mathbb{E}_k\big[\|x^{k+1} - x^*\|^2 + 2\gamma\eta^2(f(x_f^{k+1}) - f(x^*))\big]$$

$$\le (1 - \beta/2)\big[\|x^k - x^*\|^2 + 2\gamma\eta^2(f(x_f^k) - f(x^*))\big]$$

$$+ \frac{81}{4}\beta^2\mu^{-2}\Big((1 + 2/\beta) \cdot 2C_1\tau M^{-1} + 208C_1\tau \log_2 M\Big)\sigma^2$$

Finally, we perform the recursion and substitute $\beta = \sqrt{4\mu\gamma/9}$:

$$\mathbb{E}\big[\|x^N - x^*\|^2 + 2\gamma\eta^2(f(x_f^N) - f(x^*))\big]$$

$$\le \left(1 - \sqrt{\frac{\mu\gamma}{9}}\right)^N \big[\|x^0 - x^*\|^2 + 2\gamma\eta^2(f(x_f^0) - f(x^*))\big]$$

$$+ \frac{81}{2}\beta\mu^{-2}\Big((1 + 2/\beta) \cdot 2C_1\tau M^{-1} + 208C_1\tau \log_2 M\Big)\sigma^2$$

$$\le \exp\left(-\sqrt{\frac{\mu\gamma N^2}{9}}\right)\big[\|x^0 - x^*\|^2 + 2\gamma\eta^2(f(x_f^0) - f(x^*))\big]$$

$$+ \frac{81\sqrt{\gamma}}{\mu^{3/2}}C_1\tau\Big(1 + 104\log_2 M\Big)\sigma^2.$$

Substituting of $\eta = \sqrt{\frac{9}{\mu\gamma}}$ concludes the proof. $\square$

Corollary 1 (*Step tuning for Theorem 1*). *Under the conditions of Theorem 1, choosing γ as*

$$\gamma \lesssim \min\left\{\frac{\mu^3}{\delta^4\tau^2}; \frac{1}{L}; \frac{1}{\mu N^2}\ln^2\left(\frac{\mu^2 N[\|x^0 - x^*\|^2 + 18\mu^{-1}(f(x_f^0) - f(x^*))]}{\tau\sigma^2}\right)\right\},$$

in order to achieve ϵ-approximate solution (in terms of $\mathbb{E}\big[\|x^N - x^*\|^2\big] \lesssim \epsilon$) it takes

$$\tilde{\mathcal{O}}\left(\left(\sqrt{\frac{L}{\mu}} + \frac{\tau\delta^2}{\mu^2}\right)\log\left(\frac{1}{\epsilon}\right) + \frac{\tau\sigma^2}{\mu^2\epsilon}\right) \quad \text{oracle calls.}$$

4   Numerical Experiments

In this section, we present numerical experiments that compare the proposed method and the existing approaches for the problem of finding consensus in distributed network.

4.1   Problem Formulation

Let us consider the next problem. Assume that we have $\{x_i\}_{i=1}^d$, where $x_i \in \mathbb{R}$. Also we get a communication network, where i^{th} agent stores x_i. Moreover, the communication graph can be described as $G_k = (V, E_k)$, where the set of edges depends on the k – the current moment. The task is formulated as a consensus search, i.e., to find $\bar{x} = \frac{1}{d}\sum_{i=1}^{d} x_i$ – the average value of the agents.

To formalize our problem, we introduce the Laplacian matrix of the graph G_k: $W_k = D_k - A_k$ (here D_k is the diagonal matrix with degrees of nodes, A_k – adjacency matrix) and its properties:

1. $[W_k]_{i,j} \neq 0$ if and only if $(i,j) \in E_k$ or $i = j$,
2. $\ker W_k \supset \{(x_1,\ldots,x_d) \in \mathbb{R}^d : x_1 = \ldots = x_d\}$,
3. range $W_k \subset \left\{(x_1,\ldots,x_d) \in \mathbb{R}^d : \sum_{i=1}^{d} x_i = 0\right\}$.

If we consider $x = (x_1,\ldots,x_d)^\top$, then, because of second property, one can obtain

$$x_1 = \ldots = x_n \Leftrightarrow W_k x = 0.$$

Moreover, it is known that

$$W_k x = 0 \Leftrightarrow \sqrt{W_k} x = 0.$$

Hence, the problem of finding the consensus on the moment k can be reformulated as

$$\min_{x \in \mathbb{R}} \left[f(x) := \frac{1}{2} x^\top W_k x \right]. \tag{13}$$

It is important that the problem formulations (13) for each k have the same optimal point x^*, which is equal to consensus.

The classic approaches to find a consensus is a gossip protocol [36]. In terms of problem (13) the method can be formulated as a gradient descent:

$$x^{k+1} = x^k - \gamma W_k x^k = (1 - \gamma W_k) x^k.$$

This iteration sequence gives the consensus since the third property is fulfilled – it allows to keep the sum of coordinates of x^k the same, preventing the departure from the desired optimal point.

As mentioned above, the problem changes over time as the set of edges specifying the communication system changes. This situation occurs quite often in practice – when additional resources are available to improve the network, edges may be added to speed up processes, and in some system failures, communications between agents may be disconnected due to crashes and overloads. Therefore, it is natural to assume that the changes in the graphs G_k occur according to the Markovian law, since the changes are confined only to the current state of the communication system.

Since for the problem (13) the gradient is equal to $W_k x$, we have

$$\|W_k x - \mathbb{E}(W_k) x\|^2 = \|W_k x - W_k x^* - \mathbb{E}(W_k) x + \mathbb{E}(W_k) x^*\|^2$$
$$\leq \lambda_{max}^2 (W_k - \mathbb{E}(W_k)) \|x - x^*\|^2,$$

where $\mathbb{E}(W_k)$ is an expectation of Laplacian matrix of a graph G_k taking into account the stochasticity responsible for the changes in the graph (more detailed description see later). Consequently, the considered problem satisfies Assumption 4, what means that the theoretical analysis of our paper is applicable to (13).

4.2 Setup

In numerical experiments, we consider the problem described above on different topologies with certain Markovian stochastisity.

Brief Description. We design the experiments in the following way. Suppose we have some starting, or base topology. Then we modify it according to some Markovian law, during which we cannot affect the base graph (i.e., discard edges from it). Based on these changes, we compare two methods: proposed and classic one.

Topologies. As a base topologies we consider two types of graphs – cycle-graph and star-graph. For each starting network we conducted numerical experiments for problems with different dimensions: 10, 100, 1000.

Markovian Stochasticity. The network changes in time in the certain way. On each moment k with probability $\frac{1}{2}$ the random edge can be added to the topology, but if it already exists in the graph, then nothing happens. At the same time, with the same probability the random edge can be removed from the network. Nevertheless, if this edge is in the base topology or communication topology does not contain this edge, we keep the graph in the same condition.

4.3 Results

We performed numerical experiments with different base topologies (see Figs. 1 and 2) with $d = 10$ (see Figs. 1a, 2a), 100 (see Figs. 1b, 2b) and 1000 (see Figs. 1c,

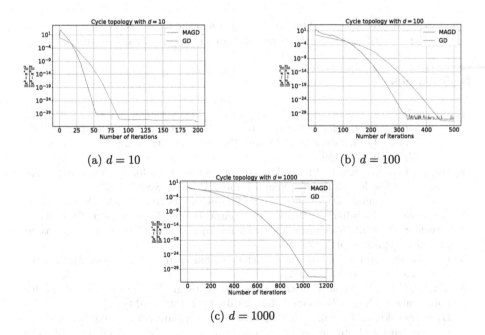

(a) $d = 10$

(b) $d = 100$

(c) $d = 1000$

Fig. 1. Comparison of MAGD and GD for the consensus problem (13) on the cycle topology with different dimensions.

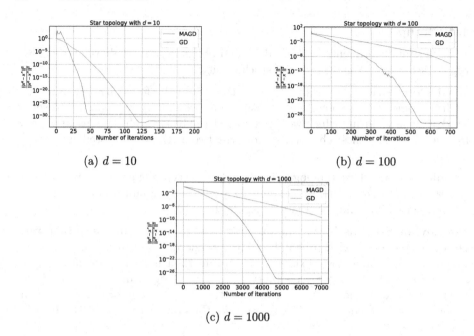

(a) $d = 10$ (b) $d = 100$

(c) $d = 1000$

Fig. 2. Comparison of MAGD and GD for the consensus problem (13) on the star topology with different dimensions.

2c). As a result, the proposed method outperform the classic approach [36] showing a faster rate of convergence, especially for the high-dimension problem.

References

1. Beznosikov, A., Samsonov, S., Sheshukova, M., Gasnikov, A., Naumov, A., Moulines, E.: First order methods with markovian noise: from acceleration to variational inequalities. Adv. Neural Inform. Process. Syst. **36** (2022). https://doi.org/10.48550/arXiv.2305.15938
2. Hu, C., Pan, W., Kwok, J.: Accelerated gradient methods for stochastic optimization and online learning. Adv. Neural Inform. Process. Syst. **22** (2009)
3. Cotter, A., Shamir, O., Srebro, N., Sridharan, K.: Better mini-batch algorithms via accelerated gradient methods. Adv. Neural Inform. Process. Syst. **24** (2011)
4. Devolder, O., et al.: Stochastic first order methods in smooth convex optimization. Technical report, CORE (2011)
5. Lan, G.: An optimal method for stochastic composite optimization. Math. Program. **133**(1–2), 365–397 (2012)
6. Lin, Q., Chen, X., Pena, J.: A smoothing stochastic gradient method for composite optimization. Optim. Methods Softw. **29**(6), 1281–1301 (2014)
7. Dvurechensky, P., Gasnikov, A.: Stochastic intermediate gradient method for convex problems with stochastic inexact oracle. J. Optim. Theory Appl. **171**, 121–145 (2016)

8. Gasnikov, A.V., Nesterov, Y.E.: Universal method for stochastic composite optimization problems. Comput. Math. Math. Phys. **58**, 48–64 (2018)
9. Vaswani, S., Bach, F., Schmidt, M.: Fast and faster convergence of SGD for overparameterized models and an accelerated perceptron. In: The 22nd International Conference on Artificial Intelligence and Statistics, pp. 1195–1204. PMLR (2019a)
10. Taylor, A., Bach, F.: Stochastic first-order methods: non-asymptotic and computeraided analyses via potential functions. In: Conference on Learning Theory, pp. 2934–2992. PMLR (2019)
11. Aybat, N.S., Fallah, A., Gurbuzbalaban, M., Ozdaglar, A.: A universally optimal multistage accelerated stochastic gradient method. Adv. Neural Inform. Process. Syst. **32** (2019)
12. Gorbunov, E., Danilova, M., Shibaev, I., Dvurechensky, P., Gasnikov, A.: Nearoptimal high probability complexity bounds for non-smooth stochastic optimization with heavy-tailed noise. arXiv preprint arXiv:2106.05958 (2021)
13. Woodworth, B.E., Srebro, N.: An even more optimal stochastic optimization algorithm: minibatching and interpolation learning. Adv. Neural. Inf. Process. Syst. **34**, 7333–7345 (2021)
14. Duchi, J.C., Agarwal, A., Johansson, M., Jordan, M.I.: Ergodic mirror descent. SIAM J. Optim. **22**(4), 1549–1578 (2012)
15. Kingma, D.P., Ba, J.: Adam: A method for stochastic optimization. *arXiv preprint*arXiv:1412.6980 (2014)
16. Sutskever, I., Martens, J., Dahl, G., Hinton, G.: On the importance of initialization and momentum in deep learning. In: Dasgupta, S., McAllester, D. (eds.) Proceedings of the 30th International Conference on Machine Learning, vol. 28. Proceedings of Machine Learning Research, pp. 1139–1147, Atlanta, Georgia, USA, 17–19 June. PMLR (2013). https://proceedings.mlr.press/v28/sutskever13.html
17. Lan, G.: First-order and Stochastic Optimization Methods for Machine Learning (Jan 2020). https://doi.org/10.1007/978-3-030-39568-1, ISBN 978-3-030-39567-4
18. Dorfman, R., Levy, K.Y.: Adapting to mixing time in stochastic optimization with markovian data. In: International Conference on Machine Learning, pp. 5429–5446. PMLR (2022)
19. Sun, T., Sun, Y., Yin, W.: On Markov chain gradient descent. Adv. Neural Inform. Process Syst. **31** (2018)
20. Beznosikov, A., Samokhin, V., Gasnikov, A.: Distributed saddle-point problems: Lower bounds, optimal and robust algorithms. arXiv preprint arXiv:2010.13112 (2020)
21. Doan, T.T.: Finite-time analysis of markov gradient descent. IEEE Trans. Autom. Control **68**(4), 2140–2153 (2023). https://doi.org/10.1109/TAC.2022.3172593
22. Doan, T.T., Nguyen, L.M., Pham, N.H., Romberg, J.: Convergence rates of accelerated markov gradient descent with applications in reinforcement learning. arXiv preprint arXiv:2002.02873 (2020)
23. Liu, C., Belkin, M.: Accelerating sgd with momentum for over-parameterized learning. arXiv preprint arXiv:1810.13395 (2018)
24. Sun, T., Li, D., Wang, B.: Adaptive random walk gradient descent for decentralized optimization. In: Chaudhuri, K., Jegelka, S., Song, L., Szepesvari, C., Niu, G., Sabato, S. (eds.)Proceedings of the 39th International Conference on Machine Learning, vol. 162. Proceedings of Machine Learning Research, pp. 20790–20809. 17–23. PMLR (Jul 2022). https://proceedings.mlr.press/v162/sun22b.html
25. Even, M.: Stochastic gradient descent under Markovian sampling schemes. arXiv preprint arXiv:2302.14428 (2023)

26. Wang, P., Lei, Y., Ying, Y., Zhou, D.-X.: Stability and generalization for markov chain stochastic gradient methods. arXiv preprint arXiv:2209.08005 (2022)
27. Nagaraj, D., Wu, X., Bresler, G., Jain, P., Netrapalli, P.: Least squares regression with markovian data: fundamental limits and algorithms. Adv. Neural. Inf. Process. Syst. **33**, 16666–16676 (2020)
28. Hsieh, Y.-G., Iutzeler, F., Malick, J., Mertikopoulos, P.: Explore aggressively, update conservatively: stochastic extragradient methods with variable stepsize scaling. Adv. Neural. Inf. Process. Syst. **33**, 16223–16234 (2020)
29. Gorbunov, E., Berard, H., Gidel, G., Loizou, N.: Stochastic extragradient: General analysis and improved rates. In: International Conference on Artificial Intelligence and Statistics, pp. 7865–7901. PMLR (2022)
30. Iusem, A.N., Jofré, A., Oliveira, R.I., Thompson, P.: Extragradient method with variance reduction for stochastic variational inequalities. SIAM J. Optim. **27**(2), 686–724 (2017). https://doi.org/10.1137/15M1031953
31. Douc, R., Moulines, E., Priouret, P., Soulier, P.: Springer Series in Operations Research and Financial Engineering. Springer (2018). ISBN 978-3-319-97703-4
32. Gorbunov, E., Danilova, M., Gasnikov, A.: Stochastic optimization with heavy-tailed noise via accelerated gradient clipping. Adv. Neural Inform. Process. Syst. **34** (2020). https://doi.org/10.48550/arXiv.2005.10785
33. Robbins, H., Monro, S.: A stochastic approximation method. Ann. Math. Statist. **22**(3), 400–407 (1951). https://doi.org/10.1214/aoms/1177729586
34. Nemirovski, A.S., Yudin, D.B.: Cesari convergence of the gradient method of approximating saddle points of convex-concave functions. Doklady Akademii Nauk **239** pp. 1056–1059. Russian Academy of Sciences (1978)
35. Nemirovsky, A.S., Yudin, D.B.: Problem complexity and method efficiency in optimization (1983)
36. Bertsekas, D., Tsitsiklis, J.: Parallel and distributed computation: numerical methods (2015)
37. Nesterov, Y.E.: A method for solving the convex programming problem with convergence rate $O(1/k^2)$. Dokl. Akad. Nauk SSSR **269**(3), 543–547 (1983). ISSN 0002-3264
38. Nesterov, Y.: Efficiency of coordinate descent methods on huge-scale optimization problems. SIAM J. Optim. **22**(2), 341–362 (2012). https://doi.org/10.1137/100802001
39. Lan, G.: Efficient Methods for Stochastic Composite Optimization. https://api.semanticscholar.org/CorpusID:15780105
40. Schmidt, M., Le Roux, N.: Fast Convergence of Stochastic Gradient Descent under a Strong Growth Condition (2013). https://arxiv.org/pdf/1308.6370.pdf
41. Wang, H., Gürbüzbalaban, M., Zhu, L., Şimşekli, U., Erdogdu, M.A.: Convergence rates of stochastic gradient descent under infinite noise variance. Adv. Neural Inform. Process. Syst. **34** (2021). https://doi.org/10.48550/arXiv.2102.10346
42. Qu, G., Li, N.: Accelerated distributed nesterov gradient descent. IEEE Trans. Autom. Control, 2566–2581 (2020). https://doi.org/10.1109/TAC.2019.2937496
43. Devolder, O.: Stochastic first order methods in smooth convex optimization. CORE Discussion Paper; 2011/70 (2011)

Combinatorial Optimization

Tabu Search for a Service Zone Clustering Problem

Davydov Ivan[1]([✉]) [iD], Gabdullina Aliya[2], Shevtsova Margarita[1,2],
and Arkhipov Dmitry[1,2] [iD]

[1] Sobolev Institute of Mathematics, Novosibirsk, Russia
vann.davydov@gmail.com, shevtsova.ma@phystech.edu
[2] National Research University Higher School of Economics, Moscow, Russia

Abstract. Network maintenance by service engineers (SE) involves a range of activities to ensure that the network is functioning optimally and providing reliable service to users. To optimize the network maintenance process, company need to find suitable places for service offices, decide how to distribute engineers between them, and find a partition of sites into office responsibility zones. These decisions should take into account multiple factors: routes from offices and sites, fair workload distribution, zone topologies, etc. Such a problem can be considered as a generalization of a well-known facility location problem with additional geometric and workload constraints.

In this study, we consider the following formulation of the network maintenance optimization problem. There is a set of sites, a set of potential service office locations, and a limited number of engineers. The site workload is defined by a vector of integers. The objective is to decide which offices should be open, distribute engineers between offices, and find a site partition. With respect to the referenced industrial scenarios, different types of zone topologies are considered. Two criteria are considered to be minimized: total traveling time between offices and sites in a related zone and minimal deviation of workload per worker. An original multi-stage heuristic, which includes greedy and Tabu-search approaches, is proposed. The algorithm includes geometry-based search steps to obtain solutions with star and convex zone topologies. Numerical experiments on industrial instances with up to 10000 sites and 20 offices demonstrate the efficiency of the proposed approach. The obtained results outperformed the CPLEX solver and demonstrated algorithm scalability and obtained solution quality.

Keywords: service engineer · clustering · cellular networks · facility location · Tabu search

1 Introduction

A proper maintenance system of a complex infrastructure plays an important role in functioning of many different services. Wind farms, cellular base stations,

A. Eremeev et al. (Eds.): MOTOR 2024, LNCS 14766, pp. 89–102, 2024.
https://doi.org/10.1007/978-3-031-62792-7_6

water pumps and many other service providing machinery is able to run automatically, but still require a regular maintenance procedures to be performed by human personnel. Optimal use of the resources requires to allocate the staff offices in a vise manner so as to minimize the traveling time to a servicing object on the one hand and at the same time to distribute the whole set of equipment to be maintained evenly between servicing stuff. In this study we consider a new version of a service maintenance zones clustering problem, which comes from a real-life application. In this problem we are given a set of geographical points, where sites with infrastructure are located and a set of potential locations for engineer offices. Each site should be assigned to exactly one office to be serviced from. An office is open and is able to service sites only if at least one engineer is assigned to it. A company seeks for an optimal assignment of a limited group of service engineers to offices, minimizing total traveling time between sites and offices. In this work we present a mathematical model for this problem with three different type of geometry constraints for site clusters. We present an original Tabu-search based heuristic to tackle the problem and provide the results of computational experiments on real-life data instances with up to 10000 sites. The obtained results are compared with CPLEX [1] solver and shows that the proposed approach is capable of obtaining high quality solutions in a short time and demonstrates high scalability.

2   Problem Statement

We consider a new version of a service engineers (SE) clustering problem in the following setting. We are given a set of sites, representing servicing points and a set of potential locations for engineer offices. Both sets of points are associated with corresponding geographical coordinates. Each site is associated with a set of values, representing the workload, induced on an office when servicing this site. Several workload metrics are given. Each site should be assigned to exactly one office to be serviced from. An office is open and is able to provide service to sites only if at least one SE is assigned to it. Each office has a limit on the number of engineers that can be assigned to it. A company seeks for an optimal assignment of a limited group of service engineers to offices, minimizing total costs. The costs are calculated as a sum of euclidean distances from sites to their corresponding servicing offices. In order to maintain fairness in engineer workload, it is demanded that the total workload of each engineer in each type of workload metrics does not deviate from the average by no more than a certain value. This constraint is considered as a soft one in this study, and the corresponding penalty value is added to the goal function in case of violation. The problem also contain a set of nontrivial geometry constraints. First, for each site we are given a set of offices, which considered as preferable ones. It means, that if at least one of these offices is opened, the site cannot be attached to another, unpreferable office. Another group of geometry constraints imposes some restrictions on the geometry form of a set of sites, serviced by one office. In this work we consider three types of such constraints that will be discussed later in details.

The considered problem has a structure, similar to a well-known p-median problem and its generalizations.

3   Problem Formulation

Below we present necessary notation and formulate the problem as a MIP. In a Service Engineer (SE) Clustering Problem (SECP) we are given with:

- J is a set of sites (sites) to be served by the SEs.
- I is a set of potential facility (office) locations, where SEs can be located.
- d_{ij} and t_{ij} are the distance and travel time correspondingly between facility $i \in I$ and site $j \in J$. Distance and travel time can be calculated in different ways, e.g. as an Earth distance, by KNN algorithm, etc.
- For each facility $i \in I$ by $R_i^{\max}$ we denote the maximum number of SEs, which can work in facility $i \in I$.
- W is a set of workload metrics.
- V_w^{lb} and V_w^{ub} are the lower and upper bounds on SE workload $w \in W$ respectively.
- v_{jw} is a workload of metric $w \in W$ induced by serving site $j \in J$.
- p is the maximum number of SEs to be distributed among the open facilities.
- $I_j \subset I$ is a subset of facilities, which are preferable for site $j \in$, i.e. site j can be served from other facilities, only when all the facilities form I_j are closed.

The SECP consists of finding a subset of open facilities, assigning to them p SEs and sites (clusters) with respect to different requirements, which will be discussed further. The main aim is to minimize the sum of distanced between sites and their servicing facilities.

3.1   MIP Formulation

To formulate the problem as a MIP, let us consider the following variables:

$$x_{ij} = \begin{cases} 1, & \text{if site } j \text{ is served form facility } i \\ 0, & \text{otherwise} \end{cases} \quad \forall i \in I, j \in J$$

$$z_i = \begin{cases} 1, & \text{if facility } i \text{ is open} \\ 0, & \text{otherwise} \end{cases} \quad \forall i \in I$$

y_i is an non-negative integer variable, which denote the number of SEs in facility $i \in I$, $0 \le y_i \le R_i^{\max}$.

1. **Objective function** is to minimize the travel time of serving all the sites and sum of workload deviations:

$$\min \sum_{i \in I} \sum_{j \in J} t_{ij} x_{ij} + \omega_1 \sum_{w \in W} u_w \qquad (1)$$

2. **Assignment constraints.** All the sites must be served:

$$\sum_{i \in I} x_{ij} = 1 \qquad\qquad \forall j \in J \qquad\qquad (2)$$

3. **Variable Upper Bounding (VUB) constraints.** sites can be served only from an open facility:

$$x_{ij} \leq y_i \qquad\qquad \forall i \in I, j \in J \qquad\qquad (3)$$

4. **Open facility constraint.** SE can be located only in an open facility:

$$y_i \leq R_i^{\max} z_i \qquad\qquad \forall i \in I \qquad\qquad (4)$$

5. **Number of workers** must be equal p :

$$\sum_{i \in I} y_i = p \qquad\qquad (5)$$

6. **Workload constraints:**

$$\sum_{j \in J} v_{jw} x_{ij} + u_w \geq V_w^{lb} y_i \qquad\qquad \forall i \in I, w \in W \qquad\qquad (6)$$

$$\sum_{j \in J} v_{jw} x_{ij} - u_w \leq V_w^{ub} y_i \qquad\qquad \forall i \in I, w \in W \qquad\qquad (7)$$

We are interested in solutions, where workloads of SE's are balanced, so the lower and upper bounds are close to average with respect to 0.8 and 1.2.

7. **Preference constraints.** site must be served from preferable facilities first:

$$\sum_{l \in I_j} x_{lj} \geq z_i \qquad\qquad \forall j \in J, i \in I_j \qquad\qquad (8)$$

The preferable set of facilities are defined as those which are located closer to the site than a given threshold D, i.e.

$$I_j = \{i \in I : \ d_{ij} < D\}.$$

8. **Geometry constraint.** We are interested in solutions that satisfy certain geometry constraints. We consider three types of such constraints:
   (a) "Free" constraint. NO additional geometry constraints are imposed.
   (b) Convex hulls of clusters can not intersect. These are the most strict constraints. They can be presented by Voronoi's constraints using Voronoi diagram:

$$h_{i_2} - h_{i_1} + M x_{i_1 j} + M z_{i_1} + M z_{i_2} \leq d_{i_2 j}^2 - d_{i_1 j}^2 + 3M$$
$$\forall i_1, i_2 \in I : i_1 \neq i_2, j \in J \qquad (9)$$

where h_i are auxiliary non-negative integer variables, $i \in I$, M is a sufficiently big constant. For example in our case we consider $M = \max_{i \in I, j \in J} (d_{ij})^2$.

(c) "Star" constraints. Softer constraints that forbid segments connecting site with it's serving office to cross. These constraint can be modeled using the vector product of corresponding segments. x_i and y_i values represent the coordinates of the corresponding object (office or site) on the plane. An example of such constraints is shown in the Fig. 1.

$$x_{ik}p_{ikjl} + M_1(1 - u_{1ikjl}) \geq 0 \quad \forall i, j \in I : i \neq i; k, l \in J : k \neq l \quad (10)$$

$$x_{jl}p_{jlik} + M_1(1 - u_{2ikjl}) \geq 0 \quad \forall i, j \in I : i \neq i; k, l \in J : k \neq l \quad (11)$$

$$u_{1ikjl} + u_{2ikjl} \leq 1 \quad \forall i, j \in I : i \neq i; k, l \in J : k \neq l \quad (12)$$

where p_{ikjl} is defined as $p_{ikjl} = ((x_k - x_i)(y_j - y_i) - (y_k - y_i)(x_j - x_i))((x_k - x_i)(y_l - y_i) - (y_k - y_i)(x_l - x_i))$

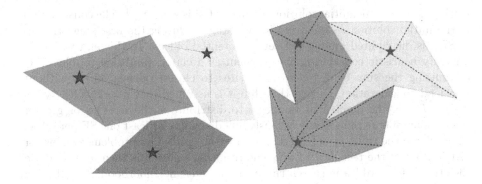

Fig. 1. Convex hull (left) and star(right) constraints

The resulting model can be considered as a generalization of the capacitated p-median problem and is NP-hard in a strong sense even in the case of free geometry. Modeling of the geometry constraints forces us to use the BigM constraints, which in turn become a serious obstacle for MIP solvers to perform rapidly. In the following section we propose a heuristic approach to tackle the problem and describe it's modifications, capable of dealing with various geometry constraints.

4   Bi-level Tabu Search Approach

In this chapter we propose a bi-level heuristic approach to tackle the problem efficiently. The description of the approach can be divided into three phases: construction of initial solution, lower-level stochastic local search approach for the site assignment problem and upper-level Tabu Search for SE distribution optimization. The idea of Tabu Search was presented and formalized by Glover [4–6]. Comprehensive surveys about the approach details and its numerous applications can be found in [3,7]. The general oultine of the approach can be presented as follows.

Initialization: $X, Y = \emptyset$;
tabuList$\emptyset$;
Use greedy approach to construct initial solution;
while *Stopping Condition does not hold* **do**
    Run Upper-level TS approach, seek for optimal SE distribution;
    for *Every Y', SE allocation considered* **do**
        Use lower-level SLS to evaluate the goal function;
    end
    Update *tabuList*;
end
return X^*, Y^*.

Algorithm 1: Bi-level TS approach

4.1   Initial Solution Generation

During this step, the initial solution for the SECP is generated. The construction of the initial solution is organized in two stages. During the first stage the set of SE's is distributed among the set of potential offices. To this end we apply a greedy heuristic considering the problem as a classic p-median problem, i.e. minimizing the sum of distances from sites to closest opened offices without workload consideration. During this step it is assumed that $p < |I|$. Otherwise, if $p \geq |I|$, i.e. the number of engineers to distribute is the same as or greater than the number of offices, we obviously will distribute at least one SE per office. Thus we use the value $p_1 = |I| \bmod p$ to solve the p-median problem, seeking for distribution of the rest part of SE. The resulting solution is then improved via the straightforward Tabu search run over the *move* neighborhood. Neighboring solutions here are obtained from the incumbent by moving one SE from one office to another one. The Tabu list contains the set of offices, that has received at least one new SE during the last t iterations. I.e. if an engineer has been moved to the new office, the number of SE in that office do not decrease for at least t subsequent iterations. The procedure is terminated after a 5*|I| subsequent iterations without improvement of the goal function. Finally we distribute the rest (if the case) $p - p_1$ engineers uniformly, among all offices. Namely, each office receive one new engineer (if possible, i.e. satisfying constraint (4)) per round. If any SE's are still not distributed, the new round is launched. In order to finalize the construction, we assign each site to the closest opened office to be serviced from. We note the resulting solution does not violate the geometry constraints of the second type (star constraints). In order to fulfill the convex geometry constraints, the corresponding procedure have to be implemented.

During the construction of the initial solution we do not take into account the workload constraints, thus the resulting solution may appear to be unwise from this point of view. In order to improve the assignment, we apply a lower-level stochastic Tabu Search to the resulting solution.

Initialization $Y = \emptyset$;
for $a \leftarrow 1$ **to** p_1 **do**
    for $i \in I | y_i = 0$ **do**
        if $F(Y \cup y_i) \leq F^*$ **then**
          | $F^* = F(Y \cup y_i), y^* = y_i$
        end
    end
    $X \leftarrow Y \cup y^*$
end
Result: $Y, |Y| = p_1$

Algorithm 2: Initial solution construction

Initialization $Y, TL = \emptyset$;
for $a \leftarrow 1$ **to** $Iter$ **do**
    $Y^* \leftarrow \text{argmax } F(N_{move}(Y) \backslash TL)$;
    update TL;
end

Algorithm 3: Tabu search for initial solution

4.2 Convex Geometry Constraints

In order to satisfy convex geometry constraint, the we apply an additional procedure to the incumbent solution. It is assumed that at the beginning of the procedure, the solution satisfy star geometry constraints. For each cluster of sites, served by the same office we calculate a convex hull. If a pair of clusters is found, such that their convex hulls intersect, a serie of local steps is applied in order to fix the topology. During each local step we choose one site to be reassigned to the opposite cluster. To this end, we consider those sites, which represent the convex hull vertices of the corresponding clusters and violate the constraints. The site, which reassignment leads to the smallest increase in the goal function is chosen and is reassigned. The procedure is applied to each couple of clusters, one by one, until no more violations are observed.

4.3 Lower-Level Stochastic Local Search

The approach, presented in this section is aimed at solving the problem of optimal assignment of sites to offices. Thus, it is assumed that the distribution of SEs among offices is fixed and given as an input data. In order to run the procedure efficiently, first we calculate some preliminary data to reduce the search space.

1. Half-plane unassignment constraints. These constraints allows us to significantly reduce the set of feasible assignments for each site. For each pair of opened offices a and b we construct a line, connecting these offices and two perpendiculars p_a and p_b, passing through a and b respectively. Any site that is located in the half-plane, opposite to point a with respect to line p_b can not be assigned to office a. The same rule is applied to office b. See an illustrative

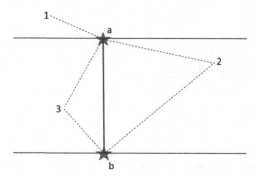

Fig. 2. Half-plane unassignment constraints

example in Fig. 2. Here sites 2 and 3 can be assigned both to offices a and b, while site 1 is located in the half-plane opposite to the office b with respect to line p_a, and thus can not be assigned to office b.

2. Based on half-plane unassignments constraints, for each pair of opened offices a and b we obtain a list of shared sites SH_{ab}, i.e. sites that can be assigned to both offices. Such lists will be used to check and satisfy the star geometry constraint.

As the preliminary data is calculated, we run the local search process. The search here is conducted over a flip neighborhood. Neighboring solution in this neighborhood is obtained from the incumbent one by reassigning one site from it's current servicing office to another one. Only feasible assignments are considered. For each possible move we also check the geometry constraints violation, using SH_{ab} data. The exploration of the neighborhood is conducted in a randomized manner. During one iteration we evaluate only a small fraction f of the whole neighborhood. The exact size of this fraction is defined empirically, and as a rule belongs to the interval $[0.05, 0.20]$. This modification brings several benefits to the procedure. First it obviously reduces the computational time, needed for exploration of the neighborhood. Secondly it allows the search process to avoid being stuck in a local optima, as the probability of return moves is reduced. Most of them will not be included into the neighborhood on a subsequent move. It is known that such a trick allows one to sufficiently reduce the calculation time without any significant reduction in solution quality [2]. The *first improvement* rule is applied to speed up the search process, i.e. if a flip move, improving the incumbent solution is found, the neighborhood exploration is terminated, the corresponding flip is applied and a new iteration starts. During the search process each feasible solution is evaluated with respect to the original goal function. This function calculates the sum of distances from sites to the serving offices and the overall penalty for workload constraints violation.

The general scheme of the approach can be presented as follows:

Initialization X ;
Calculate trivial geometry constraints;
Obtain lists of feasible assignments SH_{ab};
for $i \leftarrow 1$ **to** $Iter$ **do**
    for $k \leftarrow 1$ **to** $|J| * f$ **do**
        $s \leftarrow RAND(1, |J|)$;
        $s^* \leftarrow \operatorname{argmax} F(Ns_{flip}(X))$;
    end
    if $F(X(s^*)) \leq F^*$ **then**
        $F(X(s^*)) = F^*$
    end
end

Algorithm 4: Lower-level stochastic local search

The resulting solution, obtained as an output of Algorithm 4, is optimized according to workload constraints while the star geometry constraints are fully satisfied.

4.4 Upper-Level Tabu Search

The upper-level approach is devoted to the search of an optimal distribution of maintenance stuff among offices. In order to find an optimal distribution of SE we apply an adapted TabuSearch approach. The search is conducted over the *move* neighborhood, i.e. we obtain neighboring solutions by reallocating one staff member from one office to another. We note that as a result of such move initial office might be closed, if there was only one unit of staff or remain operating, otherwise. An office, to which the staff member was relocated might become open, if there were no staff, or remains operating otherwise. We use the Tabu List, that contains the indices of offices, that has been modified during previous steps of the search process. To evaluate the neighboring solution, each time a lower-level search is applied. We note that the lower-level search has to start from a feasible solution. If a neighboring solution is obtained from the current one by closing one office (i.e. $y_i = 0$), we have to reassign sites, served by this office to other, still opened offices. This is done in a greedy manner, satisfying the geometry constraints. The original goal function is applied during this step.

The general scheme of the approach can be presented as follows:

5 Computational Experiments

In this section we present the results of computational experiments, performed on two sets of instances. The first set contains three instances of a medium size, corresponding to a real data of a network maintenance company. We will denote them as "Test1", "Test2" and "Test3". Main metrics of these instances are presented in Table 1. Clus and Site columns refer here to the number of available offices to open and the number of sites to be serviced correspondingly.

Initialization X ;
Calculate trivial geometry constraints;
Obtain lists of feasible assignments SH_{ab};
for $i \leftarrow 1$ to *Iter* do
    for $k \leftarrow 1$ to $|J| * f$ do
        $s \leftarrow RAND(1, |J|)$;
        $s^* \leftarrow$ first_improve $F(Ns_{flip}(X))$;
    end
    if $F(X(s^*)) \leq F^*$ then
        $F(X(s^*)) = F^*$
    end
end

Algorithm 5: Lower-level stochastic local search

Res.max value presents the maximal possible number of service engineers to be distributed among offices. Two last columns, Nvar and Nconstr, demonstrate the size of the corresponding MILP model, with convex geometry constraints. For each of these instance we have considered 9 cases with different values of "p", the number of staff to be allocated, equal to 50, 75 and 100% of the total possible number of engineers (indicated in "perc" column in subsequent tables). We also consider different penalty coefficient for workload constraints violation, $\omega_1 = \{0, 10^3, 10^6\}$. We note that the distances between objects in these instances are measured in meters, and vary from 100 to 100000. Thus, these penalty values can be described as follows: $\omega_1 = 0$, no penalty, thus the workload constraints become non-relevant, and the problem transforms to the p-median one in therms of goal function, yet with specific geometry constraints; $\omega_1 = 10^3$, moderate penalty, workload constraints might be violated in the case of a good gain in distance; $\omega_1 = 10^6$, high penalty, workload constraints are more preferable, violation have to be minimized.

Second test package contains sufficiently large instances with the number of sites reaching 10000, artificially generated based on a real data;

All the instances were tested with the CPLEX v22.1.1 package with a calculation time limited to 3600 sec. Test were carried out on a PC with Intel(R) Core(TM) i7-12700 CPU 2.7GHz and 64 GB of RAM. Two versions of the geometry constraints has been applied, the (1) and the (2) in these tests. The second version of geometry constraints appeared to be rather hard to be tackled by a

Table 1. Instance metrics

Name	Clus	Site	Res. max	Nvar	Nconstr
Test1	6	886	8	5337	32836
Test2	17	1633	18	27815	473712
Test3	24	1461	25	35139	843202

Table 2. CPLEX Results, convex topology

Name	Perc.	Res.	w1	LB	UB	Gap	Time
Test1	50	4	0	7,928,981.00	7,928,981	0.00	0.46
Test2	50	9	0	8,985,817.00	8,985,817	0.00	13.18
Test3	50	13	0	12,169,029.00	12,169,029	0.00	26.54
Test1	75	6	0	7,151,098.00	7,151,098	0.00	0.52
Test2	75	14	0	8,640,836.00	8,640,836	0.00	12.18
Test3	75	19	0	11,886,876.00	11,886,876	0.00	21.37
Test1	100	8	0	7,151,098.00	7,151,098	0.00	0.36
Test2	100	18	0	8,639,057.00	8,639,057	0.00	11.04
Test3	100	25	0	11,872,506.00	11,872,506	0.00	16.22
Test1	50	4	1000	8,371,424.02	8,372,254	0.01	157.18
Test2	50	9	1000	9,624,089.28	30,526,258	68.47	3600
Test3	50	13	1000	12,627,483.44	42,361,029	70.19	3600
Test1	75	6	1000	8,294,828.90	9,012,052	7.96	3600
Test2	75	14	1000	9,099,622.43	24,389,913	62.69	3600
Test3	75	19	1000	12,806,517.19	46,158,876	72.26	3600
Test1	100	8	1000	13,578,445.88	13,579,801	0.01	109.45
Test2	100	18	1000	11,163,320.55	26,509,105	57.89	3600
Test3	100	25	1000	32,562,567.00	55,408,765	41.23	3600
Test1	50	4	1000000	8,371,482.49	8,372,316	0.01	409.3
Test2	50	9	1000000	9,656,078.77	17,859,215,491	99.95	3600
Test3	50	13	1000000	12,627,483.44	29,532,170,808	99.96	3600
Test1	75	6	1000000	8,373,348.90	8,509,472	1.60	3600
Test2	75	14	1000000	9,090,542.65	12,690,800,816	99.93	3600
Test3	75	19	1000000	12,962,035.43	34,115,886,903	99.96	3600
Test1	100	8	1000000	5,533,532,226.13	5,534,070,323	0.01	171.32
Test2	100	18	1000000	12,997,776.30	19,490,639,057	99.93	3600
Test3	100	25	1000000	16,077,045,575.00	43,835,872,654	63.32	3600

Table 3. CPLEX Results, star topology

Test1	50	4	0	7,628,981.00	7,628,981.00	0.00	177.07
Test1	75	6	0	7,151,098.00	7,151,098.00	0.00	174.09
Test1	100	8	0	7,151,098.00	7,151,098.00	0.00	156.25
Test1	50	4	1000	8,209,955.00	8,210,673.00	0.01	191.5
Test1	75	6	1000	7,585,897.15	7,586,568.00	0.01	308.78
Test1	100	8	1000	7,711,384.08	7,712,154.00	0.01	264.59
Test1	50	4	1000000	8,209,955.00	8,210,673.00	0.01	190.58
Test1	75	6	1000000	7,585,854.26	7,586,568.00	0.01	301.59
Test1	100	8	1000000	7,737,456.73	7,738,228.00	0.01	216.55

solver due to extremely high number of constraints, especially in "Test2" and "Test3" instances.

Table 4. Tabu Search Results, convex topology

Name	Perc.	Res.	w1	Objective	Time	CPLEX gap
Test1	50	4	0	7,928,981	10	0.00
Test2	50	9	0	8,985,817	10	0.00
Test3	50	13	0	12,169,029	10	0.00
Test1	75	6	0	7,151,098	10	0.00
Test2	75	14	0	8,640,836	10	0.00
Test3	75	19	0	11,886,876	10	0.00
Test1	100	8	0	7,151,098	10	0.00
Test2	100	18	0	8,639,057	10	0.00
Test3	100	25	0	11,872,506	10	0.00
Test1	50	4	1000	8,373,392	60	0.01
Test2	50	9	1000	9,721,293	60	1.01
Test3	50	13	1000	12,814,370	60	1.48
Test1	75	6	1000	8,476,486	60	2.19
Test2	75	14	1000	9,267,965	60	1.85
Test3	75	19	1000	12,928,179	60	0.95
Test1	100	8	1000	13,578,717	60	0.01
Test2	100	18	1000	11,240,347	60	0.69
Test3	100	25	1000	329,695,990	60	1.25
Test1	50	4	1000000	8,372,320	60	0.01
Test2	50	9	1000000	9,885,893	60	2.38
Test3	50	13	1000000	12,986,104	60	2.84
Test1	75	6	1000000	8,436,986	60	0.76
Test2	75	14	1000000	9,247,809	60	1.73
Test3	75	19	1000000	13,205,722	60	1.88
Test1	100	8	1000000	5,534,085,579	60	0.01
Test2	100	18	1000000	13,106,958	60	0.84
Test3	100	25	1000000	16,200,838,826	60	0.77

As we can observe from Table 2, the CLPEX solver can easily be applied in case of absence of workload constraints, providing optimal solutions is less than 30 sec. As soon as workload balancing appears to be necessary, in becomes rather hard for a solver to tackle the problem, especially when the number of allocated personnel is high. Table 3 shows the results of CPLEX under the star topology constraints. In this case, the number of corresponding geometry constraints doesn't allow to apply solver to instances with more than 1300 sites. Instance Test1 is solved to optimality in less than 10 min. Tables 4 and 5 shows the result of the Tabu Search approach on the same instances. The proposed

Table 5. Tabu Search Results, star topology

Name	Perc.	Res.	w1	Objective	Time	CPLEX gap
Test1	50	4	0	7,628,981.00	60	0.00
Test1	75	6	0	7,151,098.00	60	0.00
Test1	100	8	0	7,151,098.00	60	0.00
Test1	50	4	1000	8,318,490.00	60	1.30
Test1	75	6	1000	7,848,065.00	60	3.34
Test1	100	8	1000	7,874,341.00	60	2.07
Test1	50	4	1000000	8,384,260.00	60	2.08
Test1	75	6	1000000	7,656,577.00	60	0.92
Test1	100	8	1000000	7,941,725.00	60	2.57

approach is capable of finding optimal solutions under both types of topology constraints when balancing constraints are switched off. In the balanced case, the resulting gap between the lower bound, provided by CPLEX and the solution found is within 3% in 60 s of computational time. The proposed approach was also tested on high-dimensional instances with up to 10000 sites and 20 office locations and showed good scalability, providing solutions with high accuracy in a reasonable computational time (600 s). Comparison of the results, obtained under different topology constraints shows that in a complicated setting, i.e. many SEs to allocate and high penalty for balance violation, convex topology appears to be too hard, leading to high penalty values. On the other hand, star topology sometimes provide solutions, inconvenient for a real user to implement.

6   Conclusion

In this paper we consider a new network maintenance optimization problem aiming at minimizing the total traveling time between offices and sites in a related maintenance zone and minimal deviation of workload per worker. We propose an integer programming formulation of this problem with two types of zone topology constraints. As real-life problems instances turn out to be hardly tractable by general MIP solvers, we developed a fast two-phase solution approach based on a randomized Tabu-search heuristic. The proposed approach showed capability to solve big dimension instances in a reasonable computational time with high accuracy.

Our future research may be focused on extending the proposed problem formulation by incorporating new topology constraints and considering non-homogeneous personnel assignment.

Acknowledgement. The study of the first author was carried out within the framework of the state contract of the Sobolev Institute of Mathematics (project FWNF-2022-0019).

References

1. Cplex, I.I.: V12. 1: user's manual for cplex. Int. Bus. Mach. Corp. **46**(53), 157 (2009)
2. Davydov, I., Kochetov, Y., Dempe, S.: Local search approach for the competitive facility location problem in mobile networks. Internat. J. Artif. Intell. **16**(1), 130–143 (2018)
3. Gendreau, M., Potvin, J.Y.: Tabu search. In: Burke, E.K., Kendall, G. (eds.) Search Methodologies: Introductory Tutorials in Optimization and Decision Support Techniques, pp. 165–186. Springer, Boston (2005). https://doi.org/10.1007/0-387-28356-0_6
4. Glover, F.: Future paths for integer programming and links to artificial intelligence. Comput. Oper. Res. **13**(5), 533–549 (1986). https://doi.org/10.1016/0305-0548(86)90048-1
5. Glover, F.: Tabu search-part II. ORSA J. Comput. **2**(1), 4–32 (1990)
6. Glover, F.: Tabu search-part I. ORSA J. Comput. **1**(3), 190–206 (1989)
7. Prajapati, V.K., Jain, M., Chouhan, L.: Tabu search algorithm (TSA): a comprehensive survey. In: Proceedings of the 3rd International Conference on Emerging Technologies in Computer Engineering: Machine Learning and Internet of Things, pp. 1–8 (2020). https://doi.org/10.1109/ICETCE48199.2020.9091743

Clustering Complexity
and an Approximation Algorithm
for a Version of the Cluster Editing
Problem

Artyom Il'ev[1]([✉]) [ID] and Victor Il'ev[1,2] [ID]

[1] Sobolev Institute of Mathematics SB RAS, Omsk, Russia
artyom_iljev@mail.ru
[2] Dostoevsky Omsk State University, Omsk, Russia

Abstract. In graph clustering problems, one has to partition the vertex set of a given undirected graph into pairwise disjoint subsets (clusters). Vertices of the graph correspond to some objects, edges connect the pairs of similar objects. In cluster editing (CE) problems the goal is to find a nearest to a given graph $G = (V, E)$ cluster graph, i.e., a graph on the same vertex set V each connected component of which is a complete graph. The distance between graphs is understood as the number of their non-coinciding edges. The distance between a graph G and a nearest to G cluster graph is called clustering complexity of G.

We consider a version of CE problem in which the size of each cluster is bounded from above by a positive integer s. This problem is NP-hard for any fixed $s \geqslant 3$. In 2015, Puleo and Milenkovic proposed a 6-approximation algorithm for this problem.

For the version of the problem with $s = 5$ we propose a polynomial-time approximation algorithm with better performance guarantee and prove an upper bound on clustering complexity of a graph that is better than earlier known one.

Keywords: Graph cluster editing · Approximation algorithm · Performance guarantee · Clustering complexity

1 Introduction

Clustering is the problem of partitioning a given set of objects into several pairwise disjoint subsets (clusters) taking into account only similarity of objects. In graph clustering problems similarity relation on the set of objects is given by an undirected graph whose vertices are in one-to-one correspondence with objects and edges correspond to pairs of similar of objects. A version of this problem is known as the Graph Approximation problem [1,5,7,15,17]. In this problem, the goal is to find a nearest to a given graph $G = (V, E)$ cluster graph, i.e., a graph on the vertex set V each connected component of which is a complete graph. The distance between graphs is understood as the number of their non-coinciding edges.

A. Eremeev et al. (Eds.): MOTOR 2024, LNCS 14766, pp. 103–115, 2024.
https://doi.org/10.1007/978-3-031-62792-7_7

Later, the Graph Approximation problem was repeatedly and independently rediscovered and studied under various names (Correlation Clustering, Cluster Editing, etc. [3,4,14]). As a rule, now the unweighted version of the problem is named the Cluster Editing problem, whereas the Correlation Clustering problem refers to statements with arbitrary weights of edges [13,14,16].

In different traditional statements of the Cluster Editing problem the number of clusters may be given, bounded, or undefined. We focus our attention on a relatively new version of this problem in which the size of every cluster is bounded from above by a positive integer s.

We consider only ordinary graphs, i.e., the graphs without loops and multiple edges. An ordinary graph $G = (V, E)$ is called a *cluster graph* if every connected component of G is a complete graph [14]. Let $\mathcal{M}(V)$ be the family of all cluster graphs on the set of vertices V, let $\mathcal{M}_k(V)$ be the family of all cluster graphs on the vertex set V having exactly k connected components, and let $\mathcal{M}_{\leqslant k}(V)$ be the family of all cluster graphs on V having at most k connected components, $2 \leqslant k \leqslant |V|$.

If $G_1 = (V, E_1)$ and $G_2 = (V, E_2)$ are ordinary graphs on the set of vertices V, then the *distance* $d(G_1, G_2)$ between them is defined as

$$d(G_1, G_2) = |E_1 \Delta E_2| = |E_1 \setminus E_2| + |E_2 \setminus E_1|,$$

i.e., $d(G_1, G_2)$ is the number of non-coinciding edges in G_1 and G_2.

In the 1960–1980s, the following three Graph Approximation problems were under study. They can be considered as different formalizations of the Cluster Editing problem [1,5,7,15,17].

Problem CE. Given a graph $G = (V, E)$, find a graph $M^* \in \mathcal{M}(V)$ such that

$$d(G, M^*) = \min_{M \in \mathcal{M}(V)} d(G, M) \overset{dn}{=} \tau(G).$$

Problem $\mathbf{CE_k}$. Given a graph $G = (V, E)$ and an integer k, $2 \leqslant k \leqslant |V|$, find a graph $M^* \in \mathcal{M}_k(V)$ such that

$$d(G, M^*) = \min_{M \in \mathcal{M}_k(V)} d(G, M) \overset{dn}{=} \tau_k(G).$$

Problem $\mathbf{CE_{\leqslant k}}$. Given a graph $G = (V, E)$ and an integer k, $2 \leqslant k \leqslant |V|$, find a graph $M^* \in \mathcal{M}_{\leqslant k}(V)$ such that

$$d(G, M^*) = \min_{M \in \mathcal{M}_{\leqslant k}(V)} d(G, M) \overset{dn}{=} \tau_{\leqslant k}(G).$$

All the problems are NP-hard, problems $\mathbf{CE_k}$ and $\mathbf{CE_{\leqslant k}}$ are NP-hard for any fixed $k \geqslant 2$. Main results on computational complexity and approximation algorithms with performance guarantees for these problems can be found in surveys [9,16].

In this paper, we consider the following statement of the problem.

Let $\mathcal{M}^{\leq s}(V)$ be the family of all cluster graphs on V such that the size of each connected component is at most some integer s, $2 \leq s \leq |V|$.

Problem CE$^{\leq s}$. Given a graph $G = (V, E)$ and an integer s, $2 \leq s \leq |V|$, find $M^* \in \mathcal{M}^{\leq s}(V)$ such that

$$d(G, M^*) = \min_{M \in \mathcal{M}^{\leq s}(V)} d(G, M) \stackrel{dn}{=} \tau^{\leq s}(G).$$

In 2011, Il'ev and Navrotskaya [11] proved that Problem **CE$^{\leq s}$** is NP-hard for any fixed $s \geq 3$, whereas Problem **CE$^{\leq 2}$** is polynomially solvable. In 2015, Puleo and Milenkovic [13] proposed a 6-approximation algorithm for Problem **CE$^{\leq s}$**.

In [10] an approximation algorithm for this problem was presented that is 3-approximation in the case $s = 3$ and 5-approximation in the case $s = 4$. In [12] a $\frac{5}{3}$-approximation algorithm for Problem **CE$^{\leq 3}$** and in [8] a 2-approximation algorithm for Problem **CE$^{\leq 4}$** were proposed.

Together with theoretical interest, cases of small cluster size are also interesting from a practical point of view (e.g., allocation of bulky loads onto transport trucks of bounded capacity).

The quantities $\tau(G)$, $\tau_k(G)$, $\tau_{\leq k}(G)$ and $\tau^{\leq s}(G)$ characterize the clustering complexity of a graph G in Problems **CE**, **CE$_k$**, **CE$_{\leq k}$** and **CE$^{\leq s}$**, respectively.

In the 1970–80s tight upper bounds on $\tau(G)$, $\tau_{\leq k}(G)$ and $\tau_k(G)$ were obtained [5,7,15]. It was proved that for every n-vertex graph G each of them does not exceed $\left\lfloor \frac{(n-1)^2}{4} \right\rfloor$ (the last for $n \geq 5(k-1)$).

But for $\tau^{\leq s}(G)$ this bound is not true. In [2,6] tight upper bounds on $\tau^{\leq 3}(G)$ and $\tau^{\leq 4}(G)$ were obtained. As a corollary it was proved that for any $s \geq 4$, $n \geq 5$ and for every n-vertex graph G

$$\tau^{\leq s}(G) \leq \frac{n(n-1)}{2} - 6 \left\lfloor \frac{n}{4} \right\rfloor.$$

This paper is organized as follows. In Sect. 2, we propose a new greedy-type algorithm for the case $s = 5$ with better tight performance guarantee than earlier known one. In Sect. 3, a new bound on clustering complexity of an arbitrary graph in Problem **CE$^{\leq s}$** is proved for $s \geq 5$.

2   An Approximation Algorithm for Problem CE$^{\leq 5}$

In this section, we offer a polynomial-time $\frac{7}{2}$-approximation algorithm for Problem **CE$^{\leq 5}$**.

Let L_4 be either the clique K_4, or a 4-vertex graph obtained from K_4 by removing one edge, and L_5 be either K_5, or a 5-vertex graph obtained from K_5 by removing one edge or two non-adjacent edges (i.e., edges without common vertices). We will refer to these removed edges as *missing edges* in L_4 and L_5.

Lemma 1. *Let $G = (V, E)$ be an arbitrary graph. There is an optimal solution to Problem* $\mathbf{CE}^{\leqslant 5}$ *every connected component of which either is a clique of the graph G, or is a clique obtained from a subgraph $L_i \neq K_i$ of G by adding missing edges, $i = 4, 5$.*

Proof. Consider an optimal solution M^* to Problem $\mathbf{CE}^{\leqslant 5}$. Obviously, if a clique $C = K_2 \subseteq M^*$, then $C \subseteq G$. Suppose that a clique $C = K_3 \subseteq M^*$ is obtained from some 3-vertex subgraph H of G by adding one edge e. Then instead of adding e we can remove from H another edge and obtain another cluster graph $M' \in \mathcal{M}^{\leqslant 5}(V)$ with the same value of the objective function: $d(G, M') = d(G, M^*)$. Let now a clique $C = K_4 \subseteq M^*$ be obtained from some 4-vertex subgraph H of G by adding at least 2 edges. Then instead of their adding we can remove from H at most 2 edges and obtain another cluster graph $M' \in \mathcal{M}^{\leqslant 5}(V)$ such that $d(G, M') \leqslant d(G, M^*)$. Finally, suppose that a clique $C = K_5 \subseteq M^*$ is obtained from some 5-vertex subgraph H of G by adding either 2 adjacent edges, or by adding $i \geqslant 3$ edges. In the first case instead of adding these 2 edges we can remove from H 2 edges and obtain another cluster graph $M' \in \mathcal{M}^{\leqslant 5}(V)$ with the same value of the objective function: $d(G, M') = d(G, M^*)$. In the second case instead of adding these i edges we can remove from H at most i edges and obtain another cluster graph $M' \in \mathcal{M}^{\leqslant 5}(V)$ such that $d(G, M') \leqslant d(G, M^*)$.

In any case the cluster graph M' is obtained from M^* by replacing the component C by 2 components of smaller size that are subgraphs of the graph G. The only exception is the case when 3 adding to H edges are pairwise adjacent and form a triangle. In this case instead of adding these 3 edges we can add to H one of them and remove from H 2 edges to obtain another cluster graph $M' \in \mathcal{M}^{\leqslant 5}(V)$ in which the component C is replaced by components $C_1 = K_1$ and $C_2 = K_4$ that is obtained from the subgraph L_4 of G. In this case $d(G, M') = d(G, M^*)$.

Lemma 1 is proved.

Further, we will consider only optimal solutions satisfying Lemma 1. Consider the following approximation algorithm for Problem $\mathbf{CE}^{\leqslant 5}$.

Algorithm 1

Input: a graph $G = (V, E)$.
Output: a cluster graph $M \in \mathcal{M}^{\leqslant 5}(V)$.
0   $G' \leftarrow G$, $M \leftarrow (V, \emptyset)$.
1   **While** there is a L_5-subgraph H in G' **do**
    add $\bar{H}$ to M and remove H from G' with all incident edges.
    /* $\bar{H}$ is the complete graph on the vertex set of H */
2   **While** there is a L_4-subgraph H in G' **do**
    add $\bar{H}$ to M and remove H from G' with all incident edges.
3   **While** there is a K_3-subgraph H in G' **do**
    add H to M and remove H from G' with all incident edges.
4   **While** there is a K_2-subgraph H in G' **do**

add H to M and remove H from G' with all incident edges.
End.

Theorem 1. *Let $G = (V, E) \notin \mathcal{M}^{\leqslant 5}(V)$ be an arbitrary graph. Then*

$$\frac{d(G, M)}{d(G, M^*)} \leqslant \frac{7}{2}, \tag{1}$$

where M^* is an optimal solution to Problem $\mathbf{CE}^{\leqslant 5}$ on the graph G, M is the cluster graph constructed by Algorithm 1.

Proof. We can think of Algorithm 1 as sequetially finding L_5-, L_4-, K_3- and K_2-subgraphs in G'. Number these subgraphs in the order of their finding by Algorithm 1.

Let k be the total number of steps of Algorithm 1 and H_i be the subgraph found by Algorithm 1 at i-th step. Denote by

- G_i the graph whose vertex-set is V and whose edges are all edges of G' incident with vertices of H_i;
- M_i the graph whose vertex-set is V and edge-set coinsides with edge-set of $\bar{H}_i$;
- M_i^* the graph whose vertex-set is V and whose edges are all edges of M^* incident with vertices of H_i and not belonging to M_j^* for $j < i$.

Then the graphs G, M, and M^* can be presented as unions of pairwise edge-disjoint subgraphs: $G = \cup_{i=1}^k G_i$, $M = \cup_{i=1}^k M_i$, and $M^* = \cup_{i=1}^k M_i^*$. Note that

$$d(G, M) = \sum_{i=1}^k d(G_i, M_i), \quad d(G, M^*) = \sum_{i=1}^k d(G_i, M_i^*).$$

We claim that $\forall i \in \{1, 2, \ldots, k\}$ (unless $d(G_i, M_i) = d(G_i, M_i^*) = 0$)

$$\frac{d(G_i, M_i)}{d(G_i, M_i^*)} \leqslant \frac{7}{2}. \tag{2}$$

To prove this, we consider all possible cases for H_i. Denote by $m \in \{0, 1, 2\}$ the number of missing edges in H_i, and by r the number of edges of G incident with vertices of H_i that do not belong to H_i and any component $C \subseteq M^*$.

1. Let $H_i = L_5$. The following cases are possible.
   1.1. All vertices of H_i belong to different components of M^*. The ratio $\frac{d(G_i, M_i)}{d(G_i, M_i^*)}$ reachs maximum value when all these components of M^* are K_5, i.e., each vertex of H_i is incident with exactly 4 edges of $G_i \cap M_i^*$. Hence $d(G_i, M_i) \leqslant 20 + m + r$ and $d(G_i, M_i^*) \geqslant 10 - m + r$.
   1.2. Two vertices of H_i belong to the same component, and the others belong to different three components of M^*. We have $d(G_i, M_i) \leqslant 18 + m + r$ and $d(G_i, M_i^*) \geqslant 9 - m + r$.

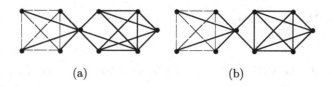

Fig. 1. Extremal (a) and impossible (b) subgraphs H_i for case 1.6. Edges of the graph G_i are thick. The graph H_i is L_5 located in the center (a).

1.3. Two vertices of H_i belong to the same component, another two vertices belong to another component, and the fifth vertex belongs to the third component of M^*. We have
$$d(G_i, M_i) \leqslant 16 + m + r \text{ and } d(G_i, M_i^*) \geqslant 8 - m + r.$$

1.4. Two vertices of H_i belong to different two components, and the remaining three vertices belong to the third component of M^*. We have
$$d(G_i, M_i) \leqslant 14 + m + r \text{ and } d(G_i, M_i^*) \geqslant 7 - m + r.$$

1.5. Two vertices of H_i belong to the same component, and the other three vertices belong to another component of M^*. We have
$$d(G_i, M_i) \leqslant 12 + m + r \text{ and } d(G_i, M_i^*) \geqslant 6 - m + r.$$

1.6. One vertex of H_i belongs to one component, and the others belong to another component of M^*. If $m \in \{0, 1\}$, then
$$d(G_i, M_i) \leqslant 8 + m + r \text{ and } d(G_i, M_i^*) \geqslant 4 - m + r.$$
If $m = 2$, then we have $d(G_i, M_i) \leqslant 10 + r$ and $d(G_i, M_i^*) \geqslant 3 + r$ (Fig. 1(a)), since otherwise H_i would contain two adjacent missing edges (Fig. 1(b)).

1.7. All vertices of H_i belong to the same component of M^*. Then
$$M_i = M_i^* \text{ and } d(G_i, M_i) = d(G_i, M_i^*).$$
It is not difficult to verify that $\frac{d(G_i, M_i)}{d(G_i, M_i^*)}$ reachs maximum value $\frac{14}{4}$ in case 1.5 when $m = 2$, $r = 0$ and both components of M^* are K_5 (Fig. 2(a)). Thus, if $H_i = L_5$, then inequality (2) is true.

2. Let $H_i = L_4$. Note that in this case $m \neq 2$ by definition L_4. After completion of step 1 of Algorithm 1 the graph G' does not contain subgraphs L_5. So, the following cases are possible.

2.1. All vertices of H_i belong to different components of M^*. The ratio $\frac{d(G_i, M_i)}{d(G_i, M_i^*)}$ reachs maximum value when all these components of M^* are K_4, i.e., each vertex of H_i is incident with exactly 3 edges of $G_i \cap M_i^*$. Hence
$$d(G_i, M_i) \leqslant 12 + m + r \text{ and } d(G_i, M_i^*) \geqslant 6 - m + r.$$

2.2. Two vertices of H_i belong to the same component, and the other two vertices belong to different components of M^*. We have
$$d(G_i, M_i) \leqslant 10 + m + r \text{ and } d(G_i, M_i^*) \geqslant 5 - m + r.$$

2.3. Two vertices of H_i belong to the same component, and two other vertices belong to another component of M^*. We have
$$d(G_i, M_i) \leqslant 8 + m + r \text{ and } d(G_i, M_i^*) \geqslant 4 - m + r.$$

2.4. Three vertices of H_i belong to the same component and the fourth vertex belongs to another component of M^*. We have
$d(G_i, M_i) \leqslant 6 + m + r$ and $d(G_i, M_i^*) \geqslant 3 - m + r$.

2.5. All vertices of H_i belong to the same component of M^*. Then
$M_i = M_i^*$ and $d(G_i, M_i) = d(G_i, M_i^*)$.

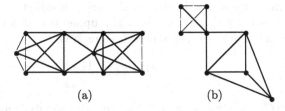

(a)   (b)

Fig. 2. Extremal subgraphs H_i for cases 1.5 and 2.4. Edges of the graph G_i are thick. The graph H_i is L_5 (a) and L_4 (b) located in the center.

It is not difficult to verify that $\frac{d(G_i, M_i)}{d(G_i, M_i^*)}$ reachs maximum value $\frac{7}{2}$ in case 2.4 when $m = 1$, $r = 0$ and both components of M^* are K_4 (Fig. 2(b)). Thus, if $H_i = L_4$, then inequality (2) is true.

3. Let $H_i = K_3$. After completion of steps 1, 2 of Algorithm 1 the graph G' does not contain L_5 and L_4 as subgraphs. So, the following cases are possible.
   3.1. All vertices of H_i belong to different components of M^*. The ratio $\frac{d(G_i, M_i)}{d(G_i, M_i^*)}$ reachs maximum value when all these components of M^* are K_3, i.e., each vertex of H_i is incident with exactly 2 edges of $G_i \cap M_i^*$. Hence
   $d(G_i, M_i) \leqslant 6 + r$ and $d(G_i, M_i^*) \geqslant 3 + r$.
   3.2. Two vertices of H_i belong to the same component of M^* and the third vertex belongs to another component. We have
   $d(G_i, M_i) \leqslant 4 + r$ and $d(G_i, M_i^*) \geqslant 2 + r$.
   3.3. All vertices of H_i belong to the same component of M^*. Then
   $d(G_i, M_i) = d(G_i, M_i^*)$.
   Thus, if $H_i = K_3$, then inequality (2) is true.

4. Let $H_i = K_2$. After completion of steps 1, 2 and 3 of Algorithm 1 the graph G' does not contain L_5, L_4 and K_3 as subgraphs. The following cases are possible.
   4.1. Both vertices of H_i belong to different components of M^*. Then
   $d(G_i, M_i) \leqslant 2 + r$ and $d(G_i, M_i^*) \geqslant 1 + r$.
   4.2. Vertices of H_i belong to the same component of M^*. Then
   $d(G_i, M_i) = d(G_i, M_i^*)$.
   Thus, if $H_i = K_2$, then inequality (2) is true as well.

We obtain
$$\frac{d(G, M)}{d(G, M^*)} = \frac{\sum_{i=1}^{k} d(G_i, M_i)}{\sum_{i=1}^{k} d(G_i, M_i^*)} \leqslant \frac{7}{2}.$$

Theorem 1 is proved.

Remark 1. Bound (1) is tight.

For example, inequality (1) becomes equality for two graphs in Fig. 2.

3 Bound on Clustering Complexity

In this section, we prove a new bound on clustering complexity of a graph in Problem $\mathbf{CE}^{\leqslant s}$ when $s \geqslant 5$. Earlier, the tight upper bound for $s = 3, 4$ was obtained: for any $n \geqslant s + 1$ and every n-vertex graph G

$$\tau^{\leqslant s}(G) \leqslant \frac{n(n-1)}{2} - \frac{s(s-1)}{2}\left\lfloor \frac{n}{s} \right\rfloor.$$

However, for $\tau^{\leqslant 5}(G)$ a similar bound on clustering complexity is incorrect in the general case. The simple counterexample is the complete bipartite graph $K_{3,3}$ for which $\tau^{\leqslant 5}(K_{3,3}) = 6$, whereas the right-hand side of the inequality above equals 5 for $n = 6$ and $s = 5$. So, we prove the following theorem.

Theorem 2. *For any $n \geqslant 6$ and every n-vertex graph G*

$$\tau^{\leqslant 5}(G) \leqslant \frac{(n-2)(n-3)}{2}. \tag{3}$$

Proof. Further, we use **Algorithm 2**, similar to Algorithm 1, which finds in G subgraphs K_5 instead of L_5 and K_4 instead of L_4. Thus, Algorithm 2 finds a cluster subgraph M of G. We devide the proof of the theorem into two steps.

Step 1. Let Algorithm 2 find the cluster graph $M \subseteq G$. We express $d(G, M)$ and the right-hand side of (3) in terms of m_i, where m_i is the number of cliques K_i in the graph M, $i = 1, 2, 3, 4, 5$. We note, that in the cluster subgraph M of the graph G found by Algorithm 2 the number of removed edges between clusters has different upper bounds in each of the possible cases:

- there are at most 25 edges between K_5 and K_5;
- there are at most 20 edges between K_5 and K_4;
- there are at most 15 edges between K_5 and K_3;
- there are at most 10 edges between K_5 and K_2;
- there are at most 5 edges between K_5 and K_1;
- there are at most 12 edges between K_4 and K_4;
- there are at most 9 edges between K_4 and K_3;
- there are at most 6 edges between K_4 and K_2;
- there are at most 3 edges between K_4 and K_1;
- there are at most 6 edges between K_3 and K_3;
- there are at most 4 edges between K_3 and K_2;
- there are at most 2 edges between K_3 and K_1;
- there are at most 2 edges between K_2 and K_2;
- there is at most 1 edge between K_2 and K_1;
- there is no edge between K_1 and K_1.

Bounds on the number of edges for clusters K_5 are defined in each case as the maximum possible number of edges connecting these clusters with others in ordinary graph. In all other cases, we carry out the following reasonings. For example, if clusters K_4 and K_4 have more than 12 connecting edges, then Algorithm 2 would find a clique K_5 instead of the first of them. Also, if clusters K_3 and K_3 have more than 6 connecting edges, then Algorithm 2 would find a clique K_4 instead of the first of them. And so on. Therefore,

$$d(G, M) \leqslant \frac{25m_5(m_5 - 1)}{2} + 20m_5m_4 + 15m_5m_3 + 10m_5m_2 + 5m_5m_1 +$$

$$+ \frac{12m_4(m_4 - 1)}{2} + 9m_4m_3 + 6m_4m_2 + 3m_4m_1 + \frac{6m_3(m_3 - 1)}{2} +$$

$$+ 4m_3m_2 + 2m_3m_1 + \frac{2m_2(m_2 - 1)}{2} + m_2m_1 =$$

$$= \frac{25}{2}m_5^2 + 6m_4^2 + 3m_3^2 + m_2^2 + 20m_5m_4 + 15m_5m_3 + 10m_5m_2 +$$

$$+ 5m_5m_1 + 9m_4m_3 + 6m_4m_2 + 3m_4m_1 + 4m_3m_2 +$$

$$+ 2m_3m_1 + m_2m_1 - \frac{25}{2}m_5 - 6m_4 - 3m_3 - m_2.$$

We note that $n = 5m_5 + 4m_4 + 3m_3 + 2m_2 + m_1$ and so

$$\frac{(n - 2)(n - 3)}{2} = \frac{25}{2}m_5^2 + 8m_4^2 + \frac{9}{2}m_3^2 + 2m_2^2 + \frac{1}{2}m_1^2 + 20m_5m_4 + 15m_5m_3 +$$

$$+ 10m_5m_2 + 5m_5m_1 + 12m_4m_3 + 8m_4m_2 + 4m_4m_1 + 6m_3m_2 +$$

$$+ 3m_3m_1 + 2m_2m_1 - \frac{25}{2}m_5 - 10m_4 - \frac{15}{2}m_3 - 5m_2 - \frac{5}{2}m_1 + 3.$$

Then

$$\frac{(n - 2)(n - 3)}{2} - d(G, M) \geqslant 2m_4^2 + \frac{3}{2}m_3^2 + m_2^2 + \frac{1}{2}m_1^2 + 3m_4m_3 + 2m_4m_2 + m_4m_1 +$$

$$+ 2m_3m_2 + m_3m_1 + m_2m_1 - 4m_4 - \frac{9}{2}m_3 - 4m_2 - \frac{5}{2}m_1 + 3.$$

Thus, if the graph M found by Algorithm 2 satisfies the following inequality

$$2m_4^2 + \frac{3}{2}m_3^2 + m_2^2 + \frac{1}{2}m_1^2 + 3m_4m_3 + 2m_4m_2 + m_4m_1 + 2m_3m_2 +$$

$$+ m_3m_1 + m_2m_1 - 4m_4 - \frac{9}{2}m_3 - 4m_2 - \frac{5}{2}m_1 + 3 \geqslant 0, \tag{4}$$

then the graph M satisfies the inequality:

$$\frac{(n - 2)(n - 3)}{2} - d(G, M) \geqslant 0. \tag{5}$$

We clarify for which values of $m_4, m_3, m_2, m_1 \in \mathbb{Z}_+$ inequality (4) is true.

When $m_4 \geqslant 2$ and $m_3, m_2, m_1 \geqslant 0$, the left-hand side of (4) is at least

$$\frac{3}{2}m_3^2 + m_2^2 + \frac{1}{2}m_1^2 + \frac{3}{2}m_3 - \frac{1}{2}m_1 + 3 \geqslant 0.$$

When $m_3 \geqslant 2$ and $m_4, m_2, m_1 \geqslant 0$, the left-hand side of (4) is at least

$$2m_4^2 + m_2^2 + \frac{1}{2}m_1^2 + 2m_4 - \frac{1}{2}m_1 \geqslant 0.$$

When $m_2 \geqslant 3$ and $m_4, m_3, m_1 \geqslant 0$, the left-hand side of (4) is at least

$$2m_4^2 + \frac{3}{2}m_3^2 + \frac{1}{2}m_1^2 + 2m_4 + \frac{3}{2}m_3 + \frac{1}{2}m_1 \geqslant 0.$$

When $m_1 \geqslant 3$ and $m_4, m_3, m_2 \geqslant 0$, the left-hand side of (4) is at least

$$2m_4^2 + \frac{3}{2}m_3^2 + m_2^2 - m_4 - \frac{3}{2}m_3 - m_2 \geqslant 0.$$

Thus, for $m_4 \geqslant 2$ or $m_3 \geqslant 2$ or $m_2 \geqslant 3$ or $m_1 \geqslant 3$ inequality (4) is always true.

Step 2. Let us find the values of $m_4 \leqslant 1$, $m_3 \leqslant 1$, $m_2 \leqslant 2$, $m_1 \leqslant 2$ for which inequality (4) fails, i.e., Algorithm 2 finds a cluster graph M such that $d(G, M)$ can exceed bound (3) on clustering complexity $\tau^{\leqslant 5}(G)$. All these values are enumerated in the following table.

Table 1. All values of variables when inequality (4) fails.

m_4	m_3	m_2	m_1	Edges*	Left-hand side of (4)
0	0	1	1	1	-1
0	0	1	2	2	-1
0	0	2	0	2	-1
0	0	2	1	4	-1
0	1	0	1	2	-1
0	1	0	2	4	-1
0	1	1	0	4	-1
0	1	1	1	7	-1

*The maximum number of edges between all K_4, K_3, K_2 and K_1 in M.

We consider all cases in which inequality (4) fails. In each case we prove that there is another cluster graph $M' \subseteq G$ such that $d(G, M') < d(G, M)$, and for the graph M' inequality (5) is true.

- Consider the case $m_4 = 0$, $m_3 = 0$, $m_2 = 1$, $m_1 = 1$. Since $n \geqslant 6$, then the graph G must contain at least one clique K_5. Hence $m_5 \geqslant 1$ in the cluster graph M founded by Algorithm 2.

We consider the 8-vertex subgraph G' in which Algorithm 2 at last steps finds a clique K_5, then a clique K_2, and after that one isolated vertex remains in G'. We remind that in M there are at most 10 edges between K_5 and K_2, at most 5 edges between K_5 and K_1, and at most 1 edge between K_2 and every K_1. Thus, altogether at most 16 edges were removed from G' by Algorithm 2.

Since cliques K_5 earlier found by Algorithm 2 don't affect to validity of inequality (4), then violation of this inequality occurs precisely at last steps of the algorithm. In this case $d(G, M)$ exceeds bound (3) on clustering complexity $\tau^{\leqslant 5}(G)$ by 1. If in the subgraph G' the number of removed edges is at most 15, then for the graph M inequality (5) and therefore bound (3) are valid.

So we further assume that the graph G' has 16 edges between cliques $K_5 = \langle v_1, v_2, v_3, v_4, v_5 \rangle$, $K_2 = \langle v_6, v_7 \rangle$ and $K_1 = \langle v_8 \rangle$. Then in the graph G' there are another clusters: $K_5 = \langle v_1, v_2, v_3, v_6, v_7 \rangle$ and $K_3 = \langle v_4, v_5, v_8 \rangle$, to obtain which one needs to remove from G' 2 fewer edges than to obtain clusters of the graph M. Thus, the graph G contains another cluster subgraph M' such that $d(G, M') = d(G, M) - 2$. Therefore, inequality (5) is valid for the graph M', and hence bound (3) on clustering complexity $\tau^{\leqslant 5}(G)$ is correct.

- Consider the cases $m_4 = 0$, $m_3 = 0$, $m_2 = 1$, $m_1 = 2$ and $m_4 = 0$, $m_3 = 0$, $m_2 = 2$, $m_1 = 0$. We assume that the subgraph G' has 20 edges between clique $K_5 = \langle v_1, v_2, v_3, v_4, v_5 \rangle$ and other vertices, including $K_2 = \langle v_6, v_7 \rangle$. Then in the graph G' there are another clusters: $K_5 = \langle v_1, v_2, v_3, v_4, v_8 \rangle$, $K_3 = \langle v_5, v_6, v_7 \rangle$ and $K_1 = \langle v_9 \rangle$, to obtain the cluster graph M' with which one needs to remove from G' 2 or 1 fewer edges respectively than to obtain clusters of the graph M.

- Consider the case $m_4 = 0$, $m_3 = 1$, $m_2 = 0$, $m_1 = 1$. We assume that the subgraph G' has 20 edges between clique $K_5 = \langle v_1, v_2, v_3, v_4, v_5 \rangle$ and other vertices, including $K_3 = \langle v_6, v_7, v_8 \rangle$. Then in the graph G' there are another clusters: $K_5 = \langle v_1, v_2, v_6, v_7, v_8 \rangle$ and $K_4 = \langle v_3, v_4, v_5, v_9 \rangle$, to obtain the cluster graph M' with which one needs to remove from G' 3 fewer edges than to obtain clusters of the graph M.

- Consider the case $m_4 = 0$, $m_3 = 0$, $m_2 = 2$, $m_1 = 1$. We assume that the subgraph G' has 25 edges between clique $K_5 = \langle v_1, v_2, v_3, v_4, v_5 \rangle$ and other vertices, including $K_2 = \langle v_6, v_7 \rangle$ and $K_2 = \langle v_8, v_9 \rangle$. Then in the graph G' there are another clusters: $K_5 = \langle v_1, v_2, v_3, v_6, v_7 \rangle$, $K_4 = \langle v_4, v_5, v_8, v_9 \rangle$ and $K_1 = \langle v_{10} \rangle$, to obtain the cluster graph M' with which one needs to remove from G' 4 fewer edges than to obtain clusters of the graph M.

- Consider the cases $m_4 = 0$, $m_3 = 1$, $m_2 = 0$, $m_1 = 2$ and $m_4 = 0$, $m_3 = 1$, $m_2 = 1$, $m_1 = 0$. We assume that the subgraph G' has 25 edges between clique $K_5 = \langle v_1, v_2, v_3, v_4, v_5 \rangle$ and other vertices, including $K_3 = \langle v_6, v_7, v_8 \rangle$. Then in the graph G' there are another clusters: $K_5 = \langle v_1, v_2, v_6, v_7, v_8 \rangle$, $K_4 = \langle v_3, v_4, v_5, v_9 \rangle$ and $K_1 = \langle v_{10} \rangle$, to obtain the cluster graph M' with which one needs to remove from G' 3 or 2 fewer edges respectively than to obtain clusters of the graph M.

- The case $m_4 = 0$, $m_3 = 1$, $m_2 = 1$, $m_1 = 1$ is special. Here we need to consider the 6-vertex subgraph G' with 11 edges in which Algorithm 2 at last steps finds a clique K_3, then a clique K_2, and after that one isolated vertex remains in G'. We notice that the minimum degree of a vertex in G' with this number of edges can be only 3, because if this minimum equals 1 or 2, then Algorithm 2 must find a clique K_4 in G', and inequality (4) is valid (see Table 1).

Let the minimum degree of a vertex in G' is 3. Then we can isolate this vertex and add 2 edges to the subgraph on the remaining 5 vertices. Thus, we obtain another cluster graph M' with cliques K_5 and K_1, such that $d(G, M') = d(G, M) - 2$. Therefore, inequality (5) is valid for the graph M', and hence bound (3) on clustering complexity $\tau^{\leqslant 5}(G)$ is correct.

Thus, for every n-vertex graph G ($n \geqslant 6$) there is a cluster graph M such that
$$\frac{(n-2)(n-3)}{2} - d(G, M) \geqslant 0.$$

Therefore,
$$\tau^{\leqslant 5}(G) \leqslant \frac{(n-2)(n-3)}{2}.$$

Theorem 2 is proved.

Remark 2. Bound on clustering complexity (3) is tight.

For example, inequality (3) becomes equality for the complete bipartite graph $G = K_{3,3}$.

It follows from Theorem 2 that for any $s \geqslant 5$, $n \geqslant 6$ and for an arbitrary n-vertex graph G
$$\tau^{\leqslant s}(G) \leqslant \frac{(n-2)(n-3)}{2}.$$

4   Conclusion

A version of the graph clustering problem is considered. In this version sizes of all clusters don't exceed a given positive integer s. This problem is NP-hard for every fixed $s \geqslant 3$. We propose a new greedy-type algorithm for the case $s = 5$ with better tight performance guarantee than earlier known one. We also prove a new bound on clustering complexity of an arbitrary graph in Problem $\mathbf{CE}^{\leqslant s}$ when $s \geqslant 5$.

Acknowledgement. The research of A.V. Il'ev was supported by the Russian Science Foundation, project No. 22-11-20019. The research of V.P. Il'ev was funded in accordance with the state task of the IM SB RAS, project FWNF-2022-0020.

References

1. Ageev, A.A., Il'ev, V.P., Kononov, A.V., Talevnin, A.S.: Computational complexity of the graph approximation problem. Diskretnyi Analiz i Issledovanie Operatsii. Ser. 1. **13**(1), 3–11 (2006). (in Russian). English transl. in: J. of Applied and Industrial Math. **1**(1), 1–8 (2007)

2. Baldzhanova, R.V., Il'ev, A.V., Il'ev, V.P.: On the clustering complexity of a graph in the problem with bounded sizes of clusters. Prikl. Diskretn. Mat. **60**, 76–84 (2023). (in Russian)
3. Bansal, N., Blum, A., Chawla, S.: Correlation clustering. Mach. Learn. **56**, 89–113 (2004)
4. Ben-Dor, A., Shamir, R., Yakhimi, Z.: Clustering gene expression patterns. J. Comput. Biol. **6**(3–4), 281–297 (1999)
5. Fridman, G.Š.: Investigation of a classifying problem on graphs. Methods of Modelling and Data Processing (Nauka, Novosibirsk). 147-177 (1976). (in Russian)
6. Il'ev, A.V., Il'ev, V.P.: Bounds for the clustering complexity in a graph clustering problem with clusters of bounded size. J. Math. Sci. **275**(1), 78–84 (2023)
7. Il'ev, V.P., Fridman, G.Š.: On the problem of approximation by graphs with a fixed number of components. Dokl. Akad. Nauk SSSR. **264**(3), 533–538 (1982). (in Russian). English transl. in: Sov. Math. Dokl. **25**(3), 666–670 (1982)
8. Il'ev, V., Il'eva, S.: Approximation algorithms for graph cluster editing problems with cluster size at most 3 and 4. In: Khachay, M., et al. (eds.) MOTOR 2023, CCIS, vol. 1881, pp. 134–145. Springer, Cham (2023). https://doi.org/10.1007/978-3-031-43257-6_11
9. Il'ev, V., Il'eva, S., Kononov, A.: Short survey on graph correlation clustering with minimization criteria. In: Kochetov, Y., Khachay, M., Beresnev, V., Nurminski, E., Pardalos, P. (eds.) DOOR 2016. LNCS, vol. 9869, pp. 25–36. Springer, Cham (2016). https://doi.org/10.1007/978-3-319-44914-2_3
10. Il'ev, V.P., Il'eva, S.D., Navrotskaya, A.A.: Graph clustering with a constraint on cluster sizes. Diskretn. Anal. Issled. Oper. **23**(3), 5–20 (2016). (in Russian). English transl. in: J. Appl. Indust. Math. **10**(3), 341–348 (2016)
11. Il'ev, V.P., Navrotskaya, A.A.: Computational complexity of the problem of approximation by graphs with connected components of bounded size. Prikl. Diskretn. Mat. **3**(13), 80–84 (2011). (in Russian)
12. Kononov, A., Il'ev, V.: On cluster editing problem with clusters of small sizes. In: Olenev, N., et al. (eds.) OPTIMA 2023, LNCS, vol. 14395, pp. 316–328. Springer, Cham (2023). https://doi.org/10.1007/978-3-031-47859-8_23
13. Puleo, G.J., Milenkovic, O.: Correlation clustering with constrained cluster sizes and extended weights bounds. SIAM J. Optim. **25**(3), 1857–1872 (2015)
14. Shamir, R., Sharan, R., Tsur, D.: Cluster graph modification problems. Discrete Appl. Math. **144**(1–2), 173–182 (2004)
15. Tomescu, I.: La reduction minimale d'un graphe à une reunion de cliques. Discrete Math. **10**(1–2), 173–179 (1974)
16. Wahid, D.F., Hassini, E.: A literature review on correlation clustering: cross-disciplinary taxonomy with bibliometric analysis. Oper. Res. Forum **3**(47), 1–42 (2020)
17. Zahn, C.T.: Approximating symmetric relations by equivalence relations. J. Soc. Ind. Appl. Math. **12**(4), 840–847 (1964)

A Learning-Augmented Algorithm
for the Parking Permit Problem
with Three Permit Types

Kharchenko Yaroslav[1]($\boxtimes$) ⓘ and Kononov Alexander[2] ⓘ

[1] Novosibirsk State University, Pirogova 2, 630090 Novosibirsk, Russian Federation
kharchenko.yar@gmail.com
[2] Sobolev Institute of Mathematics, RAS, Akademika Koptyuga 4,
630090 Novosibirsk, Russian Federation

Abstract. We consider the parking permit problem with three permit types. First, we prove the tight lower bound on the competitiveness of any deterministic online algorithm. Next, we present a learning-augmented algorithm and show its' consistency and robustness.

Keywords: Learning-augmented · Online algorithm · Prediction

1 Introduction

Online algorithms are algorithms that have their input gradually revealed to them during their execution, and as such are forced to make irrevocable decisions based only on already revealed data. These algorithms do not make any assumptions about unrevealed data, and are able to process all possible inputs. However, since the input can not be analyzed before algorithm execution, online algorithms design is often focused on worst-case scenarios, possibly delivering sub-optimal performance on inputs that are encountered more frequently in actual applications. Thus, many works consider models that are provided with some information about the uncertain part of the input [6,21].

One type of such models are prediction-augmented algorithms. These algorithms are provided with some prediction regarding the unknown part of the input, however, this information is not guaranteed to be correct. Latest developments in machine learning has allowed to use ML models as a source of such predictions [2,4,10,14,20]. Prediction-augmented algorithms have better competitive bounds than similar algorithms without predictions, if provided predictions are correct. Even if the predictions are incorrect, competitive bounds are worsened only by a constant factor.

This paper considers an online parking permit problem – a generalization of the ski rental problem that was first introduced in [19]. Unlike the ski rental problem, in the parking permit problem days that require a permit ("rainy"') can be interleaved with days that do not ("sunny"). Also, multiple permit types with different costs and durations are available.

© The Author(s), under exclusive license to Springer Nature Switzerland AG 2024
A. Eremeev et al. (Eds.): MOTOR 2024, LNCS 14766, pp. 116–126, 2024.
https://doi.org/10.1007/978-3-031-62792-7_8

In this work we present a parameterized deterministic prediction-augmented algorithm and provide its' consistency and robustness bounds. We also provide a lower boundary for deterministic algorithms without predictions.

2   Preliminaries

2.1   Problem Definition

The following problem is considered in [19].

We are given a set of K permit types. Each permit type k is assigned a cost C_k and duration D_k days. We are also given a sequence of days called a schedule, each day in a schedule is marked as either sunny or rainy. The schedule is revealed one day at a time. When a day is revealed, the algorithm may purchase arbitrary permits of given types. A purchased permit of type k is valid from the day of purchase for D_k consequtive days. Before the next day is revealed, the algorithm must have purchased such a set of permits, that any day marked as rainy is within the duration of some permit. Purchased permits can not be refunded or switched for another permit on a later day. The total cost of purchased permits must be minimized.

For any deterministic algorithm there exists such a schedule that the algorithm will find a suboptimal solution for it. Thus we will minimize worst-case ratio between the cost paid by algorithm and cost of the optimal solution. This ratio is known as the algorithm's competitive ratio.

We consider the interval variant of the parking permit problem, and denote this formulation as $\mathcal{P}$. In the original formulation of the parking permit problem the permit bought on some day is valid the day of purchase and during the following D_k days. In the interval variant for each k the schedule is split into fixed intervals of length D_k. The permit is only valid for the interval it was purchased on. Additionally we suppose that any interval of length D_k is composed of an integer number of intervals with length D_{k-1}. Meyerson [19] shows that any ρ-competitive algorithm for $\mathcal{P}$ is 2ρ-competitive for the corresponding original problem.

Instance I of $\mathcal{P}$ is comprised of fixed values of K, C_k, D_k for $k = 1..K$ and a schedule. For an instance I, we denote by $ALG(I)$ the cost of solution obtained by algorithm Υ, and let $OPT(I)$ be the cost of the optimal offline solution. We call $R(\Upsilon) = \max_I(\frac{ALG(I)}{OPT(I)})$ the competitive ratio of algorithm Υ.

If some algorithm Υ has a competitive ratio $R(\Upsilon) = \gamma$, and any other algorithm Υ' holds $R(\Upsilon') \geq \gamma$, we call the algorithm Υ competitive-best.

2.2   Prediction-Augmented Algorithms

If Π is a statement that is true for some schedules and false for other schedules, we call this statement a prediction. Prediction Π is a strict prediction if any schedule for which Π is true has same optimal cost $c(\Pi)$. In this work we only consider strict predictions. Prediction type is such a set of predictions that for each possible schedule exactly one prediction is true.

Suppose we are given a prediction type and an algorithm that receives a prediction Π along the problem input.

If for any input I and any prediction Π

$$\frac{ALG(I, \Pi)}{OPT(I)} \leq \beta,$$

we say the algorithm is β-robust.

If for any input I and any prediction Π

$$\frac{ALG(I, \Pi)}{c(\Pi)} \leq \sigma,$$

we say the algorithm is σ-consistent.

Consistency and robustness are respective measures of the algorithm's ability to utilize predicted information and compensating for incorrect predictions. Consistency shows that algorithmic solution is close to the optimal for the predicted schedule. Robustness measures worst-case actual performance for all predictions, including incorrect ones.

2.3   Related Work

The ski rental problem is the simplest and most well-studied online problem. A trivial deterministic 2-competitive online algorithm has been known for a long time [11]. Later, Karlin et al. [12] presented a randomized $\frac{e}{e-1}$-competitive algorithm. Both competitive ratio are optimal for deterministic and randomized algorithms, respectively, Recently, this problem has been extensively studied in various augmented learning settings [2,8,14]. For the ski rental problem, many generalizations are known, such as the multi-option ski rental problem, the bahncard problem [7], and others [10,20].

In our paper, we consider the parking permit problem [19], which is also generalization of the ski rental problem. Meyerson [19] considered the parking permit problem with k different types of permits. He presented k-competitive deterministic online algorithm and $O(\log k)$-competitive randomized online algorithm. Meyerson also showed that these algorithms are asymptotically competitive best i.e., any deterministic algorithm has competitive ratio at least $k/3$ and any randomized algorithm has expected competitive ratio at least $(\log k)/2$.

In our paper, we analyze the parking permit problem with three types of permits. Note that in the case of small k, Meyerson's result does not provide a reasonable lower bound for the competition ratio of online algorithms for the parking permit problem. We assume that the lower bound on the competitiveness of any deterministic online algorithm with k permits is equal to k and prove it for $k = 3$.

Lykouris and Vasilvitsky [18] described a general framework for incorporating predictions into online algorithms and developed the first learning-augmented algorithm to solve the online caching problem. A preliminary version of their paper appeared in the proceedings of the 35th International Conference on Machine Learning in 2018. Over the next five years, prediction-aware algorithms were developed for online problems in scheduling [3,9,14–16], covering [2], dynamic TCP acknowledgement [2,10], and others [5,13]. We focus here only on the related work for ski rental problems.

For the ski rental problem with a predictor, Kumar et al. [14] presented a deterministic online algorithm that is $\frac{1+\lambda}{\lambda}$-robust and $(1+\lambda)$-consistent for $\lambda \in (0,1)$. The authors used a prediction based on the total number of vacation days. They next improved these bounds by obtaining a randomized algorithm that is $\frac{1}{1-e^{-\lambda}}$-robust and $\frac{\lambda}{1-e^{-\lambda}}$-consistent. Bamas et al. [2] obtained similar results by incorporating predictions into the primal-dual method for online algorithms and showed that the obtained consistency-robustness trade-off is optimal. They also applied the primal-dual method with predictions to the Bahncard problem, which is a generalization of the ski rental problem.

3   Lower Bound for Competitive Ratio

In this work we consider the case of the interval parking permit problem with $K = 3$ permit types. Suppose w.l.o.g. that

- $C_1 = 1, D_1 = 1$,
- $C_2 = B, D_2 = D$, we call this interval type a week,
- $C_3 = A, D_3 = nD$, we call this interval type a year.

For this problem we can prove the following theorem.

Theorem 1. *No deterministic algorithm can obtain a competitive ratio of less than 3 for the parking permit problem with three permit types.*

Proof. Assign numbers to days, starting with day zero. We suppose that a day is rainy if and only if the algorithm has not yet bought a permit that covers this day. Thus we consider a single adversarial schedule for each algorithm.

Firstly we note that an algorithm which never buys the year permit can have arbitrarily large competitive ratio. Consider such an algorithm. Let $k > 1$ be an arbitrary number. Consider a schedule that has n weeks where $n > k \lceil \frac{A}{B} \rceil$. Then total cost paid by the algorithm is greater than kA. Since the optimal solution for such a schedule costs $OPT = A$, we get $R > k$.

Consider an arbitrary deterministic algorithm Υ. We will show that for any real $\varepsilon > 0$ there exists an instance I, such that the competitive ratio of Υ on instance I is greater than $3 - \varepsilon$. Let $A > D\frac{(1-\varepsilon)}{\varepsilon} + 2D$ and $B > 3D$.

Let $ALG(t)$ denote the total cost paid up to the start of day t in the solution produced by the algorithm Υ. Similarly, consider a schedule that only consists of days $0..t-1$ and denote the cost of optimal solution for this schedule as $OPT(t)$.

Assume that algorithm Υ buy the year parking permit on day t_0. Denote by $N = \lfloor \frac{t_0}{D} \rfloor$ the number of whole weeks before the day t_0. We will show that $\frac{ALG(ND)}{OPT(ND)} \geq 2$.

We can split the set of weeks into three disjoint subsets based on the number of the day that the algorithm purchased weekly permit on. Let E be the set of weeks, such that on week I_e algorithm buys weekly permit on day number $B - e(I_e)$. Similarly, on weeks $I_l \in L$ algorithm buys weekly permit on day number $B + l(I_l)$. Finally, let T be the number of weeks on which the algorithm purchases the weekly permit exactly on day number B. Then $ALG(ND) = 2NB - \sum_{I_e \in E} e(I_e) + \sum_{I_l \in L} l(I_l)$. Since $OPT(ND) = NB - \sum_{e \in E} e(I_e)$,

$$\frac{ALG(ND)}{OPT(ND)} = 1 + \frac{NB + \sum_{l \in L} l}{NB - \sum_{e \in E} e} \geq 2.$$

Note that $\frac{ALG(ND)}{OPT(ND)} = 2$ when $E = L = \emptyset$.

The competitive ratio R equals to $\frac{ALG(t_0)}{OPT(t_0)}$. Using algorithm cost and optimal cost on day ND, we can express the total cost paid by the algorithm as

$$ALG(t_0 + 1) = ALG(ND) + t_0 - ND + A,$$

$$OPT(t_0 + 1) = \min\{(OPT(ND) + t_0 - ND, OPT(ND) + B, A\}.$$

Consider possible values of $OPT(t_0 + 1)$. Let $OPT(t_0 + 1) = A$. Then $OPT(ND) + t_0 - ND \geq A > D\frac{(1-\varepsilon)}{\varepsilon} + 2D = D\left(\frac{(1-\varepsilon)}{\varepsilon} + 2\right) = D\frac{(1+\varepsilon)}{\varepsilon}$. By definition of N it follows, that $t_0 - ND < D$. Thus, $OPT(ND) + t_0 - ND \geq (t_0 - ND)\frac{(1+\varepsilon)}{\varepsilon}$, which is equivalent to $\frac{t_0 - ND}{OPT(ND) + t_0 - ND} \leq \frac{\varepsilon}{1+\varepsilon} < \varepsilon$.

Competitive ratio

$$R = \frac{ALG(ND) + t_0 - ND + A}{A} = 1 + \frac{ALG(ND) + t_0 - ND}{A}$$

$$\geq 1 + \frac{ALG(ND) + t_0 - ND}{OPT(ND) + t_0 - ND} \geq 3 - \frac{t_0 - ND}{OPT(ND) + t_0 - ND} > 3 - \varepsilon.$$

Consider the competitive ratio for other possible values of $OPT(t_0 + 1)$.

1. $R = \frac{ALG(ND) + t_0 - ND + A}{OPT(ND) + t_0 - ND} \geq \frac{2OPT(ND) + t_0 - ND + A}{OPT(ND) + t_0 - ND} = 2 + \frac{A - (t_0 - ND)}{OPT(ND) + t_0 - ND} > 3 - \frac{t_0 - ND}{OPT(ND) + t_0 - ND}$

2. $R = \frac{ALG(ND) + t_0 - ND + A}{OPT(ND) + B} \geq \frac{2OPT(ND) + t_0 - ND + A}{OPT(ND) + t_0 - ND} = 2 + \frac{A - (t_0 - ND)}{OPT(ND) + t_0 - ND} > 3 - \frac{t_0 - ND}{OPT(ND) + t_0 - ND}$.

In both cases $R \geq 3 - \frac{t_0 - ND}{OPT(ND) + t_0 - ND}$.

Suppose that $\frac{t_0 - ND}{OPT(ND) + t_0 - ND} > \varepsilon => t_0 - ND > \varepsilon(OPT(ND) + t_0 - ND) => OPT(ND) < (t_0 - ND)\frac{(1-\varepsilon)}{\varepsilon} < D\frac{(1-\varepsilon)}{\varepsilon}$, thus $OPT(ND) + t_0 - ND < A - (t_0 - ND)$. Note that in both cases $R \geq \frac{2OPT(ND) + t_0 - ND + A}{OPT(ND) + t_0 - ND}$. By substituting $A - (t_0 - ND) > OPT(ND) + t_0 - ND$ into the numerator, we get $R > 3$.

If the above assumption is incorrect, we have $\frac{t_0 - ND}{OPT(ND) + t_0 - ND} < \varepsilon$ and we obtain $R \geq 3 - \varepsilon$.

Let $\gamma \in \mathcal{R}, \gamma < 3$. Then there exists a problem instance I, such that $\frac{ALG(I)}{OPT(I)} > \gamma$, thus the algorithm is not γ-competitive. Since the algorithm is arbitrary, this proves the theorem.

4   Prediction-Augmented Algorithm with Two Prediction Types

We consider the following prediction type. For each week a prediction states, whether the week will contain at least B rainy days. For each year the prediction states, whether the cost of covering the year only using daily and weekly permits is greater than or equal to A. Thus the prediction is strict, and optimal solution cost for the corresponding schedule can be obtained trivially.

To adjust algorithm performance with regard to prediction errors we will use two parameters $\lambda, \mu \in [0, 1]$. Intuitively, they reflect the degree of mistrusting the prediction the algorithm should employ. Lesser values make the algorithm more consistent to prediction, yet less robust to prediction errors. By raising the values the algorithm can be made more robust to worst-case scenarios, but makes less use of provided information.

Formally, together with the usual input algorithm receives the prediction in form of binary values P_3, P_2^i, $i = 1, \ldots, n$, and controlling parameters $\lambda, \mu \in [0, 1]$. $P_3 = 1$ if the prediction states that buying yearly permit is optimal, else $P_3 = 0$. Similarly, $P_2^i = 1$ if prediction states that week i has at least B rainy days, else $P_2^i = 0$.

Denote cost of optimal solution if the schedule only consisted of already seen days by OPT_{cur}, same cost if the solution only consisted of daily and weekly permits by $OPT_{cur}^{(2)}$.

For an arbitrary instance of $\mathcal{P}$ and an arbitrary prediction we introduce the error of the weekly prediction. Let η be equal the number of weeks for which the prediction is incorrect, that is, either $P_2^i = 1$ and week i has less than B rainy days, or $P_2^i = 0$ and week i has at least B rainy days. Let $\alpha = \frac{\eta}{n}$ be the fraction of incorrectly predicted weeks.

Suppose that from day 0 to day t_0 algorithm and the optimal solution were not allowed to purchase the yearly permit. Denote respective paid costs by $ALG_2(t_0), OPT_2(t_0)$.

We split the set of weeks into disjoint subsets according to predictions P_2^i and actual numbers of rainy days x_i.

1. $D_0 = \{i | P_2^i = 1 \text{ and } x_i < \lambda B\}$,
2. $D_1 = \{i | P_2^i = 1 \text{ and } x_i \in [\lambda B, B]\}$,
3. $D_2 = \{i | P_2^i = 1 \text{ and } x_i > B\}$,
4. $Y_0 = \{i | P_2^i = 0 \text{ and } x_i < B\}$,
5. $Y_1 = \{i | P_2^i = 0 \text{ and } x_i \in [B, \frac{B}{\lambda}]\}$,
6. $Y_2 = \{i | P_2^i = 0 \text{ and } x_i > \frac{B}{\lambda}\}$.

Algorithm 1 Prediction-Augmented Algorithm

1: $i \leftarrow 0$
2: **if** current day is first day of the week **then**
3:     $sum_2 \leftarrow 0$
4:     $i \leftarrow i + 1$
5: **end if**
6: **if** current day is rainy and no purchased permit is currently valid **then**
7:     Find optimal offline solution for already seen days, denote its' cost by OPT_{cur}.
8:     **if** $P_3 = 1$ and $OPT_{cur} \geq \mu A$ **then**
9:         purchase yearly permit, go to 25.
10:     **end if**
11:     **if** $P_3 = 0$ and $OPT_{cur}^{(2)} \geq \frac{A}{\mu}$ **then**
12:         purchase yearly permit, go to 25.
13:     **end if**
14:     **if** $P_2^i = 1$ and $sum_2 \geq \lambda B$ **then**
15:         purchase weekly permit.
16:     **else if** $P_2^i = 0$ and $sum_2 \geq \frac{B}{\lambda}$ **then**
17:         purchase weekly permit.
18:     **else**
19:         purchase daily permit, $sum_2 \leftarrow sum_2 + 1$
20:     **end if**
21: **end if**
22: **if** schedule is not empty **then**
23:     yield next day
24: **else**
25:     Finish.
26: **end if**

The Prediction-Augmented Algorithm paid

$$ALG_2(t_0) = \sum_{i \in D_0} x_i + D_1(\lambda+1)B + D_2(\lambda+1)B + \sum_{i \in Y_0} x_i + \sum_{i \in Y_1} x_i + Y_2(\frac{1}{\lambda}+1)B.$$

The optimal solution has a cost

$$OPT_2(t_0) = \sum_{i \in D_0} x_i + \sum_{i \in D_1} x_i + D_2 B + \sum_{i \in Y_0} x_i + Y_1 B + Y_2 B.$$

Let $R_2(t_0) = \frac{ALG(t_0)}{OPT(t_0)}$. Then we have

$$R_2(t_0) = \frac{\sum\limits_{i \in D_0} x_i + D_1(\lambda+1)B + D_2(\lambda+1)B + \sum\limits_{i \in Y_0} x_i + \sum\limits_{i \in Y_1} x_i + Y_2(\frac{1}{\lambda}+1)B}{\sum\limits_{i \in D_0} x_i + \sum\limits_{i \in D_1} x_i + D_2 B + \sum\limits_{i \in Y_0} x_i + Y_1 B + Y_2 B}.$$

Since $\frac{A+x}{B+x} \leq \frac{A}{B}$ for $A \geq B > 0$ and $x \geq 0$, we obtain

$$R_2(t_0) \leq \frac{D_1(\lambda+1)B + D_2(\lambda+1)B + \sum\limits_{i \in Y_1} x_i + Y_2(\frac{1}{\lambda}+1)B}{\sum\limits_{i \in D_1} x_i + D_2 B + Y_1 B + Y_2 B}. \tag{1}$$

By definition of D_1 we have $x_i > \lambda B, i \in D_1$. By definition of Y_1 we have $x_i < \frac{1}{\lambda}B$. Substituting into (1) and reducing by B we get

$$R_2(t_0) \leq \frac{D_1(\lambda + 1) + D_2(\lambda + 1) + Y_1 \frac{1}{\lambda} + Y_2(\frac{1}{\lambda} + 1)}{D_1\lambda + D_2 + Y_1 + Y_2}.$$

Suppose we change the schedule by replacing every week in Y_1 by an arbitrary week that belongs into Y_2. The competitive ratio for the new schedule will be not less than the original, because the denominator stays the same and the numerator diminishes by $|Y_1|$. Thus,

$$R_2(t_0) \leq \frac{D_1(\lambda + 1) + D_2(\lambda + 1) + Y_2(\frac{1}{\lambda} + 1)}{D_1\lambda + D_2 + Y_2}.$$

Adding and subtracting $D_1(1 - \lambda)$ in the denominator, we get

$$R_2(t_0) \leq \frac{D_1(\lambda + 1) + D_2(\lambda + 1) + Y_2(\frac{1}{\lambda} + 1)}{n - D_1(1 - \lambda)}. \tag{2}$$

Consider the prediction error. Let η_0 be the total number of such weeks i that $P_i = 0$ and $x_i \geq B$. Let η_1 be the total number of such weeks i, that $P_i = 1$ and $x_i < B$. Let $\alpha_0 = \frac{\eta_0}{n}$, $\alpha_1 = \frac{\eta_1}{n}$, $\alpha = \alpha_0 + \alpha_1 = \frac{\eta_0 + \eta_1}{n}$. Then $D_1 = \eta_1 = \alpha_1 n = (\alpha - \alpha_0)n$, $Y_2 = \eta_0 = \alpha_0 n$ and $D_2 = n - \alpha n$. Substituting into (??) and reducing by n we get

$$R_2(t_0) \leq \frac{(\lambda + 1)(1 - \alpha_0) + \alpha_0(\frac{1}{\lambda} + 1)}{1 - \alpha_1(1 - \lambda)} = \frac{\alpha_0(\frac{1}{\lambda} - \lambda) + \lambda + 1}{1 - \alpha_1(1 - \lambda)}.$$

Let $x = \alpha_0$. Then $\alpha_1 = \alpha - x$. Consider the function

$$f(x) = \frac{x(\lambda^{-1} - \lambda) + \lambda + 1}{1 - (\alpha - x)(1 - \lambda)}, x \in [0, \alpha]$$

Derivative of the function

$$f'(x) = \frac{(\lambda^{-1} - \lambda)(1 - (\alpha - x)(1 - \lambda)) - (1 - \lambda)(x(\lambda^{-1} - \lambda) + \lambda + 1)}{(1 - (\alpha - x)(1 - \lambda))^2}.$$

Multiplying the numerator and denominator by $\lambda > 0$ and transforming the numerator, we get

$$(1 - \lambda^2)(1 - (\alpha - x)(1 - \lambda)) - (1 - \lambda)(x(1 - \lambda^2) + \lambda^2 + \lambda) =$$
$$= (1 - \lambda^2)(1 + (x - \alpha)(1 - \lambda)) - (1 - \lambda)(x(1 - \lambda)(1 + \lambda) + \lambda(\lambda + 1)) =$$
$$= (1 - \lambda)((1 + \lambda)(1 + (x - \alpha)(1 - \lambda)) - (1 + \lambda)(x(1 - \lambda) + \lambda)) =$$
$$= (1 - \lambda^2)((1 + (x - \alpha)(1 - \lambda)) - (x(1 - \lambda) + \lambda)) =$$
$$= (1 - \lambda^2)(1 + x - \alpha - \lambda x + \lambda\alpha - x + \lambda x - \lambda) =$$
$$= (1 - \lambda^2)(1 - \alpha + \lambda\alpha - \lambda) = (1 - \lambda^2)(1 - \alpha)(1 - \lambda) \geq 0 \tag{3}$$

Since both numerator and denominator are non-negative, $f(x)$ has a global maximum at $x = \alpha$. Thus, the competitive ratio

$$R_2(t_0) \le \frac{\alpha(\lambda^{-1} - \lambda) + \lambda + 1}{1 - (\alpha - \alpha)(1 - \lambda)} = \lambda(1 - \alpha) + \alpha\lambda^{-1} + 1.$$

Denote the last day the algorithm makes any purchase by j. **Case 1:** $P_3 = 0$.

1. $OPT_2(j) < A$. Since $OPT(j) \le OPT_2(j)$ it follows that $OPT(j) < A$. Neither the algorithm nor the optimal solution purchase the yearly permit, thus, $ALG = ALG_2(j)$, $OPT = OPT_2(j)$ and

$$R = R_2(j) \le \lambda(1 - \alpha) + \alpha\lambda^{-1} + 1.$$

2. $OPT_2(j) \in [A, \frac{A}{\mu})$. Then

$$ALG(j) = ALG_2(j) \le (\lambda(1-\alpha)+\alpha\lambda^{-1}+1)OPT_2(j) < (\lambda(1-\alpha)+\alpha\lambda^{-1}+1)\frac{A}{\mu}.$$

When $OPT_2 \ge A$ we have $OPT = A$. Thus

$$R < (\lambda(1 - \alpha) + \alpha\lambda^{-1} + 1)\frac{1}{\mu}$$

3. $OPT_2(j) \ge \frac{A}{\mu}$. $ALG = ALG(j) = ALG(j - 1) + A$. By the definition of j we have $ALG(j - 1) = ALG_2(j - 1)$ and $OPT_2(j - 1) < \frac{A}{\mu}$. Thus, $ALG < (\lambda(1 - \alpha) + \alpha\lambda^{-1} + 1)\frac{A}{\mu} + A$, and

$$R < \frac{(\lambda(1 - \alpha) + \alpha\lambda^{-1} + 1)\frac{A}{\mu} + A}{A} = (\lambda(1 - \alpha) + \alpha\lambda^{-1} + 1)\frac{1}{\mu} + 1$$

Case 2: $P_3 = 1$.

1. $OPT < \mu A$. Thus, $OPT = OPT_2$. Since line 12 of the algorithm was not reached, $ALG(j) = ALG_2(j)$ and

$$R \le \lambda(1 - \alpha) + \alpha\lambda^{-1} + 1.$$

2. $OPT \in [\mu A, A)$.

$$ALG(j) = ALG(j - 1) + A = ALG_2(j - 1) + A \le$$
$$\le (\lambda(1 - \alpha) + \alpha\lambda^{-1} + 1)OPT_2(j - 1) + A \le$$
$$\le (\lambda(1 - \alpha) + \alpha\lambda^{-1} + 1)A + A. \quad (4)$$

Thus,

$$R \le \frac{(\lambda(1 - \alpha) + \alpha\lambda^{-1} + 1)A + A}{\mu A} = (\lambda(1 - \alpha) + \alpha\lambda^{-1} + 1)\frac{1}{\mu}.$$

3. $OPT = A$. Since line 12 was not reached on the previous day, $OPT_2(j-1) < \mu A$. Thus,

$$
\begin{aligned}
ALG = ALG(j) &= ALG(j-1) + A \leq \\
&\leq (\lambda(1-\alpha) + \alpha\lambda^{-1} + 1)OPT_2(j-1) + A < \\
&< (\lambda(1-\alpha) + \alpha\lambda^{-1} + 1)\mu A + A \quad (5)
\end{aligned}
$$

It follows, that
$$
R < (\lambda(1-\alpha) + \alpha\lambda^{-1} + 1)\mu + 1.
$$

Thus for any prediction

$$
R \leq (\lambda(1-\alpha) + \alpha\lambda^{-1} + 1)\frac{1}{\mu} + 1.
$$

This bound increases if α increases, and since $\alpha \in [0,1]$,

$$
R \leq (\lambda^{-1} + 1)\frac{1}{\mu} + 1.
$$

Suppose that the prediction is correct. This corresponds to cases 1.1 and 2.3 with $\alpha = 0$. In case 1.1 we have $R \leq \lambda + 1$. In case 2.3 we have $R < (\lambda + 1)\mu + 1$. This proves the following theorem.

Theorem 2. *Algorithm 1 is $((1+\frac{1}{\lambda})\frac{1}{\mu}+1)$-robust and $\max\{\lambda+1, (\lambda+1)\mu+1\}$-consistent.*

5   Conclusion

We have studied the parking permit problem with three permit types. We have proven the tight lower bound on the competitiveness of any deterministic online algorithm and presented a deterministic learning-augmented algorithm. The algorithm is $((1+\frac{1}{\lambda})\frac{1}{\mu}+1)$-robust and $\max\{\lambda+1, (\lambda+1)\mu+1\}$-consistent. In the full version of the article we plan to present a randomized learning-augmented algorithm based on the primal-dual method. We note that a randomized primal-dual algorithm has competitive ratio of $\frac{e+1}{e-1} < 3$ for the parking permit problem with three permit types.

References

1. Aamand, A., Chen, J., Indyk, P.: (Optimal) online bipartite matching with degree information. In: Advances in Neural Information Processing Systems, vol. 35, pp. 5724–5737 (2022)
2. Bamas, E., Maggiori, A., Svensson, O.: The primal-dual method for learning augmented algorithms. In: Advances in Neural Information Processing Systems, vol. 33, pp. 20083–20094 (2020)

3. Bampis, E., Dogeas, K., Kononov, A.V., Lucarelli, G., Pascual, F.: Scheduling with untrusted predictions. In: Proceedings of the Thirty-First International Joint Conference on Artificial Intelligence, IJCAI 2022, Vienna, Austria, pp. 23–29 (2022)
4. Buchbinder, N., Jain, K., Naor, J.S.: Online primal-dual algorithms for maximizing ad-auctions revenue. In: Arge, L., Hoffmann, M., Welzl, E. (eds.) ESA 2007. LNCS, vol. 4698, pp. 253–264. Springer, Heidelberg (2007). https://doi.org/10.1007/978-3-540-75520-3_24
5. Eberle, F., Lindermayr, A., Megow, N., Nölke, L., Schlöter, J.: Robustification of online graph exploration methods. In: Proceedings of the AAAI Conference on Artificial Intelligence **36**(9), 9732–9740 (2022)
6. Fiat, A., Woeginger, G. (eds.): Online Algorithms - The State of the Art. Springer, Heidelberg (1998). https://doi.org/10.1007/BFb0029561
7. Fleicher, R.: On the Bahncard problem. Theoret. Comput. Sci. **268**(1), 161–174 (2001)
8. Gollapudi, S., Panigrahi, P.: Online algorithms for rent-or-buy with expert advice. In: Proceedings of the 36th International Conference on Machine Learning, volume 97 of Proceedings of Machine Learning Research, pp. 2319–2327 (2019)
9. Im, S., Kumar, R., Montazer Qaem, M., Purohit, M.: Non-clairvoyant scheduling with predictions. In: SPAA, pp. 285–294. ACM (2021)
10. Karlin, A.R., Kenyon, C., Randall, D.: Dynamic TCP acknowledgement and other stories about e/(e-1). In: Proceedings of the thirty-third annual ACM symposium on Theory of Computing (STOC 2001), pp. 502–509 (2001)
11. Karlin, A.R., Manasse, M.S., Rudolph, L., Sleator, D.D.: Competitive snoopy cashing. Algorithmica **3**, 79–119 (1988)
12. Karlin, A.R., Manasse, M.S., McGeoch, L.A., Owicki, S.: Competitive randomized algorithms for nonuniform problems. Algorithmica **11**(6), 542–571 (1994)
13. Kodialam, M., Lakshman, T.V.: Prediction augmented segment routing. In: 2021 IEEE 22nd International Conference on High Performance Switching and Routing (HPSR), pp. 1–6 (2021)
14. Kumar, R., Purohit, M., Svitkina, Z.: Improving online algorithms via ML predictions. In: Advances in Neural Information Processing Systems, vol. 31, pp. 9661–9670 (2018)
15. Lattanzi, S., Lavastida, T., Moseley, B., Vassilvitskii, S.: Online scheduling via learned weights. In: Proceedings of the Fourteens Annual ACM-SIAM Symposium on Discrete Algorithms, pp. 1859–1877 (2020)
16. Lindermayr A., Megow, N.: Permutation predictions for non-clairvoyant scheduling. In: SPAA '22: 34th ACM Symposium on Parallelism in Algorithms and Architectures, pp. 357-368. ACM (2022)
17. Lykouris, T., Vassilvitskii, S.: Competitive caching with machine learned advice. In: Proceedings of the 35th International Conference on Machine Learning, ICML 2018, pp. 3302–3311 (2018)
18. Lykouris, T., Vassilvitskii, S.: Competitive cashing with machine learned advice. J. ACM **68**(4), 1–25 (2021)
19. Meyerson, A.: The parking permit problem. In: 46th Annual IEEE Symposium on Foundations of Computer Science (FOCS 2005), pp. 274–282. IEEE (2005)
20. Shin, Y., et al.: Improved learning-augmented algorithms for the multi-option Ski Rental problem via best-possible competitive analysis. arXiv preprint arXiv:2302.06832 (2023)
21. Special Semester on Algorithms and Uncertainty (2016). https://simons.berkeley.edu/programs/uncertainty2016

One Optimization Problem Induced by the Segregation Problem for the Sum of Two Quasiperiodic Sequences

Liudmila Mikhailova$^{(\boxtimes)}$ (iD)

Sobolev Institute of Mathematics, 4 Koptyug Ave., 630090 Novosibirsk, Russia
mikh@math.nsc.ru

Abstract. An unexplored discrete optimization problem of summing the elements of three given numerical sequences is considered. This problem is a core, within the framework of a posteriori approach, of the noise-proof segregation problem for two independent unobservable quasiperiodic sequences, i.e., the sequences that include some non-intersecting subsequences-fragments having the predetermined characteristic properties with the limitations from below and above on the interval between two successive fragments. The segregation problem is to restore the unobservable sequences on the base of their noisy sum. In the current paper, all the fragments in a single sequence are assumed to be identical and coinciding with the given reference fragment, at that, the information about the number of fragments in it is unavailable.

It is shown constructively that, despite the exponentially-sized set of possible solutions to the optimization problem under consideration, as well as in the segregation problem, both these problems are polynomially solvable. Some numerical simulation results are given for illustration.

Keywords: Discrete optimization problem · Polynomial-time solvability · Quasiperiodic sequence · Segregation · Detection · One-microphone signal separation

1 Introduction

An unexplored discrete optimization problem of summing the elements of three numerical sequences is studied. The research goal is to prove the polynomial-time solvability of the problem and construct an algorithm guaranteeing the solution optimality. The research is motivated by the absence of efficient (polynomial-time) algorithms solving this problem with theoretical guarantees of quality (accuracy and complexity).

The interest to this optimization problem is explained by its relevance to one problem of noise-proof segregation of the sum of two quasiperiodic sequences

The study was supported by the Russian Academy of Science (the Program of basic research), project FWNF-2022-0015.

(restoring the unobservable quasiperiodic sequences on the base of their noisy sum). By quasiperiodic sequence we mean any sequence that includes some non-intersecting subsequences-fragments with predetermined characteristic properties while the interval between two consecutive fragments is limited from below and above by the given constants. The segregation problem is stated for the case when every sequence to be restored includes an unknown number of identical fragments coinciding with the reference sequence given as a part of the problem input. This problem is actual for applied problems of distant monitoring of natural objects (underwater, aerospace, bio-medical, and so on) having quasiperiodically repeated states in the case when it is impossible to isolate a signal from a single object and a result of monitoring is a mixed signal from two close objects (see Sect. 5 for more details). It is shown that in the framework of a posteriori approach, the segregation problem reduces to an unexplored discrete optimization problem, being a particular case of the subject of the study.

The paper has the following structure. Sect. 2 includes the statement of the discrete optimization problem and a brief overview of the related problems. In Sect. 3, the polynomial-time solvability of Problem 1 is established. The main result of the current research, Algorithm $\mathcal{A}$ solving Problem 1, is written down and justified in Sect. 4. An implementation of Algorithm $\mathcal{A}$ to the actual applied segregation problem is considered in Sect. 5. Sect. 6 contains the numerical simulation results. The summary is given in Sect. 7.

2 Optimisation Problem and Related Problems

The discrete optimization problem under consideration is

Problem 1. *Given:* the numerical sequences $(f_0^{(1)}, \ldots, f_{K-1}^{(1)})$, $(f_0^{(2)}, \ldots, f_{K-1}^{(2)})$, the numerical sequence h_k, $k = 0, \pm 1, \pm 2, \ldots$, such that $h_k = 0$, if $|k| \geq p$, positive integers $L_{\min}^{(1)}, L_{\max}^{(1)}, L_{\min}^{(2)}, L_{\max}^{(2)}$, such that

$$2p \leq L_{\min}^{(i)} \leq L_{\max}^{(i)} < K, \ i \in \{1, 2\}. \tag{1}$$

Find: the collections $\mathcal{K}^{(1)} = \{k_1^{(1)}, \ldots, k_m^{(1)}, \ldots\} \subset \{0, \ldots, K-1\}$ and $\mathcal{K}^{(2)} = \{k_1^{(2)}, \ldots, k_m^{(2)}, \ldots\} \subset \{0, \ldots, K-1\}$, their sizes M_1 and M_2, that minimize the objective function

$$F(\mathcal{K}^{(1)}, \mathcal{K}^{(2)}) = \sum_{m=1}^{M_1} f_{k_m^{(1)}}^{(1)} + \sum_{s=1}^{M_2} f_{k_s^{(2)}}^{(2)} + \sum_{m=1}^{M_1} \sum_{s=1}^{M_2} h_{k_m^{(1)} - k_s^{(2)}} \tag{2}$$

under the following constraints on the elements of the collections $\mathcal{K}^{(1)}$, $\mathcal{K}^{(2)}$:

$$L_{\min}^{(i)} \leq k_m^{(i)} - k_{m-1}^{(i)} \leq L_{\max}^{(i)}, \ k = 2, \ldots, M_i, \ i \in \{1, 2\}. \tag{3}$$

An interest to this previously unexplored optimization problem is connected with an important particular case of Problem 1 considered in Sect. 5. This particular case arises within a posteriori approach when solving the actual for applications noise-proof segregation problem for the sum of two independent quasiperiodic sequences in the case when every sequence contains some unknown number of identical subsequences-fragments.

The problem under consideration has the already solved particular case [1] with $(f_0^{(2)}, \ldots, f_{K-1}^{(2)}) = (0, \ldots, 0)$ and $h_k = 0$, $k = -p+1, \ldots, p-1$. A modification of this particular case, where M_1 is a part of the input data, has been considered in [2]. In these papers, the algorithms that allow obtaining optimal solutions in time $\mathcal{O}(L_{\max}^{(1)} - L_{\min}^{(1)} + 1)K$ and $\mathcal{O}(L_{\max}^{(1)} - L_{\min}^{(1)} + 1)KM_1$, respectively, are proved. These optimization problems are the core of algorithms solving such applied quasiperiodic sequences processing problems as detecting identical subsequences-fragments [1,2], simultaneous detection and identification of subsequences-fragments or their parts in a quasiperiodic sequence [3,4] and some others. A variant of the segregation problem, when all information about one of the sequences is available, also reduces to this particular case.

A modification of Problem 1, when the sizes M_1 and M_2 of the collections to be found are the parts of the input, has been studied in [5]. It follows from the compatibility of (3) that these values are bound by the constraints:

$$1 \le M_i \le M_{\max}^{(i)} = \lfloor (K-1)/L_{\min}^{(i)} \rfloor + 1, \ i \in \{1, 2\}. \tag{4}$$

The main result of this paper is

Lemma 1. *[5] Let the conditions of Problem 1 hold and positive integers M_1 and M_2 satisfying (4) are given, then there exists an algorithm $\mathcal{A}_1$ that finds the exact solution to the modification of Problem 1 in time $\mathcal{O}(\Delta_{\max} M_1 M_2 K^2)$, where $\Delta_{\max} = \max\{L_{\max}^{(1)} - L_{\min}^{(1)} + 1, L_{\max}^{(2)} - L_{\min}^{(2)} + 1\}$.*

3   Polynomial-Time Solvability of the Problem

Let $\mathcal{S}$ be the set of all possible solutions to Problem 1. In order to estimate its size, consider the sets $\Omega^{(1)}$ and $\Omega^{(2)}$ of collections $\{k_1^{(i)}, \ldots, k_m^{(i)}, \ldots\} \subset \{0, \ldots, K-1\}$, $i \in \{1, 2\}$, satisfying (3). It follows from combinatorial considerations, that except for trivial case $L_{\max}^{(i)} = 2p$, every set $\Omega^{(1)}$ and $\Omega^{(2)}$ is exponentially-sized:

$$|\Omega^{(i)}| > 2^{M_{\max}^{(i)}/2} = 2^{\mathcal{O}(K)}.$$

Since the collections $\mathcal{K}^{(1)}$ and $\mathcal{K}^{(2)}$ are independent, $|\mathcal{S}| = |\Omega^{(1)}| \times |\Omega^{(2)}|$. So, $\mathcal{S}$ is also exponentially-sized. Despite this, there exists an exact polynomial-time algorithm solving Problem 1.

Corollary 1. *Let the conditions of Problem 1 hold. Let $\hat{F}_{M_1, M_2}$, $\hat{\mathcal{K}}_{M_1, M_2}^{(1)}$, $\hat{\mathcal{K}}_{M_1, M_2}^{(2)}$, $M_1 \in \{1, \ldots, M_{\max}^{(1)}\}$, $M_2 \in \{1, \ldots, M_{\max}^{(2)}\}$, be the optimal value of the objective function and the optimal collections in modified Problem 1 with M_1 and M_2 being the parts of the input. Then the optimal value $\hat{F}$ of the objective function and the optimal collections $\hat{\mathcal{K}}^{(1)}$ and $\hat{\mathcal{K}}^{(2)}$ in Problem 1 are*

$$\hat{F} = \hat{F}_{\hat{M}_1, \hat{M}_2}, \quad \hat{\mathcal{K}}^{(1)} = \hat{\mathcal{K}}_{\hat{M}_1, \hat{M}_2}^{(1)}, \quad \hat{\mathcal{K}}^{(2)} = \hat{\mathcal{K}}_{\hat{M}_1, \hat{M}_2}^{(2)},$$

where $(\hat{M}_1, \hat{M}_2) = \arg \min_{(M_1,M_2)\in\{1,...,M_{\max}^{(1)}\}\times\{1,...,M_{\max}^{(2)}\}} \hat{F}_{M_1,M_2}.$

Proof. Fix M_1 and M_2 satisfying (4) and denote by $\mathcal{S}_{M_1,M_2}$ the set of admissible solutions to modified Problem 1, with M_1 and M_2 being the part of the input. It is easy to see that $\mathcal{S}_{M_1,M_2} \subseteq \mathcal{S}$. Moreover, we have

$$\mathcal{S} = \bigcup_{M_1=1}^{M_{\max}^{(1)}} \bigcup_{M_2=1}^{M_{\max}^{(2)}} \mathcal{S}_{M_1,M_2}. \tag{5}$$

The validity of this corollary follows from the chain of equalities based on (5) and the definition of an optimal value:

$$\hat{F} = \min_{(\mathcal{K}^{(1)},\mathcal{K}^{(2)})\in\mathcal{S}} F(\mathcal{K}^{(1)},\mathcal{K}^{(2)}) = \min_{(\mathcal{K}^{(1)},\mathcal{K}^{(2)})\in\cup_{M_1=1}^{M_{\max}^{(1)}}\cup_{M_2=1}^{M_{\max}^{(2)}}\mathcal{S}_{M_1,M_2}} F(\mathcal{K}^{(1)},\mathcal{K}^{(2)}) =$$

$$\min_{M_1\in\{1,...,M_{\max}^{(1)}\}} \min_{M_2\in\{1,...,M_{\max}^{(2)}\}} \min_{(\mathcal{K}^{(1)},\mathcal{K}^{(2)})\in\mathcal{S}_{M_1,M_2}} F(\mathcal{K}^{(1)},\mathcal{K}^{(2)}) =$$

$$\min_{M_1\in\{1,...,M_{\max}^{(1)}\}} \min_{M_2\in\{1,...,M_{\max}^{(2)}\}} \hat{F}_{M_1,M_2}.$$

It follows from the results of Corollary 1 and Lemma 1 that the time needed to find the exact solution to Problem 1 is estimated as

$$\sum_{M_1=1}^{M_{\max}^{(1)}} \sum_{M_2=1}^{M_{\max}^{(2)}} \mathcal{O}(\Delta_{\max}M_1 M_2 K^2) = \mathcal{O}(\Delta_{\max}(M_{\max}^{(1)})^2(M_{\max}^{(2)})^2 K^2).$$

Remark 1. Since $\Delta_{\max} \leq K$, $M_{\max}^{(1)} \leq K$, and $M_{\max}^{(2)} \leq K$, the exact solution to Problem 1 can be found in time $\mathcal{O}(K^7)$.

Corollary 1 proves that Problem 1 is polynomial-time solvable. But the complexity of the suggested procedure is too large, making it inapplicable. The goal of the current paper is to construct and justify a less time-complicated algorithm without trying all possible sizes of the collections $\mathcal{K}^{(1)}$ and $\mathcal{K}^{(2)}$ to be found.

4   Algorithm $\mathcal{A}$

In order to construct an algorithm solving Problem 1, we need the following result and subsequent designations.

Lemma 2. *Let the conditions of Problem 1 hold. Then the optimal value $\hat{F}$ of the objective function of the problem is given by the formula*

$$\hat{F} = \min_{n\in\{0,...,K-1\}} \min_{s\in\{0,...,K-1\}} F(n,s), \tag{6}$$

and the values of $F(n,s)$ are calculated with the recurrent formula

$$F(n,s) = h_{n-s} +$$
$$\begin{cases} f_n^{(1)} + f_s^{(2)}, & \text{if } n \in \omega_0^{(1)}, \ s \in \omega_0^{(2)}; \\ f_n^{(1)} + f_s^{(2)} + \min\{0, \min\limits_{j \in \gamma^{(2)}(s)} G^{(2)}(j|n)\}, & \text{if } n \in \omega_0^{(1)}, \ s \in \omega^{(2)}; \\ f_n^{(1)} + f_s^{(2)} + \min\{0, \min\limits_{j \in \gamma^{(1)}(n)} G^{(1)}(j|s)\}, & \text{if } n \in \omega^{(1)}, \ s \in \omega_0^{(2)}; \\ \min \begin{cases} f_n^{(1)} + f_s^{(2)} + G_{\min}(n,s); \\ f_s^{(2)} + \min\limits_{j \in \gamma^{(2)}(s)} F(n,j); \end{cases} & \text{if } n \in \omega^{(1)}, \ s \in \omega^{(2)}, \ s \geq n; \\ \min \begin{cases} f_n^{(1)} + f_s^{(2)} + G_{\min}(n,s); \\ f_n^{(1)} + \min\limits_{j \in \gamma^{(1)}(n)} F(j,s); \end{cases} & \text{if } n \in \omega^{(1)}, \ s \in \omega^{(2)}, \ s < n; \end{cases}$$
$$(7)$$

here

$$\omega_0^{(i)} = \{0, \ldots, L_{\min}^{(i)} - 1\}, \quad \omega^{(i)} = \{L_{\min}^{(i)}, \ldots, K - 1\},$$
$$\gamma^{(i)}(n) = \left\{\max\{0, n - L_{\max}^{(i)}\}, \ldots, n - L_{\min}^{(i)}\right\}, \quad n \in \omega^{(i)}, \quad i \in \{1, 2\}, \quad (8)$$

and for all $n \in \omega^{(1)} \cup \omega_0^{(1)}$, $s \in \omega^{(2)} \cup \omega_0^{(2)}$,

$$G^{(1)}(n|s) = \begin{cases} f_n^{(1)} + h_{n-s}, & \text{if } n \in \omega_0^{(1)}; \\ f_n^{(1)} + h_{n-s} + \min\{0, \min\limits_{j \in \gamma^{(1)}(n)} G^{(1)}(j|s)\}, & \text{if } n \in \omega^{(1)}, \end{cases} \quad (9)$$

$$G^{(2)}(s|n) = \begin{cases} f_s^{(2)} + h_{n-s}, & \text{if } s = \omega_0^{(2)}; \\ f_s^{(2)} + h_{n-s} + \min\{0, \min\limits_{j \in \gamma^{(2)}(s)} G^{(2)}(j|n)\}, & \text{if } s \in \omega^{(2)}, \end{cases} \quad (10)$$

$$G_{\min}(n,s) = \min\{0, \min\limits_{j \in \gamma^{(2)}(s)} G^{(2)}(j|n), \min\limits_{j \in \gamma^{(1)}(n)} G^{(1)}(j|s)\}. \quad (11)$$

In this lemma, equalities (8) define the sets of admissible values for the components of $\mathcal{K}^{(1)}$ and $\mathcal{K}^{(2)}$: $k_1^{(i)} \in \omega_0^{(i)} \cup \omega^{(i)}$, $k_m^{(i)} \in \omega^{(i)}$, $m = 2, \ldots, M_{\max}^{(i)}$, $i \in \{1, 2\}$. If $k_m^{(i)} = k$, $k \in \omega^{(i)}$, $m = 2, \ldots, M_{\max}^{(i)}$, then $k_{m-1}^{(i)} \in \gamma^{(i)}(k)$. The proof of this lemma is based on the direct derivation of the recurrent formulas. The finiteness of h_k plays a key role when constructing these formulas. Namely, this property, combined with (1) and (3), guarantees that there is not greater than one non-zero summand in the double sum $\sum_{m=1}^{M_1} \sum_{s=1}^{M_2} h_{k_m^{(1)} - k_s^{(2)}}$ in (2).

In order to find the optimal collections $\hat{\mathcal{K}}^{(1)}$ and $\hat{\mathcal{K}}^{(2)}$ we need three auxiliary functions: $I^{(1)}(n,s)$, $I^{(2)}(n,s)$, and $J(n,s)$, $n \in \{0, \ldots, K-1\}$, $s \in \{1, \ldots, K-1\}$. Consider them as a collection

$$\left\{\left(I^{(1)}(n,s), I^{(2)}(n,s), J(n,s)\right), n = 0, \ldots, K-1, \ s = 0, \ldots, K-1\right\}$$

of triples and calculate together for all $n \in \{0, \ldots, K-1\}$, $s \in \{1, \ldots, K-1\}$:

$$(I^{(1)}(n,s), I^{(2)}(n,s), J(n,s)) =$$

$$
\begin{cases}
(-1,-1,0), & \text{if } n \in \omega_0^{(1)}, \ s \in \omega_0^{(2)}; \\[2mm]
(-1,-1,0), & \text{if } n \in \omega_0^{(1)}, \ s \in \omega^{(2)}, \ \min\limits_{j \in \gamma^{(2)}(s)} G^{(2)}(j|n) > 0; \\[2mm]
(n, \arg\min\limits_{j \in \gamma^{(2)}(s)} G^{(2)}(j|n), 2), \\
\quad\quad\quad \text{if } n \in \omega_0^{(1)}, \ s \in \omega^{(2)}, \ \min\limits_{j \in \gamma^{(2)}(s)} G^{(2)}(j|n) \leq 0; \\[2mm]
(-1,-1,0), & \text{if } n \in \omega^{(1)}, \ s \in \omega_0^{(2)}, \ \min\limits_{j \in \gamma^{(1)}(n)} G^{(1)}(j|s) \geq 0; \\[2mm]
(\arg\min\limits_{j \in \gamma^{(1)}(n)} G^{(1)}(j|s), s, 1), \\
\quad\quad\quad \text{if } n \in \omega^{(1)}, \ s \in \omega_0^{(2)}, \ \min\limits_{j \in \gamma^{(1)}(n)} G^{(1)}(j|s) < 0; \\[2mm]
(-1,-1,0), & \text{if } n \in \omega^{(1)}, \ s \in \omega^{(2)}, \ n \leq s, \\
\quad f_1(n) + G_{\min}(n,s) < \min\limits_{j \in \gamma^{(2)}(s)} F(n,j), \ G_{\min}(n,s) = 0; \\[2mm]
(n, \arg\min\limits_{j \in \gamma^{(2)}(s)} G^{(2)}(j|n), 2), & \text{if } n \in \omega^{(1)}, \ s \in \omega^{(2)}, \ n \leq s, \\
\quad f_1(n) + G_{\min}(n,s) < \min\limits_{j \in \gamma^{(2)}(s)} F(n,j), \\
\quad G_{\min}(n,s) = \min\limits_{j \in \gamma^{(1)}(n)} G^{(1)}(j|s); \\[2mm]
(\arg\min\limits_{j \in \gamma^{(1)}(n)} G^{(1)}(j|s), s, 1), & \text{if } n \in \omega^{(1)}, \ s \in \omega^{(2)}, \ n \leq s, \\
\quad f_1(n) + G_{\min}(n,s) < \min\limits_{j \in \gamma^{(2)}(s)} F(n,j), \\
\quad G_{\min}(n,s) = \min_{j \in \gamma^{(1)}(n)} G^{(1)}(j|s); \\[2mm]
(n, \arg\min\limits_{j \in \gamma^{(2)}(s)} F(n,j), 2), & \text{if } n \in \omega^{(1)}, \ s \in \omega^{(2)}, \ n \leq s, \\
\quad f_1(n) + G_{\min}(n,s) \geq \min\limits_{j \in \gamma^{(2)}(s)} F(n,j); \\[2mm]
(-1,-1,0), & \text{if } n \in \omega^{(1)}, \ s \in \omega^{(2)}, \ n > s, \\
\quad f_2(s) + G_{\min}(n,s) < \min\limits_{j \in \gamma^{(1)}(n)} F(j,s), \ G_{\min}(n,s) = 0; \\[2mm]
(n, \arg\min\limits_{j \in \gamma^{(2)}(s)} G^{(2)}(j|n), 2), & \text{if } n \in \omega^{(1)}, \ s \in \omega^{(2)}, \ n > s, \\
\quad f_2(s) + G_{\min}(n,s) < \min\limits_{j \in \gamma^{(1)}(n)} F(j,s), \\
\quad G_{\min}(n,s) = \min\limits_{j \in \gamma^{(2)}(s)} G^{(2)}(j|n); \\[2mm]
(\arg\min\limits_{j \in \gamma^{(1)}(n)} G^{(1)}(j|s), s, 1), & \text{if } n \in \omega^{(1)}, \ s \in \omega^{(2)}, \ n > s, \\
\quad f_2(s) + G_{\min}(n,s) < \min\limits_{j \in \gamma^{(1)}(n)} F(j,s), \\
\quad G_{\min}(n,s) = \min\limits_{j \in \gamma^{(1)}(n)} G^{(1)}(j|s); \\[2mm]
(\arg\min\limits_{j \in \gamma^{(1)}(n)} F(j,s), s, 1), & \text{if } n \in \omega^{(1)}, \ s \in \omega^{(2)}, \ n > s, \\
\quad f_2(s) + G_{\min}(n,s) \geq \min\limits_{j \in \gamma^{(1)}(n)} F(j,s).
\end{cases}
\tag{12}
$$

Using (12), calculate the components of four auxiliary collections $(\nu_1^{(1)}, \ldots, \nu_m^{(1)}, \ldots)$ $(\nu_1^{(2)}, \ldots, \nu_m^{(2)}, \ldots)$, $(\mu_1^{(1)}, \ldots, \mu_m^{(1)}, \ldots)$, and $(\mu_1^{(2)}, \ldots,$

$\mu_m^{(2)}, \ldots)$, of positive integers having the same size. The first components of these collections are

$$\nu_1^{(1)} = \arg \min_{n \in \{0,\ldots,K-1\}} \left\{ \min_{s \in \{0,\ldots,K-1\}} F(n,s) \right\},$$
$$\nu_1^{(2)} = \arg \min_{s \in \{0,\ldots,K-1\}} F(\nu_1^{(1)}, s), \tag{13}$$
$$\mu_1^{(1)} = 1, \quad \mu_1^{(2)} = 1,$$

while the rest components are obtained by the following recurrent formulas for $m = 1, \ldots, M-1$, where

$$M = \min\{m \in \{0, \ldots, 2K-1\} \ : \ J(\nu_m^{(1)}, \nu_m^{(2)}) = 0\}, \tag{14}$$

$$\begin{aligned}
\nu_{m+1}^{(1)} &= I^{(1)}(\nu_m^{(1)}, \nu_m^{(2)}), \\
\nu_{m+1}^{(2)} &= I^{(2)}(\nu_m^{(1)}, \nu_m^{(2)}), \\
\mu_{m+1}^{(1)} &= \begin{cases} \mu_m^{(1)} + 1, & \text{if } J(\nu_m^{(1)}, \nu_m^{(2)}) = 1, \\ \mu_m^{(1)}, & \text{if } J(\nu_m^{(1)}, \nu_m^{(2)}) = 2, \end{cases} \\
\mu_{m+1}^{(2)} &= \begin{cases} \mu_m^{(2)}, & \text{if } J(\nu_m^{(1)}, \nu_m^{(2)}) = 1, \\ \mu_m^{(2)} + 1, & \text{if } J(\nu_m^{(1)}, \nu_m^{(2)}) = 2. \end{cases}
\end{aligned} \tag{15}$$

Corollary 2. *The optimal solution to Problem 1 are the collections* $\hat{\mathcal{K}}^{(1)} = (\hat{k}_1^{(1)}, \ldots, \hat{k}_{\hat{M}_1}^{(1)})$ *and* $\hat{\mathcal{K}}^{(2)} = (\hat{k}_1^{(2)}, \ldots, \hat{k}_{\hat{M}_2}^{(2)})$ *of the sizes*

$$\hat{M}_1 = \max_{m=1,\ldots,M} \mu_m^{(1)}, \quad \hat{M}_2 = \max_{m=1,\ldots,M} \mu_m^{(2)}, \tag{16}$$

obtained as

$$\hat{k}_m^{(1)} = \nu_{\kappa_m^{(1)}}^{(1)}, m = 1, \ldots, \hat{M}_1,$$
$$\hat{k}_m^{(2)} = \nu_{\kappa_m^{(2)}}^{(2)}, m = 1, \ldots, \hat{M}_2, \tag{17}$$

where

$$\kappa_m^{(i)} = \max\{j \in \{1, \ldots, M\} \ : \ \mu_j^{(i)} = \hat{M}_i - m + 1\}, \ m = 1, \ldots, \hat{M}_i, \ i \in \{1,2\}. \tag{18}$$

The proof of Corollary 1 is based on the analysis of variants in (7).

Algorithm $\mathcal{A}$ presented below finds the exact solution to Problem 1.

Algorithm $\mathcal{A}$.

INPUT: numerical sequences $(f_0^{(1)}, \ldots, f_{K-1}^{(1)})$, $(f_0^{(2)}, \ldots, f_{K-1}^{(2)})$, a numerical sequence h_k, $k = 0, \pm 1, \pm 2, \ldots$, positive integers $L_{\min}^{(1)}, L_{\max}^{(1)}, L_{\min}^{(2)}, L_{\max}^{(2)}$.

STEP 1. Check that the conditions of Problem 1 hold for this input.

Forward pass.

STEP 2. Compute $G^{(1)}(n|s)$, $G^{(2)}(s|n)$, and $G_{\min}(n,s)$, $n \in \{0, \ldots, K-1\}$, $s \in \{0, \ldots, K-1\}$, using (9), (10), and (11).

STEP 3. Compute $F(n,s)$, $n \in \{0, \ldots, K-1\}$, $s \in \{0, \ldots, K-1\}$, by (7).

STEP 4. Compute $\hat{F}$ by (6). Put $F_{\mathcal{A}} = \hat{F}$.

Backward pass.

STEP 5. Compute $\{(I^{(1)}(n,s), I^{(2)}(n,s), J(n,s)), n \in \{0,\ldots,K-1\}, s \in \{1,\ldots,K-1\}\}$ by (12). Find $(\nu_1^{(1)},\ldots,\nu_m^{(1)},\ldots)$, $(\nu_1^{(2)},\ldots,\nu_m^{(2)},\ldots)$, $(\mu_1^{(1)},\ldots,\mu_m^{(1)},\ldots)$, and $(\mu_1^{(2)},\ldots,\mu_m^{(2)},\ldots)$ and their size M by (13), (14), and (15).

STEP 6. Find $\hat{\mathcal{K}}^{(1)}$, $\hat{\mathcal{K}}^{(2)}$, their sizes $\hat{M}_1$ and $\hat{M}_2$ by (16), (17), (18). Put $\mathcal{K}_{\mathcal{A}}^{(1)} = \hat{\mathcal{K}}^{(1)}$ and $\mathcal{K}_{\mathcal{A}}^{(2)} = \hat{\mathcal{K}}^{(2)}$.

OUTPUT: the collections $\mathcal{K}_{\mathcal{A}}^{(1)}$, $\mathcal{K}_{\mathcal{A}}^{(2)}$ and the value $F_{\mathcal{A}}$.

Theorem 1. *Algorithm $\mathcal{A}$ finds the exact solution to Problem 1 in $\mathcal{O}(\Delta_{\max}K^2)$ time.*

The optimality of the solution obtained by Algorithm $\mathcal{A}$ follows from Lemma 2 and Corollary 2. The step-by-step representation of the algorithm makes it possible to estimate its time-complexity.

Remark 2. Since $\Delta_{\max} \leq K$, the running time of Algorithm $\mathcal{A}$ is $\mathcal{O}(K^3)$. If $\Delta_{\max}$ is fixed (bounded by a constant), then the running time of Algorithm $\mathcal{A}$ is $\mathcal{O}(K^2)$.

Remark 3. Algorithm $\mathcal{A}$, obtained as a main result of the current paper, is significantly less time-complicated as compared with Algorithm from Corollary 1. The advantage can be estimated as $\mathcal{O}((M_{\max}^{(1)})^2 (M_{\max}^{(2)})^2) = \mathcal{O}(K^4)$ times, according to (4), Corollary 1, and Theorem 1.

5   Segregation Problem

As mentioned above, Problem 1 is relevant to one actual problem of noise-proof remote monitoring of natural objects. Let's consider in detail this applied problem and show that its solving reduces to solving a particular case of Problem 1.

Assume that an object to be monitored has two states: active and passive. These states alternate in such a way that the interval between two successive active states is limited from below and above by the given constants. The number of active states within the observation period is unknown but obviously limited. The active state duration is fixed, while the passive state duration varies, remaining longer than the duration of the active state. This type of state repeatability is said to be quasiperiodic; it is typical for natural objects. In the current paper, we assume additionally that the behaviour of the object is the same for all active states. Within the period of monitoring, this object generates some signal in the form of a series of chronologically ordered numerical values. In this signal, subsequences-fragments corresponding to active states are identical since the behaviour doesn't vary and each of them coincides with the given reference sequence. In the passive state, the signal is zero. According to the definition given in the introduction, the signal to be monitored is a quasiperiodic sequence, with the subsequences-fragments corresponding to active states.

In the case of a single object, the result of monitoring is a quasiperiodic sequence distorted by some noise. But if we have two close objects generating such signals, it may be impossible to isolate the signal generated by a single object. In such a case, the result of monitoring is a sum of signals distorted by some inevitable noise — a noisy sum of two quasiperiodic sequences. The segregation problem is to restore the signals (the quasiperiodic sequences) on the base of their noisy sum. Since the behaviour of each object is the same for all active states, it is sufficient to find the collection of the beginning instants of active states for every object in order to restore the signals.

The stated segregation problem can be interpreted as the blind signal separation problem (BSS) (see, for example, [6–8], and so on). The BSS problem is to restore initial unobserved signals using some observations of their mix, with a lack of information about the source signals or the mixing process. If we have only one observation of the mixture in the BSS problem, as in the segregation problem under consideration, the problem is called a one-microphone signal or source separation problem. The existing algorithms solving this problem implement a traditional (multistage) approach. This approach implies the division of the solving process into some stages, implementing via suitable well-known signal-processing techniques: statistical principles, spectral analysis, time-frequency mask estimation, and so on [9,10], accompanied, as a rule, by filtering out noise with subsequent analysis of the received unmixed signals (detecting separate fragments of every unmixed signal, analyzing and identifying these fragments). Algorithms implementing the sequential approach are quick, but they don't guarantee the optimality of the resulting solution, even if each stage is solved optimally.

In an alternative a posteriori approach, all the stages of signal processing are combined into one optimization process. The main disadvantage of this approach as compared with the traditional (multistage) one is the necessity of solving a unique, as a rule, previously unexplored discrete optimization problem. There are a lot of algorithms for qiuasiperiodic sequences processing (detection, identification, recognition, and so on) with various assumptions about a sequence structure and fragment properties in the framework of a posteriori approach (see, for example, [1–5,11,12]). But all of them, except for [5], deal with a single quasiperiodic sequence, which reflects the monitoring of a single isolated object. But even in the simplest case of identical subsequences-fragments in every sequence and only two quasiperiodic sequences to be summed up, the resulting sum is no longer quasiperiodic, and all these algorithms are inapplicable.

Let's give a formal statement of the segregation problem under consideration and make sure that a posteriori approach to solving this problem leads to an unexplored discrete optimization problem being a particular case of Problem 1. Let we have two not-coinciding sequences $U^{(1)} = (u_0^{(1)}, \ldots, u_{q-1}^{(1)})$ and $U^{(2)} = (u_0^{(2)}, \ldots, u_{q-1}^{(2)})$ of the same length q, and two pairs of positive integers $T_{\min}^{(1)}$, $T_{\max}^{(1)}$, and $T_{\min}^{(2)}$, $T_{\max}^{(2)}$ such that

$$2q < T_{\min}^{(i)} \le T_{\max}^{(i)}, \quad i \in \{1,2\}. \tag{19}$$

Put $u_n^{(i)} = 0$, $i \in \{1,2\}$, if $n \notin \{0,\ldots,q-1\}$, and consider two sequences $X^{(1)} = (x_0^{(1)},\ldots,x_{N-1}^{(1)})$ and $X^{(2)} = (x_0^{(2)},\ldots,x_{N-1}^{(2)})$ of the same length N with the general term given for every $n = 0,\ldots,N-1$, by the rule

$$x_n^{(i)} = x_n^{(i)}(\mathcal{M}^{(i)}) = x_n^{(i)}(\mathcal{M}^{(i)}|U^{(i)},T_{\min}^{(i)},T_{\max}^{(i)}) = \sum_{m=1}^{M_i} u_{n-n_m^{(i)}}^{(i)}, \quad i \in \{1,2\}, \quad (20)$$

where $\mathcal{M}^{(1)} = \{n_1^{(1)},\ldots,n_m^{(1)},\ldots\}$ and $\mathcal{M}^{(2)} = \{n_1^{(2)},\ldots,n_m^{(2)},\ldots\}$ are the collections of positive integers having unknown sizes M_1 and M_2, and their components are bounded by the restrictions:

$$T_{\min}^{(i)} \le n_m^{(i)} - n_{m-1}^{(i)} \le T_{\max}^{(i)}, \quad m = 2,\ldots,M_i, \quad i \in \{1,2\}. \quad (21)$$
$$n_{M_i}^{(i)} \le N - q,$$

Call the sequences $X^{(1)}$ and $X^{(2)}$ quasiperiodic sequences engendered by the reference sequences $U^{(1)}$ and $U^{(2)}$. In (20) $(x_{n_m^{(i)}}^{(i)},\ldots,x_{n_m^{(i)}+q-1}^{(i)}) = U^{(i)}$, $m = 1,\ldots,M_i$, $i \in \{1,2\}$, while the remaining values are zero. Restrictions (21) reflect the quasiperiodicity of states. Inequalities $2q \le T_{\min}^{(i)}$, $i \in \{1,2\}$ guarantees that the duration of the active state is not greater than the duration of a passive one.

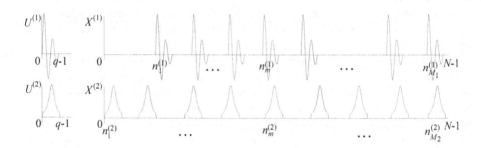

Fig. 1. Example of quasiperiodic sequences

Fig. 1 presents two examples of quasiperiodic sequences engendered by the reference sequences depicted in the left part. The first sequence in this figure includes $M_1 = 7$ subsequences-fragments and the second one — $M_2 = 9$.

Assume that we observe a sequence $Y = (y_0,\ldots,y_{N-1})$ being the element-wise sum of $X^{(1)}$, $X^{(2)}$, and some numerical sequence $E = (e_0,\ldots,e_{N-1})$, which reflects possible noise distortion:

$$y_n = x_n^{(1)} + x_n^{(2)} + e_n, \quad n = 0,\ldots,N-1.$$

Figs. 2 and 3 present examples of the unobservable sum $X^{(1)} + X^{(2)}$, where the summands are presented in Fig. 1, and the observable sequence Y.

The *segregation problem for the sum of two quasiperiodic sequences* is to restore the unobservable sequences $X^{(1)}$ and $X^{(2)}$ when processing their

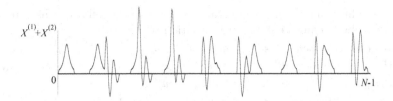

$X^{(1)}{+}X^{(2)}$

0

N-1

Fig. 2. Example of an unobservable sum of two quasiperiodic sequences

Y

0

N-1

Fig. 3. Example of the obeservable sequence

noisy sum Y. It means that we have to find two collections $\mathcal{M}^{(1)}$ and $\mathcal{M}^{(2)}$ satisfying (21) such that $X^{(1)} = X(\mathcal{M}^{(1)}|U^{(1)}, T_{\min}^{(1)}, T_{\max}^{(1)})$ and $X^{(2)} = X(\mathcal{M}^{(2)}|U^{(2)}, T_{\min}^{(2)}, T_{\max}^{(2)})$, having Y, $U^{(1)}$, $U^{(2)}$, $T_{\min}^{(1)}$, $T_{\max}^{(1)}$, $T_{\min}^{(2)}$, and $T_{\max}^{(2)}$ as the problem input.

Let $\mathcal{X}(U^{(i)}, T_{\min}^{(i)}, T_{\max}^{(i)})$, $i \in \{1, 2\}$, be the set of all N-length sequenses satisfying (20). Consider $X^{(1)}$, $X^{(2)}$, and Y as vectors in N-dimensional space and formulate the segregation problem in the form of the approximation problem:

$$\|Y - (X^{(1)} + X^{(2)})\|^2 \longrightarrow \min_{\mathcal{X}(U^{(1)}, T_{\min}^{(1)}, T_{\max}^{(1)}), \mathcal{X}(U^{(2)}, T_{\min}^{(2)}, T_{\max}^{(2)})} . \tag{22}$$

Since $U^{(i)}$, $T_{\min}^{(i)}$, $T_{\max}^{(i)}$, $i \in \{1, 2\}$, are fixed, $X^{(i)} = X(\mathcal{M}^{(i)}|U^{(i)}, T_{\min}^{(i)}, T_{\max}^{(i)})$, $i \in \{1, 2\}$. For this reason, we can rewrite (22) in the following equivalent form:

$$\|Y - (X^{(1)} + X^{(2)})\|^2 = $$
$$\left\| Y - \left(X(\mathcal{M}^{(1)}|U^{(1)}, T_{\min}^{(1)}, T_{\max}^{(1)}) + X(\mathcal{M}^{(2)}|U^{(2)}, T_{\min}^{(2)}, T_{\max}^{(2)}) \right) \right\|^2 \tag{23}$$
$$\longrightarrow \min_{\mathcal{M}^{(1)}, \mathcal{M}^{(2)}} .$$

Finally, by transforming $\|Y - (X^{(1)} + X^{(2)})\|^2$ with (20), we have

$$\|Y - (X^{(1)} + X^{(2)})\|^2 = \sum_{n=0}^{N-1} y_n^2 + \sum_{m=1}^{M_1} \sum_{i=0}^{q-1} \left\{ (u_i^{(1)})^2 - 2u_i^{(1)} y_{n_m^{(1)}+i} \right\} +$$
$$+ \sum_{m=1}^{M_2} \sum_{j=0}^{q-1} \left\{ (u_j^{(2)})^2 - 2u_j^{(2)} y_{n_m^{(2)}+j} \right\} + 2 \sum_{m=1}^{M_1} \sum_{s=1}^{M_2} \sum_{i=0}^{q-1} u_i^{(1)} u_{i+n_m^{(1)}-n_s^{(2)}}^{(2)}.$$

Here the first summand is a constant, so we have the following discrete optimization problem equivalent to approximation problem (23).

Problem 2 *Given*: Positive integers N, q, $T_{\min}^{(1)}$, $T_{\min}^{(2)}$, $T_{\max}^{(1)}$, and $T_{\max}^{(2)}$ such that (19) are valid, the numerical sequences $Y = (y_0, \ldots, y_{N-1})$, $U^{(1)} = (u_0^{(1)}, \ldots, u_{q-1}^{(1)})$, and $U^{(2)} = (u_0^{(2)}, \ldots, u_{q-1}^{(2)})$.

Find: the collections $\mathcal{M}^{(1)} = \{n_1^{(1)}, \ldots, n_m^{(1)}, \ldots\} \subset \{0, \ldots, N-1\}$ and $\mathcal{M}^{(2)} = \{n_1^{(2)}, \ldots, n_m^{(2)}, \ldots\} \subset \{0, \ldots, N-1\}$ of indices of the sequence Y and their sizes M_1 and M_2 that minimize the objective function

$$
\begin{aligned}
G(\mathcal{M}^{(1)}, \mathcal{M}^{(2)}) = &\sum_{m=1}^{M_1} \sum_{i=0}^{q-1} \left\{ (u_i^{(1)})^2 - 2u_i^{(1)} y_{n_m^{(1)}+i} \right\} + \\
&\sum_{m=1}^{M_2} \sum_{j=0}^{q-1} \left\{ (u_j^{(2)})^2 - 2u_j^{(2)} y_{n_m^{(2)}+j} \right\} + 2 \sum_{m=1}^{M_1} \sum_{s=1}^{M_2} \sum_{i=0}^{q-1} u_i^{(1)} u_{i+n_m^{(1)}-n_s^{(2)}}^{(2)},
\end{aligned}
\tag{24}
$$

under the constraints (21) on the elements of $\mathcal{M}^{(1)}$ and $\mathcal{M}^{(2)}$.

The following result confirms that Problem 2 is a particular case of Problem 1.

Corollary 3. *Let the conditions of Problem 2 hold. If in the conditions of Problem 1 put $K = N - q + 1$, $p = q$, $L_{\min}^{(i)} = T_{\min}^{(i)}$, $L_{\max}^{(i)} = T_{\max}^{(i)}$, $i \in \{1, 2\}$,*

$$
f_k^{(i)} = \sum_{j=0}^{q-1} \left\{ (u^{(i)})^2 - 2u_j^{(1)} y_{k+j} \right\}, \quad k = 0, \ldots, K-1, \quad i \in \{1, 2\},
\tag{25}
$$

$$
h_k = 2 \sum_{i=0}^{q-1} u_i^{(1)} u_{i+k}^{(2)}, \quad k = 0, \pm 1, \ldots,
$$

then Algorithm $\mathcal{A}$ gives an exact solution to Problem 2 in $\mathcal{O}(\max\{T_{\max}^{(1)} - T_{\min}^{(1)} + 1, T_{\max}^{(2)} - T_{\min}^{(2)} + 1\})N^2$ time.

To prove this corollary, it is sufficient to substitute the input data into the conditions of Problem 1: restrictions (1) and (3) coincide with (19) and (21), and the objective function (2) coincides with (24). Since $u_j^{(i)} = 0$, if $j \notin \{0, \ldots, q-1\}$, $i \in \{1, 2\}$, function h_k given by (25) satisfy the conditions of Problem 1.

6   Numerical Simulation

Theorems 1 and Corollary 3 prove that with the appropriate choice of input data, Algorithm $\mathcal{A}$ gives the exact solution to the segregation problem. So, the numerical simulation results play an illustration role only. There are two examples of solving the segregation problem for the modelled quasiperiodic sequences in this section. The reference sequences are chosen among signals having abstract geometrical form (such as a sinusoid half-period, a step, a decaying sinusoid, and so on) in order to illustrate the potential applicability of the algorithm as one of the steps to constructing noise resistant segregation algorithms without focusing on specific features of possible applications.

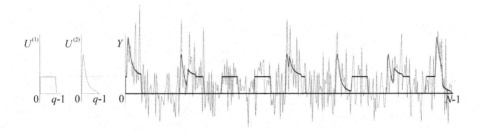

Fig. 4. Example 1. Input data and unobservable sequence

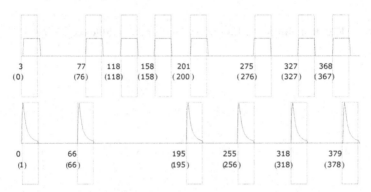

Fig. 5. Example 1. Processing result

Figure 4 presents an example of input data for the segregation problem. There are reference sequences $U^{(1)}$ and $U^{(2)}$ in its left part and a modelled noisy sequence Y in the right part. The sequence Y to be processed is the element-wise sum of the program-generated sequence $X^{(1)} + X^{(2)}$ plotted in the same figure by a thicker line and the sequence of independent, identically distributed Gaussian random variables with zero mathematical expectation. It should be emphasized that only $U^{(1)}$, $U^{(2)}$, and Y are included in the problem input. The unobservable sequence $X^{(1)} + X^{(2)}$ is given for illustration only; this data is not available. Figure 5 shows the result of the algorithm operation — two quasiperiodic sequences $X_{\mathcal{A}}^{(1)}$ and $X_{\mathcal{A}}^{(2)}$ restored according to (20) on the base of collections $\mathcal{M}_{\mathcal{A}}^{(1)}$ and $\mathcal{M}_{\mathcal{A}}^{(2)}$ being Algorithm $\mathcal{A}$ output. The elements of these collections are plotted below the graph. For the convenience of visual comparison, the locations of fragments in the unobservable sequences are marked by frames, and their beginning indices are written in brackets. The example is computed for $N = 400$, $q = 20$, $T_{\min}^{(1)} = 40$, $T_{\max}^{(1)} = 200$, $T_{\min}^{(2)} = 60$, $T_{\max}^{(2)} = 300$, the maximum amplitude pulse value is 118, and the noise level $\sigma = 60$.

Figures 6 and 7, in a similar manner, give another numerical processing example. This example is computed for $N = 600$, $q = 20$, $T_{\min}^{(1)} = 40$, $T_{\max}^{(1)} = 135$, $T_{\min}^{(2)} = 40$, $T_{\max}^{(2)} = 135$, the maximum amplitude pulse value is 127, and the noise level $\sigma = 60$.

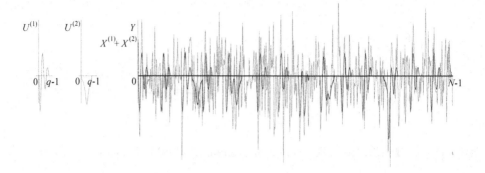

Fig. 6. Example 2. Input data and unobservable sequence

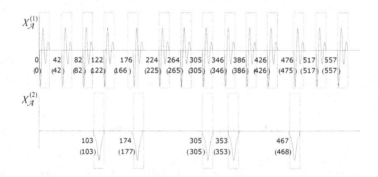

Fig. 7. Example 2. Processing result

The numerical simulation examples show that the algorithm presented allows processing data in the form of the sum of two quasiperiodic sequences with quite acceptable quality. Firstly, the number of fragments in every summand is evaluated correctly. Secondly, a comparison of fragments beginning indices (unobservable and found) shows only insignificant deviations.

7   Conclusion

The main result of this paper is the polinomial-time algorithm that guarantees the optimality of the solution to previously unexplored discrete optimization Problem 1. The algorithm constructed finds the exact solution to the problem in $\mathcal{O}(K)^3$ time when processing K-length sequences; if all integers in the input are bounded by some constant, time complexity is $\mathcal{O}(K^2)$. It is also proved that the previously unexplored applied segregation problem for the sum of two quasiperiodic sequences, including an unknown number of identical subsequences-fragments, reduces (in the framework of a posteriori approach) to the particular case of the considered problem. So, the segregation problem is polynomially solvable as well. The numerical simulation results have demonstrated that the proposed algorithm can serve as a suitable tool to solve the problem of noise-resistant segregation for the sum of two quasiperiodic sequences.

A modification of Problem 1, where the size of the first sought collection is given while the size of the second one is to be found, remains to be studied. Optimization problems induced by segregation problems for quasiperiodic sequences having a more complicated structure (for example, including fragments from the given alphabet or generated by a collection of fragments according to some rule), as well as problems of joint segregation and recognition, are also of the interest. The nearest plans are connected with exploring these problems.

References

1. Kel'manov, A., Khamidullin, S., Mikhailova, L.: A posteriori detection of identical subsequences in a quasiperiodic sequence. Pattern Recognit Image Anal. **12**(4), 438–447 (2002)
2. Kel'manov, A., Khamidullin, S.: Posterior detection of a given number of identical subsequences in a quasi-periodic sequence. Comput. Math. and Math. Phys. **41**, 762–774 (2001)
3. Kel'manov, A.V., Okol'nishnikova, L.V.: A posteriori simultaneous detection and discrimination of subsequences in a quasiperiodic sequence. Pattern Recognit Image Anal. **11**(3), 505–520 (2001)
4. Kel'manov, A.V., Khamidullin, S.A.: A posteriori concurrent detection and identification of quasiperiodic fragments in a sequence from their pieces. Pattern Recognit Image Anal. **16**(4), 599–613 (2006)
5. Mikhailova, L.: One segregation problem for the sum of two quasiperiodic sequences. In: Olenev, N., Evtushenko, Y., Jacimovic, M., Khachay, M., Malkova, V. (eds.): Optimization and Applications. OPTIMA 2023. LNCS, vol 14395. Springer, Cham (2023). https://doi.org/10.1007/978-3-031-47859-8_11
6. Comon, P., Jutten, C. (eds.): Handbook of Blind Source Separation, Independent Component Analysis and Applications. Academic Press (2010)
7. Chien, J.-T.: Source Separation and Machine Learning. Academic Press (2019)
8. Schobben, D.W.E.: Blind Signal Separation, An Overview. In: Real-time Adaptive Concepts in Acoustics. Springer, Dordrecht (2001)
9. Cardoso, J.-F.: Blind signal separation: statistical principles. Proc. IEEE **86**(10), 2009–2025 (1998). https://doi.org/10.1109/5.720250
10. Bach, F.R., Jordan, M.I.: Blind one-microphone speech separation: a spectral learning approach. In: Saul, L.K., Weiss, Y., Bottou, L. (eds.) Advances in Neural Information Processing Systems, vol. 17, pp. 65–72. MIT Press, Cambridge (2005)
11. Kel'manov, A.V., Mikhailova, L.V., Ruzankin, P.S., et al.: Recognition of a quasiperiodic sequence containing an unknown number of nonlinearly extended reference subsequences. Comput. Math. and Math. Phys. **61**, 1153–1161 (2021)
12. Kel'manov, A.V., Mikhailova, L.V.: A posteriori joint detection of reference fragments in a quasi-periodic sequence. Comput. Math. Math. Phys. **48**, 850–865 (2008)

On 1-Skeleton of the Cut Polytopes

Andrei V. Nikolaev[✉]ⓘ

P.G. Demidov Yaroslavl State University, Yaroslavl, Russia
a.nikolaev@uniyar.ac.ru

Abstract. Given an undirected graph $G = (V, E)$, the cut polytope CUT(G) is defined as the convex hull of the incidence vectors of all cuts in G. The 1-skeleton of CUT(G) is a graph whose vertex set is the vertex set of the polytope, and the edge set is the set of geometric edges or one-dimensional faces of the polytope.

We study the diameter and the clique number of 1-skeleton of cut polytopes for several classes of graphs. These characteristics are of interest since they estimate the computational complexity of the max-cut problem for certain computational models and classes of algorithms.

It is established that while the diameter of the 1-skeleton of a cut polytope does not exceed $|V| - 1$ for any connected graph, the clique number varies significantly depending on the class of graphs. For trees, cacti, and almost trees (2), the clique number is linear in the dimension, whereas for complete bipartite and k-partite graphs, it is superpolynomial.

Keywords: Max-cut problem · Cut polytope · 1-skeleton · Diameter · Clique number · Connected component · Chromatic number

1 Introduction

The *1-skeleton* of a polytope P is a graph whose vertex set is the vertex set of P and edge set is the set of geometric edges or one-dimensional faces of P. We are interested in 1-skeletons of polytopes arising from linear and integer programming models since some of their characteristics help to estimate the time complexity for different computation models and classes of algorithms.

The *diameter* of a graph G, denoted by $d(G)$, is the maximum edge distance between any pair of vertices. The diameter of 1-skeleton is known as a lower bound on the number of iterations of the simplex method and similar algorithms. Indeed, let the shortest path between a pair of vertices $\mathbf{u}$ and $\mathbf{v}$ of a polytope P consist of $d(P)$ edges. If the simplex method chooses $\mathbf{u}$ as the initial solution, and the optimal solution is $\mathbf{v}$, then no matter how successfully the algorithm chooses the next adjacent vertex of the 1-skeleton, the number of iterations cannot be less than $d(P)$ (see, for example, Dantzig [17]).

The *clique number* of a graph G, denoted by $\omega(G)$, is the number of vertices in the largest complete subgraph. The clique number of 1-skeleton serves as a lower bound for computational complexity in a class of *direct-type* algorithms

Table 1. Summary table of properties of the 1-skeleton of the cut polytope CUT(G) for some classes of graphs

	Max-cut complexity	Graph diameter	Clique number
Tree	P [21,28]	$|V| - 1$ [26]	2 [26]
Cactus	P [21,28]	$\lfloor \frac{|V|}{2} \rfloor \leq d \leq |V| - 1$	$\leq 2^{\lceil \log_2(|E|+1) \rceil} \leq 2|E|$
Almost tree (2)	P [21,28]	$\lfloor \frac{|V|}{3} \rfloor \leq d \leq |V| - 1$	$\leq 2^{\lceil \log_2(|E|) \rceil + 1} \leq 4|E|$
Complete bipartite graph	NP-hard [25]	2 [26]	$\geq 2^{\min\{|V_1|,|V_2|\}-1}$
Complete k-partite graph	NP-hard [6]	2	$\geq 2^{\text{2nd largest}\{|V_1|,...,|V_k|\}-1}$
Complete graph	NP-hard [22]	1 [3]	$2^{|V|-1}$ [3]

based on linear decision trees. The idea is that direct-type algorithms cannot, in the worst case, discard more than one of the pairwise adjacent solutions in one iteration. This class includes, for example, some classical dynamic programming, branch-and-bound and other algorithms. See Bondarenko [8,9] for more details. Besides, for all known cases, the clique number of 1-skeleton of a polytope is polynomial for polynomially solvable problems (Bondarenko and Nikolaev [7,10, 27]) and superpolynomial for intractable problems (Bondarenko et al. [8,11,12], Padberg [29], and Simanchev [30]).

In this paper, we consider 1-skeletons of cut polytopes associated with the maximum cut problem for several classes of graphs, in particular for trees, cacti, almost trees (2), complete bipartite, and k-partite graphs. The results of the research are summarized in Table 1. New results are highlighted in bold.

2 Cut Polytopes

We consider the classical max-cut problem in an undirected weighted graph.

Maximum cut problem (max-cut).
INSTANCE. Given an undirected graph $G = (V, E)$ with an edge weight function $w : E \to \mathbb{Z}$.
QUESTION. Find a subset of vertices $S \subseteq V$ such that the sum of the weights of the edges from E with one endpoint in S and another in $V \backslash S$ (cut) is as large as possible.

Max-cut is a well-known NP-hard problem (Karp [22], Garey and Johnson [19]) that arises, for example, in cluster analysis (Boros and Hammer [13]), the Ising model in statistical physics (Barahona et al. [2]), the VLSI design (Chen at al. [15]), social network analysis (Agrawal et al. [1]), and the image segmentation (de Sousa at al. [31]).

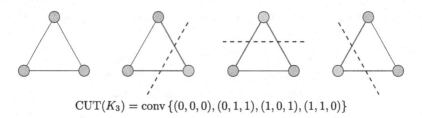

$$\mathrm{CUT}(K_3) = \mathrm{conv}\left\{(0,0,0),(0,1,1),(1,0,1),(1,1,0)\right\}$$

Fig. 1. An example of constructing a cut polytope for K_3

The cut polytopes were introduced by Barahona and Mahjoub in their seminal paper [3]. Let $G = (V, E)$ be an undirected graph with node set V and edge set E. With each subset $S \subseteq V$ we associate the incidence $0/1$−vector $\mathbf{v}(S) \in \{0,1\}^{|E|}$:

$$v(S)_e = \begin{cases} 1, & \text{if } e \in \delta(S), \\ 0, & \text{otherwise,} \end{cases}$$

where $\delta(S) \subseteq E$ is the *cut-set* of edges with exactly one endpoint in S.

The *cut polytope* $\mathrm{CUT}(G)$ is defined as the convex hull of all incidence vectors:

$$\mathrm{CUT}(G) = \mathrm{conv}\left\{\mathbf{v}(S) : S \subseteq V\right\} \subset \mathbb{R}^{|E|}.$$

An example of constructing the cut polytope $\mathrm{CUT}(K_3)$ for the complete graph K_3 on 3 vertices is shown in Fig. 1.

Note that since $\delta(S) = \delta(V \backslash S)$, then any graph $G = (V, E)$ contains exactly $2^{|V|-1}$ cuts, and the cut polytope $\mathrm{CUT}(G)$ will have the same number of vertices.

The cut polytope and its various relaxations often serve as the linear programming models for the maximum cut problem. See, for example, the polynomial time algorithm by Barahona for the max-cut on graphs not contractible to K_5 [4]. For more details, see the monograph by Deza and Laurent [18].

3   Adjacency

The 1-skeletons of cut polytopes were first studied by Barahona and Mahjoub [3]. In particular, they described a criterion for the adjacency of vertices in a cut polytope.

We denote by $X \triangle Y$ the symmetric difference of the sets X and Y. It is easy to check that for any cuts $X, Y \subseteq V$ we have $\delta(X) \triangle \delta(Y) = \delta(X \triangle Y)$.

Theorem 1 (Barahona and Mahjoub [3]). Let $G = (V, E)$ be a connected graph. Let $\mathbf{v}(X)$ and $\mathbf{v}(Y)$ be extreme points of $\mathrm{CUT}(G)$, where $X, Y \subseteq V$ are the corresponding cuts. Then $\mathbf{v}(X)$ and $\mathbf{v}(Y)$ are adjacent in $\mathrm{CUT}(G)$ if and only if the graph $H_{X \triangle Y} = (V, E \backslash (\delta(X \triangle Y)))$ has two connected components.

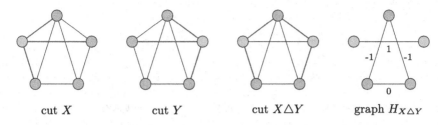

cut X cut Y cut $X \triangle Y$ graph $H_{X \triangle Y}$

Fig. 2. An example of adjacent cuts according to Theorem 1

A short summary of the proof is as follows. Let $G_i = (V_i, E_i)$, $i = 1, 2$ be two connected components of $H_{X \triangle Y} = (V, E \backslash (\delta(X \triangle Y)))$, and let T_i be a spanning tree of G_i, $i = 1, 2$. Then we can consider vector $\mathbf{c} \in \mathbb{R}^{|E|}$:

$$c_e = \begin{cases} 1, & \text{if } e \in T_i \cap (\delta(X) \cap \delta(Y)), \ i = 1, 2, \\ -1, & \text{if } e \in T_i \backslash (\delta(X) \cap \delta(Y)), \ i = 1, 2, \\ 0, & \text{otherwise,} \end{cases}$$

such that $\mathbf{v}(X)$ and $\mathbf{v}(Y)$ are the only two extreme points that maximize $\mathbf{c}^T \mathbf{v}$ over $\mathrm{CUT}(G)$ (see Fig. 2), therefore they are adjacent in the 1-skeleton.

Let now the graph $H_{X \triangle Y}$ contain more than two connected components (see Fig. 3), then at least one of these components defines a non-empty cut L, such that $\delta(L) \subset \delta(X \triangle Y)$, and

$$\mathbf{v}(X \triangle L) + \mathbf{v}(Y \triangle L) = \mathbf{v}(X) + \mathbf{v}(Y).$$

A line segment connecting vertices $\mathbf{v}(X)$ and $\mathbf{v}(Y)$ intersects with a line segment connecting two other vertices, therefore, $\mathbf{v}(X)$ and $\mathbf{v}(Y)$ cannot be adjacent.

Theorem 1 directly implies the following statement.

Corollary 1 (Barahona and Mahjoub [3]). If G is a complete graph, then $\mathrm{CUT}(G)$ has diameter one.

The cut polytope of a complete graph is a classic example of a combinatorial polytope, any two vertices of which are pairwise adjacent. Note also that the clique number of the 1-skeleton here is equal to $2^{|V|-1}$.

However, the situation may be fundamentally different for other classes of graphs. In this paper, we follow Neto [26], who described the diameters of the 1-skeletons of cut polytopes for trees and complete bipartite graphs.

The study of cut polytopes for different classes of graphs is also of interest since the adjacency of vertices in the 1-skeleton is inherited from the subgraphs.

Theorem 2 (Neto [26]). Let $G' = (V, E')$ be a connected subgraph of $G = (V, E)$, and let X, Y denote two distinct node subsets of V. Then, if the vertices $\mathbf{v}(X)$ and $\mathbf{v}(Y)$ are adjacent in $\mathrm{CUT}(G')$, they are also adjacent in $\mathrm{CUT}(G)$.

Indeed, if the graph $H'_{X \triangle Y}$ for G' has only 2 connected components, then these same components will remain connected in the graph $H_{X \triangle Y}$ for G. Thus, the 1-skeleton of $\mathrm{CUT}(G')$ is a subgraph of the 1-skeleton of $\mathrm{CUT}(G)$.

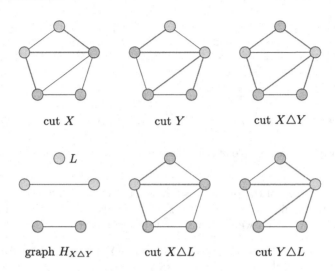

Fig. 3. An example of non-adjacent cuts according to Theorem 1

4   Trees, Cacti, and Almost Trees

In this section, we study cut polytopes for connected graphs with a small number of edges. In particular, trees, cacti, and almost trees (2). Let us recall the definitions of the corresponding classes of graphs, following Brandstädt et al. [14]:

- a *tree* is a connected graph without cycles;
- a *cactus* is a connected graph in which every edge belongs to at most one simple cycle;
- a connected graph is an *almost tree* (k) if every *biconnected component* (maximal subgraph that remains connected when any vertex is removed) has the property that there are at most k edges not in a spanning tree of this biconnected component.

It's easy to see that G is a cactus if and only if G is an almost tree (1). Some problems that are NP-hard for general graphs are polynomially solvable on almost trees with small parameter k (see Ben-Moshe et al. [5], Coppersmith and Vishkin [16], Gurevich et al. [20]). As for the max-cut problem, trees, cacti, and almost trees (2) are planar graphs, and it is well known that the maximum cut in a planar graph can be found in polynomial time (Orlova and Dorfman [28], Hadlock [21]), and, in general, on graphs not contractible to K_5 (Barahona [4]).

4.1   Trees

Cut polytopes for trees were first considered by Neto [26] as an example of graphs with the largest possible value of the diameter of the corresponding cut polytopes. Here we provide an alternative proof of this result.

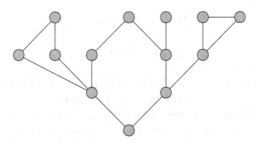

Fig. 4. Example of a cactus graph

The *Hamming distance* between two equal-length strings of symbols $\mathbf{x}$ and $\mathbf{y}$, denoted by $\Delta(\mathbf{x}, \mathbf{y})$, is defined as the number of positions at which the entries of $\mathbf{x}$ and $\mathbf{y}$ are different.

Theorem 3 (Neto [26]). If $G = (V, E)$ is a tree, then the cut polytope $\mathrm{CUT}(G)$ is a unit hypercube $[0, 1]^{|V|-1}$.

Proof. Note that if we remove k edges from the tree, then it splits into $k + 1$ connected components. Therefore, for some two distinct cuts X and Y the graph $H_{X \triangle Y}$ has 2 connected components if and only if $|\delta(X \triangle Y)| = 1$. Thus, the incidence vectors of cuts in a tree are $0/1$ vectors of length $|E| = |V| - 1$, and two vectors are adjacent if and only if the Hamming distance between them is exactly 1. Clearly, this is a unit hypercube $[0, 1]^{|V|-1}$.

Corollary 2. *If $G = (V, E)$ is a tree, then the diameter $d(\mathrm{CUT}(G)) = |V| - 1$, and the clique number $\omega(\mathrm{CUT}(G)) = 2$.*

Thus, taking into account Theorem 2, cut polytopes for trees, among all connected graphs, have the smallest clique number and the largest diameter of the 1-skeleton.

4.2 Cacti

Here we consider cacti as a simple generalization of trees (see Fig. 4).

Lemma 1. *Let $G = (V, E)$ be a cactus. If for two cuts $X, Y \subseteq V$ the vertices $\mathbf{v}(X)$ and $\mathbf{v}(Y)$ are adjacent in $\mathrm{CUT}(G)$, then $|\delta(X \triangle Y)| \leq 2$.*

Proof. Note that if $G = (V, E)$ is a cactus, then for any cut $S \subseteq V$, where $|\delta(S)| \geq 3$, the graph $H_S = (V, E \backslash \delta(S))$ has at least three connected components. Indeed, each cactus edge is either a *bridge* (an edge whose removal would disconnect the graph) or belongs to a unique cycle. Since each edge belongs to at most one simple cycle, any cut-set can contain either an even number of edges from some cycles or one or more bridges. However, removing any bridge or any two edges from the cycle breaks the cactus into different connected components. Thus, by Theorem 1, if the vertices $\mathbf{v}(X)$ and $\mathbf{v}(Y)$ are adjacent, then $\delta(X \triangle Y)$ contains at most two edges.

Theorem 4. *Let $G = (V, E)$ be a cactus, then*

$$\left\lfloor \frac{|V|}{2} \right\rfloor \leq d(\mathrm{CUT}(G)) \leq |V| - 1.$$

Proof. An upper bound on the diameter, common to all connected graphs, is achieved on a tree, which is a special case of a cactus.

For the lower bound, consider the spanning tree of the graph $G = (V, E)$. Since every tree is a bipartite graph, we can always construct a non-empty cut $S \subset V$ such that the cut-set $\delta(S)$ contains at least $|V| - 1$ edges. Therefore, by Lemma 1, the shortest path between $\mathbf{v}(S)$ and the empty cut $\mathbf{v}(\emptyset)$ for the cactus graph has length of at least $\lceil \frac{|V|-1}{2} \rceil = \lfloor \frac{|V|}{2} \rfloor$. The exact value is achieved, for example, on any cycle graph.

Theorem 5. *Let $G = (V, E)$ be a cactus, then*

$$\omega(\mathrm{CUT}(G)) \leq 2^{\lceil \log_2(|E|+1) \rceil} \leq 2|E|.$$

Proof. We bound the clique number of the 1-skeleton of a cut polytope from above by the *chromatic number*, i.e. the smallest number of colors needed to color the vertices of G so that no two adjacent vertices share the same color:

$$\omega(\mathrm{CUT}(G)) \leq \chi(\mathrm{CUT}(G)).$$

To color the 1-skeleton of a cut polytope for a cactus, we use the idea of binary representation matrices by Wan [32] (see also Linial et al. [24]). The *binary representation matrix* $\mathrm{BRM}(k)$ is a $k \times \lceil \log_2 k \rceil$ matrix, in which row i ($1 \leq i \leq k$) is the 0/1-vector of length $\lceil \log_2 k \rceil$ obtained from the binary expansion of i, adding leading zeroes as necessary. For example, the 3th row of the matrix $\mathrm{BRM}(5)$ has the form $(0, 1, 1)$ since $3 = 0 \cdot 2^2 + 1 \cdot 2^1 + 1 \cdot 2^0$. To each vertex $\mathbf{v}(S)$ of the cut polytope $\mathrm{CUT}(G)$ we assign the color

$$\mathrm{color}(\mathbf{v}(S)) = \mathbf{v}(S)\mathrm{BRM}(|E| + 1),$$

where all arithmetic operations are performed modulo 2.

Let us show that this is a proper graph coloring. Consider two distinct cuts $X, Y \subseteq V$ and the corresponding incidence vectors $\mathbf{v}(X), \mathbf{v}(Y)$, then

$$\mathbf{v}(X)\mathrm{BRM}(|E| + 1) - \mathbf{v}(Y)\mathrm{BRM}(|E| + 1) = (\mathbf{v}(X) - \mathbf{v}(Y))\mathrm{BRM}(|E| + 1). \quad (1)$$

If the extreme points $\mathbf{v}(X)$ and $\mathbf{v}(Y)$ are adjacent, then they differ by no more than two coordinates. Therefore, equation (1) is either some row of the matrix $\mathrm{BRM}(|E| + 1)$, or the difference of two rows. In both cases (1) cannot be a zero row since all rows of the matrix $\mathrm{BRM}(|E| + 1)$ are distinct and non-zero. Thus,

$$\mathbf{v}(X)\mathrm{BRM}(|E| + 1) \neq \mathbf{v}(Y)\mathrm{BRM}(|E| + 1),$$

and no two adjacent vertices share the same color.

It remains to note that each color is uniquely represented by a binary string of length $\lceil \log_2(|E| + 1) \rceil$, from which we obtain the required upper bound

$$\omega(\mathrm{CUT}(G)) \leq \chi(\mathrm{CUT}(G)) \leq 2^{\lceil \log_2(|E|+1) \rceil} \leq 2|E|.$$

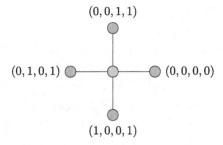

$(0,0,1,1)$

$(0,1,0,1)$ $(0,0,0,0)$

$(1,0,0,1)$

Fig. 5. A Hamming ball of radius 1 centered around $(0,0,0,1)$ (red vertex) (Color figure online)

To estimate how tight the upper bound of the Theorem 5, we consider the cycle graph $C_n = (V, E)$. A cycle is split into two connected components if exactly two edges are removed from it. Then, by Theorem 1, the incidence vectors $\mathbf{v}(X)$ and $\mathbf{v}(Y)$ of the cuts $X, Y \subseteq V$ are adjacent in $\mathrm{CUT}(C_n)$ if and only if $|\delta(X \triangle Y)| = 2$ (or $\Delta(\mathbf{v}(X), \mathbf{v}(Y)) = 2$).

Consider a *Hamming ball* of radius 1 centered around some vector $\mathbf{c} \in \{0,1\}^n$ with an odd number of 1's, i.e. set of vectors $\mathbf{v} \in \{0,1\}^n$ such that $\Delta(\mathbf{c}, \mathbf{v}) = 1$ (see Fig. 5). This ball contains n binary vectors, the Hamming distance between any pair of which equals 2, therefore the corresponding incidence vectors are pairwise adjacent in $\mathrm{CUT}(C_n)$. For example, we can consider the cuts $\{\{v_1\}, \{v_1, v_2\}, \{v_1, v_2, v_3\}, \ldots, \{V\}\}$.

Thus, if $G = (V, E)$ is a cactus, then the clique number of the 1-skeleton of $\mathrm{CUT}(G)$ can be equal to $|E|$, while Theorem 5 provides an upper bound of $2|E|$, which is quite tight. Note also that the 1-skeleton of the cut polytope $\mathrm{CUT}(C_3) = \mathrm{CUT}(K_3)$ (Fig. 1) is a complete graph whose clique number is 4.

4.3 Almost Trees (2)

Now we consider almost trees (2), i.e. such graphs that there are at most 2 edges not in a spanning tree of each biconnected component (see Fig. 6).

Lemma 2. *Let $G = (V, E)$ be an almost tree (2). If for two cuts $X, Y \subseteq V$ the vertices $\mathbf{v}(X)$ and $\mathbf{v}(Y)$ are adjacent in $\mathrm{CUT}(G)$, then $|\delta(X \triangle Y)| \leq 3$.*

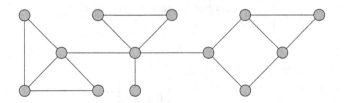

Fig. 6. Example of an almost tree (2)

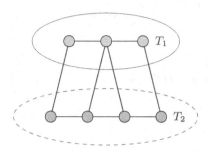

Fig. 7. The case with 4 edges in a cut-set separating two connected components

Proof. Similar to Lemma 1, if $G = (V, E)$ is an almost tree (2), then for any cut $S \subseteq V$, where $|\delta(S)| \geq 4$, the graph $H_S = (V, E \backslash \delta(S))$ has at least three connected components.

Suppose to the contrary that the cut S splits G into exactly two connected components. Let us consider the spanning trees $T_1 = (V_1, E_1)$ and $T_2 = (V_2, E_2)$ of these components. Let's add to them edges from the cut-set $\delta(S)$. Then the resulting subgraph $G' = (V, E_1 \cup E_2 \cup \delta(S))$ of the original graph G contains a biconnected component whose 3 edges do not belong to the spanning tree (see Fig. 7). Thus, we have a contradiction.

Theorem 6. *Let $G = (V, E)$ be an almost tree (2), then*

$$\left\lfloor \frac{|V|}{3} \right\rfloor \leq d(\mathrm{CUT}(G)) \leq |V| - 1.$$

Proof. The reasoning is completely similar to the proof of Theorem 4.

Theorem 7. *Let $G = (V, E)$ be an almost tree (2), then*

$$\omega(\mathrm{CUT}(G)) \leq 2^{\lceil \log_2(|E|) \rceil + 1} \leq 4|E|.$$

Proof. We again bound the clique number of the 1-skeleton of a cut polytope by the chromatic number. This time we are adapting the idea of hypercube coloring by Kim et al. [23].

Let $m = \lceil \log_2 k \rceil + 1$. The *modified binary representation matrix* $\mathrm{BRM}^*(k)$ is a $k \times m$ matrix, in which row i $(1 \leq i \leq k)$ is the 0/1-vector of the form (b_i, n_i), where b_i is the $(m - 1)$-dimensional binary expansion of integer $i - 1$, adding leading zeroes as necessary, and

$$n_i = \begin{cases} 1, & \text{if } b_i \text{ contains an even number of 1's;} \\ 0, & \text{otherwise.} \end{cases}$$

For example,

$$\mathrm{BRM}^*(5) = \begin{pmatrix} 0\,0\,0\,1 \\ 0\,0\,1\,0 \\ 0\,1\,0\,0 \\ 0\,1\,1\,1 \\ 1\,0\,0\,0 \end{pmatrix}.$$

To each vertex $\mathbf{v}(S)$ of the cut polytope $\mathrm{CUT}(G)$ we assign the color

$$\mathrm{color}(\mathbf{v}(S)) = \mathbf{v}(S)\mathrm{BRM}^*(|E|),$$

where all arithmetic operations are performed modulo 2. Again, note that we use 2^m colors in total.

Consider two cuts $X, Y \subseteq V$ and the incidence vectors $\mathbf{v}(X), \mathbf{v}(Y)$, then

$$\mathbf{v}(X)\mathrm{BRM}^*(|E|) - \mathbf{v}(Y)\mathrm{BRM}^*(|E|) = (\mathbf{v}(X) - \mathbf{v}(Y))\mathrm{BRM}^*(|E|). \quad (2)$$

If the extreme points $\mathbf{v}(X)$ and $\mathbf{v}(Y)$ are adjacent, then they differ by no more than 3 coordinates. Therefore, equation (2) is the modulo 2 sum of at most 3 rows of the matrix $\mathrm{BRM}^*(|E|)$ that cannot be a zero row, since no two rows of $\mathrm{BRM}^*(|E|)$ are equal, and any odd number of rows of $\mathrm{BRM}^*(|E|)$ contain an odd number of 1's in total. Therefore, no two adjacent vertices share the same color. Thus, we obtain the required upper bound

$$\omega(\mathrm{CUT}(G)) \leq \chi(\mathrm{CUT}(G)) \leq 2^{\lceil \log_2(|E|)\rceil + 1} \leq 4|E|.$$

As for the tightness of this upper estimate, the cut polytope $\mathrm{CUT}(C_n)$ of the cycle graph $C_n = (V, E)$ still has a clique of size $|E|$, while Theorem 7 provides an upper bound of $4|E|$. This is not as good as the estimate of Theorem 5, but it still has the same asymptotic behavior.

5   Complete Bipartite and k-Partite Graphs

In this section, we consider cut polytopes for graphs with a large number of edges, in particular for complete bipartite and k-partite graphs. A k-*partite graph* is a graph whose vertices can be partitioned into k different independent sets $V_1, V_2, \ldots, V_k$. When $k = 2$, these are the *bipartite graphs*, and when $k = 3$, they are called the *tripartite graphs*.

The situation with the max-cut problem on a bipartite graph is a bit tricky. On the one hand, a *simple max-cut* (all edges have the same weight, for example, equal to one) and a max-cut with non-negative edges have an obvious solution: the largest cut will be the one that separates parts V_1 and V_2 since the cut-set $\delta(V_1) = \delta(V_2)$ contains all the edges of the graph. On the other hand, if the edges of the graph can take both positive and negative values, then the max-cut problem is NP-hard (McCormick et al. [25]). As for k-partite graphs, even a simple max-cut is NP-hard for tripartite graphs (Bodlaender and Jansen [6]).

5.1   Complete Bipartite Graphs

The diameter of the 1-skeleton of a cut polytope of a complete bipartite graph was first considered by Neto [26].

Theorem 8 (Neto [26]). Let $K_{n_1, n_2} = (V_1, V_2, E)$ be a complete bipartite graph, where $n_1 = |V_1| \geq 2$ and $n_2 = |V_2| \geq 2$, then the diameter of the cut polytope $d(\mathrm{CUT}(K_{n_1, n_2})) = 2$.

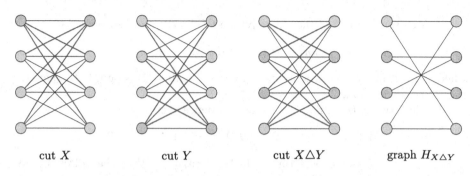

Fig. 8. Symmetric cuts of a complete bipartite graph $K_{4,4}$

Here we establish that the clique number of the 1-skeleton of a cut polytope of a complete bipartite graph is superpolynomial in dimension.

Theorem 9. *Let $K_{n_1,n_2} = (V_1, V_2, E)$ be a complete bipartite graph, where $n_1 = |V_1| \geq 2$ and $n_2 = |V_2| \geq 2$, then the clique number*

$$\omega(\mathrm{CUT}(K_{n_1,n_2})) \geq 2^{\min\{n_1,n_2\}-1}.$$

Proof. Let, without loss of generality, $n_1 = \min\{n_1, n_2\} \geq 2$.

Consider the family $\mathcal{F}$ of all possible subsets $S \subseteq V_1$ such that from each pair S and $V_1 \backslash S$ the family $\mathcal{F}$ contains only one subset. Then, clearly, $|\mathcal{F}| = 2^{n_1-1}$.

Now, for each $S \in \mathcal{F}$ we construct a symmetric cut $(S \subseteq V_1) \cup (S \subseteq V_2)$. Note that such a cut always exists since $|V_1| \leq |V_2|$.

Let X and Y be two symmetric cuts of this type, then in each of the two parts V_1 and V_2 there is at least one vertex that belongs to $X \triangle Y$, and at least one vertex that does not belong to $X \triangle Y$, since no two subsets in $\mathcal{F}$ are the same, and we previously excluded set complements.

However, K_{n_1,n_2} is a complete bipartite graph, and any two vertices from different parts are connected by an edge. Therefore, both $X \triangle Y$ and its complement $V \backslash (X \triangle Y)$ form connected components (see Fig. 8). Thus, by Theorem 1, for any two symmetric cuts X, Y of this type (any two subsets of the family $\mathcal{F}$), the vertices $\mathbf{v}(X)$ and $\mathbf{v}(Y)$ of the cut polytope $\mathrm{CUT}(K_{n_1,n_2})$ are pairwise adjacent, which gives us a clique of the required size.

5.2   Complete Tripartite and k-Partite Graphs

Theorem 10. *Let $K_{n_1,n_2,n_3} = (V_1, V_2, V_3, E)$ be a complete tripartite graph with the node set $V = V_1 \cup V_2 \cup V_3$, where $n_i = |V_i| \geq 2$, $1 \leq i \leq 3$, then the diameter of the cut polytope $d(\mathrm{CUT}(K_{n_1,n_2,n_3})) = 2$.*

Proof. We consider two cuts $X, Y \subseteq V$ and explore three possible cases.

Case 1. If $V_i \cap (X \triangle Y) \neq \emptyset$ and $V_i \backslash (X \triangle Y) \neq \emptyset$ for at least two of the three parts V_i (possibly different), then the extreme points $\mathbf{v}(X)$ and $\mathbf{v}(Y)$ are adjacent

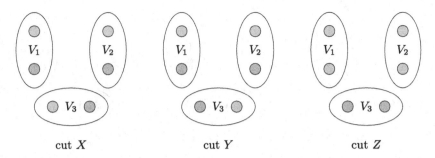

Fig. 9. Construction of the cut Z in the case of $V_1 \cap (X \triangle Y) = V_2 \cap (X \triangle Y) = \emptyset$

in the cut polytope $\mathrm{CUT}(K_{n_1,n_2,n_3})$. Indeed, both $X \triangle Y$ and its complement $V \backslash (X \triangle Y)$ form connected components (see Fig. 8).

Case 2. Let for some two parts, without loss of generality V_1 and V_2, $V_1 \cap (X \triangle Y) = V_2 \cap (X \triangle Y) = \emptyset$. Therefore, $X \triangle Y \subseteq V_3$, and if $|X \triangle Y| \geq 2$, then the symmetric difference does not form a connected component, and the vectors $\mathbf{v}(X)$ and $\mathbf{v}(Y)$ are not adjacent.

We construct a cut $Z \subseteq V$ such that $|V_1 \cap (Z \triangle X)| = 1$ and $|V_2 \cap (Z \triangle X)| = 1$ (see Fig. 9). This is always possible since both V_1 and V_2 contain at least two vertices. It is enough to add or subtract a vertex from X in each of the parts. We obtain that $Z \triangle X$ has exactly one vertex each in V_1 and V_2. Then, $Z \triangle X$ and $V \backslash (Z \triangle X)$ form connected components. The same is true for $Z \triangle Y$. Therefore, the vector $\mathbf{v}(Z)$ is adjacent to both $\mathbf{v}(X)$ and $\mathbf{v}(Y)$.

Case 3. Let now for some two parts, without loss of generality V_1 and V_2, $V_1 \backslash (X \triangle Y) = \emptyset$ and $V_2 \backslash (X \triangle Y) = \emptyset$. Then, both cuts X and Y cannot simultaneously contain all vertices from V_1 and V_2. Without loss of generality, we assume that $X \cap V_1 \neq V_1$ and $Y \cap V_2 \neq V_2$.

Let us choose vertices $v_1 \in V_1 \backslash X$ and $v_2 \in V_2 \backslash Y$ and construct a cut $Z \subseteq V$ (see Fig. 10) such that

$$Z \cap V_1 = (X \cap V_1) \cup \{v_1\} \text{ and } Z \cap V_2 = (Y \cap V_2) \cup \{v_2\}.$$

It only remains to note that

- $V_1 \cap (X \triangle Z) = \{v_1\}$ and $V_2 \cap (X \triangle Z) = V_2 \backslash \{v_2\}$, hence the vectors $\mathbf{v}(X)$ and $\mathbf{v}(Z)$ are adjacent;
- $V_1 \cap (Y \triangle Z) = V_1 \backslash \{v_1\}$ and $V_2 \cap (Y \triangle Z) = \{v_2\}$, hence the vectors $\mathbf{v}(Y)$ and $\mathbf{v}(Z)$ are adjacent.

Thus, for any pair of non-adjacent vertices $\mathbf{v}(X)$ and $\mathbf{v}(Y)$ of a cut polytope $\mathrm{CUT}(K_{n_1,n_2,n_3})$ of a complete tripartite graph, we can construct a third vertex $\mathbf{v}(Z)$ adjacent to both. Therefore, $d(\mathrm{CUT}(K_{n_1,n_2,n_3})) = 2$.

Theorem 11. *Let* $K_{n_1,\ldots,n_k} = (V_1,\ldots,V_k,E)$ *be a complete k-partite graph with the node set* $V = V_1 \cup \ldots \cup V_k$, *where* $n_i = |V_i| \geq 2$, $1 \leq i \leq k$, *then the diameter of the cut polytope* $d(\mathrm{CUT}(K_{n_1,\ldots,n_k})) = 2$.

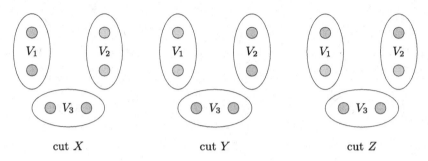

Fig. 10. Construction of the cut Z in the case of $V_1 \backslash (X \triangle Y) = \emptyset$ and $V_2 \backslash (X \triangle Y) = \emptyset$

Proof. The cases of $k = 2$ and $k = 3$ have already been considered above, and for the remaining k, the proof is similar to complete tripartite graphs.

Theorem 12. *Let* $K_{n_1, n_2, \dots, n_k} = (V_1, V_2, \dots, V_k, E)$ *be a complete k-partite graph, where* $n_1 = |V_1| \geq n_2 = |V_2| \geq \dots \geq n_k = |V_k|$, *then the clique number*

$$\omega(\text{CUT}(K_{n_1, n_2, \dots, n_k})) \geq 2^{n_2 - 1}.$$

Proof. The idea is to limit the consideration to only the parts V_1 and V_2, then repeat the construction from the proof of the Theorem 9.

We again consider a family $\mathcal{F}$ of all possible subsets $S \subseteq V_2$ such that from each pair S and $V_2 \backslash S$ the family $\mathcal{F}$ contains only one subset. For each $S \in \mathcal{F}$ we construct a symmetric cut $(S \subseteq V_1) \cup (S \subseteq V_2)$. For any two symmetric cuts X, Y of this type, both $X \triangle Y$ and $V \backslash (X \triangle Y)$ form connected components, and the vectors $\mathbf{v}(X)$ and $\mathbf{v}(Y)$ of the cut polytope are pairwise adjacent, which gives us a clique of the size $|\mathcal{F}| = 2^{n_2 - 1}$.

6   Conclusion

The results of the research are summarized in Table 1. We can see that as the number of edges in the graph G grows, the diameter of the 1-skeleton of the cut polytope $\text{CUT}(G)$ decreases, and the clique number increases. This is expected since if G' is a subgraph of G, then 1-skeleton of $\text{CUT}(G')$ is a subgraph of the 1-skeleton of $\text{CUT}(G)$.

However, the clique number of the 1-skeleton undergoes a qualitative leap as we move from trees and almost trees to complete bipartite graphs. On the one hand, the max-cut problem is polynomially solvable for trees, cacti, and almost trees (2). The clique number of the 1-skeleton here is linear in the dimension. On the other hand, for complete bipartite and k-partite graphs, the max-cut problem is NP-hard, and the clique number of the 1-skeleton of the corresponding cut polytopes is superpolynomial in the dimension.

Acknowledgment. The research is supported by the Yaroslavl region grant No. 4-np/2023 and the P.G. Demidov Yaroslavl State University Project VIP-016.

Disclosure of Interests. The author has no competing interests to declare that are relevant to the content of this article.

References

1. Agrawal, R., Rajagopalan, S., Srikant, R., Xu, Y.: Mining newsgroups using networks arising from social behavior. In: Proceedings of the 12th International Conference on World Wide Web, pp. 529-535. WWW 2003. Association for Computing Machinery, New York (2003). https://doi.org/10.1145/775152.775227
2. Barahona, F., Grötschel, M., Jünger, M., Reinelt, M.: An application of combinatorial optimization to statistical physics and circuit layout design. Oper. Res. **36**(3), 493–513 (1988). https://doi.org/10.1287/opre.36.3.493
3. Barahona, F., Mahjoub, A.R.: On the cut polytope. Math. Program. **36**(2), 157–173 (1986). https://doi.org/10.1007/BF02592023
4. Barahona, F.: The max-cut problem on graphs not contractible to K_5. Oper. Res. Lett. **2**, 107–111 (1983). https://doi.org/10.1016/0167-6377(83)90016-0
5. Ben-Moshe, B., Bhattacharya, B., Shi, Q.: Efficient algorithms for the weighted 2-center problem in a cactus graph. In: Deng, X., Du, D.-Z. (eds.) ISAAC 2005. LNCS, vol. 3827, pp. 693–703. Springer, Heidelberg (2005). https://doi.org/10.1007/11602613_70
6. Bodlaender, H.L., Jansen, K.: On the complexity of the maximum cut problem. In: Enjalbert, P., Mayr, E.W., Wagner, K.W. (eds.) STACS 1994. LNCS, vol. 775, pp. 769–780. Springer, Heidelberg (1994). https://doi.org/10.1007/3-540-57785-8_189
7. Bondarenko, V., Nikolaev, A.: On graphs of the cone decompositions for the min-cut and max-cut problems. Int. J. Math. Math. Sci. **2016**, 7863650 (2016). https://doi.org/10.1155/2016/7863650
8. Bondarenko, V.A.: Nonpolynomial lower bounds for the complexity of the traveling salesman problem in a class of algorithms. Autom. Remote. Control. **44**(9), 1137–1142 (1983)
9. Bondarenko, V.A.: Estimating the complexity of problems on combinatorial optimization in one class of algorithms. Physics-Doklady **38**(1), 6–7 (1993)
10. Bondarenko, V.A., Nikolaev, A.V.: On the skeleton of the polytope of pyramidal tours. J. Appl. Ind. Math. **12**(1), 9–18 (2018). https://doi.org/10.1134/S1990478918010027
11. Bondarenko, V.A., Nikolaev, A.V., Shovgenov, D.A.: 1-skeletons of the spanning tree problems with additional constraints. Autom. Control. Comput. Sci. **51**(7), 682–688 (2017). https://doi.org/10.3103/S0146411617070033
12. Bondarenko, V.A., Nikolaev, A.V., Shovgenov, D.A.: Polyhedral characteristics of balanced and unbalanced bipartite subgraph problems. Autom. Control. Comput. Sci. **51**(7), 576–585 (2017). https://doi.org/10.3103/S0146411617070276
13. Boros, E., Hammer, P.L.: On clustering problems with connected optima in euclidean spaces. Discret. Math. **75**(1), 81–88 (1989). https://doi.org/10.1016/0012-365X(89)90080-0
14. Brandstädt, A., Le, V.B., Spinrad, J.P.: Graph classes: a survey. Society for Industrial and Applied Mathematics (1999). https://doi.org/10.1137/1.9780898719796

15. Chen, R.W., Kajitani, Y., Chan, S.P.: A graph-theoretic via minimization algorithm for two-layer printed circuit boards. IEEE Trans. Circ. Syst. **30**(5), 284–299 (1983). https://doi.org/10.1109/TCS.1983.1085357
16. Coppersmith, D., Vishkin, U.: Solving NP-hard problems in 'almost trees': vertex cover. Discret. Appl. Math. **10**(1), 27–45 (1985). https://doi.org/10.1016/0166-218X(85)90057-5
17. Dantzig, G.B.: Linear Programming and Extensions. Princeton University Press (1991)
18. Deza, M.M., Laurent, M.: Geometry of Cuts and Metrics. Springer-Verlag Berlin Heidelberg (1997). https://doi.org/10.1007/978-3-642-04295-9
19. Garey, M.R., Johnson, D.S.: Computers and Intractability: A Guide to the Theory of NP-Completeness (Series of Books in the Mathematical Sciences). Freeman, W. H (1979)
20. Gurevich, Y., Stockmeyer, L., Vishkin, U.: Solving NP-hard problems on graphs that are almost trees and an application to facility location problems. J. ACM **31**(3), 459–473 (1984). https://doi.org/10.1145/828.322439
21. Hadlock, F.: Finding a maximum cut of a planar graph in polynomial time. SIAM J. Comput. **4**(3), 221–225 (1975). https://doi.org/10.1137/0204019
22. Karp, R.M.: Reducibility among combinatorial problems. In: Miller, R.E., Thatcher, J.W., Bohlinger, J.D. (eds.) Complexity of Computer Computations. The IBM Research Symposia Series, pp. 85–103. Springer US, Boston, MA (1972). https://doi.org/10.1007/978-1-4684-2001-2_9
23. Kim, D.S., Du, D.Z., Pardalos, P.M.: A coloring problem on the n-cube. Discret. Appl. Math. **103**(1), 307–311 (2000). https://doi.org/10.1016/S0166-218X(99)00249-8
24. Linial, N., Meshulam, R., Tarsi, M.: Matroidal bijections between graphs. J. Combinatorial Theory Ser.B **45**(1), 31–44 (1988). https://doi.org/10.1016/0095-8956(88)90053-6
25. McCormick, S.T., Rao, M., Rinaldi, G.: Easy and difficult objective functions for max cut. Math. Program. **94**, 459–466 (2003). https://doi.org/10.1007/s10107-002-0328-8
26. Neto, J.: On the diameter of cut polytopes. Discret. Math. **339**(5), 1605–1612 (2016). https://doi.org/10.1016/j.disc.2016.01.002
27. Nikolaev, A.V.: On 1-skeleton of the polytope of pyramidal tours with step-backs. Siberian Electr. Math. Rep. **19**, 674–687 (2022). https://doi.org/10.33048/semi.2022.19.056
28. Orlova, G.I., Dorfman, Y.G.: Finding maximum cut in a graph. Eng. Cybern. **10**(3), 502–506 (1972)
29. Padberg, M.: The boolean quadric polytope: some characteristics, facets and relatives. Math. Program. **45**(1), 139–172 (1989). https://doi.org/10.1007/BF01589101
30. Simanchev, R.Y.: On the vertex adjacency in a polytope of connected k-factors. Trudy Inst. Mat. i Mekh. UrO RAN **24**, 235–242 (2018). https://doi.org/10.21538/0134-4889-2018-24-2-235-242
31. de Sousa, S., Haxhimusa, Y., Kropatsch, W.G.: Estimation of distribution algorithm for the max-cut problem. In: Kropatsch, W.G., Artner, N.M., Haxhimusa, Y., Jiang, X. (eds.) GbRPR 2013. LNCS, vol. 7877, pp. 244–253. Springer, Heidelberg (2013). https://doi.org/10.1007/978-3-642-38221-5_26
32. Wan, P.J.: Near-optimal conflict-free channel set assignments for an optical cluster-based hypercube network. J. Comb. Optim. **1**, 179–186 (1997). https://doi.org/10.1023/A:1009759916586

Temporal Bin Packing Problems with Placement Constraints: MIP-Models and Complexity

Pavel Borisovsky[ID], Anton Eremeev[ID], Artem Panin[(✉)][ID], and Maksim Sakhno

Sobolev Institute of Mathematics SB RAS, Novosibirsk, Russia
{pborisovsky,eremeev,sakhno}@ofim.oscsbras.ru, aapanin1988@gmail.com

Abstract. In this paper, we investigate new problem statements, generalizing the Temporal Bin Packing Problem (TBPP) with possible applications in cloud computing. We suppose that items are organized into *batches*. All items in the same batch are placed simultaneously. In cloud computing, items correspond to virtual machines (VMs) and batches correspond to user requests for VM placement. In addition, cloud users can create *placement groups* consisting of VMs united by a single placement constraint named *cluster*: at any moment in time, VMs from the same placement group must be hosted on the same rack of servers. In this paper, we consider servers as one-dimensional bins.

We investigate the computational complexity and inapproximability of different formulations of the TBPP with cluster placement constraint and suggest mixed integer programming models for them.

Keywords: Temporal Bin Packing · Placement groups · Virtual machines · Complexity · Inapproximability · MIP model

1 Introduction

In the classical Bin Packing Problem (BPP) [11], a finite set of items should be placed into the minimum number of identical bins. This problem arises mainly when optimizing container filling and usage. A natural and relevant generalization of BPP is the Temporal Bin Packing problem (TBPP) [2,5,6,15,17], which is directly related to cloud computing where virtual machines (VMs) are placed on servers. In cloud computing, decisions are made online. Each VM or item is assigned a time interval. The left boundary of the interval represents the time when it is placed or packed, and the right boundary represents the time when it is deleted. The objective is to determine the minimum number of servers (bins) required to host VMs throughout the entire planning horizon.

We investigate new problem statements generalizing TBPP and arising primarily in cloud computing. A cloud has a hierarchical structure that basically consists of regions, network zones, racks and servers. Note that in transportation logistics a similar problem is known as a Multi-Level Bin Packing Problem [4]. For simplicity, in this study we consider the two-level hierarchy in which sets of servers (bins) are united into racks. Items are organized into *batches* that correspond to user requests for virtual machine placement. A user request occurs at one time, so all virtual machines in the same request are placed at once. In addition, users can create *placement groups* consisting of virtual machines united by a single *placement rule (constraint)* that reflects the dependency between VMs. For example, certain types of applications such as high performance scientific computing or big data processing have to be run on several VMs and produce a large traffic between them and so require a high network bandwidth (such a dependency is known as *affinity*) [3,13,18]. To reduce the latency, such VMs are considered as a placement group, and at any moment in time, virtual machines from the same placement group must be hosted on the same network domain or even on the same rack. According to [14], this type of constraints is referred to as *cluster*. Note that there are other forms of dependencies, for example if some data has to be replicated in order to provide a fault tolerance then the corresponding VMs must be placed on different network domains, but in this papers, we concentrate only on the cluster constraints. In our problem settings, input consists of batches of items. These items belong to groups that have the cluster type constraint at the rack level. Each rack consists of a given number of bins, see Fig. 1.

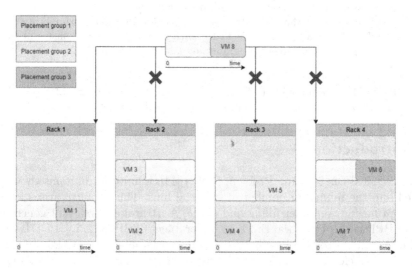

Fig. 1. Illustration of a feasible solution with respect to the cluster placement constraint. Here the lifetime of each VM is shown as an interval on time axis. Let the placement groups be PG1={VM1,VM4}, PG2 = {VM2,VM3,VM5}, and PG3 = {VM6,VM7}. Suppose VM1 and VM8 intersect in time. The crosses mark the racks where VM8 can not be placed due to the cluster placement constraint, because it must be put together with VM1.

A common approach to handle a large number of technical requirements and preferences consists in development a multi-objective model, in which among the minimization of active servers there are such additional criteria as power consumption, resource wastage, network traffic, number of rejections and migrations of virtual machines etc. [10,16,20]. Comprehensive surveys that cover these and other technological and economical issues of virtual machines scheduling can be found in [9,12,19]. In order to keep the models simple and to investigate certain aspects individually, we restrict ourselves to the two VM placement scenarios. In the first one, the set of virtual machines is given and the number of required servers to pack all of them is to be minimized. In the second scenario, there are a fixed number of servers and a sequence of requests, which must be fulfilled without skips until the first placement failure. The goal is to maximize the number of satisfied requests. This formulation can be useful for estimation of the time moment when the datacenter needs to be expanded.

The main purpose of this study is to investigate the complexity of the problem under different settings. Since the basic VM packing problem is a generalization of the classic BPP it is clearly NP-hard. However, there could be particular cases, in which this reasoning may be incorrect. For example, BPP considers items of arbitrary sizes, but the cloud providers usually offer a limited set of VM types, and it is proven in [8] that BPP with the limited number of item types is polynomially solvable. We show that in presence of cluster constraints the problem is NP-hard even for only one VM type. Besides, we consider other practically interesting cases and prove that under certain settings the solutions cannot be approximated with any ratio in polynomial time.

The paper is structured as follows. Section 2 contains detailed description of problems under consideration, together with Mixed Integer Programming (MIP) models for their solution. Section 3 shows different hardness and inapproximability results for formulations from Sect. 2. Concluding remarks are given in Sect. 4.

2   Problem Statements and MIP Formulations

Let a set of items $\mathbf{I} = \{1, \ldots, I\}$ and rack capacity C (a number of bins in one rack) be given. For each item i, the time interval $[s_i, f_i)$ is known. At the time s_i, the item i must be placed into one of the bins and remain there until the time f_i. A set of event moments $\mathbf{T}$ can be defined as the combination of all left endpoints of time intervals. We can assume $\mathbf{T} = \{1, \ldots, T\}$ without loss of generality. Let us denote the weight of item i by w_i and the bin capacity by W.

The items are divided into groups. $\mathbf{G} = \{0, \ldots, G\}$ is a set of groups of items. The group $g = 0$ comprises all items without cluster placement constraints. For each group $g \geq 1$ and each event (or time) moment t, all items of g that exist in this moment must be placed on the same rack.

2.1   Problem Statements of Minimizing the Number of Identical Bins

Consider a problem formulation that requires all items to be placed in a minimum number of identical bins, subject to placement constraints. Let $\mathbf{R} = \{1, \ldots, R\}$ be a sufficient set of racks to allocate all items (e.g., $R = \lceil I/C \rceil$). Each rack contains exactly C bins. Let $\mathbf{B}_r$ be a set of bins on a rack r. In addition, we define the following auxiliary sets:

$\mathbf{I}_t = \{i \in \mathbf{I}|s_i \le t < f_i\}$, where $t \in \mathbf{T}$;

$\mathbf{I}_{gt} = \{i \in \mathbf{I}|s_i \le t < f_i$ and i belongs to $g\}$, where $t \in \mathbf{T}$ and $g \in \mathbf{G}$.

A Boolean variable z_{rb} takes the value 1 if and only if the bin b of the rack r is used to place items. A Boolean variable x_{rbi} is equal to 1 if and only if the item i is in the bin b of the rack r. To ensure that the placement constraints are met, a Boolean variable a_{rgt} is introduced. The variable takes the value 1 if and only if the items of group g that exist at time t are placed in the rack r.

The problem of minimizing the number of bins can be written in terms of mixed integer linear programming as follows.

$$\sum_{r \in \mathbf{R}} \sum_{b \in \mathbf{B}_r} z_{rb} \to \min_{x,z,a} \tag{1}$$

$$\sum_{r \in \mathbf{R}} \sum_{b \in \mathbf{B}_r} x_{rbi} = 1, \quad i \in \mathbf{I} \tag{2}$$

$$\sum_{i \in \mathbf{I}_t} w_i x_{rbi} \le W, \quad r \in \mathbf{R}; b \in \mathbf{B}_r; t \in \mathbf{T} \tag{3}$$

$$\sum_{i \in \mathbf{I}} x_{rbi} \le I \times z_{rb}, \quad r \in \mathbf{R}; b \in \mathbf{B}_r \tag{4}$$

$$\sum_{b \in \mathbf{B}_r} \sum_{i \in \mathbf{I}_{gt}} x_{rbi} = |\mathbf{I}_{gt}| \times a_{rgt}, \quad r \in \mathbf{R}; g \in \mathbf{G} \setminus 0; t \in \mathbf{T} \tag{5}$$

Objective function (1) minimizes the number of bins used. Constraint (2) ensures that all items are placed. Constraint (3) is a limit on the bin capacity. Constraint (4) requires placing items only on used bins. Items from the same group must be on the same rack. This follows from constraint (5).

In cloud computing, it is sometimes required to minimize the number of racks. In this case, it is sufficient to replace the variable z_{rb} with a Boolean variable z_r which takes the value 1 if and only if the rack r is used to place items. Then the objective function (1) and constraint (4) are rewritten as follows:

$$\sum_{r \in \mathbf{R}} z_r \to \min_{x,z,a}$$

$$\sum_{b \in \mathbf{B}_r} \sum_{i \in \mathbf{I}} x_{rbi} \le I \times z_r, \quad r \in \mathbf{R}$$

2.2 Problem Statement of Maximizing the Number of Batches

Unlike the previous problem, the batch maximization problem aggregates items from the same left endpoint of the time interval into batches. Let a set of batches $\mathbf{U} = \{1, \dots, U\}$ be given. In fact, in the problem statement under study, the set of batches is equal to $\mathbf{T}$. A Boolean matrix γ determines whether an item belongs to a batch. The value of γ_{ui} is 1 if and only if the item i belongs to the batch u. The Boolean variable y_u determines whether items from the batch u are packed or not.

The batch maximization problem can be written in terms of mixed integer linear programming as follows.

$$\sum_{u \in \mathbf{U}} y_u \to \max_{x,y,a} \tag{6}$$

$$y_{u-1} \geq y_u, \quad 2 \leq u \leq U \tag{7}$$

$$\gamma_{ui} \times \sum_{r \in \mathbf{R}} \sum_{b \in \mathbf{B}_r} x_{rbi} = y_u, \quad u \in \mathbf{U}; i \in \mathbf{I} \tag{8}$$

$$\sum_{i \in \mathbf{I}_t} w_i x_{rbi} \leq W, \quad r \in \mathbf{R}; b \in \mathbf{B}_r; t \in \mathbf{T} \tag{9}$$

$$\sum_{b \in \mathbf{B}_r} \sum_{i \in \mathbf{I}_{gt}} x_{rbi} \leq |\mathbf{I}_{gt}| \times a_{rgt}, \quad r \in \mathbf{R}; g \in \mathbf{G} \setminus 0; t \in \mathbf{T} \tag{10}$$

Objective function (6) maximizes the number of batches packed. Constraint (7) implies that batches are placed sequentially until the first placement failure. Constraint (8) ensures that only whole batches are packed. It is not possible to place only a part of a batch. Constraints (9) and (10) are similar to constraints (3) and (5), respectively.

2.3 Allocation Rate Maximization

We have described mathematical models of the problems of minimizing the size of the resource pool (number of bins or racks) and maximizing the number of batches. Cloud computing also considers statements in which allocation rate is maximized.

Definition 1 (Allocation rate). *The allocation rate is defined as follows. For a given set of items* $\mathbf{I}$ *and for each time moment* t, *let* TW_t *be the total weight consumed by the items of* $\mathbf{I}$ *and* TW *be the total capacity available in the resource pool. Then the allocation rate is* $f = (\max_t TW_t)/TW$.

In the problem (1)–(5), all items must be placed. Therefore, we can clearly determine the time when the maximum allocation rate is reached:

$$\bar{t} = \arg\max_t \sum_{i \in \mathbf{I}_t} w_i.$$

The total weight $TW_{\bar{t}}$ is equal to $\sum_{i \in \mathbf{I}_{\bar{t}}} w_i$. The statement of maximizing allocation rate is obtained when replacing the objective function (1) by

$$TW_{\bar{t}} \ / \ \sum_{r \in \mathbf{R}} \sum_{b \in \mathbf{B}_r} W \times z_{rb} \underset{x,z,a}{\rightarrow} \max,$$

for the bins minimization case, and by

$$TW_{\bar{t}} \ / \ \sum_{r \in \mathbf{R}} W \times C \times z_r \underset{x,z,a}{\rightarrow} \max,$$

for the racks minimization case

For the problem (6)–(10), the statement of maximizing allocation rate is obtained when replacing the objective function (6) by

$$\sum_{t \in \mathbf{T}} q_t \times \sum_{i \in \mathbf{I}_t} \sum_{r \in \mathbf{R}} \sum_{b \in \mathbf{B}_r} w_i x_{rbi} \ / \ (W \times R \times C) \underset{x,y,a,q}{\rightarrow} \max$$

$$\sum_{t \in \mathbf{T}} q_t = 1,$$

where the Boolean variable q_t defines the event moment at which the maximum load of bins is reached.

3   Computational Complexity

In cloud computing, items are virtual machines. The users cannot define the configuration (type) of the virtual machine themselves. Instead, they have to choose from a set of configurations offered by the cloud provider. In practice, TBPP with placement constraints has a limited variety of item types. In our study, the type of an item is its weight. This paper explores two cases where the number of item types is a part of the problem input and there is only one item type, i.e., all items have the same weight.

3.1   The Number of Item Types as a Part of the Problem Input

The first statement corresponds to the scenario where items must be evenly distributed among bins. This approach is commonly used in partition problems. The setting of the classical partition problem is given below.

Partition problem. [7]
Input: Finite set X of items, for each $x \in X$ a weight $a_x \in Z^+$.
Question: Whether the set X can be partitioned into two subsets X^1 and X^2 such that $\sum_{x \in X^1} a_x = \sum_{x \in X^2} a_x$?

The presence of placement groups allows to use the Partition problem ideology for complexity evaluation. This leads to the following statement.

Theorem 1. *The problem of finding a feasible solution for the problem (1)–(5) is NP-hard even in the case of one placement group, one event moment, and any rack capacity $C \geq 2$.*

Proof. Consider an arbitrary instance of Partition problem. Let us construct an instance of the problem (1)–(5) with one placement group and one event moment as follows. The set of items I is exactly equivalent to the set of items X and the weight of each item is equal to corresponding a_x. The group $g = 1$ consists of all items. The capacity of the bin is equal to half the total weight of all items. The rack contains exactly two bins.

In a feasible solution of the constructed instance of the problem (1)–(5), the placement group must be allocated entirely to a single rack. Therefore, the set of items must be distributed into two bins, which corresponds to a positive answer in the NP-complete Partition problem.   □

Corollary 1. *The problems of finding a feasible solution for the problems (1)–(5) with the rack minimization criterion and the allocation rate maximization criterion are NP-hard even in the case of one placement group and one event moment.*

Let us turn to the hardness of finding approximate solutions with guaranteed approximation ratio ρ, which is also called *the performance ratio* [1]. Let F_x^* be the optimal value of the objective function for input x (initial data) and $F_x(y)$ be the value of the objective function for solution y obtained by an approximate algorithm.

Definition 2. *An approximate polynomial-time algorithm for a minimization problem is called a ρ-approximation algorithm if it finds a solution y such that $F_x(y) \leq \rho F_x^*$ for any input x.*

Definition 3. *An approximate polynomial-time algorithm for a maximization problem is called a ρ-approximation algorithm if it finds a solution y such that $F_x(y) \geq F_x^*/\rho$ for any input x.*

Implicitly the performance ratio ρ may be a function of x and we can write $\rho(x)$ to emphasize this. So, for an arbitrary minimization problem with input x, the inequality $F_x^* \geq \frac{F_x(y)}{\rho(x)}$ holds. In the case of maximization, we have $F_x^* \leq \rho(x)F_x(y)$.

Corollary 2. *For the problem (1)–(5) with criteria of minimizing the number of racks or bins and maximizing the allocation rate, there are no polynomial-time approximation algorithms with any performance ratio, unless $P = NP$.*

Corollary 2 follows from the fact that the approximation algorithm constructs a feasible solution in polynomial time. Therefore, for the problem (1)–(5), there are no algorithms satisfying Definition 2 or 3. The obtained results show that the presence of placement constraints strongly complicates the classical Bin Packing-type formulations. For the BPP, polynomial-time approximation algorithms with performance ratio $3/2$ exist [1].

Theorem 2. *For the problem (6)–(10), there is no ρ-approximation algorithm for any constant ρ unless $P = NP$.*

Proof. Consider a decision problem P, in which there is a single time moment and one batch consisting of n items is to be allocated on the given set of bins of one large rack. No placement constraints are assumed. It is required to decide whether all of the n items can be packed or not. This problem is NP-complete since it is the corresponding decision problem of the BPP.

Assume for simplicity that ρ is integer. Consider the problem Q to maximize the number of batches that represents ρ copies of problem P. Namely, the first copy of the batch occupies only the time moment $t = 1$, similarly the second copy of the batch exists in the time moment $t = 2$, and so on up to the moment $t = \rho$. Since ρ is a constant, the size of Q is bounded by a polynomial in n.

Suppose there is a ρ-approximation algorithm for Q. If an instance of the decision problem P has a positive answer, then all the batches of problem Q can be packed, i.e. the optimal objective value of problem Q is $F^* = \rho$. In this case, the approximation algorithm will find a solution with $F \geq \frac{F^*}{\rho} = 1$, i.e. it will pack the first batch and therefore solve P. On the other hand, if an instance of problem P has the negative answer, then necessarily $F = 0$.

This allows to correctly decide the problem P in polynomial time, which is impossible unless $P = NP$.   □

Now let us consider the optimization problem, where the solution quality is the maximum allocation rate of packed batches until the first placement failure.

Naturally, for a given solution, the allocation rate is calculated over the batches that are packed in this solution. Note that the optimal solution to the problem that maximizes the number of placed batches until the first failure is an optimal solution to the problem of allocation rate maximization, but not vice versa.

Denote the optimal allocation rate by f^*.

Theorem 3. *For the problem (6)–(10) with allocation rate maximization criterion, there is no R-approximation algorithm for any constant number of racks R and any rack capacity $C \geq 2$, unless $P = NP$.*

Proof. Consider a decision problem P, in which there is a single time moment and a set of n items to be allocated on the given set of bins of one rack. Only one cluster placement group with all items is given. It is required to decide whether all of the n items can be packed or not. This problem is NP-hard even if the total demand of all n items equals to the total resource of a rack (i.e. the allocation rate is equal to 1) because Partition problem reduces to it as a special case (see Theorem 1). Let us limit ourselves to considering only such instances of the problem P.

Consider an optimization problem Q that represents R copies of problem P combined into R batches with different lifetime and different cluster groups, assuming that the resource pool consists of R identical racks as in the problem P. The first copy of items from P make up the first batch and the first group, the second copy of items (indexed from $n+1$ to $2n$) makes up the second batch and group, and so on, up to the batch R and group number R with item indices $n(R-1)+1,\ldots,nR$. Items of batch 1 exist at time period from 1 to R, items of batch 2 exist at time period from 2 to R, etc. Items of batch R exist only at time moment R. Since R is a constant, the size of Q is bounded by a polynomial in n.

Suppose there is a R-approximation algorithm for Q. If an instance of the decision problem P has a positive answer, then all the items of problem Q can be packed, i.e. the allocation rate of problem Q is $f^* = 1$. In this case, the R-approximation algorithm will find a solution with an allocation rate $f \geq \frac{f^*}{R} = \frac{1}{R}$, i.e. it will pack at least the first batch. On the other hand, if an instance of problem P has the negative answer, then necessarily $f < \frac{1}{R}$.

This allows to correctly decide the problem P in polynomial time, which is impossible unless P = NP.   □

3.2   The Single Item Type Case

Theorems 1–3 above characterize the computational complexity of the problems under study in the case of unlimited number of item types. The following statement describes the complexity in the case of a single type (and also applies to the cases with any number of types upper-bounded by a constant).

The proof of the following statement relies on a decision version of Graph Coloring Problem.

K-coloring problem. [7]
Input: Graph $G = (V,E)$ and integer $K \leq |V|$.
A coloring of G into K colors is a partition of V into K independent sets in G.
Question: Can we color G with no more than K colors?

Theorem 4. *The problem (6)–(10) is NP-hard, even in the case of one item type, any fixed number of racks $R \geq 3$, any fixed rack capacity.*

Proof. We show that the K-coloring problem with any fixed number of colors K reduces in polynomial time to the problem (6)–(10) with the number of racks $R = K$ and one bin per rack.

Let us match placement groups subject to the cluster constraint with the vertices of the given graph $G = (V,E), |V| = n, |E| = m$. In our reduction, we will match utilized racks with colors in the K-coloring problem. Placing items from a placement group on a rack will indicate coloring of a vertex that corresponds to this group.

Let the bin capacity be $2n$ and the planning horizon **T** be $\{1,\ldots,m+1\}$. For each placement group corresponding to vertex i $(i = 1,\ldots,n)$, let's add an item with lifetime from 1 to $m+1$ and weight equal to 1. This way there will be at least one item in each placement group. Since this item exists for the entire time period, the other items in this placement group cannot be packed into the other racks. Given the weights of these items, they can be packed into any bins, even may all be packed into the same bin. For each edge $e_\ell = (i, j), \ell = 1, \ldots, m$, let us add to the placement groups i, j, corresponding to the incident vertices, n items of the same type as described above and set their start time to ℓ and the end time to $\ell + 1$. Assume that each item makes up a separate batch, so the total number of batches is $U = nm + n$.

Figure 2 illustrates the reduction. For example, edge $e_1 = (V1, V2)$ corresponds to four VMs in PG 1 and four VMs in PG 2. Since $l = 1$, they have the time interval $[1,2)$, i.e. they are active at time moment $t = 1$.

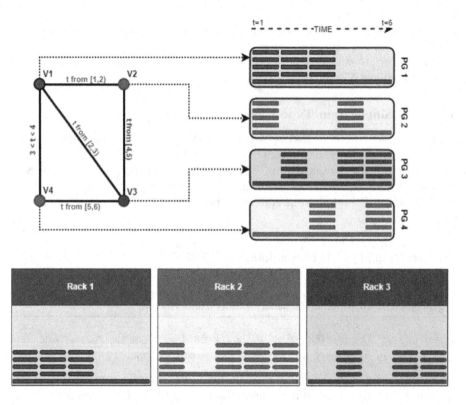

Fig. 2. Illustration of a comparison of instances and solutions of the graph coloring problem and the problems under study in the case of one type of items and unit capacity of racks. A graph with four vertices and five edges is colored in three colors. Each vertex is assigned a placement group. Each color corresponds to a rack.

Note that for any edge $e_\ell = (i, j)$, items from different groups i, j cannot be packed simultaneously on the same rack because together they require $2n + 2$

resource units, while the rack capacity is only $2n$. On the other hand, if two vertices v_i, v_j of the graph can be colored in the same color then groups i, j can be packed into the same rack, because groups i, j have no items living at the same time except for long-lived items, one in each group, since there are no edges between the vertices v_i, v_j in the graph.

The K-coloring problem corresponds one-to-one to the feasible solutions of the constructed instance with $R = K$ racks, and the number of the allocated batches in such solutions is equal to the total number of batches U, i.e. the objective value is equal to U in these solutions. Thus, answering the question: "Is the optimum to the problem (6)–(10) equal to at least U batches?" allows to solve the K-coloring problem. So the latter problem reduces to the problem (6)–(10) with $R = K$ in polynomial time.

Since the reduction yields only instances in which numerical parameters are at most polynomial in n and the K-coloring problem is NP-complete [7] for any fixed number of colors $K \geq 3$, the considered problem is NP-hard.   □

Corollary 3. *The problem (1)–(5) is NP-hard with all optimization criteria considered in the paper, even in the case of one item type, any maximum number of racks $R \geq 3$.*

Proof. Placing items in no more than K racks corresponds to a positive answer in the K-coloring problem.   □

Corollary 4. *For the problem (6)–(10) even in the case of one item type, any maximum number of racks $R \geq 3$, in the presence of cluster constraints, there is no ρ-approximation algorithm for any constant ρ unless $P = NP$.*

Proof. It is enough to assume that the problem P in the proof of Theorem 2 has cluster constraints and a given number of time moments T (and so it is NP-complete), the problem Q is constructed as ρ copies of P with ρT moments, then the rest of the proof can be reproduced.   □

4   Conclusions

In this paper, we have shown that cluster placement constraint significantly complicates the Temporal Bin Packing problem. These are the theoretical results, but our preliminary experiments with modern MIP solvers also indicate that TBPP with placement constraints requires much more CPU time and memory. We expect that future research will provide a detailed experimental analysis of the behaviour of MIP models and algorithms on such problems. Also, we hope that certain special cases will be identified, where TBPP with placement constraints admits constant-factor approximation algorithms.

Acknowledgement. The research was carried out in accordance with the state task of the IM SB RAS (projects FWNF-2022-0020 and FWNF-2022-0019).

References

1. Ausiello, G., Crescenzi, P., Gambosi, G., et al.: Complexity and Approximation: Combinatorial Optimization Problems and Their Approximability Properties. Springer-Verlag, Berlin (1999)
2. de Cauwer, M., Mehta, D., O'Sullivan, B.: The temporal bin packing problem: an application to workload management in data centres. In: 2016 IEEE 28th International Conference on Tools with Artificial Intelligence (ICTAI), San Jose, CA, pp. 157–164. IEEE (2016)
3. Chen, J., et al.: Joint affinity aware grouping and virtual machine placement. Microprocess. Microsyst. **52**, 365–380 (2017)
4. Chen, L., Tong, X., Yuan, M., Zeng, J., Chen, L: A data-driven approach for multi-level packing problems in manufacturing industry. In: 25th ACM SIGKDD International Conference on Knowledge Discovery & Data Mining (KDD 2019), Anchorage, USA, pp. 1762–1770. ACM (2019)
5. Dell'Amico, M., Furini, F., Iori, M.: A branch-and-price algorithm for the temporal bin packing problem. Comput. Operat. Res. **114**, 104825 (2020)
6. Furini, F., Shen, X.: Matheuristics for the temporal bin packing problem. In: Amodeo, L., Talbi, E.-G., Yalaoui, F. (eds.) Recent Developments in Metaheuristics. ORSIS, vol. 62, pp. 333–345. Springer, Cham (2018). https://doi.org/10.1007/978-3-319-58253-5_19
7. Garey, M.R., Johnson, D.S.: Computers and Intractability: A Guide to the Theory of NP-Completeness. W. H. Freeman & Co., San Francisco, Calif (1990)
8. Goemans, M.X., Rothvoss, T.: Polynomiality for bin packing with a constant number of item types. J. ACM **67**(6), 1–21 (2020)
9. Grushin, D.A., Kuzyurin, N.N.: On effective scheduling in computing clusters. Program. Comput. Softw. **45**(7), 398–404 (2019)
10. Guo, X.: Multi-objective task scheduling optimization in cloud computing based on fuzzy self-defense algorithm. Alex. Eng. J. **60**(6), 5603–5609 (2021)
11. Johnson, D. Near-optimal bin packing algorithms. Ph. D. Thesis, Dept. of Mathematics, M.I.T., Cambridge, MA (1973)
12. Mann, Z.: Allocation of virtual machines in cloud data centers - a survey of problem models and optimization algorithms. ACM Comput. Surv. **48**(1), 1–34 (2015)
13. Pachorkar, N., Ingle, R.: Multi-dimensional affinity aware VM placement algorithm in cloud computing. Inter. J. Adv. Comput. Res. **3**(4), 121–125 (2013)
14. Placement groups. (n.d.). Amazon Elastic Compute Cloud. https://docs.aws.amazon.com/AWSEC2/latest/UserGuide/placement-groups.html
15. Ratushnyi, A., Kochetov, Y.: A column generation based heuristic for a temporal bin packing problem. In: Pardalos, P., Khachay, M., Kazakov, A. (eds.) MOTOR 2021. LNCS, vol. 12755, pp. 96–110. Springer, Cham (2021). https://doi.org/10.1007/978-3-030-77876-7_7
16. Regaieg, R., Koubàa, M., Osei-Opoku, E., Aguili, T.: Multi-objective mixed integer linear programming model for VM placement to minimize resource wastage in a heterogeneous cloud provider data center. In: 2018 Tenth International Conference on Ubiquitous and Future Networks (ICUFN), Prague, Czech Republic, pp. 401–406. IEEE (2018)
17. Sakhno, M.: A grouping genetic algorithm for the temporal vector bin packing problem. In: 19th International Asian School-Seminar on Optimization Problems of Complex Systems (OPCS), pp. 94–99, IEEE (2023)

18. Su, K., Xu, L., Chen,C., Chen W., Wang,Z.: Affinity and Conflict-Aware Placement of Virtual Machines in Heterogeneous Data Centers. In: 2015 IEEE Twelfth International Symposium on Autonomous Decentralized Systems, Taichung, Taiwan, pp. 289–294. IEEE (2015)
19. Talebian, H., Gani, A., Sookhak, M., et al.: Optimizing virtual machine placement in IaaS data centers: taxonomy, review and open issues. Cluster Comput. **23**, 837–878 (2020)
20. Zheng, Q., et al.: Virtual machine consolidated placement based on multi-objective biogeography-based optimization. Futur. Gener. Comput. Syst. **54**, 95–122 (2016)

Branching Algorithms for the Reliable Production Process Design Problem

Roman Rudakov[1,2] , Yuri Ogorodnikov[2(✉)] , and Michael Khachay[2]

[1] Ural Federal University, Ekaterinburg, Russia
[2] Krasovsky Institute of Mathematics and Mechanics, Ekaterinburg, Russia
{yogorodnikov,mkhachay}@imm.uran.ru

Abstract. In the well-known Subgraph Homeomorphism Problem (SHP), it is required to make a homeomorphic embedding of some pattern digraph Π into the given target digraph G. Such an embedding is performed by some one-to-one map f defined on the node set $V(\Pi)$ such that, for each arc (v, u) of Π, there exists an elementary $f(v)$-$f(u)$-path in G, and all such paths are vertex-disjoint. In this paper we consider the proposed recently Reliable Production Process Design Problem (RPPDP), which generalizes the SHP in the following way:

(i) graph Π is supposed to be acyclic, while G is edge- and node-weighted, which assigns costs to the considered embeddings;

(ii) the mutual uncrossing constraint for $f(v)$-$f(u)$-paths is slightly relaxed;

(iii) for a given number k, we are requested to find k vertex-disjoint homeomorphic images of the pattern Π, such that the largest cost of the obtained images has the minimum value.

The RPPDP has applications in production management planning, where the decision maker is aimed to propose a family of plans tolerant to possible faults of manufacturing units and supply chains. We propose the first branch-and-bound and branch-and-price algorithms for the RPPDP. The results of numerical experiments demonstrate high performance and mutual complementarity of the proposed algorithms.

Keywords: Subgraph Homeomorphism Problem · reliable supply chains · branch-and-price

1 Introduction

The Subgraph Homeomorphism Problem (SHP) is the well-known combinatorial optimization problem, where, for a given pattern digraph Π and target digraph G, it is required to find a one-to-one embedding f defined on the node set $V(\Pi)$ that maps an arbitrary arc (v, u) from Π to an elementary $f(v)$-$f(u)$-path in G, such that all these paths are vertex-disjoint. As it was shown in [9], some special cases of the SHP induced by specific patterns Π can be solved in polynomial time, while in general this problem is NP-complete.

A. Eremeev et al. (Eds.): MOTOR 2024, LNCS 14766, pp. 170–186, 2024.
https://doi.org/10.1007/978-3-031-62792-7_12

We consider a generalization of the SHP known as Reliable Production Process Design Problem (RPPDP) introduced in recent paper [23]. In this paper, we follow the similar motivation based on prospective importance of the RPPDP in practical applications of operations research. In particular, optimal and even close to optimal solutions of the RPPDP can make design of distributed production systems more tolerant to possible faults. It is known, that there can be some troubles regarding insufficient reliability of supply chains [26]. Traditional approach to overcome such issues employs construction of special stochastic models [5], which perform well in the case of prior information about possible faults. In the same time, this approach becomes inefficient in the case of unexpected faults occurrence. In this context, minmax deterministic models that give guaranteed results, one of them is considered in this paper, seem to be more reliable.

The RPPDP is aimed to find a minimum cost fault-tolerant design of an abstract distributed production process. By *production process* we refer to a set of m manufacturing operations ordered by a given precedence constraints. Each operation can be done in one of *manufacturing units* specialized to perform this single operation. Combining these units, which assumed to be mutually interchangeble, we obtain m *production clusters*. According to the given precedence (partial) order, commodity produced in cluster-predecessor should be transmitted to the cluster-successor for the further production.

Production process is convenient to describe in terms of a transportation network, a directed graph whose node set is partitioned onto manufacturing clusters and a set of *transportation hubs* whose aim are to become a transfer station for sending production products between clusters. In addition, it contains two artificial nodes s and t corresponding to the start and finish of the production process. Each hub has two attributes: its *opening cost* that is paid when hub is included to the production process, and transportation *capacity* bound for the number of routes serviced byt this hub. The transportation costs are modeled by weights assigned to the arcs of the given graph.

Further, by a *production plan* we call a subgraph of the transportation network that contains nodes s and t, a single representative for each manufacturing cluster, and directed routes connecting these nodes with respect to the given precedence order and hub capacities. For each route, its cost is defined by a sum of weights for all the incident arcs and opening costs of all the visited hubs. In turn, *cost* of a production plan is equal to the total cost of all its routes. By definition, the cost of any family of production plans coincides with the maximum cost for the contained plans.

Similarly to in the classic SHP, each production plan is a homeomorphic image of some pattern graph Π. The main difference between the problems is as follows. Unlike the SHP, in the RPPDP setting we should construct $k \geq 2$ vertex-disjoint homeomorphic images of the graph Π. In addition, the pattern digraph Π is assumed to be acyclic, the target digraph G should be edge- and node-weighted, and $f(v)$-$f(u)$-paths can cross each other by some more general way. Finally, the RPPDP has a slightly more general optimization criterion.

2   Related Work

To the best of our knowledge, the SHP appears to be the mostly closest to the RPPDP. In a special case of the SHP, the problem of finding k node-disjoint paths (k-DP) [9], it is required to construct a set of vertex-disjoint s_i-t_i-paths for a given set of node pairs $(s_i, t_i), i = \overline{1, k}$. As it was shown in [4], k-DP is NP-complete for an arbitrary fixed $k \geq 2$.

Another similar to the RPPDP problem is Constrained Shortest Path Tour Problem (CSPTP) [7]. An arbitrary instance of the CSPTP is given by edge- and node-weighted digraph $G = (V \cup \{s, t\}, A)$, an ordered set of mutually disjoint node subsets $\mathfrak{T} = \{T_1, \ldots, T_k\}$, and weight functions $c, q \colon A \to \mathbb{R}^+$ defining for each arc $a \in A$ its transportation cost and capacity. The goal is to find the shortest s-t-route that passes some hitting set of the family $\mathfrak{T}$ accoding to the given linear order. The CSPTP is NP-hard [7] and has several exact and heuristic algorithms. Among them are efficient Greedy Randomized Adaptive Search (GRASP) based algorithm, several versions of branch-and-bound techniques [6], and a recent branch-and-price algorithm based on improved Mixed Integer Linear Programming (MILP) model [20],

The Shortest Simple Path Problem with k Must-Pass Nodes (SSPP-k-MPN) [24] is another combinatorial problem close to the RPPDP. In the SSPP-k-MPN, we are given by an edge- and node-weighted digraph $G = (V, E, c)$, a subset $T \subset V, |T| = k$, and dedicated departure and arrival nodes s and t. It is required to find the shortest elementary s-t-path that visits each node from T exactly once in an arbitrary order. Paper [1] investigated several MILP models for the problem. Further, the authors of [19] shown that the problem is strongly NP-hard for any fixed $k \geq 1$ and proposed the first problem-specific branch-and-bound algorithm [19].

Finally, the well-known Precedence Constrained Generalized Traveling Salesman Problem (PCGTSP) [13] has a lot of common with the RPPDP considered in this paper. An instance of the PCGTSP is given by a edge- and node-weighted digraph G augmented with a partition $\{C_1, \ldots, C_m\}$ of its node set into the mutually disjoint clusters. Each cluster must be visited by an arbitrary feasible tour in a single node according to the partial order specified by an auxiliary acyclic graph Π. The goal is to find the shortest tour starting and finishing in the initial cluster C_1. The PCGTSP problem is strongly NP-hard [21] in general case and polynomially solvable, if $m = O(\log n)$ [18]. There are known efficient approximation algorithms developed for several specific precedence orders [3,15,17], a general PCGLNS solver [16] based on the well-known Adaptive Large Neighborhood Search (ALNS) meta heuristic, dynamic programming and branch-and-bound techniques [18,25]. Evolving the seminal results of [2,11], and [10] leads to the state-of-the-art branch-and-cut algorithm for the PCGTSP [14], which managed to significantly increase the progress in solving instances of the public PCGTSPLIB library [25].

As it follows from the presented overview, the existing techniques allow to find efficiently shortest paths or tours in transportation networks with respect to various additional constraints. However, none of these aforementioned combinatorial optimization problems can be used to model fault tolerant production processes directly.

On the other hand, the RPPDP introduced in [23] is designed intentionally to model the requested reliability. In that initial work, along with the statement of the RPPDP, the authors proved its strongly NP-hardness and proposed the first compact MILP model and simple ALNS-based heuristic. Exploratory numerical experiments performed in that work showed that out-of-the-box Gurobi MIP-solver equipped with the mentioned model and heuristic managed to solve small instances of RPPDP, in quite a reasonable time, within very small gaps or even to optimality. Nevertheless, these first proof-of-concept results still need further extension, both in algorithmic and experimental directions, which was the main motivation of this paper.

Our contribution is three-fold. In this paper we

(i) improve our branch-and-bound and propose a novel brach-and-price algorithm to tackle instances of moderate and large size, respectively
(ii) propose a novel library of syntetic RPPDP instances derived from the public PCGTSPLIB test library
(iii) perform much more extended numerical evaluation of both algorithms, which shows their high performance and mutual complementarity.

The rest of the paper is organized as follows. In Sect. 3, we provide formal description of the RPPDP. Section 4 presents the compact MILP model, which our improved branch-and-bound method is relied on. Section 5 introduces a route packing formulation of the RPPDP and describes the novel branch-and-price algorithm. In Sect. 6, we report on numerical evaluation results for the both algorithms. Finally, in Sect. 7, we conclude the paper and discuss some open questions.

3   Problem Statement

An instance of the RPPDP is specified by a triple (G, Π, k), where

(i) $G = (V, E, c, q, C)$ is an edge- and node-weighted digraph supplemented with a partition $V = \mathfrak{M}_0 \cup \ldots \cup \mathfrak{M}_{m+1} \cup \mathfrak{H}$, such that
  – each cluster subset $\mathfrak{M}_j$, $j = \overline{1, m}$, consists of nodes that model manufacturing units performing the same operation
  – clusters $\mathfrak{M}_0 = \{s\}$ and $\mathfrak{M}_{m+1} = \{t\}$ consists of artificial nodes s and t specifying start and termination states of an arbitrary production plan
  – all the other nodes (from the subset $\mathfrak{H}$) are called transportation hubs
  – the weighting function $c \colon E \to \mathbb{R}_+$ assigns to each arc $e \in E$ the corresponding transportation cost, while functions $q, C \colon \mathfrak{H} \to \mathbb{R}_+$ define capacity and opening costs of the hubs, respectively;
(ii) $\Pi = (\{0, 1, \ldots, m+1\}, A)$ is an auxiliary acyclic digraph that encodes the precedence constraints defined on the clusters $\mathfrak{M}_j$, where 0 and $m+1$ are assumed to be the minimum and maximum elements, respectively;

(iii) $k \geq 2$ is the number of production plans to be constructed.

The goal is to construct a k-element family of the node-disjoint production plans $\mathfrak{P} = \{P_1, P_2, \ldots, P_k\}$, such that

(i) each plan P_r is a digraph homeomorphic to Π such that each node i of the digraph Π is represented by some node $v \in \mathfrak{M}_i$, while each arc $(i, j) \in A$ is replaced by a route $v_i, h_1, \ldots, h_p, v_j$ all whose interior nodes are transportation hubs visited with respect to their capacities;

(ii) $\mathfrak{P} = \arg\min\{\max\{\text{cost}(P_r) \colon r = \overline{1, k}\}\}$, where

$$\text{cost}(P_r) = \sum_{e \in P_r} c_e + \sum_{h \in P_r} C_h.$$

A simple example of the RPPDP is presented on Fig. 1. For the sake of brevity, we skip transportation costs.

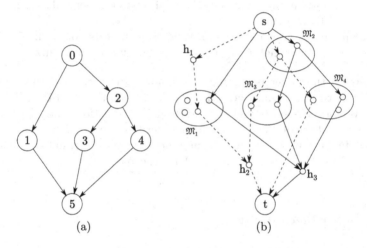

$$\text{(a)} \qquad\qquad\qquad\qquad \text{(b)}$$

Fig. 1. A simple RPPDP instance for $m = 4$ and $k = 2$. (1a) order graph Π (1b) corresponding graph G. Solid and dashed lines specify first and second plans of some feasible solution, respectively.

4   Compact Formulation for RPPDP

We start with the MILP model for the RPPDP introduced in [23]. The model operates with plan-to-arcs incidence indicators $x_e^{r,a} \in \{0, 1\}$, ($r = \overline{1, k}$, $a \in A$, $e \in E$), plan-to-node indicators $y_v^r \in \{0, 1\}$, ($r = \overline{1, k}$, $v \in V$), and real-valued variable $cost$. An indicator $x_e^{r,a} = 1$ if and only if the arc $e \in E$ is included into the route of the plan r encoding the arc $a \in A$. Similarly, $y_v^r = 1$ if and only if the node $v \in V$ is visited by some route of the plan r. Thus, we obtain the following MILP model with at most polynomial number of variables and constraints, with respect of size of the initial problem formulation. We call this model compact.

$$\min \quad cost \tag{1}$$

$$\text{s.t.} \quad \sum_{a \in A, e \in E} c_e x_e^{r,a} + \sum_{h \in \mathfrak{H}} C_h y_h^r - cost \leq 0 \qquad \left(r = \overline{1,k} \right) \tag{2}$$

$$\sum_{u \in \mathfrak{M}_j \cup \mathfrak{H} : \; (v,u) \in E} x_{(v,u)}^{r,(i,j)} - y_v^r = 0 \qquad \begin{pmatrix} r = \overline{1,k}, \\ (i,j) \in A, \\ v \in \mathfrak{M}_i \end{pmatrix} \tag{3}$$

$$\sum_{v \in \mathfrak{M}_i \cup \mathfrak{H} : \; (v,u) \in E} x_{(v,u)}^{r,(i,j)} - y_u^r = 0 \qquad \begin{pmatrix} r = \overline{1,k}, \\ (i,j) \in A, \\ u \in \mathfrak{M}_j \end{pmatrix} \tag{4}$$

$$\sum_{v \in \mathfrak{M} \cup \mathfrak{H} : \; (v,u) \in E} x_{(v,u)}^{r,(i,j)} = 0 \qquad \begin{pmatrix} r = \overline{1,k}, \\ (i,j) \in A, \\ u \in \mathfrak{M} \setminus \mathfrak{M}_j \end{pmatrix} \tag{5}$$

$$\sum_{u \in \mathfrak{M} \cup \mathfrak{H} : \; (v,u) \in E} x_{(v,u)}^{r,(i,j)} = 0 \qquad \begin{pmatrix} r = \overline{1,k}, \\ (i,j) \in A, \\ v \in \mathfrak{M} \setminus \mathfrak{M}_i \end{pmatrix} \tag{6}$$

$$\sum_{v \in \mathfrak{M}_i \cup \mathfrak{H} : \; (v,h) \in E} x_{(v,h)}^{r,(i,j)} - \sum_{u \in \mathfrak{M}_j \cup \mathfrak{H} : \; (h,u) \in E} x_{(h,u)}^{r,(i,j)} = 0 \qquad \begin{pmatrix} r = \overline{1,k}, \\ (i,j) \in A \\ h \in \mathfrak{H} \end{pmatrix} \tag{7}$$

$$\sum_{(i,j \in A)} \sum_{v \in \mathfrak{M}_i \cup \mathfrak{H} : \; (v,h) \in E} x_{(v,h)}^{r,(i,j)} - q_h y_h^r \leq 0 \qquad \begin{pmatrix} r = \overline{1,k} \\ h \in \mathfrak{H} \end{pmatrix} \tag{8}$$

$$\sum_{v \in \mathfrak{M}_i} y_v^r = 1 \qquad \begin{pmatrix} r = \overline{1,k} \\ i = \overline{1,m} \end{pmatrix} \tag{9}$$

$$\sum_{r=1}^{k} y_v^r \leq 1 \qquad \left(v \in V \setminus \{s,t\} \right) \tag{10}$$

$$x_e^{r,a} \in \{0,1\}, \; y_v^r \in \{0,1\} \tag{11}$$

Here, equations (1) and (2) form the objective that is to minimize the maximum cost among the k desired production plans. Equations (3) and (4) set the start and destination nodes for any route assigned to any arc (i,j) of the digraph Π and each plan r. Equations (5) and (6) guarantee that each interior point of any such route is a transportation hub. The connectivity of the constructed digraph is ensured by flow conservation constraint (7). Equation (8) restricts an ingoing edges for any transportation hub according to its capacity bound. Finally, equations (9) and (10) make sure that an arbitrary cluster $\mathfrak{M}_i$ is visited by any plan r exactly once and all the plans are vertex-disjoint, respectively.

Model (1)–(11) is compact and can be easily equipped into any modern MIP solver. In paper [23], we proposed a simple branch-and-bound algorithm relying on this model implemented over Gurobi user callback framework employing the proposed ALNS-based heuristic as the only primal heuristic. The first exploratory experiments performed in [23] show that this algorithm outperforms Gurobi augmented with default settings. In Sect. 6, we use an improved version of this algorithm (we call it A_1) as a main baseline.

5   Branch-and-Price Algorithm for the RPPDP

As we have found in numerical experiments, branch-and-bound algorithm based on the model discussed in Sect. 4 performs well on instances of small and moderate size. But its performance degrades fast as the size of instances increases. To tackle large instances, we introduce another model based on a route packing and develop the appropriate branch-an-price algorithm.

5.1   Route Formulation Model

As it mentioned in Sect. 3, an arbitrary route p implementing an arc $(i, j) \in A$ departs at some node of cluster $\mathfrak{M}_i$, optionally passes through a number of transportation hubs respecting their capacity and arrives to some node of cluster $\mathfrak{M}_j$. We define indicator variables $\lambda_p^r \in \{0, 1\}$ such that $\lambda_p^r = 1$ if route p is included to the plan r and $\lambda_p^r = 0$ otherwise. In addition, we use plan-to-node indicators y_v^r defined in Sect. 4. Finally, to speed-up branching procedure, we introduce real-valued variables $cost_r = \mathrm{cost}(P_r)$.

$$\min \ cost_1 \tag{12}$$

$$\sum_{p \in \mathcal{P}} \sum_{e \in E} E_{e,p} c_e \cdot \lambda_p^r + \sum_{h \in \mathfrak{H}} C_h y_h^r - cost_r \leq 0, \qquad (r = \overline{1,k}) \tag{13}$$

$$\sum_{p \in \mathcal{P}_{(i,j),v}^+} \lambda_p^r - y_v^r = 0, \qquad \begin{pmatrix} r = \overline{1,k} \\ (i,j) \in A \\ v \in \mathfrak{M}_i \end{pmatrix} \tag{14}$$

$$\sum_{p \in \mathcal{P}_{(i,j),v}^-} \lambda_p^r - y_v^r = 0, \qquad \begin{pmatrix} r = \overline{1,k} \\ (i,j) \in A \\ v \in \mathfrak{M}_j \end{pmatrix} \tag{15}$$

$$\sum_{p \in \mathcal{P}} B_{v,p} \lambda_p^r - q_v y_v^r \leq 0, \qquad \begin{pmatrix} r = \overline{1,k} \\ v \in \mathfrak{H} \end{pmatrix} \tag{16}$$

$$\sum_{p \in \mathcal{P}_{(i,j)}} B_{v,p} \lambda_p^r - y_v^r \leq 0, \qquad \begin{pmatrix} r = \overline{1,k} \\ (i,j) \in A \\ v \in \mathfrak{H} \end{pmatrix} \tag{17}$$

$$\sum_{p \in \mathcal{P}_{(i,j)}} \lambda_p^r = 1, \quad \begin{pmatrix} r = \overline{1,k} \\ (i,j) \in A \end{pmatrix} \tag{18}$$

$$\sum_{r=1}^{k} y_v^r \leq 1, \quad (v \in V \setminus \{s,t\}) \tag{19}$$

$$\sum_{v \in \mathfrak{M}_i} y_v^r = 1, \quad \begin{pmatrix} r = \overline{1,k} \\ i = \overline{1,m} \end{pmatrix} \tag{20}$$

$$cost_r \geq cost_{r+1}, \quad (r = \overline{1,k-1}) \tag{21}$$

$$\lambda_p^r \in \{0,1\}, \ y_v^r \in \{0,1\}. \tag{22}$$

Equation (12) defines an objective function similarly to the model (1)–(11). For each plan, its cost is defined by equation (13), where $E_{e,p}$ is an incidence matrix between the arcs $e \in E$ and all the possible routes p. Equations (14)–(15) ensure the connectivity of the subgraph to be constructed. Here $\mathcal{P}_{(i,j),u}^+$ and $\mathcal{P}_{(i,j),v}^-$ are sets of routes, each of them implements the arc $(i,j) \in A$ and departs from and arrives to the node $u \in \mathfrak{M}_i$ and $v \in \mathfrak{M}_j$, respectively. Equation (16) provides the capacity constraint for each transportation hub, while equation (17) establishes the incidence between any hub and the routes implementing an arbitrary arc $(i,j) \in A$ where $B_{v,p}$ is an indicator denoting whether node v belongs to path p or not. Equation (18) makes sure that any arc $(i,j) \in A$ is implemented by one route of an arbitrary plan r exactly. Equation (19) guarantees that all the plans are vertex-disjoint, while equation (20) ensures that each cluster is visited by an arbitrary plan exactly once. Finally, equation (21) is included to the model to reduce possible symmetry.

The proposed model (12)–(22) induces a master problem in the column generation scheme of our branch-and-price algorithm.

5.2 Column Generation and Pricing Problem

We proceed with the LP-relaxation of model (12)–(22) that can be obtained by replacing (22) by

$$\lambda_p^r \geq 0 \quad (p \in \mathcal{P}, \ r = \overline{1,k}) \tag{23}$$

$$y_v^r \geq 0 \quad (v \in V, \ r = \overline{1,k}) \tag{24}$$

Since this LP-relaxation relies on exponential number of variables, to obtain its optimal solutions, we apply the well-known column generation framework. Evidently, each column (variable) in the primal linear program corresponds to the constraint in the dual one. By tradition, any LP-relaxation induced by a subset $\mathcal{P}' \subset \mathcal{P}$ of columns is called a *Restricted Master Problem* (RMP). To ensure the feasibility of RMP, we initialize it with a set of 'artificial' routes, each of them is penalized in the objective function. The new columns are generated iteratively by the following scheme.

Denote by $\alpha_r, \beta_{v,(i,j)}^r, \gamma_{v,(i,j)}^r, \xi_v^r, \sigma_{(i,j),v}^r, \delta_{(i,j)}^r, \eta_v, \mu_i^r, \psi^r$ the dual variables associated with equations (13)–(21), respectively. Suppose, we managed to find optimal solutions $(\bar\lambda, \bar y)$ and $(\bar\alpha_r, \bar\beta_{v,(i,j)}^r, \bar\gamma_{v,(i,j)}^r, \bar\xi_v^r, \bar\sigma_{(i,j),v}^r, \bar\delta_{(i,j)}^r, \bar\eta_v, \bar\mu_i^r, \bar\psi^r)$ of RMP($\mathcal{P}'$) and its dual DP($\mathcal{P}'$), respectively.

If the obtained dual solution appears to be feasible in DP($\mathcal{P}$) as well, then $(\bar\lambda, \bar y)$ is an optimal solution of RMP($\mathcal{P}$), which coincides with the initial LP-relaxation. Otherwise, there is a route $p \in \mathcal{P} \setminus \mathcal{P}'$, $p = (v, h_1, ..., h_l, u)$, that satisfies inequality:

$$\sum_{e \in E} E_{e,p} c_e \bar\alpha_r + \bar\beta_{v,(i,j)_p}^r + \bar\gamma_{u,(i,j)_p}^r + \sum_{h \in \mathfrak{H}: \, h \in p} B_{h,p} \left(\bar\xi_h^r + \bar\sigma_{(i,j)_p,h}^r \right) < \bar\delta_{(i,j)_p}^r, \quad (25)$$

where $(i,j)_p$ denotes an arc $a \in A$ represented in the graph G by the route p.

In order to find such a route p, for each arc $(i,j) \in A$ and each plan r, we solve the corresponding *pricing problem* in an auxiliary subgraph. If at least one of such pricing problems produce a route p satisfying (25) (or, in other words, having *negative reduced cost*), we add p to $\mathcal{P}'$ and proceed with the column generation procedure.

To solve the pricing problem, we construct the auxiliary subgraph $G_{i,j}^r = (V_{i,j}^r, E_{i,j}^r)$, $V_{i,j}^r = \mathfrak{M}_i \cup \mathfrak{M}_j \cup \mathfrak{H} \cup \{s_i^r, t_j^r\}$, where s_i^r and t_j^r are artificial nodes connected with all the nodes of $\mathfrak{M}_i$ and $\mathfrak{M}_j$ respectively by zero-weight arcs. Besides that, $E_{i,j}^r = \{(u,v): u \in \mathfrak{M}_i \cup \mathfrak{H}, \, v \in \mathfrak{M}_j \cup \mathfrak{H}, \, (u,v) \in E\}$. For each arc $e \in E_{i,j}^r$ its weight is defined as $c_e \cdot \bar\alpha_r$. Then, any node from $V_{i,j}^r$ except s_i^r and t_j^r is transformed as follows:

- any $u \in \mathfrak{M}_i$ is split onto two nodes u' and u'' that are connected by an arc of the weight $\bar\beta_{u,(i,j)}^r$
- any $v \in \mathfrak{M}_j$ is split onto two nodes v' and v'' that are connected by an arc of the weight $\bar\gamma_{v,(i,j)}^r$
- any $h \in \mathfrak{H}$ is split onto two nodes h' and h'' that are connected by an arc of the weight $\bar\xi_h^r + \bar\sigma_{(i,j),h}^r$.

Figure 2 illustrates construction of the graph $G_{i,j}^r$.

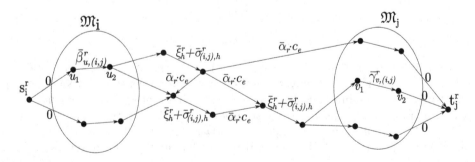

Fig. 2. Example of an auxiliary subgraph $G_{i,j}^r$ specifying the pricing problem

The goal is to find a shortest s_i^r-t_j^r-path in the constructed graph $G_{i,j}^r$. Notice that, by construction, all weights assigned to the edges of this graph except maybe $\bar{\beta}_{v,(i,j)}^r$ and $\bar{\gamma}_{v,(i,j)}^r$ are non-negative. In addition, the only predecessor of any arc of weight $\bar{\beta}_{u,(i,j)}^r$ is the node s_i^r. In a similar way, the node t_j^r appears to be the only successor of arcs with weight $\bar{\gamma}_{v,(i,j)}^r$. Therefore, the graph $G_{i,j}^r$ contains no cycles of negative weight, and the pricing problem appears to be the classic Shortest Path Problem that can be solved to optimality in polynomial time. In this paper, we use the well-known Ford-Bellman algorithm to find such path.

5.3   Branching Scheme

Since an optimal solution $(\bar{\lambda}, \bar{y})$ of the LP-relaxation of model (12)–(22) is most likely fractional, in order to solve the initial mixed integer linear program to optimality, we employ a branching scheme based on the following strategies.

Branching on the 'most fractional' variable y_h^r for $h \in \mathfrak{H}$. Equations (19) and (22) imply that $0 \leq \bar{y}_h^r \leq 1$ for each r and h. Therefore, for the branching, we take the variable y_h^r with minimum value $|\bar{y}_h^r - 0.5|$ and produce two branches. If $y_h^r = 1$, all variables $y_h^{r'}$ for $r' \neq r$ are fixed to zero automatically.

Branching on the fractional set of cluster nodes. Suppose, in the current node of the branching tree, a subset $\bar{\mathfrak{M}}_i^r \subset \mathfrak{M}_i$ consists of the nodes that are prohibited to be included to some plan r i.e., there is an additional constraint $y_v^r = 0$ for any $v \in \bar{\mathfrak{M}}_i^r$. Consider the set

$$\mathfrak{M}_i^r = \{v \colon \bar{y}_v^r > 0, v \in \mathfrak{M}_i \setminus \bar{\mathfrak{M}}_i^r\}.$$

Obviously, for an arbitrary integer solution, $|\mathfrak{M}_i^r| = 1$. On the other hand, for a fractional solution, it can be not a singleton set. In this case, we take the largest one, split it at random into two 'half-subsets' F_1 and F_2 and branch as follows. For the first child node, we set $y_v^r = 0$ for each $v \in F_1$ and extend $\bar{\mathfrak{M}}_i^r$ by the subset F_1 while, for the second child, we proceed with F_2 in a similar way.

At each node of the branching tree, we try to use the first strategy if there are fraction y_h^r for $h \in \mathfrak{H}$. Otherwise, we use the second strategy. Notice that, any time when some branching strategy succeeds, we remove all the routes violating the appended branching constraints.

5.4   Algorithm Description

The proposed algorithm follows the classic branch-and-price framework (see, e.g. [27]) that consists of combining the well-known column generation technique and the aforementioned branching scheme in order to obtain an optimal or close to optimal approximate solution with an accuracy bound (see Fig. 3).

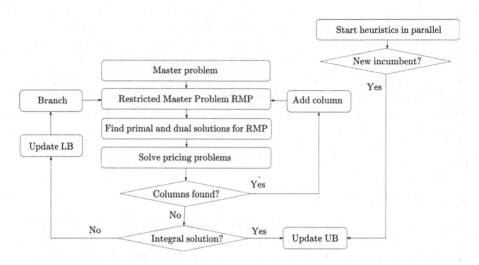

Fig. 3. An overall scheme of the branch-and-price framework

At first, we solve the LP-relaxation of model (12)–(22) using column generation technique. In the case of fractional solution, we proceed with brachnig applying strategies described in Subsect. 5.3.

In order to obtain an upper bound and speed up the overall convergence of the scheme, we apply the following primal heuristics in parallel. Each time, when some of them manage to find an incumbent solution, we update current value of the upper bound.

Start Greedy Heuristic. The goal of this simple heuristic is to provide a start feasible (incumbent) solution to set up the upper bound. In our method, we consequently construct $r = \overline{1,k}$ plans in a following way. Considering arcs of A in a topological order, we find the shortest feasible route p implementing arc $a \in A$. Each time when such a route is found, we add it to the current plan, update properties of the visited nodes and proceed with the further data. Otherwise, we move to another arc of the graph Π.

Local search. This heuristic runs each time when column generation provides a new route. We take routes p for which $\lambda_p^r > 0$ and try to find an integer solution by Gurobi by restricting model (12)–(22) to these routes.

Feasibility Pump (FP). This classic heuristic introduced by Fischetti in his seminal paper [8]. This heuristic tries to convert iteratively current fractional solution to an integer one feasible for the initial mixed integer linear problem.

Adaptive Search. Key idea of the classic ALNS meta heuristic (see. e.g. [22]) is to iteratively apply *ruin* and *recreate* procedures with online learning and simulated annealing stopping criterion. In this paper, we employ an improved version of our ALNS-based heuristic proposed initially in [23].

6   Numerical Evaluation

In our experiments, we are focused on significant extension of the evaluation performed in our initial paper [23]. In this paper, we compare our branch-and-price algorithm A_2 introduced in Sect. 5 with branch-and-bound algorithm A_1, which is exploited in the subsequent experiments as a baseline.

6.1   Instance Benchmark

To produce a test-bench for our experiments, we take PCGTSP instances from the well-known library PCGTSPLIB [14] and transform each of them in a following way:

- the start and destination points s and t are inherited from the initial PCGTSP instance;
- all the clusters of size k or greater are inherited as well;
- any smaller cluster is contracted into a single hub, whose inbound and outgoing arcs are assigned by average weights of the initial arcs incident to this cluster;
- to any obtained hub h, we assign a tuple (q_h, C_h), each its entry is taken at random from $[1, q]$ and $[1, C]$, respectively, where q and C are predefined bounds of hub capacity and opening cost;
- an order graph of PCGTSP instance is updated with respect to the aforementioned transformations and inherited as well.

The described procedure produces a pair of graph (G, Π) for which we generate RPPDP instances (G, Π, k) for the different number of production plans k.

6.2   Experimental Setup

In this paper, we restrict ourselves to the following values of the outer parameters: $k = 2, 3$, $q = 1$, $C = C_{avg}$, where C_{avg} is an average weight of all arcs of the initial PCGTSP instance. For each triple (G, Π, k) we generate 10 random RPPDP testing instances.

Our experiment consists in two stages. At the first exploratory stage, we set time limit to ten hours (36000 sec) and solve the generated RPPDP instances by the baseline algorithm A_1 implemented over Gurobi user callback framework [12]. As we observed, this method solved all the instances of small and medium size to optimality. However, its performance degrades as the instances size grows. Thus, in the second stage, we enlarge time limit to 20 h and try to solve not well solved instances by both A_1 and A_2 algorithms.

The experiments were carried out on the 'Uran' supercomputer of the N.N. Krasovskii Institute of Mathematics and Mechanics of the Ural Branch of the Russian Academy of Sciences http://parallel.uran.ru, Intel(R) Xeon(R) 16 core CPU E5-2697 v4 @2.30 GHz 256GB RAM.

Table 1. Exploratory stage results: instances solved to optimality by A_1. Iname – identifier of the initial PCGTSP instance, N – number of nodes; $|A|$ – number of arcs in the graph Π; m, $|H|$, and k – number of clusters, hubs, and plans to construct, respectively

Iname	N	$|A|$	m	$|H|$	k	Time, sec Avg	Min	Max	Iname	N	$|A|$	m	$|H|$	k	Time, sec Avg	Min	Max
ESC07	38	9	7	2	2	0.05	0.02	0.09	ft70.4	343	84	50	21	2	45.67	16.26	139.37
					3	0.1	0.07	0.13						3	3380.87	1226.77	15908.33
ESC12	64	16	11	3	2	0.09	0.07	0.12	kro124p.1	508	138	78	23	2	41.52	19.0	82.99
					3	0.25	0.21	0.31						3	163.41	59.69	293.21
ESC25	133	38	23	4	2	0.14	0.11	0.29	kro124p.2	508	138	78	23	2	23.29	12.61	39.12
					3	0.55	0.42	0.67						3	23.29	12.61	39.12
ESC47	241	75	42	7	2	1.68	1.41	2.39	kro124p.3	527	139	82	19	2	18.68	12.93	25.94
					3	2.86	2.31	3.59						3	320.6	237.62	390.53
ESC63	348	128	57	8	2	2.54	1.83	3.48	kro124p.4	516	134	80	21	2	319.66	87.65	2226.8
					3	20.7	16.39	30.0						3	5988.87	3285	14902.23
ESC78	405	87	62	18	2	31.8	2.53	66.57	p43.1	198	31	54	13	2	3.55	1.56	6.78
					3	104.17	41.07	254.34						3	10.48	7.04	22.41
br17.10	87	21	14	4	2	0.12	0.07	0.19	p43.2	193	53	34	10	2	1.28	0.46	1.59
					3	1.15	0.19	3.58						3	12.72	6.66	26.98
br17.12	91	22	16	2	2	0.08	0.05	0.14	p43.4	201	50	33	11	2	4.87	1.8	15.74
					3	1.5	0.62	2.9						3	2035.1	669.11	5165.67
ft53.1	277	76	43	11	2	4.23	1.33	11.06	prob.100	497	126	78	22	2	31.25	7.84	73.41
					3	21.11	7.45	38.25						3	115.67	24.9	192.91
ft53.2	270	67	41	13	2	7.1	1.98	12.31	prob.42	206	56	32	10	2	1.27	1.03	1.79
					3	33.25	6.8	49.8						3	2.77	1.32	3.94
ft53.3	278	72	45	9	2	1.95	1.24	3.09	ry48p.1	255	71	41	8	2	2.4	0.53	5.48
					3	16.67	10.26	44.12						3	10.74	4.05	17.81
ft53.4	271	73	45	9	2	5.42	1.97	8.87	ry48p.2	246	64	39	10	2	4.27	1.21	8.7
					3	216.91	65.68	606.87						3	12.98	4.86	23.15
ft70.1	342	103	58	13	2	6.52	2.45	18.01	ry48p.3	252	62	38	11	2	8.92	3.59	13.06
					3	34.52	4.8	59.63						3	26.33	7.17	54.67
ft70.2	346	97	58	13	2	21.6	2.71	42.4	ry48p.4	243	64	39	10	2	6.93	2.54	10.89
					3	55.43	10.04	119.92						3	229.67	51.94	486.36
ft70.3	336	92	52	19	2	8.18	5.7	10.61	p43.3	209	50	34	10	2	4.33	2.24	7.26
					3	444.29	37.52	835.57	rbg048a	252	123	40	10	2	662.53	172.62	1333.06

6.3 Results and Discussion

We report the results of the numerical evaluation of the first stage in Tables 1 and 2. All instances presented in Table 1 were solved to optimality within the given time limit of 10 h. Therefore, in this table we report only running times for each group of test instances.

On the other hand, as it follows from Table 2, A_1 does not manage to obtain optimal solutions even for some instances of moderate size. While it solves instances p43.3, rbg048a for 3 plans and rbg050c for both 2 and 3 plans within average gap less than 2, for the instances rbg109a, rbg150a, rbg174a the gap appears to be twice as large. Further, for all the larger instances of our library including the instances rbg253a, rbg323a, rbg341a, rbg358a, rbg378a algorithm

Table 2. Exploratory stage results: instances solved by A_1 within positive gap

Iname	N	$|A|$	m	$|H|$	k	Gap, % Avg	Min	Max	Iname	N	$|A|$	m	$|H|$	k	Gap, % Avg	Min	Max
p43.3	209	50	34	10	3	1.0	0.0	2.85	rbg109a	563	395	86	25	2	2.69	2.49	2.9
rbg048a	252	123	40	10	3	0.15	0.0	0.5						3	3.36	2.63	7.6
rbg050c	255	203	43	9	2	0.47	0.16	0.75	rbg150a	855	626	120	32	2	6.97	3.33	8.23
					3	1.63	1.37	2.03						3	7.9	7.86	7.99
									rbg174a	944	753	144	32	2	5.78	2.84	7.59
														3	7.8	7.77	7.84

A_1 did not managed to find even a feasible solution in 10 h. For the sake of brevity, we do not include these results to Table 2.

In the second stage of our numerical evaluation, we compare performance of A_1 and A_2 on instances $rbg109a$, $rbg150a$, $rbg174a$, $rbg253a$, $rbg323a$, $rbg341a$, $rbg358a$, $rbg378a$. Since both A_1 and A_2 did not find optimal solutions within the given time limit (of 20 h), in Table 3 we report only on average, minimum and

Table 3. Second stage results: comparison of A_1 and A_2 on large instances

		Explored Nodes			UB			LB			
A_1											
Iname	k	Avg	Min	Max	Avg	Min	Max	Avg	Min	Max	Gap, %
rbg109a	2	9321	7512	10395	5880.5	5872	5889	5721.13	5714	5731	**2.66**
	3	9271.67	8239	10424	5907	5894	5929	5735.16	5726	5740.98	**2.85**
rbg150a	2	7736.83	6448	10061	9587	9574	9596	9271.83	9266	9277	**3.29**
	3	479.67	478	482	10101	10101	10101	9301.44	9298	9304.08	7.9
rbg174a	2	4510.17	3118	5619	11742.67	11716	11763	11386.5	11381	11394	**3.03**
	3	16.4	2	30	15628.5	12384	25362	11658	11413	12384	7.82
A_2											
rbg109a	2	149.67	149	150	6118	6118	6118	5718.35	5714.36	5724.85	6.81
	3	64.67	63	67	6157	6157	6157	5734.89	5726.28	5743.28	6.53
rbg150a	2	96.6	96	98	9995.8	9968	10014	9270.54	9264.81	9275.1	7.24
	3	47.5	47	48	**10007**	**10007**	**10007**	9302.67	9300.43	9305.3	**6.66**
rbg174a	2	41.5	41	22	12109	12064	12151	11385.85	11377.7	11395.2	6.22
	3	19.36	19	22	**12192.6**	**12044**	**12237**	11416.34	11409.21	11421.21	**6.76**
rbg253a	2	12.33	11	13	19004	19004	19004	17705.03	17764.86	17780.86	**6.44**
	3	4.67	4	5	19037	19037	19037	17833.33	17832.26	17833.92	**6.34**
rbg323a	2	4.5	4	5	31089	31089	31089	29164.25	29160.25	29168.25	**6.19**
	3	—	—	—	31187	31187	31187	—	—	—	—
rbg341a	2	4.5	4	5	28291.33	28288	28294	26508.74	26504	26511.76	**6.39**
	3	—	—	—	28350	28350	28350	—	—	—	—
rbg358a	2	—	—	—	38803	38803	38803	—	—	—	—
	3	—	—	—	38872	38872	38872	—	—	—	—
rbg378a	2	—	—	—	28681	28681	28681	—	—	—	—
	3	—	—	—	28681	28681	28681	—	—	—	—

maximum values of the explored nodes of search trees, upper and lower bounds, and average values of obtained gaps.

As it follows from Table 3, algorithm A_1 demonstrates better performance on instances $rbg109a$, $k = \overline{2,3}$ and $rbg150a$, $rbg174a$, $k = 2$. On the other side, for the larger instances $rbg150a$, $rbg174a$, $k = 3$ algorithm A_2 finds approximate solutions of better average gap (highlighted by bold). Emphasize that A_2 obtains the presented results exploring significantly less number of nodes in searching tree. While lower bounds obtained by both algorithms are close to each other, algorithm A_2 managed to find the better upper bounds for instances $rbg150a$ and $rbg174a$, $k = 3$.

Finally, A_2 provides approximate solutions for the instances $rbg253a$ and $rbg323a$, $rbg341a$ for $k = 2$ and incumbents for $rbg323a$, $rbg341a$ for $k = 3$ and for $rbg358a$ and $rbg378a$ as well. For all these instances algorithm A_1 found neither upper nor lower bounds during the same time limit. For the sake of brevity, we omit presentation of these data in Table 3.

7   Conclusion

In this paper, we extended our algorithmic analysis of the Reliable Production Process Design Problem. To date, we have branch-and-bound algorithm A_1 based on compact MILP model and implemented over the Gurobi user callback framework, and the novel branch-and-price algorithm A_2 employing route packing model and the shortest path pricing problem. Results of the comparative numerical experiments show that algorithms A_1 and A_2 together managed to find optimal or close to optimal approximate solutions for the most instances of the testing library forked from the public PCGTSPLIB and complement well to each other. In the same time, for large size instances, performance of both algorithms still needs further improvement, which can be considered as a subject of the future work.

Acknowledgements. This work is part of research carried out in the Ural Mathematical Center with the financial support of the Ministry of Science and Higher Education of the Russian Federation (Agreement no. 075-02-2024-1377).

References

1. Andrade, R.C.D.: New formulations for the elementary shortest-path problem visiting a given set of nodes. Eur. J. Oper. Res. **254**(3), 755–768 (2016). https://doi.org/10.1016/j.ejor.2016.05.008
2. Balas, E., Fischetti, M., Pulleyblank, W.: The precedence-constraint asymmetric traveling salesman polytope. Math. Program. **68**, 241–265 (1995). https://doi.org/10.1007/BF01585767
3. Chentsov, A.G., Khachai, M.Y., Khachai, D.M.: An exact algorithm with linear complexity for a problem of visiting megalopolises. Proc. Steklov Instit. Math. **295**(1), 38–46 (2016). https://doi.org/10.1134/S0081543816090054

4. Eilam-Tzoreff, T.: The disjoint shortest paths problem. Discret. Appl. Math. **85**(2), 113–138 (1998). https://doi.org/10.1016/S0166-218X(97)00121-2

5. Fan, Y., Schwartz, F., Vob, S., Woodruff, D.L.: Catastrophe insurance and flexible planning for supply chain disruption management: a stochastic simulation case study. Int. J. Prod. Res. **62**(4), 1108–1125 (2023). https://doi.org/10.1080/00207543.2023.2176179

6. Ferone, D., Festa, P., Guerriero, F.: An efficient exact approach for the constrained shortest path tour problem. Optim. Methods Softw. **35**(1), 1–20 (2020). https://doi.org/10.1080/10556788.2018.1548015

7. Ferone, D., Festa, P., Guerriero, F., Laganà, D.: The constrained shortest path tour problem. Comput. Oper. Res. **74**, 64–77 (2016). https://doi.org/10.1016/j.cor.2016.04.002

8. Fischetti, M., Glover, F., Lodi, A.: The feasibility pump. Math. Program. **104**, 91–104 (2005). https://doi.org/10.1007/s10107-004-0570-3

9. Fortune, S., Hopcroft, J., Wyllie, J.: The directed subgraph homeomorphism problem. Theoret. Comput. Sci. **10**(2), 111–121 (1980). https://doi.org/10.1016/0304-3975(80)90009-2

10. Gouveia, L., Pesneau, P., Ruthmair, M., Santos, D.: Combining and projecting flow models for the (precedence constrained) asymmetric traveling salesman problem. Networks **71**(4), 451–465 (2018). https://doi.org/10.1002/net.21765

11. Gouveia, L., Ruthmair, M.: Load-dependent and precedence-based models for pickup and delivery problems. Comput. Oper. Res. **63**, 56–71 (2015). https://doi.org/10.1016/j.cor.2015.04.008

12. Gurobi Optimization, L.: Gurobi optimizer reference manual (2024). https://www.gurobi.com/documentation/9.5/refman/index.html

13. Gutin, G., Punnen, A.P.: The Traveling Salesman Problem and Its Variations. Springer, US, Boston, MA (2007). https://doi.org/10.1007/b101971

14. Khachai, D., Sadykov, R., Battaia, O., Khachay, M.: Precedence constrained generalized traveling salesman problem: polyhedral study, formulations, and branch-and-cut algorithm. Eur. J. Oper. Res. **309**(2), 488–505 (2023). https://doi.org/10.1016/j.ejor.2023.01.039

15. Khachai, M.Y., Neznakhina, E.D.: Approximation schemes for the generalized traveling salesman problem. Proc. Steklov Instit. Math. **299**(1), 97–105 (2017). https://doi.org/10.1134/S0081543817090127

16. Khachay, M., Kudriavtsev, A., Petunin, A.: PCGLNS: a heuristic solver for the precedence constrained generalized traveling salesman problem. In: Olenev, N., Evtushenko, Y., Khachay, M., Malkova, V. (eds.) OPTIMA 2020. LNCS, vol. 12422, pp. 196–208. Springer, Cham (2020). https://doi.org/10.1007/978-3-030-62867-3_15

17. Khachay, M., Neznakhina, K.: Complexity and approximability of the Euclidean generalized traveling salesman problem in grid clusters. Ann. Math. Artif. Intell. **88**(1), 53–69 (2020). https://doi.org/10.1007/s10472-019-09626-w

18. Khachay, M., Ukolov, S., Petunin, A.: Problem-specific branch-and-bound algorithms for the precedence constrained generalized traveling salesman problem. LNCS **13078**, 136–148 (2021). https://doi.org/10.1007/978-3-030-91059-4_10

19. Kudriavtsev, A., et al.: The shortest simple path problem with a fixed number of must-pass nodes: a problem-specific branch-and-bound algorithm. In: Simos, D.E., Pardalos, P.M., Kotsireas, I.S. (eds.) LION 2021. LNCS, vol. 12931, pp. 198–210. Springer, Cham (2021). https://doi.org/10.1007/978-3-030-92121-7_17

20. Martin, S., Magnouche, Y., Juvigny, C., Leguay, J.: Constrained shortest path tour problem: branch-and-price algorithm. Comput. Oper. Res. **144**, 105819 (2022). https://doi.org/10.1016/j.cor.2022.105819
21. Papadimitriou, C.: Euclidean TSP is NP-complete. Theoret. Comput. Sci. **4**, 237–244 (1977)
22. Ropke, S., Pisinger, D.: An adaptive large neighborhood search heuristic for the pickup and delivery problem with time windows. Transp. Sci. **40**, 455–472 (2006). https://doi.org/10.1287/trsc.1050.0135
23. Rudakov, R., Khachai, D., Ogorodnikov, Y., Khachay, M.: Reliable production process design problem: compact MILP model and ALNS-based primal heuristic. LNCS **14395**, 174–188 (2023). https://doi.org/10.1007/978-3-031-47859-8_13
24. Saksena, J.P., Kumar, S.: The routing problem with 'k' specified nodes. Oper. Res. **14**(5), 909–913 (1966)
25. Salman, R., Ekstedt, F., Damaschke, P.: Branch-and-bound for the precedence constrained generalized traveling salesman problem. Oper. Res. Lett. **48**(2), 163–166 (2020). https://doi.org/10.1016/j.orl.2020.01.009
26. Schilling, L., Seuring, S.: Linking the digital and sustainable transformation with supply chain practices. Int. J. Prod. Res. **62**(3), 949–973 (2023). https://doi.org/10.1080/00207543.2023.2173502
27. Vanderbeck, F.: Branching in branch-and-price: a generic scheme. Math. Program. **130**(2), 249–294 (2011). https://doi.org/10.1007/s10107-009-0334-1

The Problem of Planning Investment Projects with Lending

Svetlana A. Malakh and Vladimir V. Servakh[✉]

Sobolev Institute of Mathematics, Siberian Branch of the Russian Academy of
Sciences, Omsk, Russia
svv_usa@rambler.ru

Abstract. We are working on an approach that can be used to implement large-scale projects. First, the general formulation of the resource-constrained scheduling problem (RCPSP) is described. The next option for solving the problem with the criterion of maximizing net present value (NPV) from the entire project. The only type of resource used is financial, since all other resources are exchanged in monetary terms, and cash flows can be exchanged at any time during the execution of work. Next, we consider the problem of scheduling an investment project with the possibility of attracting borrowed capital, in which the net present value of the entire project will be maximum.

The use of investment or borrowed funds makes it possible not to consider resource limitations, but you have to pay for their use. With this approach, any schedule consistent with a partial order becomes valid. The computational complexity of the problem is researched, and a pseudopolynomially solvable case is found.

Keywords: Scheduling · Investment projects · Optimization

1 Introduction

A project will be a set of technologically interrelated works $V = 1, 2, \ldots, n$, the implementation of which is aimed at achieving a specific goal. Typically, jobs depend on each other in at least two ways. First, they compete for the resources needed to get the job done. Second, priority constraints between pairs of jobs require that each such pair be executed in a predetermined order. The relationship is determined by the partial order E, and the project itself will be specified by the graph $G = (V, E)$. The durations of each job p_i are known, $i \in V$. The problem of minimizing the overall project completion time was successfully solved in 1958 during the implementation of the project to create the Polaris missile system [7]. The project, consisting of 60 thousand works, was completed two years ahead of schedule. Around the same time, when planning work to

The work was carried out within the framework of the state assignment of the IM SB RAS, project FWNF-2022-0020.

modernize DuPont plants, the critical path method was proposed [17]. These developments have received wide practical application due to the simplicity, clarity and efficiency of their use. They made it possible to calculate resource requirements at each stage of the project.

Difficulties in scheduling work arose when projects had to be implemented under resource constraints. This problem is called the resource constrained project scheduling problem (RCPSP) and consists in scheduling work taking into account their priority and resource limitations, while keeping the execution time to a minimum. This has become a classic problem in the context of project planning, attracting the attention of numerous researchers who have developed both precise and heuristic planning procedures. However, it is a rather formal model with assumptions that are too narrow to capture many real-world needs. Consequently, various extensions to the basic problem of resource-constrained project planning have been developed.

A mathematical model representing RCPSP was developed by Pritsker, Watters, and Wolfe (1969) [25]. Blazewicz, Lenstra, and Rinnooy Kan (1983) [6] showed that RCPSP belongs to the class of strongly NP-hard problems.

2   Scheduling Problem

Let us present the classical formulation of the problem [1]. Let's consider two types of resources: renewable and stored. The first includes equipment, workers, specialists, production facilities, and the stored ones include consumables, raw materials, finances, and so on. Let W^r and W^a be the number of types of renewable and stored resources, respectively. The presence of a resource of the form $w \in W^r U W^a$ in the interval $[t-1, t)$ will be denoted by $K^w(t)$. For renewable resources, these resources are available throughout the entire interval; for stored resources, this volume is available at the beginning of the specified period. Without loss of generality, it is assumed that the durations of all jobs are integer. Each job $i \in V$ is characterized by a discrete resource consumption function $k_i^w(\tau)$ in the interval $[\tau-1, \tau)$, where $\tau = 1, 2, \ldots, p_i$, counting from the beginning of this work. It is necessary to find s_i - the start time of work $i \in V$.

Let us denote by $N_t = \{i \in V | \ s_i < t \le s_i + p_i\}$ the set of jobs performed on the interval $[t-1, t)$. Schedule $(s_1, s_2, \ldots, s_n)$ is called admissible if the technological sequence of work is followed:

$$s_i + p_i \le s_j, \quad (i, j) \in E;$$

restrictions on stored type resources are observed:

$$\sum_{\tau=1}^{t} \sum_{i \in N_\tau} k_i^w(\tau - s_i) \le \sum_{\tau=1}^{t} K^w(\tau), \quad w \in W^a, t = 1, 2, \ldots;$$

restrictions on renewable resources are observed:

$$\sum_{i \in N_\tau} k_i^w(\tau - s_i) \le K^w(t), \quad w \in W^r, \quad t = 1, 2, \ldots.$$

If there are only resources of the stored type, then the problem of minimizing the total completion time of all jobs is polynomially solvable [1]. For other criteria, the problem is NP-hard in the strong sense. If resources are renewable, then even finding a feasible solution is a highly NP-hard problem. In this case, it is easy to construct examples when there is at least one feasible solution, but it is difficult to find it. To do this, it is enough to take an arbitrary work schedule and calculate exactly how many resources are required to implement it. Record this minimum required level of resources as input data. Next, remove information about the schedule. As a result, we obtain an example with a non-empty set of feasible solutions. Finding a feasible schedule for such an example is a very difficult task, since the resource restrictions are very strict. Such examples are the most difficult in discrete optimization.

The literature provides many ideas on how resource constraints can be relaxed. The trade-off problem is well known, in which the duration of a job depends on the resources allocated to it [12]: it is possible to redistribute limited resources in favor of critical jobs. On the other hand, if there is a lack of resources at some point in time, you can increase the duration of the work and meet the restrictions. Another important approach aimed at working more flexibly with resource constraints is resource acquisition. The problem of minimizing the cost of purchasing resources was first considered in the work of Mohring (1984) [23]. Among the latest studies, we note the articles by Rodrigues, Yamashita (2010) and Romanova (2018) [5,26]. In the works of [9,35], in addition to drawing up a schedule, they solved a problem in which it was necessary to determine additional parameters: at what time, from whom and in what quantity to order materials for the project. Also, the works of [5,35] also took into account discounts on the number of orders. The disadvantage of this approach is the need to optimize the second objective function - the cost of purchased resources.

Currently, ideas about resources have expanded significantly. First of all, let us note the concept of resource regeneration [22,28]. Each job performed consumes a certain number of resource units. This is interpreted as an expense. And reproduces another number of units of the resource after its completion, which is interpreted as income. For renewable resources, consumption is equal to income, and for stored resources, income is 0. Resource regeneration assumes the presence of a β coefficient for resource recovery after work is completed. Previously, two values of this coefficient were considered: if $\beta = 1$, then the resource is renewable, if $\beta = 0$ the resource is stored. More generally, the value β can take any value. As an example, consider the articles by Kaplan (1986) and Amir (1988) [15,16], which examine the model of intermediate relocation of residents as part of a city redevelopment project.

This problem differs from the classical scheduling problem with limited resources in that the income of any completed work can be not only positive or zero, but also negative. The authors of the work [13] consider a special case with pairs of jobs, when the first job occupies a resource at the time of its start, while the same amount of capacity is released upon completion of the second job. Bound resources are called take-and-give resources.

This approach can be generalized by suggesting the reproduction of a resource not only after completion of the work, but also during its execution, as well as with a time lag after its completion. The concepts of partially renewable resources, as well as the concept of progressively renewable resources, were introduced. In [24] the authors use partially renewable resources in a multi-project scheduling problem. [34] considers a problem with partially renewable resources and minimum and maximum time delays.

Let us note a few more aspects related to [11] resources. This is the concept of a shared or aggregate resource, where a subset of jobs uses a common resource. Another direction is related to resources that have multiple skills, and each job requires a certain set of these skills. Some researchers consider the problem of multiple skills with learning or fatigue effects. But research into these types of resources is beyond the scope of this work.

Although RCPSP, as stated above, is already a powerful model, it cannot cover all situations that arise in practice. Therefore, many researchers have developed more general project planning problems, often using the standard RCPSP as a starting point. Since the 1990s, several review articles on project planning have been published. Most of these works focus on standard RCPSP methods (see Hartmann and Kolisch (2000); Kolisch and Hartmann (1999, 2006); Pellerin, Perrier, and Berthaut (2020)) and basic variants (see Brucker (2002), Brucker, Drexl , Mohring, Neumann, and Pesch (1999), Herroelen, de Reyck, and Demeulemeester (1998), Herroelen (2005), Kolisch and Padman (2001), Ozdamar and Ulusoy (1995), and Tavares (2002)). The books by Schwindt and Zimmermann (2015a,b) contain more than 60 articles covering many important project planning models and methods. Hartmann and Briskorn (2022) provided a broad overview of variations and extensions of RCPSP that have been proposed in the literature.

3   Criteria Based on Net Present Value

One of the important options for the problem under consideration is the planning of investment projects, the main goal of which is aimed at generating profit from the implementation of a set of technologically interrelated works. In this setting, cash flows arise, and to compare money at different points in time, the discounting operation is used. Since when implementing financial transactions it is assumed that there is the possibility of an alternative risk-free liquid placement of capital at the rate of r_0, then having placed capital K_0 at the time t_0 at the rate of r_0, by the time t it will increase to the value $K_t = K_0(1 + r_0)^{t-t_0}$. Thus, capital K_t at time t is equivalent to capital $\frac{K_t}{(1+r_0)^{t-t_0}}$ at time t_0. The operation of reduction to the initial point in time is called discounting and allows you to compare money at different points in time.

The literature discusses different approaches to modeling investment projects. Cash outflows are caused by the completion of work and the use of resources; cash inflows occur upon completion of certain parts of the project. This leads to the goal of maximizing the net present value (NPV) of the project. In addition to standard priority and resource limitations, deadline constraints are taken into account. RCPSP for NPV was studied by Gu, Schutt, Stuckey, Wallace, and Chu (2015) [10], Leyman and Vanhoucke (2015) [19], Thiruvady, Wallace, Gu, and Schutt (2014) [31] and Vanhoucke (2010) [32]. These studies are based on continuous compounding, that is, cash flows are discounted by a factor of $e^{-\beta t}$. Fink and Homberger (2013) [8] consider the same situation, but use period compounding with a discount factor of $(1+\alpha)^{-t}$. Note, however, that these two types of compounding are not significantly different as they can be converted into each other. Leyman and Vanhoucke (2017) [21] extend the RCPSP objective to the NPV objective by considering cash outflows either at the beginning or end of the activity, or over the entire duration of the activity (cash inflows occur only at the end of the period). The limit ensures that the capital never goes negative.

Leyman et al. (2019) [18] take a similar approach under DTCTP, where the three payment models are also applied to cash inflows. Leyman and Vanhoucke (2016) [20] consider payments at regular and irregular times, as well as activity-related payments, which are included in both single-mode and multimode RCPSP.

Shahsawar, Niaki, and Najafi (2010) [27] propose a net present value-based objective that takes into account post-completion cash inflows, resource capacity costs, and bonuses and penalty payments based on project completion with respect to time frame execution. The inflation rate is also taken into account.

Resource capacity costs apply only to the time interval during which the resource is actually used. Walig'ora and Weglarz (2014) [33] consider RCPSP with the goal of maximizing net present value and an additional single continuous resource.

Tirkolai et al. (2019) [30] aims to maximize NPV under MRCPSP. Khoshjahan, Najafi and Afshar-Nadjafi (2013) [14] minimize the net present value of early and late action penalties in RCPSP with action deadlines.

Our work proposes an approach that can be used to model large-scale projects and includes most of the concepts described above. In addition, the model manages to avoid multicriteria.

Let's consider a single type of resource - financial, and replace all other resources with a monetary equivalent. Having a financial resource, workers can be hired, equipment can be rented, leased or bought with its subsequent sale, the necessary premises can be rented, etc. In the end, it all comes down to the flow of payments $(c_i(0), c_i(1), \ldots, c_i(p_i), \ldots)$, where $c_i(\tau)$ is the balance of payments at time $\tau = 0.1, \ldots, p_i$, that is, the difference between receipts and expenses. If $c_i(\tau) < 0$, costs are greater than revenues. If $c_i(\tau) > 0$ - revenues are greater than costs. Magnitude

$$NPV_j = \sum_{\tau=0}^{p_j} \frac{c_j(\tau)}{(1+r_0)^\tau}$$

is called net profit work $j \in V$ reduced to the beginning of its execution (Net Present Value). It is assumed that every job runs without interruption. The availability of resources at time t is also specified in monetary terms by the total value $K(t)$, $t = 0, 1, 2, \ldots, T - 1$, where T – project planning horizon.

Let us denote by s_j the moment when work $j \in V$ begins. The vector $\mathbf{S} = (s_1, s_2, \ldots, s_N)$ is called the schedule execution of project work. Since p_j are integers and payment flows are discrete, it is enough to consider schedules with integer s_j. Let us denote by $N_t = \{j \in V \mid s_j \leq t < s_j + p_j\}$ set of jobs performed in interval $[t, t + 1)$, $t \in Z^+$. The schedule $\mathbf{S}$ is called valid if:
– the specified partial order of work execution is preserved:

$$s_i + p_i \leq s_j, \quad (i, j) \in E;$$

– at each point in time, taking into account reinvestment of income and placement of free capital at the rate of r_0 financial There are enough resources to complete the project work:

$$\sum_{t=0}^{t^*} \frac{K(t)}{(1 + r_0)^t} + \sum_{t=0}^{t^*} \sum_{j \in N_t} \frac{c_j(t - s_j)}{(1 + r_0)^t} \geq 0, \quad t^* = 0, 1, \ldots, T - 1.$$

It is necessary to determine an acceptable work schedule, when in which the net present value of the entire project will be maximum. To sum up the profit from all jobs, the required value is NPV_j lead to time $t = 0$. Then the objective function will be look like this:

$$NPV_{rc}(\mathbf{S}) = \sum_{j \in V} \frac{NPV_j}{(1 + r_0)^{s_j}} \rightarrow \max.$$

This problem is NP-hard in the strong sense [3]. Various versions of this model have been studied in [10, 19–21].

Note that the net present value criterion is convenient to use to achieve various project goals, in particular to minimize the total project completion time. It is enough to introduce a fictitious final work, which depends on all the works of the project, and make its income greater than the maximum possible total income of the project. Then, when optimizing the NPV criterion, this work will automatically be performed early, and the entire project will be completed as early as possible.

4   Statement of the Problem of Investment Projects with Lending

Let us generalize the model considered above. Let there be an opportunity to attract additional resources for a certain fee. In our setting with a single type of resources and NPV criterion, the mechanism for attracting additional funds is lending. The possibility of using loans expands the set of feasible solutions to the problem. Accordingly, the optimal solution should be at least no worse than without loans.

When modeling loans, a large number of additional characteristics arise, such as the type of loan, its size, term, payment scheme, and so on. Let us make some assumption that simplifies the formulation, but does not significantly affect the adequacy of the model. We will assume that at any time you can take out a loan at a fixed rate r. Then any loan can be represented as a flow of payments

$$\begin{pmatrix} t_0 & t_1 & \dots & t_k \\ \xi_0 & \xi_1 & \dots & \xi_k \end{pmatrix},$$

with the property

$$\sum_{i=0}^{k} \frac{\xi_i}{(1+r)^{t_i-t_0}} = 0,$$

where r is the interest rate on the loan. If the rate does not change over time, then a loan of any complexity can be divided into a sequence of simple loans implemented according to the scheme: "taken - returned with interest." Since our time is discrete and t takes integer values, the following statement is true [2].

Proposition 1. *With a fixed interest rate, a loan of any type can be divided into an equivalent sequence of loans with a loan term equal to a single time period.*

Indeed, with discrete time, at the next integer moment in time, the entire loan debt is returned along with the accrued interest. The loan is completely closed. And at the same moment we open a new loan for the same amount. The resulting sequence of payment flows corresponds to the original loan scheme.

This approach allows you to introduce only one additional type variables $D(t)$ – the size of the loan taken in year t, and the model with loans and reinvestment of income has the following form: build a work execution schedule $\mathbf{S} = (s_1, s_2, \dots, s_N)$, at which

- the technological order of work execution is observed

$$s_i + p_i \le s_j, \quad (i,j) \in E;$$

- at each integer moment $t^* \in Z^+$ is preserved non-negative balance of payments taking into account loans taken, payments on them, reinvestment of income and placement of free capital at an interest rate of r_0:

$$\sum_{t=0}^{t^*} \left(\frac{K(t)}{(1+r_0)^t} + \sum_{j \in N_t} \frac{c_j(t-s_j)}{(1+r_0)^t} + \frac{D(t) - (1+r)D(t-1)}{(1+r_0)^t} \right) \ge 0;$$

- net present profit taking into account payments for loans reaches its maximum value:

$$NPV_{loan}(\mathbf{S}) = \sum_{j \in V} \frac{NPV_j}{(1+r_0)^{s_j}} + \sum_{t=0}^{T} \frac{D(t) - (1+r)D(t-1)}{(1+r_0)^t} \to \max,$$

where T is the project planning horizon, $D(-1) = 0$ and $D(T) = 0$.

The use of loans expands the set of feasible solutions of the problem. Therefore, the optimal solution NPV^*_{loan} must be at least no worse than the optimal solution NPV^*_{rc} for the problem without loans. Thus, if a problem with criterion NPV_{rc} is solvable, then the following inequality holds:

$$NPV^*_{rc} \leq NPV^*_{loan}.$$

Let us highlight the features of the constructed model:

- the goal is for the investor to obtain maximum profit;
- the only resource is money;
- the time factor of the cost of funds is taken into account.
- there is a possibility of using a loan;
- the income received during the implementation of the project is reinvested.

This problem is also NP-hard in the strong sense [2]. Moreover, the number of variables has increased. Below we propose to use a different approach to modeling this problem.

The main idea is that for a given work schedule, credit borrowings are determined uniquely. To do this, at each integer point in time, the lack of financial resources is covered by the minimum necessary loan, which is returned at the next integer point in time. Next we proceed in the same way.

Magnitude

$$NPV(\mathbf{S}) = \sum_{j \in V} \frac{NPV_j}{(1 + r_0)^{s_j}}$$

in this case, it characterizes the net present value of the project when there are no restrictions on resources. If we have enough money to cover any ongoing costs of the project, then the final profit will be the specified amount.Note that for any technologically feasible schedule $\mathbf{S}$ the following holds:

$$NPV_{loan}(\mathbf{S}) \leq NPV(\mathbf{S}).$$

At this stage, you can immediately discard schedules for which $NPV(\mathbf{S}) < 0$, since taking into account resource constraints, profit will not increase. But it makes no sense to implement a project with a negative profit. Finding the schedule for which $NPV(\mathbf{S})$ takes the greatest value allows us to determine the potential capabilities of the project and construct an upper estimate of the optimum under resource constraints. However, to date it has not been possible to construct a polynomial algorithm for solving such a problem and the question of the computational complexity of the problem remains open.

Let us describe the algorithm for calculating our own net present value profit. Let's create a general flow of payments for the project for schedule S. To do this, it is enough to add up the work flows $(c_i(0), c_i(1), \ldots, c_i(p_i))$, $i = 1, 2, \ldots, n$, linking them to the start of work S_i, $i = 1, 2, \ldots, n$. As a result, we obtain the total project flow $(C_0, C_1, \ldots, C_t, \ldots, C_T)$ for a given schedule S, where $T = max_{j \in V}(S_j + p_j)$. Note that this flow of payments and revenues does not depend on how this project will be financed. Let, as before, to finance the project at time $t = 0, 1, \ldots, T-1$ there is capital K_t.

Let us denote by F_t is the current balance of payments at the moment t, $t = 0.1, \ldots, T$, taking into account the use of loans and payments on them. Initially, if $K_0 < C_0$, then it is necessary to take out a loan and $F_1 = (K_0 + C_0) \cdot (1 + r)$. If if $K_0 \geq C_0$, then we place free money under the bet r_0 and that means $F_1 = (K_0 + C_0) \cdot (1 + r_0)$. The value of F_{t+1} for $t = 1, \ldots, T-1$ is computed recursively:

$$F_{t+1} = (F_t + K_t + C_t) \cdot (1 + r_0), \text{ if } F_t + K_t + C_t \geq 0,$$
$$F_{t+1} = (F_t + K_t + C_t) \cdot (1 + r), \text{ if } F_t + K_t + C_t < 0.$$

Discounting the value $F_T + C_T$ to the initial point in time and subtracting the invested capital, we obtain our own net profit

$$NPV_{taut}(S) = \frac{F_T + C_T}{(1 + r_0)^T} - \sum_{i=0}^{T-1} \frac{K_t}{(1 + r_0)^t}.$$

Ultimately, you need to find a schedule for completing the project's work, at which your own profit will be the greatest:

$$NPV_{taut}(\mathbf{S}) \to max.$$

$$s_i + p_i \leq s_j, \quad (i, j) \in E.$$

Proposition 2. *The optimal values of the objective functions NPV_{taut}^* and NPV_{loan}^* coincide.*

The constructed model differs from the previously described model with the NPV_{loan} criterion by the additional requirement of using the minimum required volume of credit at any time. This makes it possible to exclude D_t variables from the model. Both approaches are solutions to the same problem, leading to two different versions of the model.

5   Complexity and Approaches to Solving the Problem

Let us present several results on the computational complexity of solving the problem.

Proposition 3. *The $NPV_{taut}(S)$ maximization problem is NP-hard in the strong sense.*

By virtue of Proposition 2, when we find the optimum of a problem with the criterion NPV_{taut}, we also find the optimum of the problem with the criterion NPV_{loan}. Since this problem is NP-hard in the strong sense [2], then the original problem has the same complexity.

Note that the contribution of work j to the objective function is $\frac{NPV_j}{(1+r_0)^{s_j}}$. If $NPV_j < 0$, then this work generates losses, and as s_j increases, the total profit of the project also increases. This property allows us to select a pseudopolynomially solvable case of the problem.

Theorem 1. *If the number of profitable jobs is limited by a constant m, then the problem of maximizing one's own profit is pseudopolynomially solvable.*

The proof is based on the above remark. In an optimal schedule, activities that have a negative net present value are completed as late as possible. Thus, if we fix the start times of profitable jobs S_i, then for all other jobs it is enough to find the late start times. And since $S_i \in \{0, 1, \ldots, T-1\}$, where T is the project planning horizon, the schedule enumeration algorithm has a complexity of no more than $O(T^m)$ operations , which was what needed to be proven.

The complexity $O(T^m)$ is a rough upper bound and can be reduced by using an algorithm from [4] based on a dynamic programming scheme. Its complexity depends exponentially on the maximum number of technologically independent profitable jobs. If the number of such jobs is limited by a constant, then the algorithm becomes pseudopolynomial. This strengthens the result of the theorem presented above, but its proof is beyond the scope of this work.

The advantage of the constructed model is that there are no restrictions on resources, which means that any schedule consistent with a partial order is valid. This allows the use of a wide range of metaheuristics, including evolutionary algorithms, to solve it. The development of such algorithms is the goal of our further research.

References

1. Kh, G.E., Zalyubovsky, V.V., Sevastyanov, S.V.: Polynomial solvability of scheduling problems with stored resources and deadlines. In: Discrete Analysis and Operations Research, Ser. 2, vol. 7, no. 1, pp. 9–34 (2000)
2. Kazakovtseva, E.A., Servakh, V.V.: Complexity of the scheduling problem with loans. Discrete Anal. Oper. Res. 22(4), 35–49 (2015). https://doi.org/10.17377/daio.2015.22.478 https://doi.org/10.17377/daio.2015.22.478
3. Servakh, V.V., Shcherbinina, T.A.: On the complexity of one scheduling problem with stored resources. Vestn. NSU. Ser. Mat. Mekh. Inform. 8(3), 105–112 (2008)
4. Servakh, V.V.: A Dynamic Programming Algorithm for Some Project Management Problems. In: Proceedings International Workshop "Discrete Optimization Methods in Scheduling and Computer-aided Design". Minsk, pp. 90–92 (2000)
5. Romanova, A.A.: Minimizing the cost of a project schedule with renewable resources of variable cost. In: Materials of the XII International School-Symposium AMUR. Simferopol, pp. 392–396 (2018)
6. Blazewicz, J., Lenstra, J.K., Rinnooy Kan, A.H.G.: Scheduling subject to resource constraints: classification and complexity. Discrete Appl. Math. 5(1), 11–24 (1983). https://doi.org/10.1016/0166-218X(83)90012-4
7. Fazar, W.: The Origin of PERT. Controller, 598–621 (1962)
8. Fink, A., Homberger, J.: An ant-based coordination mechanism for re- source-constrained project scheduling with multiple agents and cash flow objectives. Flex. Serv. Manuf. J. 25, 94–121 (2013)
9. Fu, F.: Integrated scheduling and batch ordering for construction project. Appl. Math. Model. 38(2), 784–797 (2014)

10. Gu, H., Schutt, A., Stuckey, P.J., Wallace, M.G., Chu, G.: Exact and heuristic methods for the resource-constrained net present value problem. In: Schwindt, C., Zimmermann, J. (eds.) Handbook on Project Management and Scheduling Vol.1. IHIS, pp. 299–318. Springer, Cham (2015). https://doi.org/10.1007/978-3-319-05443-8_14

11. Hartmann, S., Briskorn, D.: An updated survey of variants and extensions of the resource-constrained project scheduling problem. Eur. J. Oper. Res. **297**, 1–14 (2022)

12. Hazır, Ö., Haouari, M., Erel, E.: Robust optimization for the discrete time-cost tradeoff problem with cost uncertainty. In: Schwindt, C., Zimmermann, J. (eds.) Handbook on Project Management and Scheduling Vol. 2. IHIS, pp. 865–874. Springer, Cham (2015). https://doi.org/10.1007/978-3-319-05915-0_9

13. Hanzalek, Z., Sucha, P.: Time symmetry of resource constrained project scheduling with general temporal constraints and take-give resources. Ann. Oper. Res. **248**, 209–237 (2017)

14. Khoshjahan, Y., Najafi, A.A., Afshar-Nadjafi, B.: Resource constrained project scheduling problem with discounted earliness-tardiness penalties: Mathematical modeling and solving procedure. Comput. Ind. Eng. **66**(2), 293–300 (2013)

15. Kaplan, E.H.: Relocation models for public housing redevelopment programs. Plann. Des. **13**, 5–19 (1986)

16. Kaplan, E.H., Amir, A.: A fast feasibility test for relocation problems. Eur. J. Oper. Res. **38**, 201–205 (1988)

17. Kelley, J.E.: Critical-path planning and scheduling: mathematical basis. Oper. Res. **9**, 296–320 (1961)

18. Leyman, P., Van Driessche, N., Vanhoucke, M., De Causmaecker, P.: The impact of solution representations on heuristic net present value optimization in discrete time/cost trade-off project scheduling with multiple cash flow and payment models. Comput. Oper. Res. **103**, 184–197 (2019)

19. Leyman, P., Vanhoucke, M.: A new scheduling technique for the resource-constrained project scheduling problem with discounted cash flows. Int. J. Prod. Res. **53**(9), 2771–2786 (2015)

20. Leyman, P., Vanhoucke, M.: Payment models and net present value optimization for resource-constrained project scheduling. Comput. Ind. Eng. **91**, 139–153 (2016)

21. Leyman, P., Vanhoucke, M.: Capital- and resource-constrained project scheduling with net present value optimization. Eur. J. Oper. Res. **256**(3), 757–776 (2017)

22. Kononov, A., Lin, B.M.-T.: Relocation problems with multiple working crews. Discret. Optim. **3**, 366–381 (2006)

23. Mohring, R.H.: Minimizing costs of resource requirements in project networks subject to a fixed completion time. Oper. Res. **32**(1), 89–120 (1984)

24. Okubo, H., et al.: Project scheduling under partially renewable resources and resource consumption during setup operations. Comput. Ind. Eng. **83**, 91–99 (2015)

25. Alan, A., Pritsker, B., Watters, L.J., Wolfe, P.M.: Multiproject scheduling with limited resources: a zero-one programming approach. Manage. Sci. **16**(1), 93–108 (1969)

26. Rodrigues, S.B., Yamashita, D.S.: An exact algorithm for minimizing resource availability costs in project scheduling. Eur. J. Oper. Res. **206**(3), 562–568 (2010)

27. Shahsavar, M., Niaki, S.T.A., Najafi, A.A.: An efficient genetic algorithm to maximize net present value of project payments under inflation and bonus-penalty policy in resource investment problem. Adv. Eng. Softw. **41**(7), 1023–1030 (2010)

28. Sevastyanov, S.V., Lin, B.M.T., Huang, H.L.: Tight complexity analysis of the relocation problem with arbitrary release dates. Theoret. Comput. Sci. **412**, 4536–4544 (2011)
29. Tabrizi, B.H.: Integrated planning of project scheduling and material procurement considering the environmental impacts. Comput. Ind. Eng. **120**, 103–115 (2018)
30. Tirkolaee, E.B., Goli, A., Hematian, M., Sangaiah, A.K., Han, T.: Multi-objective multi-mode resource constrained project scheduling problem using pare- to-based algorithms. Computing **101**, 547–570 (2019)
31. Thiruvady, D., Wallace, M., Gu, H., Schutt, A.: A Lagrangian relaxation and ACO hybrid for resource constrained project scheduling with discounted cash flows. J. Heuristics **20**(6), 643–676 (2014). https://doi.org/10.1007/s10732-014-9260-3
32. Vanhoucke, M.: A scatter search heuristic for maximising the net present value of a resource-constrained project with fixed activity cash flows. Int. J. Prod. Res. **48**(7), 1983–2001 (2010)
33. Waligora, G., Weglarz, J.: Discrete-continuous project scheduling with dis- counted cash inflows and various payment models -a review of recent results. Ann. Oper. Res. **213**, 319–340 (2014)
34. Watermeyer, K., Zimmermann, J.: A branch-and-bound procedure for the resource-constrained project scheduling problem with partially renewable resources and general temporal constraints. OR Spectrum **42**, 427–460 (2020)
35. Zoraghi, N., Shahsavar, A., Niaki, S.: A hybrid project scheduling and material ordering problem: modeling and solution algorithms. Appl. Soft Comput. **58**, 700–713 (2017)

Stochastic Greedy Algorithms
for a Temporal Bin Packing Problem
with Placement Groups

Alexander Turnaev[1,2]([✉]) and Artem Panin[2][iD]

[1] Novosibirsk State University, 2 Pirogova St., 630090 Novosibirsk, Russia
`a.turnaev@g.nsu.ru`
[2] Sobolev Institute of Mathematics, 4 Koptyug Ave., 630090 Novosibirsk, Russia

Abstract. The paper considers a temporal bin packing problem, where bins are servers using the Non-Uniform Memory Access architecture, and items are virtual machines. Bins are grouped together into racks. The main difference from the classical bin packing problem is that items are organized into placement groups, which are subdivided into subgroups. Items from different subgroups of the same group conflict and cannot be placed on the same rack.

The additional constraints considered are relevant to cloud computing. Service reliability is essential for both service providers and their customers. This is achieved by isolating subgroups of virtual machines from the same placement group within a failure domain.

Several stochastic algorithms have been developed to solve this problem. They are based on the classical first-fit algorithm, reordering of the packing sequence, and the bisection method. The algorithms give good results even for a rather naive initial solution (ordering). Moreover, they are easily parallelizable, which allows them to have an acceptable speed even for large problems.

Keywords: Stochastic algorithms · temporal bin packing · first-fit · placement groups · placement constraints

1 Introduction

In today's digital world, cloud computing plays a central role, providing businesses, organizations, and individual users with access to advanced computing resources, large data stores, and diverse applications.

However, with the increasing demand for cloud services, new challenges arise related to the efficient management and allocation of resources in the cloud. In particular, a common problem is to minimize the number of servers used to process a given set of virtual machines (VMs) under resource constraints. This problem can be considered as a bin packing problem [6], where the bins are servers and the items are virtual machines.

In cloud computing, VMs are assigned time windows that indicate when they should be added to or removed from the server. This situation corresponds to the temporal bin packing problem [2–4]. With the growing popularity of cloud computing, the requirements are also growing, and therefore, many generalizations and variants of this problem have been formulated [1, 8–11].

This paper generalizes the temporal bin packing problem, introducing additional constraint on the placement of VMs. Specifically, VMs are organized into placement groups, which are subdivided into subgroups. VMs from different subgroups of the same group conflict and cannot be placed on the same rack. The additional constraints considered are relevant to cloud computing. Service reliability is essential for both service providers and their customers. This is achieved by isolating subgroups of VMs from the same placement group[1] within a failure domain. A failure domain is a region that may be susceptible to failures. It can be a server, a rack, or even an entire data center. In general, accounting for failure domains ensures uninterrupted operation of provided services and data safety. This paper focuses solely on the failure domain in terms of racks.

The paper proposes several stochastic algorithms to obtain upper bounds for the temporal bin packing problem, taking into account resource and failure domain constraints, minimizing the active number of racks and servers, during the scheduling period. The algorithms are based on the classical first-fit algorithm, reordering of the packing sequence, and the bisection method.

Section 2 presents the problem statement, while Sect. 3 describes the algorithms for finding lower and upper bounds. The results of numerical experiments are discussed in Sect. 4.

2 Problem Statement

In our formulation, the cloud resource pool has a hierarchical structure, as shown in Fig. 1. It consists of four levels.

1. The entire resource pool contains a potentially unlimited number of identical racks.
2. Each rack contains a given number of identical servers.
3. Each server has a Non-Uniform Memory access (NUMA) architecture [7] with a limited number of NUMA-nodes, either 2 or 4.
4. Each NUMA-node has constraints on two resources: CPU and memory. NUMA-nodes are not homogeneous and may differ in the capacity of these two resources.

Users send requests to provide them with VMs for some time periods, which sets the time of creation t_1 and the time of deletion t_2 for each VM. VMs are categorized by type, which determines:

---

[1] Placement groups. Amazon Elastic Compute Cloud. URL: https://docs.aws.amazon.com/AWSEC2/latest/UserGuide/placement-groups.html.
Using partition placement groups for large distributed and replicated workloads in Amazon EC2. URL: https://aws.amazon.com/ru/blogs/compute/using-partition-placement-groups-for-large-distributed-and-replicated-workloads-in-amazon-ec2/.

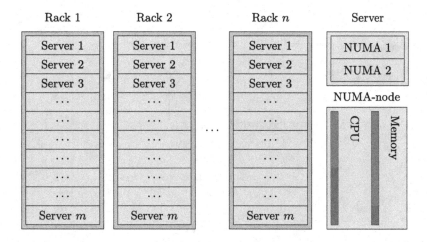

Fig. 1. Resource pool structure

1. Resource constraints: The amount of CPU and memory required to run this VM.
2. Size: the number of NUMA-nodes occupied by a VM of this type, as well as the number of parts into which this VM is divided. VM parts have the same resource constraints and must be placed within the same server, strictly on different NUMA-nodes. The paper considers two sizes: *small* (1 part) and *large* (2 parts).

After placing the VM on the server, it can no longer be moved. Each resource constraint must be satisfied at any point in the VM's time window.

The novelty of the temporal bin packing problem under study is the placement groups and placement constraints. A placement group consists of disjoint sets of VMs, called partitions. Each VM can belong to only one placement group. VMs from the same group but from different partitions conflict and cannot be placed on the same rack at the same time (see Fig. 2).

This paper discusses the problem of minimization of the number of used racks and servers, under a given set of requests (or VMs), resource constraints, and placement groups. The structure of the resource pool assumes that each rack contains a given number of servers. Therefore, minimizing the number of servers and minimizing the number of racks are equivalent criteria. However, if the number of racks is minimal, the developed algorithms will minimize the number of servers used.

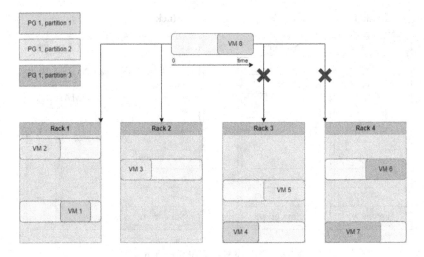

Fig. 2. Illustration of a feasible solution with respect to the placement constraint. Here the lifetime of each VM is shown as an interval on time axis. The partitions of the placement group are shown by colors. The crosses mark the racks where VM8 can not be placed due to the placement constraint. VM8 does not overlap in time with VM2 or VM3.

3   Algorithms for Computing Bounds on the Minimum Number of Racks

3.1   Lower Bound

To find the lower bound, an algorithm based on column generation [5] from the paper [9] was used. To apply it, all constraints related to placement groups are relaxed, i.e. partition conflicts are not taken into account. And since the algorithm is for a single-moment problem, we use it only for a few of the most loaded event moments.

3.2   Upper Bound

Several stochastic algorithms have been developed to calculate the upper bound. They are based on improving some initial solution obtained using the first-fit algorithm on some ordering of VMs. To begin with, we informally describe the functions that will be used later in the algorithms:

1. **Pack**(VMSeq) is a VM placement function. The argument of the function is an ordered set of VMs. The function places VMs as in the first-fit approach, checking for capacity and placement constraints. The function returns a valid VM placement (feasible solution). This placement always exists because new racks can be added in unlimited numbers.

2. **GetValue**(Sol) is a function that returns the value of the objective function, i.e., the ordered pair of racks and servers used. The argument is the solution to the problem. The values are compared in lexicographic order.

3. **RemoveVMs**(Sol,VMSeq) is a function that creates and returns a partial solution by removing the given sequence of VMs from the given partial solution Sol. That is, for each VM in the given sequence, the NUMA-node, the server, and the rack on which it is located are searched and removed from there.

4. **Shuffle**(VMSeq) is a function that randomly reorders a sequence of VMs.

5. **PartiallyPack**(Sol,VMSeq) is equivalent to **Pack**(VMSeq), but the packaging process begins from an initial resource pool state that corresponds to the feasible solution Sol.

6. **FirstHalf**(VMSeq) returns the first half of the sequence VMSeq.

7. **SecondHalf**(VMSeq) returns the second half of the sequence VMSeq.

Stochastic Local Improvement Algorithm. The first algorithm iteratively transforms a given initial sequence of VMs (InitVMSeq) and computes a feasible solution of the problem using a first-fit approach. The algorithm only improves the tail part of the sequence without changing the entire sequence. The algorithm is described by pseudocode 1.

Algorithm 1. SLI (Stochastic Local Improvements)

    Input: *InitVMSeq*
    Output: *Solution*

1: $i \leftarrow 0$, VMSeq $\leftarrow$ InitVMSeq
2: Sol $\leftarrow$ **Pack**(InitVMSeq)
3: Value $\leftarrow$ **GetValue**(Sol)
4: **while** VMSeq $\neq \varnothing$ **do**
5:     PartSol $\leftarrow$ **RemoveVMs**(Sol,VMSeq)
6:     **if** **GetValue**(PartSol) = Value **then**
7:         **break**
8:     **end if**
9:     $i \leftarrow i + 1$, $k \leftarrow 1$
10:     **while** $k \leq \varphi(i)$ **do**
11:         NewVMSeq $\leftarrow$ **Shuffle**(VMSeq)
12:         NewSol $\leftarrow$ **PartiallyPack**(PartSol,NewVMSeq)
13:         NewValue $\leftarrow$ **GetValue**(NewSol)
14:         **if** NewValue < Value **then**
15:             Sol $\leftarrow$ NewSol, Value $\leftarrow$ NewValue
16:             VMSeq $\leftarrow$ NewVMSeq
17:         **end if**
18:         $k \leftarrow k + 1$.
19:     **end while**
20:     VMSeq $\leftarrow$ **SecondHalf**(VMSeq)
21: **end while**
22: **return** Sol

At the first iteration of the algorithm (when $i = 1$), the full sequence is shuffled and the best solution is chosen in the sense of lexicographic minimization of the objective value (ordered pair of the number of occupied racks and servers). The process is repeated for the second half of the sequence, for the second half of the second half of the sequence, and so on. The function $\varphi : \mathbb{N} \to \mathbb{N}$ determines the number of random reorderings at each iteration.

Modified Stochastic Local Improvement Algorithm. In the *SLI* algorithm, the first half of the sequence is practically not affected. It would be beneficial to extend randomization to the left halves as well.

Consider the graph (V, E), which is a complete binary tree of depth h, for some $h \in \mathbb{N}$. Let v_{ij} be the j-th vertex at level i. Denote the set of vertices at level i as V^i. Each vertex in the tree is some kind of a subsequence of VMs. Define a tree of subsequences by induction:

1. V^1 contains a single vertex v_{11}, which is a complete sequence of VMs.
2. The set V^k consist of 2^{k-1} vertices $v_{k1}, \ldots, v_{kj}, \ldots v_{k(2^{k-1})}$. If j is odd, then the vertex v_{kj}, is the first half of $v_{(k-1)((j+1)/2)}$ and there is an edge between them. If j is even, then the vertex v_{kj} is the second half of $v_{(k-1)(j/2)}$ and there is an edge between them.
3. There are no other edges.

The second algorithm recursively traverses the graph, uniformly shuffling the sequence of VMs for each vertex v_{ij} $\varphi(i)$ times, propagating the best solution and ordering of VMs with each transition up and down the tree. After processing the vertex, VMs are repacked. The algorithm then compares the solutions and selects the best one. When moving left-down the tree, the second half of the sequence is removed. When moving right-down the tree, the first half of the sequence is fixed. When moving up the tree, the best solution and subsequence are passed to the top vertex for comparison with the current solution in it.

The algorithm is described by pseudocodes 2, 3. Figure 3 shows the tree of subsequences and its traversal for a sequence of VMs $(1, 2, 3, 4, 5, 6, 7, 8)$ with depth 3.

Algorithm 2. MSLI (Modified Stochastic Local Improvements)

    Input: *InitialVMSeq*
    Output: *Solution*
1: **return** ProcessNode(**Pack**(InitialVMSeq),InitialVMSeq,1)

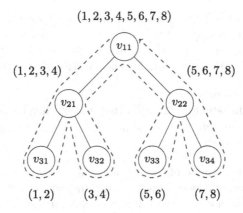

Fig. 3. Tree of subsequences and its traversal

Algorithm 3. ProcessNode

  Input: *InitSol*, *VMSeq*, *i*
  Output: *Solution*
1: Value ← **GetValue**(InitSol)
2: Sol ← InitSol
3: PartSol ← **RemoveVMs**(Sol, VMSeq)
4: $k \leftarrow 1$
5: **while** $k \leq \varphi(i)$ **do**
6:     NewVMSeq ← **Shuffle**(VMSeq)
7:     NewSol ← **PartiallyPack**(PartSol,NewVMSeq)
8:     NewValue ← **GetValue**(NewSol)
9:     **if** NewValue < Value **then**
10:         Sol ← NewSol, Value ← NewValue, VMSeq ← NewVMSeq
11:     **end if**
12:     $k \leftarrow k + 1$.
13: **end while**
14: LeftVMSeq ← **FirstHalf**(VMSeq), RightVMSeq ← **SecondHalf**(VMSeq)
15: **if** $i < h$ **then**
16:     PartSol ← **RemoveVMs**(Sol,RightVMSeq)
17:     PartSol ← **ProcessNode**(PartSol, LeftVMSeq, $i + 1$)
18:     NewSol ← **PartiallyPack**(PartSol,RightVMSeq)
19:     NewValue ← **GetValue**(NewSol)
20:     **if** NewValue < Value **then**
21:         Value ← NewValue, Sol ← NewSol
22:     **end if**
23:     Sol ← **ProcessNode**(Sol,RightVMSeq, $i + 1$)
24: **end if**
25: **return** Sol

Neighbor Modified Stochastic Local Improvements Algorithm. The previous algorithm can be improved if we sacrifice the speed. To do this, we will determine for each vertex its right neighbor and, together with the randomization of this vertex, we will pack the right neighbor with it, but without randomization. This, in a sense, guarantees that the solution will avoid falling into a bad local minimum. Let's define the neighbors for each vertex by induction:

1. The root v_{11} has no neighbors.
2. If a vertex v_{ij} is the left child of some other vertex, then its neighbor is $v_{i(j+1)}$.
3. If a vertex v_{ij} is the right child of some other vertex, then its neighbor is the neighbor of parent vertex.

Figure 4 shows an example of the neighbors for each vertex.

The changes in pseudocode 3 are insignificant, so we will describe only the main changes and will not give the full pseudocode of the algorithm:

1. The **ProcessNode** function now accepts an additional RightNodeVMSeq argument. At the very first call of this function, let RightNodeVMSeq $= \varnothing$.
2. After step 3, additionally remove RightNodeVMSeq. After step 7, additionally pack RightNodeVMSeq.
3. In steps 16 and 18, RightVMSeq is replaced by RightNodeVMSeq.
4. In step 17, the **ProcessNode** function now calls with an additional argument RightVMSeq.
   In step 23, the **ProcessNode** function now calls with an additional argument RightNodeVMSeq.

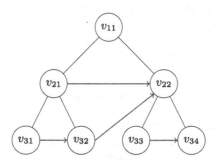

Fig. 4. Neighbors of the vertices

Additional Algorithms. In addition, two algorithms are defined for obtaining the initial ordering and complete randomization:

1. *FF*: Combine all VMs into a single sequence, starting with a set of request-ordered VMs from placement groups with partitions, followed by all other VMs also ordered by requests.
2. *FFD*: Same as FF, but the VMs are sorted by descending *area* instead of by request.

$$area = \frac{cpu + memory}{2} \cdot N \cdot (t_2 - t_1)$$

where t_1 is the creation time, t_2 is the deletion time, *cpu* and *memory* are the amount of resources needed to process a virtual machine, N is the number of VM parts. If necessary, a weighted sum is taken.
3. *FULL RAND*: Shuffle the complete sequence of VMs n times and choose the best solution.

4 Computational Experiments

Note that the loop with the search for the best subsequence in algorithms 1 and 3 can be easily parallelized. This is important because computational experiments were carried out on a parallelized version of each algorithm.

The algorithms were implemented in Python 3.7, the tests were performed on a computer with an Intel(R) Core(TM) i7-8750H 2.21 GHz processor with 16Gb of RAM and were parallelized by 6 processes using the multiprocessing Python module. The following parameters were used for the algorithms: $\varphi(i) = 10 \cdot i$, $h = 4$, and $n = 100$.

4.1 Test Data Description

An open synthetic dataset from the benchmark library "Discrete Location Problems"[2] is used to conduct the experiments. The dataset contains a total of 450 tests, which are evenly distributed according to the following parameters:

1. Types of placement groups
   (a) LPO, only large placement groups with partitions.
   (b) MP, only large and small placement groups with partitions.
   (c) DMP, placement groups with partitions and unconstrained placement groups.

---

[2] Benchmark library "Discrete Location Problems". Sobolev Institute of Mathematics. http://math.nsc.ru/AP/benchmarks/Temporal Bin Packing/binpack.html.

2. Total number of requests
   (a) Small, tests containing an average of 14,702 machines
   (b) Medium, tests containing an average of 26,545 machines
   (c) Large, tests containing an average of 51,301 machines
3. The number of NUMA-nodes used (2,4)

4.2   Results

Tables 1, 2, 3 show the average GAP values by racks and servers, the average running time of the SLI, MSLI, NMSLI algorithms with different initial solutions found by FF and FFD, the results of FF and FFD themselves, as well as the results of the full randomization algorithm on different test families. *BEST INIT* refers to selecting the best of the two solutions found by FF and FFD, respectively. The use of both initial solutions simultaneously is indicated by *ALL INIT*. In this case, the local improvement algorithm is run twice. GAP is calculated by the formula GAP = 100% (UB - LB)/LB, where LB is the lower bound obtained by the algorithm from Sect. 3.1 and UB is the upper bound obtained by one of the algorithms from Sect. 3.2.

The analysis of the results shows that each of the algorithms is able to significantly improve the initial solution. As expected, MSLI works better than SLI, and NMSLI is better than MSLI, despite the fact that the size of the tree was significantly limited for MSLI and NMSLI and can be increased if necessary.

Note that choosing an obviously dominant algorithm may increase the GAP by servers. For example, FF+SLI has a better GAP value than INIT+SLI, which includes FF+SLI. This is due to the fact that the number of racks is minimized first, and the number of servers is minimized afterwards. In fact, this situation would not have arisen if there were no constraints on placement groups.

As you can see, choosing the best initial solution improves the final result on average. However, in rare cases, selecting the inferior initial solution can result in a better outcome.

On average, the execution time of the MSLI is twice as long as the SLI, and the execution time of the NMSLI is twice as long as the MSLI. On the other hand, one can increase the height of the tree and sacrifice the number of iterations φ by reducing its growth or even making it a constant.

All developed algorithms perform best with DMP type families, large or medium size, and 2 NUMA-nodes. To improve the results, it may be necessary to reconsider the local objective function (**GetValue**).

Note that each algorithm performs better than FULL RAND. Moreover, the MSLI algorithm does almost the same amount of work as FULL RAND, as shown in Table 3. Indeed, at each level i of the subsequences tree, we randomize and pack 2^i subsequences of the same length $\varphi(i) = 10 \cdot i$ times, assuming that it takes the same amount of time on average, equal to $T/2^i$, where T is the average time spent to pack a complete sequence of VMs. Then $10 \cdot i \cdot T$ time is needed for each level or a total of $100 \cdot T$ of time, as for FULL RAND.

Table 1. Result GAP by racks

Algorithm	Type			Size			Numas		Total
	DMP	LPO	MP	large	medium	small	2	4	
FF	8.67%	11.81%	10.91%	10.05%	9.72%	11.61%	11.40%	9.52%	10.46%
FFD	4.22%	9.58%	9.84%	5.92%	6.73%	10.99%	4.17%	11.59%	7.88%
BEST INIT	4.22%	7.78%	7.10%	5.87%	5.57%	7.65%	4.16%	8.57%	6.36%
FF+SLI	3.56%	7.34%	6.94%	5.28%	5.16%	7.40%	3.80%	8.09%	5.95%
FFD+SLI	2.98%	7.32%	7.24%	4.94%	4.74%	7.86%	3.15%	8.54%	5.84%
BEST INIT+SLI	2.98%	6.90%	6.44%	4.93%	4.59%	6.80%	3.15%	7.73%	5.44%
ALL INIT+SLI	2.90%	6.74%	6.27%	4.85%	4.48%	6.58%	3.03%	7.58%	5.30%
FF+MSLI	3.38%	7.20%	6.90%	5.23%	5.08%	7.18%	3.67%	7.99%	5.83%
FFD+MSLI	2.66%	7.13%	7.07%	4.71%	4.54%	7.61%	2.93%	8.31%	5.62%
BEST INIT+MSLI	2.66%	6.69%	6.29%	4.71%	4.42%	6.51%	2.93%	7.50%	5.21%
ALL INIT+MSLI	2.58%	6.54%	6.18%	4.64%	4.31%	6.34%	2.81%	7.38%	5.10%
FF+NMSLI	3.22%	6.91%	6.57%	5.13%	4.79%	6.78%	3.44%	7.69%	5.57%
FFD+NMSLI	2.47%	6.78%	6.64%	4.61%	4.29%	6.99%	2.75%	7.84%	5.30%
BEST INIT+NMSLI	2.47%	6.43%	5.98%	4.61%	4.17%	6.10%	2.77%	7.15%	4.96%
ALL INIT+NMSLI	2.36%	6.26%	5.79%	4.51%	4.03%	5.88%	2.61%	6.99%	4.80%
FULL RAND(×100)	4.06%	7.74%	7.72%	5.47%	5.37%	8.68%	3.97%	9.04%	6.51%

Table 2. Result GAP by servers

Algorithm	Type			Size			Numas		Total
	DMP	LPO	MP	large	medium	small	2	4	
FF	8,63%	11,48%	10,32%	9,97%	9,50%	10,96%	11,31%	8,97%	10,14%
FFD	4,21%	4,49%	3,82%	4,62%	3,64%	4,26%	3,34%	5,01%	4,17%
BEST INIT	4,21%	6,15%	5,42%	4,74%	4,92%	6,12%	3,40%	7,12%	5,26%
FF+SLI	2,73%	5,81%	5,58%	4,58%	3,85%	5,68%	3,50%	5,91%	4,70%
FFD+SLI	2,91%	4,96%	4,13%	4,34%	3,25%	4,42%	2,60%	5,40%	4,00%
BEST INIT+SLI	2,91%	5,31%	4,66%	4,33%	3,47%	5,07%	2,59%	5,99%	4,29%
ALL INIT+SLI	2,76%	5,14%	4,57%	4,25%	3,32%	4,90%	2,56%	5,75%	4,16%
FF+MSLI	2,53%	5,70%	5,38%	4,45%	3,69%	5,47%	3,30%	5,77%	4,54%
FFD+MSLI	2,53%	4,77%	4,07%	4,09%	3,13%	4,15%	2,33%	5,26%	3,79%
BEST INIT+MSLI	2,53%	5,09%	4,47%	4,09%	3,30%	4,69%	2,32%	5,73%	4,03%
ALL INIT+MSLI	2,43%	5,03%	4,37%	4,07%	3,19%	4,57%	2,27%	5,61%	3,94%
FF+NMSLI	2,43%	5,43%	5,12%	4,39%	3,52%	5,08%	3,08%	5,57%	4,33%
FFD+NMSLI	2,37%	4,63%	3,94%	4,01%	2,99%	3,93%	2,22%	5,07%	3,64%
BEST INIT+NMSLI	2,37%	4,86%	4,32%	4,01%	3,15%	4,39%	2,21%	5,49%	3,85%
ALL INIT+NMSLI	2,27%	4,79%	4,20%	3,99%	3,05%	4,22%	2,17%	5,34%	3,76%
FULL RAND(×100)	2,66%	5,47%	5,05%	4,56%	3,63%	5,00%	3,41%	5,38%	4,40%

Table 3. Result time in minutes

Algorithm	Type			Size			Numas		Total
	DMP	LPO	MP	large	medium	small	2	4	
FF	0,54	0,51	0,38	1,04	0,29	0,10	0,29	0,66	0,48
FFD	0,68	0,89	0,64	1,64	0,43	0,15	0,50	0,98	0,74
BEST INIT	1,22	1,40	1,02	2,68	0,71	0,25	0,79	1,64	1,21
FF+SLI	5,94	4,40	3,53	9,66	2,96	1,25	3,33	5,92	4,62
FFD+SLI	5,63	3,62	2,76	8,12	2,90	0,99	2,11	5,90	4,00
BEST INIT+SLI	6,85	4,99	3,73	10,78	3,57	1,23	2,90	7,48	5,19
ALL INIT+SLI	11,57	8,02	6,30	17,78	5,87	2,24	5,44	11,82	8,63
FF+MSLI	14,98	12,45	9,64	23,71	8,98	4,37	9,81	14,90	12,36
FFD+MSLI	6,94	13,12	18,53	28,15	7,49	2,96	7,49	18,24	12,86
BEST INIT+MSLI	8,16	14,61	17,04	30,48	6,48	2,85	8,28	18,26	13,27
ALL INIT+MSLI	21,92	25,57	28,17	51,86	16,46	7,33	17,29	33,14	25,22
FF+NMSLI	25,75	20,14	16,03	43,48	13,17	5,28	15,64	25,64	20,64
FFD+NMSLI	26,86	16,84	13,30	38,00	13,37	5,64	11,06	26,94	19,00
BEST INIT+NMSLI	28,08	17,87	14,07	40,75	13,68	5,60	11,85	28,17	20,01
ALL INIT+NMSLI	52,62	36,99	29,33	81,48	26,54	10,91	26,70	52,58	39,64
FULL RAND(×100)	14,32	10,45	8,60	23,82	6,79	2,76	7,40	14,85	11,12

5   Conclusions

In this paper, we considered the temporal bin packing problem with additional constraints on placement groups. Several stochastic algorithms have been developed to solve this problem. They are based on the classical first-fit algorithm, reordering of the packing sequence, and the bisection method. The algorithms give good results even for a rather naive initial solution (ordering). Moreover, they are easily parallelizable, which allows them to have an acceptable speed even for large problems. To achieve better performance, algorithms can be implemented in a faster language like C++.

Further research may be related to the study of the application of algorithms to better initial solutions, the selection of optimal parameters h, φ, or even methods for making them dynamic.

Acknowledgement. The study was carried out within the framework of the state contract of the Sobolev Institute of Mathematics (project FWNF-2022-0019).

References

1. Aydın, N., Muter, İ, Birbil, Şİ: Multi-objective temporal bin packing problem: an application in cloud computing. Comput. Oper. Res. **121**, 104959 (2020). https://doi.org/10.1016/j.cor.2020.104959

2. De Cauwer, M., Mehta, D., O'Sullivan, B.: The temporal bin packing problem: an application to workload management in data centres. In: 2016 IEEE 28th International Conference on Tools with Artificial Intelligence (ICTAI), pp. 157–164. IEEE (2016). https://doi.org/10.1109/ICTAI.2016.0033

3. Dell'Amico, M., Furini, F., Iori, M.: A branch-and-price algorithm for the temporal bin packing problem. Comput. Oper. Res. **114**, 104825 (2020). https://doi.org/10.1016/j.cor.2019.104825

4. Furini, F., Shen, X.: Matheuristics for the temporal bin packing problem. Recent Developments in Metaheuristics, pp. 333–345 (2018). https://doi.org/10.1007/978-3-319-58253-5_19

5. Gilmore, P.C., Gomory, R.E.: A linear programming approach to the cutting-stock problem. Oper. Res. **9**(6), 849–859 (1961). https://doi.org/10.1287/opre.9.6.849

6. Johnson, D.S.: Near-optimal bin packing algorithms. Ph.D. thesis, Massachusetts Institute of Technology (1973)

7. Manchanda, N., Anand, K.: Non-uniform memory access (NUMA). New York University **4** (2010)

8. Pires, F.L., Barán, B.: A virtual machine placement taxonomy. In: 2015 15th IEEE/ACM International Symposium on Cluster, Cloud and Grid Computing, pp. 159–168. IEEE (2015). https://doi.org/10.1109/CCGrid.2015.15

9. Ratushnyi, A., Kochetov, Y.: A column generation based heuristic for a temporal bin packing problem. In: Pardalos, P., Khachay, M., Kazakov, A. (eds.) MOTOR 2021. LNCS, vol. 12755, pp. 96–110. Springer, Cham (2021). https://doi.org/10.1007/978-3-030-77876-7_7

10. Sakhno, M.: A grouping genetic algorithm for the temporal vector bin packing problem. In: 2023 19th International Asian School-Seminar on Optimization Problems of Complex Systems (OPCS), pp. 94–99. IEEE (2023). https://doi.org/10.1109/OPCS59592.2023.10275770

11. Shi, J., Luo, J., Dong, F., Jin, J., Shen, J.: Fast multi-resource allocation with patterns in large scale cloud data center. J. Comput. Sci. **26**, 389–401 (2018). https://doi.org/10.1016/j.jocs.2017.05.005

Fast Heuristics for a Staff Scheduling Problem with Time Interval Demand Coverage

Igor Vasilyev[1], Anton V. Ushakov[1]([✉]), Dmitry Arkhipov[4],
Ivan Davydov[2], Ildar Muftahov[5], and Maria Lavrentyeva[3]

[1] Matrosov Institute for System Dynamics and Control Theory of SB RAS,
134 Lermontov Street, 664033 Irkutsk, Russia
{vil,aushakov}@icc.ru
[2] Sobolev Institute of Mathematics of SB RAS, 4 Academic Koptyug Avenue,
630090 Novosibirsk, Russia
[3] MPG IT Solutions LLC, Novosibirsk, Russia
[4] Moscow, Russia
[5] Irkutsk, Russia

Abstract. Staff scheduling is a key component of supporting and increasing competitiveness for many service enterprises. This is of especially urgent concern for organizations that provide service on a twenty-four hour basis and often encounter significant fluctuations of demand. By scheduling personnel, the employers have also to strictly follow local laws, industrial regulations, and workload agreements that may considerably affect the final schedule. Staff preferences have also to be taken into account when planning work schedules, since it may reduce turnover and increase productivity. In this paper we consider a staff scheduling problem that arise in the industrial fields where the demand in staff is highly dynamic and varies within time intervals throughout a day. The goal is to assign each employee with a shift for each day of a planning horizon so as to minimize the sum of unsatisfied demand over all time intervals subject to hard workplace constraints. Note that each employee may have his/her day-specific set of pre-defined shifts and a set of work-rule constraints. We formulate the scheduling problem as a mixed-integer program. We develop several fast two-stage heuristic algorithms that includes a constructive step to find an initial solution followed by fast local search procedures. We demonstrate the effectiveness of the proposed approaches on a number of real-world huge-scale scheduling problems involving thousands of employees.

Keywords: staff scheduling · rostering · integer programming · local search

1 Introduction

Staff scheduling is a key component of supporting and increasing competitiveness for many service enterprises. This is of especially urgent concern for organizations

© The Author(s), under exclusive license to Springer Nature Switzerland AG 2024
A. Eremeev et al. (Eds.): MOTOR 2024, LNCS 14766, pp. 212–228, 2024.
https://doi.org/10.1007/978-3-031-62792-7_15

that provide service on a twenty-four hour basis and often encounter significant fluctuations of demand. In this setting it is very important to assign the right employee on duty at the right time, so as all customers may be provided the expected quality of service. At the same time, the local labor legislation and personnel workplace agreements strictly govern the employees' rights at work and their working conditions, that especially make staff scheduling an extremely challenging process. Finally, even if all employees are provided with a schedule feasible with respect to their workplace agreements, it is also important to meet the fairness considerations, since the latter may affect productivity and provoke turnover. Staff scheduling problems are typically highly constrained, application-oriented, and challenging optimization problems.

These problems are ones of the oldest and widely studied problems in operations research, scheduling, and combinatorial optimization communities. Staff scheduling origins are usually traced back to 1950th. By now, there are plenty of literature devoted to huge number of different variations of staff scheduling problems (e.g. for a survey, see [7,8,16,17]). Staff scheduling may have different goals and involve various decisions, which scheduling objectives and constraints are dependent on. The first classification of the personnel scheduling problems was proposed in [1]. According to [8], who proposed a more general classification, the process of scheduling is divided into six modules that are assumed to follow one another. Note that not all the modules may be required to build a schedule for a particular application area. Some modules are often combined into one. The first module is *staffing* or demand module, which is to determine the number of staff required at different time intervals. This module is heavily dependent on application. The next modules are *days off scheduling* to determine the rules of assigning rest days to staff, *shift scheduling* to select the set of shifts in order to meet demand, *line of work construction (tour scheduling)* to find out which employees may be assigned to which shifts in order to determine the work rule constraints (e.g. how shifts should follow one another). The fifth module, *task assignment*, is actually optional and is required only when there are some tasks to be carried out during each shift. These tasks may require particular staff skills or levels of seniority. The last module is *staff assignment* which involves the assignment of individual staff to the lines of work. Staff assignment is often done during construction of the work lines.

The first major decision that affects modeling process is how the demand in staff is specified. For example, under the task based demand, there may be a set of tasks that require a predefined amount of staff and skills to be completed [5,11]. Secondly, the demand can also be unknown in advance but predicted for some intervals within a day. This situation is quite common for many real-life systems, e.g. call centers, police services, toll booths etc. (for a survey, see [6]). Finally demand, may be specified by the requirements of shifts. This kind of demand is widely considered in nurse scheduling applications.

After the type of demand is determined, the staff members are scheduled to cover it. This process basically consists in scheduling days off and working shifts. There are several widespread approaches to how staff rostering is modeled [18]. The traditional one is supposed that all feasible roster sequences (or stints) of shifts and rest days are found and known, hence the problem is modeled as the set partitioning problem, i.e. one has to assign each employee to exactly one stint [4]. The second approach assumes that selection of the start and end times of shifts are also decisions to be made. The planning period is divided into time intervals (e.g. 15 minutes long), and one has to decide which intervals an employee works in [9, 19, 20]. Finally, a set of shifts, as well as their start and end times, may be pre-defined and be part of input. In this case, the traditional 3-index model consists in deciding which shift is assigned to a particular employee. This decision has to be made for each day of the planning horizon [12]. Note that in some applications (like ground staff scheduling at airports), a shift scheduled to an employ may consist of a set of operational tasks, hence an employee must further be assigned to tasks with respect to her skills [3, 13, 18].

In this paper we address a staff rostering problem where a set of shifts is already defined and the demand in staff is specified for intervals within each day of the planning horizon. Shifts may in general be specific for each particular employee and day. The problem involves a large number of hard workplace constraints and aims at minimizing understaffing over time intervals.

The large number of variations of staff scheduling problem resulted in a plenty of solution approaches developed. As these problems are highly application-oriented, many research papers addressed modeling of specific scheduling scenarios. For example, in [2], the authors addressed the problem of staff scheduling for the US post service, where there are full- and part-time shifts, and demand is given for 30 time intervals. Among others, the goal of the problem includes lunch breaks scheduling. As was noted above, flexible demand is especially inherent to call centers, hence many research papers are devoted to this application. For example, in [20], a staff scheduling problem supplemented with routing requirements was presented. It includes regular and overtime shifts, multi-skill demand requirements, and considers employees' discontent rates dependent of their satisfaction of the schedule. Both aforementioned problems were solved with an MIP solver. Due to long history and a large number of variations, almost all metaheuristic strategies and general heuristics were adapted for staff scheduling problems. Thus, the recent efforts are focused on developing a more general frameworks supporting many variations of the problem. For example, in [10], the authors developed a simulated annealing framework that can deal not only with a large number of workplace constraints but also various demand types.

2   Problem Statement

We suppose that each day in the scheduling period is divided into intervals of 15 min long, i.e. a day consists of 96 time intervals. There is demand in staff members defined for each time interval. There is also given a set of shifts. For each shift, there is a start and end time, as well as break intervals, where an employee is supposed to have lunch. Shifts may have different length, which may be used to model employees with different workplace agreements (part-time, full-time, etc.) Note that shifts may end at the next day, i.e. its end interval belongs to the next day and it is larger than 96. Hence, shifts in subsequent days may obviously overlap. The main novel feature of the problem is that each employee i may have a specific set of shifts S^{ij} for each particular day j, i.e. $\bigcap_{i \in I, j \in J} S^{ij} = \emptyset$. This property may be motivated by careful addressing employees' preferences on the shift design step. Each employee must be assigned exactly one shift or day off for each particular day of the planning horizon.

Thus, the modules related to staffing (determining demand) and shift scheduling are supposed to be already completed by the system planners and the corresponding information is given as input for the problem. Due to budget limitations and workforce shortage, the demand is usually cannot be satisfied, hence it is considered in the form of soft constraint.

The work line (or schedule) of a particular employee must meet labor agreements and government regulations, such as the required number of days off, working hours per week, etc. Note that the labor agreements are in general specific for each particular employee. For example, two employees may have different work rules for the number of working days per week, or may have different requirements for the number of rest days for the same week. Note that labor agreements and regulations cannot be violated, as they governed by laws, and must be considered as hard constraints.

It is interesting to note that the type of constraints that must be considered as hard or soft ones is dependent on application. For example, nurse rostering problems usually consider the work-rules constraints and demand satisfaction constrains as soft and hard ones, respectively. The demand in nurses with particular skills must always be satisfied for each particular shift, as this is critically important for patients' safety.

The work rule constraints are usually categorized as counters, series or successions. Counter constraints restrict the number of times a specific roster item (shift type, assignments or days-off) can occur within a certain period. Series restrict consecutive occurrences of specific roster items. These different types of constraints can be expressed as either ranged, minimum, maximum or exact. Finally, successions denote a special type of series, which restrict occurrences of specific roster items on two consecutive days [15,16]. For example a day shift cannot follow a night one, or if a particular shift is assigned in one day, the next day must be the working day, etc. For our problem, counter constraints are defined for two types of periods: week and overall planning horizon.

In the proposed problem, we also assume that some employees may have pre-assigned shift types for some particular days and these requirements must be satisfied in the schedule. For example, an employee may be allowed to assign only working shifts for some days, or she must have pre-specified rest days, etc.

In our problem we consider the following work-rules constraints:

1. Counter constraints:
   - Maximum/Minimum number of days-off per week.
   - Maximum/Minimum number of night shifts per week.
   - Maximum/Minimum number of days-off per planning period.
   - Maximum/Minimum number of working hours per week.

  – Maximum/Minimum number of working hours per planning period.
2. Series constraints:
  – Maximum/Minimum number of consecutive working days.
  – Maximum number of consecutive days-off.
  – Maximum number of consecutive shifts with the same id.
  – Maximum number of consecutive night shifts.
3. Succession constraints:
  – Minimum time between adjacent shifts.

The objective is to find a feasible schedule that meets all the work-rule constraints such that the total under-staffing over all time intervals is minimized.

3   Problem Formulation

In this section, we formulate the aforementioned staff scheduling problem as a mixed integer program. We suppose that there is a set of employees (staff) I and the planning horizon J, i.e. the number of days which need to assign shifts. Each day is divided into time intervals: each interval is 15 min long. There is also a set of shifts S, each of which has a start time, end time, and a set of rest intervals. Each shift has a unique id and may be one of the following types: day shift, night shift, and day off. Each employee can be assigned at each particular day only to some specific subset of shifts. Other problem input parameters are:

– $J_w \subseteq J$ is a set of days in week $w \in V$.
– $S_p \subset S$ – subset of shifts of type p, $s' \in S$ – day-off shift type. $S = \bigcup_{p \in P} S_p \cup \{s'\}$, (p: 1 - DAY, 2 - NIGHT).
– $W = S \setminus \{s'\}$ – working shifts.
– T – a set of all time slots.
– L_s – the number of time slots in shift s, except break intervals.
– a_{ijs} is 1 if staff i is available at day j for shift s.
   • $S^{ij} = \{s \in S : a_{ijs} = 1\}$ – a set of shifts that employee i can be assigned to on day $j \in J$.
– b_{jst} is 1 if time slot t is covered by shift s at day j.
   • $W^{jt} = \{s \in W : b_{jst} = 1\}$ is a set of working shifts that cover interval t.
– d_t – staff demand at time slot t.

To cast the staff scheduling problem as a mixed integer program, let us introduce the following binary variables:

$$x_{ijs} = \begin{cases} 1, \text{ if shift } s \text{ is assigned to staff } i \text{ on day } j; \\ 0, \text{ otherewise.} \end{cases}$$

We also introduce the integer variables u_t, $t \in T$, and λ. The variable u_t is a slack variable that defines the under-staffing rate at time interval t, whereas λ defines the maximum unsatisfied demand over all time slots $t \in T$.

With these notations the scheduling problem can be formulated as follows:

$$\min\left(w_1 \sum_{t\in T} u_t + w_2 \lambda\right), \tag{1}$$

$$\sum_{s\in S^{ij}} x_{ijs} = 1, \qquad\qquad i \in I,\ j \in J, \tag{2}$$

$$\sum_{i\in I}\sum_{j\in J}\sum_{s\in W^{jt}} x_{ijs} + u_t \geq d_t, \qquad\qquad t \in T, \tag{3}$$

$$\lambda \geq u_t, \qquad\qquad t \in T, \tag{4}$$

$$\alpha_i^1 \leq \sum_{j\in J_w} x_{ijs'} \leq \beta_i^1, \qquad\qquad i \in I,\ w \in V, \tag{5}$$

$$\alpha_i^2 \leq \sum_{j\in J_w}\sum_{s\in S^{ij}\cap S_2} x_{ijs} \leq \beta_i^2, \qquad\qquad i \in I,\ w \in V, \tag{6}$$

$$\alpha_i^3 \leq \sum_{j\in J} x_{ijs'} \leq \beta_i^3, \qquad\qquad i \in I, \tag{7}$$

$$\alpha_i^4 \leq \sum_{j\in J_w}\sum_{s\in W^{ij}} L_s x_{ijs} \leq \beta_i^4, \qquad\qquad i \in I,\ w \in V, \tag{8}$$

$$\alpha_i^5 \leq \sum_{j\in J}\sum_{s\in W^{ij}} L_s x_{ijs} \leq \beta_i^5, \qquad\qquad i \in I, \tag{9}$$

$$\sum_{j'=j}^{\beta_i^6+j} x_{ij's'} \geq 1, \qquad\qquad \begin{array}{l} i \in I, \\ j = 1,\dots |J| - \beta_i^6, \end{array} \tag{10}$$

$$x_{ijs'} + \left(\sum_{j'=1}^{k-1}\sum_{s\in W^{i(j'+j)}} x_{i(j'+j)s}\right) + x_{i(j+k)s'} \leq k, \qquad \begin{array}{l} i \in I,\ k = 2,\dots,\alpha_i^6, \\ j = 1,\dots,|J| - k, \end{array} \tag{11}$$

$$\sum_{j'=0}^{\beta_i^7}\sum_{s\in S^{i(j+j')}\cap S_2} x_{i(j+j')s} \leq \beta_i^7, \qquad\qquad \begin{array}{l} i \in I, \\ j = 1,\dots |J| - \beta_i^7, \end{array} \tag{12}$$

$$\sum_{j'=0}^{\beta_i^8} x_{i(j+j')s'} \leq \beta_i^8, \qquad\qquad \begin{array}{l} i \in I, \\ j = 1,\dots |J| - \beta_i^8, \end{array} \tag{13}$$

$$\sum_{j'=0}^{\beta_i^9} x_{i(j+j')s} \leq \beta_i^9, \qquad\qquad \begin{array}{l} i \in I, \\ j = 1,\dots |J| - \beta_i^9, \\ s \in \displaystyle\bigcap_{j'=0}^{\beta_i^9} W^{i(j+j')}, \end{array} \tag{14}$$

$$x_{ijs_1} + x_{i(j+1)s_2} \leq 1, \qquad\qquad \begin{array}{l} (s_1,s_2) \in \hat{W}(i,j), \\ i \in I, j = 1,\dots,|J| - 1, \end{array} \tag{15}$$

where the objective function is to minimize the under-staffing for all intervals t. Note that it minimizes both the overall sum of under-staffings and the maximum under-staffing over all time intervals t. (2) are assignments constraints indicating that each employee must be assigned to exactly one shift each day. Constraints (3), (4) are soft constraints for demand satisfaction. Constraints (5)–(9) are counter constraints guaranteeing that the total number of days off, night shifts, and working slots per week and overall planning horizon are within pre-defined interval. Constraints (10)–(14) are series constraints, where inequalities (10), (11) impose the maximum and minimal number of consecutive working days, respectively. At the same time, constraints (12), (14) correspondingly restrict the maximum number of consecutive night shifts, rest days, and shifts with the same ID. The last constraints ensure that in consecutive days each employ is not assigned to shifts that are not separated by a given number of time intervals, e.g. the shifts an employee is assigned to must be non-overlapping, i.e. early morning shift cannot follow night shift etc. Here $\hat{W}(i,j)$ is a set of pairs (s_1, s_2) of incompatible working shifts, i.e. (s_1, s_2): $s_1 \in W^{ij}, s_2 \in W^{ij+1}$ separated by at least β_i^9 time slots.

4   Solution Algorithms

In this section, we present two solution algorithms developed for the aforementioned staff scheduling problem. Both algorithms are based on the conventional two-stage approach. In the first stage, the algorithms construct an initial feasible solution from scratch by applying a greedy strategy. After that the initial solution found is fed into the next improvement stage. The latter is based on various local search techniques supplemented with metaheursitic strategies to avoid stucking at a local optimum.

4.1   Pattern-Driven Local Search

The core idea of the approach is to split the procedure of the assignment of shifts to employees into two phases. During the first phase, each employee is assigned to a particular week pattern – a set of seven consecutive days, where each day is associated with a day shift, night shift, or day-off. The planning horizon may contain several weeks, so the pattern assignment is made for each of them. During the second phase of the assignment, the employees are assigned particular shifts according to their week pattern, chosen on the first step. Such an approach allows us to easily control the majority of the hard constraints. During the creation of the week patterns, we check that all the hard constraints, defined for a week, are satisfied, e.g. number of consecutive working shifts, number of consecutive days-off, etc. We also take into account the sets of available shifts for each day of a particular employee during this phase. As a result, for each employee and for each week of the planning horizon, we obtain a set of feasible week patterns. Some examples of possible week patterns are presented in Fig. 1.

In order to obtain a full schedule for one employee, we have to create a chain of week patterns. To do that, we calculate the adjacency matrix that define

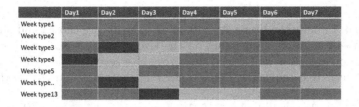

Fig. 1. Examples of different week patterns

whether a week pattern can follow or can be followed by another one. On this step, we also check whether week and month related constraints are satisfied. Note that this approach assumes that the majority of employees share the same workplace constraints, the same number of maximum/minimum working shifts, consecutive rest days etc. This assumption is quite realistic and often occurs in practical settings. If some employees have unique workplace constraints, the rows and columns of the matrix corresponding to their patterns can be obtained in the same way. Note that it is not necessary to store and generate the whole set of possible week chains for each employee, as in some cases it would take prohibitively long run time and require a large amount of RAM. The preliminary search procedure can be implemented using only a fraction of all variety of patterns, generated randomly.

During the first phase of the approach we construct an initial schedule. Although the main goal of this step is to construct a feasible schedule, we still try to reduce the unsatisfied demand during the construction. The process is based on a well-known greedy scheme:

Algorithm 1. Initial greedy algorithm

1: Initialization: Start from an empty schedule;
2: **for** $i \in I$ **do**
3:     Create a set $K_i, |K_i| > 0$ chains of feasible week patterns for employee i.
4:     **for** $k \in K_i$ **do**
5:         **for** $t = 1, .., T$ **do**
6:             Create a randomized assignment of working shifts, according to the chain pattern.
7:             Evaluate the total uncovered demand ω left after the assignment.
8:         **end for**
9:     **end for**
10:     Choose the chain and the corresponding assignment with smallest uncovered demand.
11: **end for**

The constructions starts from an empty schedule. The main loop goes through the whole set of employees. For each employee, we create a set of possible schedule patterns. Then, for each pattern, we create a number (say T) of assignments, producing a real schedule of the employee from a pattern. Each such schedule is evaluated. Finally, the best pattern and assignment are chosen and added to the incumbent solution. Depending on the size of an instance and the limits on calculation time, one might run this procedure several times with different, randomly permuted sequence of employees I in the input and choose the best output as a resulting solution.

The aim of the second phase of the approach is to improve the solution found. In order to do that, we propose a local search-based heuristic. We note that week patterns and chains assigned to each employee allows us to easily satisfy the majority of hard constraints. Thus, whenever we decide to change the schedule locally, we are still required to keep the patterns. We propose to use two basic neighborhood structures. The first neighborhood contains all solutions that can be obtained from the incumbent by changing one day assignment of one employee to another one (one day-shift to another day-shift, or one-night shift to another night-shift). This neighborhood allows us to perform a fine-tuning of the solution, choosing the best assignment of an employee withing a given set of week patterns. The second neighborhood is much bigger, and contains all solutions that can be obtained by changing the pattern of one week. During this interchange, we have to check the feasibility of the resulting week chain, using the adjacency matrix. The assignment of the employee to particular shift after the change of the pattern is done in a greedy manner. We assign a shift for each day of the changed week minimizing the resulting objective value.

4.2 Randomized Greedy and Shift-Exchange Local Search

Here we present a solution algorithm that follows another approach to both constructing an initial solution and improving it with a local search procedure. The algorithm presented here considers shifts as basic building blocks of schedule. Recall that the previous solution algorithm considers week working patterns (or stints of 7 days long) as basic stints.

The first stage of the algorithm is to generate an initial feasible solution from scratch. To do that, it follows a quite simple and effective random greedy strategy. The general outline of the construction phase is presented in Algorithm 2.

The randomized greedy algorithm considers each employee one by one. For each employee it begins with the first day in the planning horizon and consecutively assigns shifts to days uniformly at random taking into account work-rule constraints. An example of how the algorithm assigns shifts and what type of shifts are viewed as feasible in some particular cases is presented in Fig. 2. Yellow, blue, and green squares represent day, night, and day off shifts, respectively. We can see that shifts for four previous days are already assigned and the algorithm now has to decide what random shifts can be assigned for the fifth day. Let us recall that for our staff scheduling problem we deal with the *ranged* work-rule constraints, i.e. there are minimal and maximum numbers for particular roster items. The values are part of input. Let us assume that the maximal numbers of consecutive working days, night shifts, and rest days are 5, 3, and 2, respectively. In the first picture, a shift of any type can be assigned to the fifth day, hence

Algorithm 2. Randomized greedy algorithm

1: Initialization: Let r_{ij} be a schedule. Set schedule $r_{ij} = -1 \ \forall i \in I, j \in J$.
2: **for** $i \in I$ **do**
3:     Set $k \leftarrow 0$.
4:     **for** $j \in J$ **do**
5:         Select a shift $\bar{s}$ form S^{ij} uniformly at random.
6:         **if** assigning $\bar{s}$ does not violate horizontal constraints **then**
7:             $r_{ij} \leftarrow s$.
8:         **end if**
9:         **if** there is no feasible shift in S^{ij} **then**
10:            Set $j \leftarrow j - (j \mod 7) + 1$.
11:            $k \leftarrow k + 1$.
12:         **end if**
13:         **if** $k = k_{\max}$ **then**
14:            $k \leftarrow 0, i \leftarrow i - 1$.
15:         **end if**
16:         **if** $(j + 1) \mod 7 = 0$ **then**
17:            $k \leftarrow 0$.
18:         **end if**
19:     **end for**
20: **end for**

the algorithm can pick it randomly. In the next picture (top right), we can see that the previous two days are days off, hence the fifth day cannot be a day off too, because the maximal number of consecutive rest days is three. Thus, the algorithm has to choose a random working shift. In the next picture, we can see that an employee works for three consecutive night shifts, hence a night shift cannot be assigned for the current day. Note that if we suppose that seven days presented in the picture correspond to a week, then a night shift cannot also be assigned to the current day, as the number of night shifts per week must be less than or equal to three. Thus, the algorithm can assign either a day shift or day off shift, taking into account other work-rule constraints. The last picture demonstrates that there can be no more than three shifts in a row with the same id, hence the algorithm cannot select shift 35. Note that the considered examples are not the complete list and there are other possible cases that affect selecting shifts.

Once an initial solution is found, it is quite natural to try to find a better solution with an improvement heuristic. For our problem, we develop a local search algorithm based on exchanging a single shift of an employee to find a neighbor solution of better quality, i.e. better objective value. Recall that local search heuristics start from some arbitrary initial (usually random) feasible solution (incumbent) and then try to find a better solution in its neighborhood. If such a solution is found, it replaces the incumbent, and the procedure is repeated. Otherwise, the incumbent is said to be locally optimal and the heuristic halts. Thus, local search moves from a current solution to another one in the search space

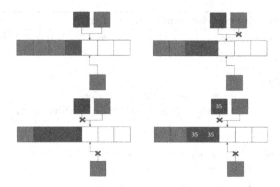

Fig. 2. Example of selecting the next shift depending to the already constructed work line

by examining its neighboring solutions. Note that our local search procedure is supposed to start with a solution found by the random greedy algorithm.

For our algorithm we adapted the following three neighborhoods:

N_1: consists of all solutions obtained from the current one by exchanging a work shift of one employee at one day by another work shift of the same type (i.e. day or night one).

N_2: contains all solutions obtained from the current one by exchanging a work shift of one employee at one day by any other work shift (taking into account work-rule constraints).

N_3: consists of all solutions obtained from the current one by exchanging work shifts of two employees at the same day (taking into account work-rule constraints).

Note that if an initial solution is feasible, then a neighbor solution in N_1-neighborhood is also feasible, if and only if the constraint for the maximal number of consecutive working shifts with the same id is not violated. This property can be checked very fast. When searching the N_2- or N_3-neighborhoods, one has to keep in mind that an exchange of shifts may cause violation of a number of work rule constraints, especially series ones. The only constraints that always remain satisfied are day off constraints (both counter and series ones). The N_2- or N_3-neighborhoods consist of only feasible solutions, hence any changes that result in violation of any work-rule constraints are discarded. Note that in case of the N_3-neighborhood, one has to ascertain the feasibility of shift exchange with respect to two employees.

In Fig. 3 one can see some examples of which shift exchange may cause infeasibility and must be discarded by the local search algorithm. For example, in the left picture we can see a schedule of an employee for a week. Six work days are interspersed with one rest day in the middle of the week. One can see that any day shift may be changed to any other day if it does not result in violating the constraint for the maximal number of consecutive working shifts with the same id. However, a day shift cannot be changed by a night shift if the next day is assigned a day shift. Indeed, in this case the succession constraints would be violated. In the right picture one can see another schedule for a week where there are three consecutive night shifts. Again, any night or day shifts can be exchanged with a shift of the same type. But, for example, a day shift cannot be assigned between two night shifts due to succession constraints. Nevertheless, it can be assigned to the day before a series of consecutive night shifts.

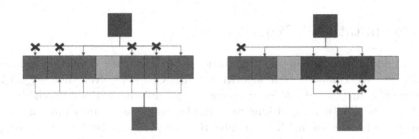

Fig. 3. Example of selecting a feasible neighbor solution

When searching for a better neighbor solution, there are several strategies of replacing incumbent with a better solution found. In best improvement local search, the best of the neighboring solutions (if it has a better objective value) replaces the current incumbent. However, in first improvement local search, the neighborhood is only explored until any improving solution is found, which then replaces the incumbent. In the case of first-improving local search, a very effective strategy for exploring a neighborhood consists in performing a circular search [14]. Let us consider how circular search works for our problem. The local search explores neighboring solutions by selecting employees one by one in ascending order $i = 1, \ldots, |I|$ and by exploring each day $j = 1, \ldots, |J|$. Under the first improvement strategy, if an improving neighbor is found for some i' and j', the incumbent would be replaced and the search would resume from the very first employee and the first day. However, since in the last iteration there was no improving move for $i \leq i'$ and $j \leq j'$, it most likely that there will still be no an improving move. That is why, a more effective strategy is to resume the search from $i = i'$ and $j = j' + 1$, instead of from $i = 1$ and $j = 1$. In this context, the move defined by $j = |J|$ is followed by that indexed by $j = 1$ and $i = |I|$ by $i = 1$ as if they were organized as a circular list. The circular local search stops at a local optimum as soon as a complete tour of this circular list is performed without any improvement in the current solution [14]. However, this stopping criterion may be intractable in case of large problem instances, hence we suppose that the algorithm stops if there is no any move after a fixed, relatively small number of local search iterations.

Another key feature of our local search algorithm is a fast procedure of computing the objective value for neighbor solutions. Indeed, as the neighborhoods are quite large it becomes crucially important to fast estimate whether the next neighbor solution is better or not. To do that, one has to determine the intervals affected by exchanging shifts and calculate the corresponding change in the objective value.

As any local search heuristic can fast stuck in a local optimum, we adapted a random-restart or repeated strategy for our approach, i.e. it iteratively runs the local search, each time starting with a random greedy solution, and keeps the best solution found.

5   Computational Experiments

Here we present the results of computational testing of the developed solution algorithms. We compared our algorithms with CPLEX solver that set up to solve the aforementioned integer programming formulation of the staff scheduling problem. The solution algorithms were implemented using Java programming language and tested on a PC with Intel(R) Core(TM) i7-8665U and 16 GB of RAM. Note that we also implemented the randomized greedy and shift-exchange local search algorithms using C++ language (compiled with GNU C++ 12.2.0 compiler).

Our test bed consists of 9 problem instances related to real-life scheduling problems. In all the problem instances we use the same set of parameters to define workplace constraints (see Table 1). However, up to 10% of employees taken at random may have some small fluctuations in this values. The problem statistics

Table 1. Constraint parameters

Constraint	Min	Max
days-off per week	1	3
night shifts per week	0	3
days-off per planning horizon	4	8
working slots per week	140	180
working slots in planning horizon	640	840
consecutive working days	3	5
consecutive night shifts	–	3
consecutive days off	–	2
consecutive shifts with the same id	–	3
time between adjacent shifts	28	–

is presented in Table 2, where column *Prob.* contains the problem names, *Staff* — the number of employees for which one has to build a schedule. The next column *Shifts* demonstrates the number of shifts in each instance. Recall that possible shifts are determined in the shift scheduling step (see above) by the system planner. The set of shifts with start time, end time, and rest intervals is part of input for our staff scheduling problem. Let us also remind that each employee for each particular day can be assigned to only some subset of shifts, defined by the so-called cover matrix. Column *Days* contains the time horizon for the required schedule. Note that it is equal to 30 days for all instances. *Time slots* contains the overall number of intervals in the time horizon. Recall that one interval is supposed to be 15 min long. The last three columns entitled as MIP indicate the characteristics of the MIP models for corresponding problem instances. For each problem, the columns demonstrate the number of variables, constraints, and non-zero elements in the constraint matrix, respectively. One can see that these characteristics are heavily dependent on the number of staff and the number of possible shifts. MIP problems are quite large, e.g. the largest problem 118g involves more than 500 thousand variables and more than 43 million constraints, which actually makes it not tractable by MIP solvers.

Table 2. Problem statistics

					MIP		
Prob.	Staff	Shifts	Days	Time slots	vars	cons	nonzeros
110g	10	97	30	2880	6781	18015	193882
111g	25	97	30	2880	12631	36330	470492
112g	50	97	30	2880	22381	67266	933926
113g	100	97	30	2880	41881	128224	1859000
114g	250	97	30	2880	177385	946290	9326054
115g	500	97	30	2880	353617	1902833	18734634
116g	1000	181	30	2880	1063843	8776706	63368648
117g	2500	181	30	2880	2654683	21935422	248659018
118g	5000	181	30	2880	5309095	43893600	316716564

The results of computational testing of the MIP model using MIP solver are presented in Table 3. We performed two series of experiments. First, we estimate the time required by the solver to find a feasible solution of the problem and report the corresponding objective value. In second series, we limit the run time of the solver to 20 h and report the best solution found. In the columns *Obj.* one can observe the value of the objective function, in *Gap*(%) – the corresponding duality gap provided by CPLEX. It is computed as $\frac{UB-LB}{UB} * 100\%$, where UB is the objective value of the found solution and LB is the best found lower bound for the objective value. The columns *Time* demonstrate the run time in seconds.

We can see that the staff scheduling problem is challenging especially in case of large problem instances. The solver was not able to find even a feasible solution for the largest problem, 118g, and spent almost two hours to find a feasible solution for 117g instance. Though for 116g and 117g, it managed to discover a feasible solution quite fast, it improved them not significantly after

Table 3. Computational results of IBM ILOG CPLEX solver

	CPLEX (first feasible)			CPLEX (20 h)		
Prob.	Obj.	Gap (%)	Time, s	Obj.	Gap (%)	Time, s
110g	21,243	52	0.2	17,721	0.01	154.6
111g	59,283	91.5	0.4	37,463	0.01	1272.3
112g	119,259	100	0.7	74,704	0.03	42692
113g	238,498	100	1.2	146,500	0.09	32785
114g	518,316	32.42	170.3	351,438	0.07	55448
115g	1,031,499	32.37	534.3	699,625	0.09	72000
116g	2,077,103	31.96	1211.9	2,071,620	31.21	20608
117g	5,182,980	32.26	6181.4	5,176,761	31.63	54000
118g	–	–	–	–	–	–

Table 4. Computational results of the developed solution algorithms

	Random Greedy + LS				Greedy pattern			Greedy+LS pattern		
Prob.	Obj.	Gap	C++	Java	Obj.	Gap	Time	Obj.	Gap	Time
110g	18, 362	3.49	1	1	18, 169	2.47	4	17, 645	−0.43	300
111g	44, 523	15.86	1	1	44, 386	15.60	5	42, 290	11.41	300
112g	89, 593	16.62	1	1	89, 184	16.24	5	84, 782	11.89	300
113g	176, 880	17.18	1	2	172, 894	15.27	7	163, 458	10.37	300
114g	436, 942	19.57	2	18	425, 909	17.49	12	391, 650	10.27	300
115g	866, 132	19.22	2	45	854, 066	18.08	22	791, 129	11.57	300
116g	1, 798, 490	−15.19	2	79	1, 710, 290	−21.13	41	1, 596, 461	−29.76	300
117g	4, 411, 390	−17.35	43	257	4, 279, 062	−21.12	95	4, 031, 003	−28.58	300
118g	9, 195, 010	–	165	546	8, 628, 296	–	200	8, 325, 944	–	300

20 h of computation. For other problems, the solver found near-optimal solutions confirmed by the tightness of gap values.

Now let us demonstrate the computational results for the developed solution algorithms that are presented in Table 4. We report the objective values, *Obj.*, Gap, and run time. Note that Gap is computed with respect to the objective value provided by the solver. In other words, it is computed as $\frac{UB-UB^{CPX}}{UB}*100\%$, where UB and UB^{CPX} are the objective values provided by our solution algorithms and the MIP solver, respectively. Thus, a negative value of Gap indicates that our algorithms found better solutions. In columns *RandomGreedy + LS* one can see the results for randomized greedy and shift-exchange local search algorithm. In *Greedy pattern* we report the results for the pattern-driven greedy procedure only and in *Greedy+LS pattern* the results for overall pattern-driven algorithm including the local search procedure. Note that for *Random Greedy + LS* we report the run time for both C++ and Java implementations (given in columns *C++* and *Java*). The run time of *Greedy+LS pattern* was limited to 300 s. After that the best found solution is returned.

One can observe that for small problem instances the developed algorithm found the solutions that are in general $10 - 20\%$ worse than the optimal ones. However, for $110g$ instance, *Greedy+LS pattern* yields a solution which is better than near-optimal one found by the solver.

On the other hand, for large problem instances $116g - 118g$ the developed algorithms considerably outperform CPLEX in both solution quality and run time. The largest improvement in quality over the MIP solver is achieved for $116g$ instance.

We can see that C++ implementation of *RandomGreedy + LS* runs significantly faster than Java one. For the largest problem, $118g$, it is more than 3 times faster. However, for small problems this superiority is not so evident.

In all cases, *Greedy+LS pattern* approach provides better solutions than *RandomGreedy + LS*. However, it runs much slower.

6 Conclusion

In this paper, we have considered a staff scheduling problem motivated by real-life applications. The main feature of the problem is that each employee may have a unique set of shifts for each particular day. We have developed fast two-stage heuristic algorithms that follow different strategies for finding solutions. We tested the algorithms on large-scale problem instances that include some specific requirements. Note that the strong application oriented nature of the problem makes it difficult to directly compare the developed algorithms with some known ones. We demonstrated the comparison of the algorithms with a commercial solver and highlight their effectiveness.

References

1. Baker, K.R.: Workforce allocation in cyclical scheduling problems: a survey. Oper. Res. Q. **27**(1), 155–167 (1976). https://doi.org/10.1057/jors.1976.30
2. Bard, J.F., Binici, C., deSilva, A.H.: Staff scheduling at the united states postal service. Comput. Oper. Res. **30**(5), 745–771 (2003). https://doi.org/10.1016/S0305-0548(02)00048-5
3. Borgonjon, T., Maenhout, B.: The impact of dynamic learning and training on the personnel staffing decision. Comput. Ind. Eng. **187**, 109784 (2024). https://doi.org/10.1016/j.cie.2023.109784
4. Brucker, P., Qu, R., Burke, E.: Personnel scheduling: models and complexity. Eur. J. Oper. Res. **210**(3), 467–473 (2011). https://doi.org/10.1016/j.ejor.2010.11.017
5. Davydov, I., Vasilyev, I., Ushakov, A.V.: Tabu search metaheuristic for the penalty minimization personnel task scheduling problem. In: Khachay, M., Kochetov, Y., Eremeev, A., Khamisov, O., Mazalov, V., Pardalos, P. (eds.) Mathematical Optimization Theory and Operations Research: Recent Trends, pp. 109–121. Springer, Cham (2023). https://doi.org/10.1007/978-3-031-43257-6_9

6. Defraeye, M., Van Nieuwenhuyse, I.: Staffing and scheduling under nonstationary demand for service: a literature review. Omega **58**, 4–25 (2016). https://doi.org/10.1016/j.omega.2015.04.002
7. Erhard, M., Schoenfelder, J., Fügener, A., Brunner, J.O.: State of the art in physician scheduling. Eur. J. Oper. Res. **265**(1), 1–18 (2018). https://doi.org/10.1016/j.ejor.2017.06.037
8. Ernst, A.T., Jiang, H., Krishnamoorthy, M., Sier, D.: Staff scheduling and rostering: a review of applications, methods and models. Eur. J. Oper. Res. **153**(1), 3–27 (2004). https://doi.org/10.1016/S0377-2217(03)00095-X, timetabling and Rostering
9. Kilincli Taskiran, G., Zhang, X.: Mathematical models and solution approach for cross-training staff scheduling at call centers. Comput. Oper. Res. **87**, 258–269 (2017). https://doi.org/10.1016/j.cor.2016.07.001
10. Kletzander, L., Musliu, N.: Solving the general employee scheduling problem. Comput. Oper. Res. **113**, 104794 (2020). https://doi.org/10.1016/j.cor.2019.104794
11. Krishnamoorthy, M., Ernst, A.T., Baatar, D.: Algorithms for large scale shift minimisation personnel task scheduling problems. Eur. J. Oper. Res. **219**(1), 34–48 (2012). https://doi.org/10.1016/j.ejor.2011.11.034
12. Lai, D., Leung, J., Dullaert, W., Marques, I.: A graph-based formulation for the shift rostering problem. Eur. J. Oper. Res. **284**(1), 285–300 (2020). https://doi.org/10.1016/j.ejor.2019.12.019
13. Maenhout, B., Vanhoucke, M.: A perturbation matheuristic for the integrated personnel shift and task re-scheduling problem. Eur. J. Oper. Res. **269**(3), 806–823 (2018). https://doi.org/10.1016/j.ejor.2018.03.005
14. Resende, M.G.C., Ribeiro, C.C.: Optimization by GRASP. Springer, New York (2016)
15. Smet, P., Bilgin, B., De Causmaecker, P., Vanden Berghe, G.: Modelling and evaluation issues in nurse rostering. Ann. Oper. Res. **218**, 303–326 (2014). https://doi.org/10.1007/s10479-012-1116-3
16. Smet, P., Brucker, P., De Causmaecker, P., Vanden Berghe, G.: Polynomially solvable personnel rostering problems. Eur. J. Oper. Res. **249**(1), 67–75 (2016). https://doi.org/10.1016/j.ejor.2015.08.025
17. Van den Bergh, J., Beliën, J., De Bruecker, P., Demeulemeester, E., De Boeck, L.: Personnel scheduling: a literature review. Eur. J. Oper. Res. **226**(3), 367–385 (2013). https://doi.org/10.1016/j.ejor.2012.11.029
18. Wang, W., Xie, K., Guo, S., Li, W., Xiao, F., Liang, Z.: A shift-based model to solve the integrated staff rostering and task assignment problem with real-world requirements. Eur. J. Oper. Res. **310**(1), 360–378 (2023). https://doi.org/10.1016/j.ejor.2023.02.040
19. Wang, Z., Liu, R., Sun, Z.: Physician scheduling for emergency departments under time-varying demand and patient return. IEEE Trans. Autom. Sci. Eng. **20**(1), 553–570 (2023). https://doi.org/10.1109/TASE.2022.3163259
20. Örmeci, E.L., Salman, F.S., Yücel, E.: Staff rostering in call centers providing employee transportation. Omega **43**, 41–53 (2014). https://doi.org/10.1016/j.omega.2013.06.003

Game Theory

Potential Game in General Transport Network with Symmetric Externalities

Julia V. Chirkova[(✉)] [iD]

Institute of Applied Mathematical Research of Karelian Research Centre of RAS,
Pushkinskaya Street 11, Petrozavodsk, Russia
julia@krc.karelia.ru
http://www.krc.karelia.ru/HP/julia

Abstract. The paper considers a model of a general transport network and BPR linear delay functions with externalities. We consider the case where the impact of channel loads to the delay is pairwise symmetric. For this case, it is proven that the game of traffic allocation among channels is potential, and the price of anarchy is limited by the value $\frac{4}{3}$.

Keywords: Wardrop equilibrium · Optimal profile · Social costs · Price of Anarchy · Symmetric externalities · BPR-functions · General network

1 Introduction

Researches exploring the efficiency and optimality of transport and communication systems are actual due to the constantly and uncontrollably growing degree of network load with traffic increase. Traffic flow equilibrium and social optimum become solution concepts that are widely used in the theory of transport systems [5,10,24]. The concept of a equilibrium is based on Wardrop's hypothesis [27] that the travel time for each flow on all its routes is the same and is less than the time on the routes, which are not used by this flow. In Wardrop models the existence of equilibria [7,16] is connected with the presence of a potential function [18,21], the minimum of which always exists and corresponds to the equilibrium. Great part of works are devoted to routing problems with traffic delay functions of a certain form. They investigate the question: how does the refusal of centralized control degrade the system, that is, how much can an equilibrium costs of the entire system exceed the costs in an optimum situation. The price of anarchy [20,23] is the metric of such worsening. An exact expressions and estimates are found for the price of anarchy in models with different characteristics. The authors of [20] show that in the Wardrop model with linear latency functions and parallel communication channels the price of anarchy is bounded above $\frac{4}{3}$, and in [23] the validity of this estimate was established for a network of an arbitrary topology with linear delays. In [22] an estimate was obtained for the price of anarchy in the case when delays have the form of polynomial functions, belonging to the class of BPR (Bureau of Public Roads) functions [19, 26], which are widely used in applications.

It is important to consider the forming of equilibrium and optimal solutions taking into account external factors, externalities, which are usually not included in the consideration of transport problems. External factors include traffic rules, speed and capacity of transport units, quality of service, passenger comfort, etc. The work [12] is one of the first to introduce the concept of externalities as external effects created by neighboring players in the network. The papers [1, 6, 11, 13] consider routing games with positive cost-sharing and negative congestion externalities. In [17] it is shown that overload externalities can cause the inefficiency of Pareto equilibria, including appearing the Braess paradox [2, 23]. Externalities of mixed type are taken into account in [15]. They include negative and positive components and influence the occurrence and characteristics of the Braess paradox in the network. In the work [9] the model of passenger transportation considers the characteristics of service companies as externalities.

The works [4, 14] examine the Wardrop model with splittable traffic in relation to a transport system with parallel channels and BPR delay functions with linear externalities. In this case the traffic delay on a channel depends not only on the flows on this channel, but also on the flows on other channels. For this model, it is shown that it is possible to find such values of externalities that ensure the optimality of the equilibrium solution. It is assumed that the system is such that there is an equilibrium and optimal solution for it. In the general case, a system with externalities may not to have a Wardrop equilibrium even for a parallel network. The work [3] shows that for a network with parallel channels in the case of symmetric externalities, where flows on the channels symmetrically influence each other in pairs, the game turns out to be potential, which means that Wardrop equilibrium always exists. In addition, for this case it is shown that the value of the price of anarchy is limited by $\frac{4}{3}$. In this paper, these results are generalized to a network of arbitrary topology.

The paper is organized as follows. In the Sect. 2 we give a motivation explaining an introducing of externalities to latencies for general network. The Sect. 3 describes a formal model of a Wardrop game with symmetric externalities in linear BPR-function latencies. In the Sect. 4 we show that this game is potential and it always has a unique Wardrop equilibrium. The Sect. 5 explores properties of social costs and in the resulting Sect. 6 we show that the price of anarchy in our game is limited by $\frac{4}{3}$.

In the paper for an ordered set $A = (a_1, \ldots, a_{|A|})$ we denote a position number of its element a as $pos(a, A)$. Also we write $a \prec b$ if $a, b \in A$ and $pos(a, A) < pos(b, A)$.

2   Externalities in the General Transport System

Our previous papers [3, 4, 14] consider externalities in linear latency BPR-functions on a parallel network. Parallel channels are interpreted as lanes on the same road, each with different speed rates and, therefore, latencies. For each lane the externality reflects the degree with which the traffic flows on all other lanes influence its latency. This influence is natural as long as flow participants move nearby and use adjacent lanes to perform maneuvers. Here we give some motivation to explain a sense of externalities (including pairwise symmetric) in the transport system of general topology. In the present paper we don't claim that a pairwise symmetry necessarily presents in the network. In further sections we show that if it presents, i.e. flows affect pairwise symmetrically to each other, then the game becomes potential and moreover the price of anarchy is limited by the value $\frac{4}{3}$, similarly to the original case of the model with linear latencies [23] without externalities.

In an original Wardrop model [10,22,23] flow latency on each edge depends only on load on this edge. Partially the model does not take into account additional delays when two or more flows are merged, especially without a semaphore dividing flows. Such delays can be significant if edges are highly loaded and flows form a traffic jam and stay waiting the possibility to join the united flow. Generally in such situation rules of road priorities, "interference on the right", etc. become ineffective, otherwise at least one of flows need to remain stopped. Drivers ask to let them join to the flow or let other participants pass ahead. Partially, often they can act symmetrically, and we observe the situation where flows symmetrically influence latencies of each other. So, it important to introduce such external delays into latency functions as they take part in forming of participants equilibrium behavior. We illustrate it on the following simple example.

Example 1. We consider a network (see Fig. 1) of topology well-known due to example of Braess paradox [23]. The traffic of a total volume X needs to be transmitted from vertex A to B. It is allocated on three routes: part x_1 of X moves on the upper way ACB, x_2 on the lower way ADB and x_3 on the way $ACDB$. The picture shows the load on each edge, that is the total amount of traffic on this edge. We suppose that traffic latencies on edges AC and DB have a form $\frac{1}{100}(1 + 1000\delta_e)$, where δ_e is a load of corresponding edge e, and on edges AD and CB they are $50(1 + \frac{1}{50}\delta_e)$. The edge connecting C and D has a latency in a form $10(1 + \frac{1}{10}\delta_{CD})$.

Each participant of traffic chooses a way where sum of latencies on edges which he uses is minimal. So they form an equilibrium flow allocation named a Wardrop equilibrium. If a total traffic amount is small, e.g. $X = 2$, the equilibrium situation is $x_1 = x_2 = 0$, $x_3 = X$, as latencies on unused ways ACB and ADB are about 70.01 and on the way $ACDB$ is about 52.02. By the way, it is also the optimum. But when the total traffic grows to e.g. $X = 4$, the optimum becomes: $x_1 = x_2 = 2$ with latencies about 72.01 on ACB and ADB and about 50.02 on the unused way $ACDB$. But selfish flow participants act as an equilibrium where $x_1 = x_2 \approx 0.308$ and $x_3 \approx 3.383$ with latencies on all ways about 87.234. We see here that in the original Wardrop model upper and lower equilibrium flows are equal with the same latencies. In real situation the flow x_2 can suffer from additional cost due to the flow x_3 joining at the vertex D, as well as x_3 suffers from x_2.

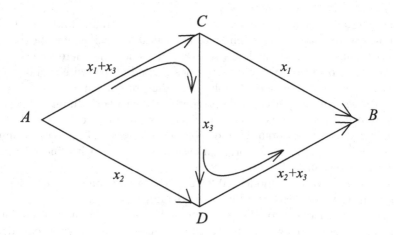

Fig. 1. Network topology.

If we add symmetric externalities into delays on edges AD and CD, their latencies become $50(1 + \frac{1}{50}\delta_{AD}) + b\delta_{CD}$ and $10(1 + \frac{1}{10}\delta_{CD}) + b\delta_{AD}$. The value b reflects a size of contribution of external flow into edge latency. For $b = 0.01$ the equilibrium allocation becomes not symmetric: $x_1 \approx 0.309$, $x_2 \approx 0.306$, $x_3 \approx 3.385$, equal latencies on ways are about 87.26.

3   A Formal Model of the Transport System

We consider a general transport network as an oriented graph $G = (V, E)$. Its edges are road parts and vertices are road intersection or points where road properties significantly change. Vertices $v \in V$ of the transport graph are source, transit and terminal points for transport flows. The size of V is n. Each edge (i, j) in the ordered edge set E means that there is a direct oneside way from i to j. A directed way or route between two vertices i and j is a sequence of edges $e_1 = (i, k_1), e_2 = (k_1, k_2), \ldots, e_m = (k_m, j)$ where all $e_k \in E$. We consider only simple routes without cycles.

A correspondence matrix $Corr[n \times n]$ defines transport flow sizes for pairs of vertices. Each its element $Corr_{ij}$ is a total traffic size which has to be transmitted from i to j through routes connecting i and j. We denote $W = \{w = (i, j) | Corr_{ij} > 0\}$ as an ordered set of all pairs which generate flows. For each such pair $w = (i, j) \in W$ we determine its route set $P_w = (r_1, \ldots, r_{|P_w|})$ as an ordered set of allowed routes connecting i and j. A traffic of each pair $w = (i, j) \in W$ has to be allocated on routes P_w to satisfy demands of the corresponding matrix. That is $\sum_{r \in P_w} x_r^w = Corr_{ij}$, where $x^w = (x_1^w, \ldots, x_{|P_w|}^w)$ is a strategy of a pair w. Each x^w is an ordered flow set, such that each $x_r^w \geq 0$ is the part of traffic of a pair w transmitting through the route $r \in P_w$. In this case a strategy profile $x = \{x_{r \in P_w}^{w \in W}\}$ is a feasible transport flow allocation in the transport graph G with the correspondence matrix $Corr$. Denote a set of all feasible profiles x as χ.

For each edge $e \in E$ its traffic load $\delta_e(x) = \sum\limits_{w \in W} \sum\limits_{r \in P_w : e \in r} x_r^w$ is a total flow on this edge, i.e. the sum of all transport flows on routes containing e.

For a flow on an edge $e \in E$ we define its latency as a linear BPR-function [4, 14, 26]

$$f_e(x) = t_e(1 + a_e \delta_e(x) + \sum_{l \in E} b_{el} \delta_l(x)),$$

where $a_e > 0$ reflects an impact of the load on the edge e to its latency, and $b_{ij} \geq 0$ are externalities. Each externality b_{el} reflects a contribution of a load on an edge l to the latency on the edge e. In the present paper we consider a case when traffic loads on edges influence latencies pairwise symmetrically. That is $t_e b_{el} = t_l b_{le}$ for all $e, l \in E$. Also we assume that all $b_{ee} := 0$. Besides we suppose that $a_e \geq \sum\limits_{l \in E} b_{el}$ for all $e, l \in E$. It means that for each edge the impact of its load to its latency is not less than the impact of loads on all other edges.

Then the latency for the traffic on the route r is $F_r(x) = \sum\limits_{e \in r} f_e(x)$. A Wardrop equilibrium [23, 27] is a profile where all routes with non-zero traffic have a minimal latency.

Definition 1. *[4] A profile x is a Wardrop equilibrium iff for each pair $w \in W$ and any routes $r, q \in P_w$ from $x_r^w > 0$ it yields that $F_r(x) = \min_{q \in P_w} F_q(x)$ and $x_r^w = 0$ implies that $F_r(x) \geq \min_{q \in P_w} F_q(x)$.*

We define a social costs function as

$$SC(x) = \sum_{e \in E} t_e \delta_e(x) \left(1 + a_e \delta_e(x) + \sum_{l \in E} b_{el} \delta_l(x) \right).$$

A socially optimal profile minimizes the social costs

$$OPT = \min_{x \in \chi} SC(x).$$

The price of anarchy for the transport system with the network G and correspondence matrix $Corr$ is a ratio of a social costs in Wardrop equilibrium and a socially optimal social costs.

4   Wardrop Equilibrium

Consider an arbitrary feasible profile $x \in \chi$. We define a function

$$\begin{aligned}
\phi(x) &= \sum_{e \in E} \int_0^{\delta_e(x)} t_e(1 + a_e y) dy + \frac{1}{2} \sum_{e, l \in E} t_e b_{el} \delta_e(x) \delta_l(x) \\
&= \sum_{e \in E} \delta_e(x) t_e(1 + \tfrac{1}{2} a_e \delta_e(x)) + \frac{1}{2} \sum_{e, l \in E} t_e b_{el} \delta_e(x) \delta_l(x).
\end{aligned} \tag{1}$$

Note that for each x_r^w

$$\frac{\partial \phi(x)}{\partial x_r^w} =$$

$$= \sum_{e \in r} t_e \left(1 + \tfrac{1}{2} a_e \delta_e(x) + \tfrac{1}{2} \sum_{l \in E} b_{el} \delta_l(x) \right) + \tfrac{1}{2} \sum_{e \in r} a_e t_e \delta_e(x) + \tfrac{1}{2} \sum_{e \in E} \delta_e(x) \sum_{l \in r} t_e b_{el}$$

$$= \sum_{e \in r} t_e \left(1 + a_e \delta_e(x) + \tfrac{1}{2} \sum_{l \in E} b_{el} \delta_l(x) \right) + \tfrac{1}{2} \sum_{e \in E} \delta_e(x) \sum_{l \in r} t_l b_{le}$$

$$= \sum_{e \in r} t_e \left(1 + a_e \delta_e(x) + \tfrac{1}{2} \sum_{l \in E} b_{el} \delta_l(x) \right) + \tfrac{1}{2} \sum_{l \in r} t_l \sum_{e \in E} b_{le} \delta_e(x)$$

$$= \sum_{e \in r} t_e \left(1 + a_e \delta_e(x) + \sum_{l \in E} b_{el} \delta_l(x) \right)$$

$$= \sum_{e \in r} f_e(x) = F_r(x).$$

The function $\phi(x)$ is defined on the convex set χ which is a Cartesian product of simplexes. If it is convex, it has a unique minimum.

Theorem 1. *The function $\phi(\cdot)$ defined by (1) is convex if $t_e b_{el} = t_l b_{le}$ and $a_e \geq \sum_l b_{el}$ for all $e, l \in E$.*

Proof. Let $\nabla^2 \phi(x)$ to be a Hessian matrix for the function $\phi(x)$. Its size is $m \times m$ where $m = \sum_{w \in W} |P_w|$.

$$\nabla^2 \phi(x) = \begin{pmatrix}
\frac{\partial^2 \phi(x)}{\partial (x_{r_1}^{w_1})^2} & \frac{\partial^2 \phi(x)}{\partial x_{r_1}^{w_1} \partial x_{r_2}^{w_1}} & \cdots & \frac{\partial^2 \phi(x)}{\partial x_{r_1}^{w_1} \partial x_{r_{|P_{w_1}|}}^{w_1}} & \frac{\partial^2 \phi(x)}{\partial x_{r_1}^{w_1} \partial x_{r_1}^{w_2}} & \cdots \\
\frac{\partial^2 \phi(x)}{\partial x_{r_2}^{w_1} \partial x_{r_1}^{w_1}} & \frac{\partial^2 \phi(x)}{\partial (x_{r_2}^{w_1})^2} & \cdots & \frac{\partial^2 \phi(x)}{\partial x_{r_2}^{w_1} \partial x_{r_{|P_{w_1}|}}^{w_1}} & \frac{\partial^2 \phi(x)}{\partial x_{r_2}^{w_1} \partial x_{r_1}^{w_2}} & \cdots \\
\cdots & \cdots & \cdots\cdots & \cdots & \cdots \\
\frac{\partial^2 \phi(x)}{\partial x_{r_{|P_{w_1}|}}^{w_1} \partial x_{r_1}^{w_1}} & \frac{\partial^2 \phi(x)}{\partial x_{r_{|P_{w_1}|}}^{w_1} \partial x_{r_2}^{w_1}} & \cdots & \frac{\partial^2 \phi(x)}{\partial (x_{r_{|P_{w_1}|}}^{w_1})^2} & \frac{\partial^2 \phi(x)}{\partial x_{r_{|P_{w_1}|}}^{w_1} \partial x_{r_1}^{w_2}} & \cdots \\
\frac{\partial^2 \phi(x)}{\partial x_{r_1}^{w_2} \partial x_{r_1}^{w_1}} & \frac{\partial^2 \phi(x)}{\partial x_{r_1}^{w_2} \partial x_{r_2}^{w_1}} & \cdots & \frac{\partial^2 \phi(x)}{\partial x_{r_1}^{w_2} \partial x_{r_{|P_{w_1}|}}^{w_1}} & \frac{\partial^2 \phi(x)}{\partial (x_{r_1}^{w_2})^2} & \cdots \\
\cdots & \cdots & \cdots\cdots & \cdots & \cdots
\end{pmatrix}$$

The elements of the Hessian matrix are

$$\frac{\partial^2 \phi(x)}{\partial (x_r^w)^2} = \sum_{e \in r} t_e \left(a_e + \sum_{l \in r} b_{el} \right),$$

$$\frac{\partial^2 \phi(x)}{\partial x_r^w \partial x_q^v} = \sum_{e \in r \cap q} t_e a_e + \sum_{e \in r} \sum_{l \in q} t_e b_{el} = \sum_{e \in r \cap q} t_e a_e + \sum_{l \in q} \sum_{e \in r} t_e b_{el}$$

$$= \sum_{e \in r \cap q} t_e a_e + \sum_{l \in q} \sum_{e \in r} t_l b_{le} = \frac{\partial^2 \phi(x)}{\partial x_q^v \partial x_r^w}.$$

One can see that $\nabla^2 \phi(x)$ is symmetric.

Let R to be a concatenation of all ordered route sets P_w, such that a route $r \in P_w$ is a route $q \in R$ where $pos(q, R) = \sum_{v \in W | v \prec w} |P_v| + pos(r, P_w)$. Moreover, for any routes $r, q \in R$ we suppose by the definition that $r \neq q$ if $pos(r, R) \neq pos(q, R)$.

Consider also y which is an arbitrary m-dimensional vector of real values such that $(y, y) > 0$. Each element $y_r \in y$ corresponds to a route $r \in R$. Then $\delta_e(y) = \sum\limits_{r \in R | e \in r} y_r$.

Then

$$(\nabla^2 \phi(\cdot)y, y) =$$

$$= \sum_{r \in R} y_r^2 \left(\sum_{e \in r} t_e(a_e + \sum_{l \in r} b_{el}) \right) + \sum_{r \in R} \sum_{q \in R | q \neq r} y_r y_q \left(\sum_{e \in r \cap q} t_e a_e + \sum_{e \in r} \sum_{l \in q} t_e b_{el} \right)$$

$$= \sum_{e \in E} t_e a_e \sum_{r \in R | e \in r} y_r^2 + \sum_{e \in E} \sum_{l \in E} t_e b_{el} \sum_{r \in R | e, l \in r} y_r^2$$

$$+ \sum_{e \in E} t_e a_e \sum_{r \in R | e \in r} \sum_{q \in R | q \neq r, e \in q} y_r y_q + \sum_{e \in E} \sum_{l \in E} t_e b_{el} \sum_{r \in R | e \in r} \sum_{q \in R | q \neq r, l \in q} y_r y_q$$

$$= \sum_{e \in E} t_e a_e \left(\sum\nolimits_{r \in R | e \in r} y_r \right)^2 + \sum_{e \in E} \sum_{l \in E} t_e b_{el} \left(\sum\nolimits_{r \in R | e \in r} y_r \right) \left(\sum\nolimits_{r \in R | l \in r} y_r \right)$$

$$= \sum_{e \in E} t_e a_e \delta_e^2(y) + \sum_{e \in E} \sum_{l \in E} t_e b_{el} \delta_e(y) \delta_l(y)$$

$$\geq \sum_{e \in E} t_e \sum_{l \in E} b_{el} \delta_e^2(y) + \sum_{e \in E} \sum_{l \in E} t_e b_{el} \delta_e(y) \delta_l(y)$$

$$= \sum_{e \in E} \sum_{l \in E | e \prec l} t_e b_{el} (\delta_e^2(y) + \delta_l^2(y)) + 2 \sum_{e \in E} \sum_{l \in E | e \prec l} t_e b_{el} \delta_e(y) \delta_l(y)$$

$$= \sum_{e \in E} \sum_{l \in E | e \prec l} t_e b_{el} (\delta_e(y) + \delta_l(y))^2 > 0.$$

Therefore, since $(\nabla^2 \phi(\cdot)y, y) > 0$ for any non-zero vector y, the Hessian matrix is positive definite [8], so the function $\phi(\cdot)$ is convex [25]. $\qquad\square$

Now we show that the function $\phi(x)$ is a potential in the game with symmetric externalities.

Theorem 2. *A profile $x \in \chi$ in the game with symmetric externalities is a Wardrop equilibrium iff $\phi(x) = \min\limits_{y \in \chi} \phi(y)$.*

Proof. Consider a Wardrop equilibrium profile $x \in \chi$ and some arbitrary feasible profile $y \in \chi$. The convexity of $\phi(x)$ yields [25]

$$\phi(y) - \phi(x) \geq \sum_{w \in W} \sum_{r \in P_w} \frac{\partial \phi(x)}{\partial x_r^w} (y_r^w - x_r^w) = \sum_{w \in W} \sum_{r \in P_w} F_r(x)(y_r^w - x_r^w).$$

In the equilibrium profile x for all $w \in W$ for all routes $r \in P_w$ such that $x_r^w > 0$ all latencies are the same and equal to $F_r(x) = \min\limits_{q \in P_w} F_q(x) = \lambda_w$, and on remaining routes q they are $F_q(x) \geq \lambda_w$. That is why

$$\phi(y) - \phi(x) \geq \sum_{w \in W} \lambda_w \sum_{r \in P_w} (y_r^w - x_r^w)$$

$$= \left(\sum_{w \in W} \lambda_w \sum_{r \in P_w} y_r^w - \sum_{w \in W} \lambda_w \sum_{r \in P_w} x_r^w \right) = 0$$

for $x, y \in \chi$. That is the equilibrium profile x provides a minimum for $\phi(\cdot)$ on χ.

Suppose now that a profile $x \in \chi$ is a minimum of the function $\phi(\cdot)$ on χ, but is not a Wardrop equilibrium. Then there are two routes $r, q \in P_w$ for some w, such that $x_r^w > 0$, but $F_r(x) > F_q(x)$. In this case we can move a small traffic part $\epsilon > 0$ of w from the route r to the route q so that $F_r(x) > F_q(x)$ remains true. It is possible due to latencies are continuous. Then we obtain a new profile $y \in \chi$, where $y_r^w = x_r^w - \epsilon$, $y_q^w = x_q^w + \epsilon$ and all other $y_p^w = x_p^w$ for all $p \in P_w$ and $p \neq r, q$. Strategies for all $v \in W$ such that $v \neq w$ remain the same: $y^v = x^v$. Then

$$\phi(x) - \phi(y) \geq \sum_{w \in W} \sum_{r \in P_w} \frac{\partial \phi(y)}{\partial y_r^w} (x_r^w - y_r^w) = \epsilon(F_r(y) - F_q(y)) > 0,$$

which contradicts with the fact that x is a minimum of $\phi(\cdot)$ on χ. □

Theorem 2 yields that the game with symmetric externalities as a potential. The potential function is convex, it has a unique minimum on χ. Therefore we obtain the following theorem.

Theorem 3. *The game of traffic allocation with symmetric externalities has a unique Wardrop equilibrium.*

5   Social Costs

Social costs are

$$SC(x) = \sum_{e \in E} \delta_e(x) t_e (1 + a_e \delta_e(x) + \sum_{l \in E} b_{el} \delta_l(x)).$$

The Hessian matrix of this function equals $2\nabla^2 \phi(x)$, so the function $SC(x)$ is also convex and has a unique minimum.

For each w such that $r \in P_w$ we define the marginal costs on the route r as

$$F_r^*(x) = \frac{\partial SC(x)}{\partial x_r^w} = \sum_{e \in r} t_e (1 + 2a_e \delta_e(x) + 2 \sum_{l \in E} b_{el} \delta_e(x)).$$

Theorem 4. *A strategy profile $x \in \chi$ is a socially optimal iff for each pair $w \in W$ and its routes $r \in P_w$ from $x_r^w > 0$ it yields $F_r^*(x) = \min_{q \in P_w} F_q^*(x)$, and $x_r^w = 0$ implies $F_r^*(x) \geq \min_{q \in P_w} F_q^*(x)$.*

Proof. Consider the profile $x \in \chi$ which is a minimum of $SC(\cdot)$ on χ. Imagine that there is some pair w with routes $r, q \in P_w$, such that $x_r^w > 0$, but $F_r^*(x) > F_q^*(x)$. Then we can to move some small traffic part $\epsilon > 0$ of the pair w from the route r to the route q so that $F_r^*(x) > F_q^*(x)$ remains true. In such a way we obtain a new profile $y \in \chi$, where $y_r^w = x_r^w - \epsilon$, $y_q^w = x_q^w + \epsilon$, and all other

$y_p^w = x_p^w$ for all $p \in P_w$ and $p \neq r, q$. Strategies for all $v \in W$ such that $v \neq w$ remain the same: $y^v = x^v$. Then

$$SC(x) - SC(y) \geq \sum_{w \in W} \sum_{r \in P_w} \frac{\partial SC(y)}{\partial y_r^w}(x_r^w - y_r^w) = \epsilon(F_r^*(y) - F_r^*(y)) > 0,$$

which contradicts with the fact that x is a minimum of $SC(\cdot)$ on χ.

Suppose now that a profile x is such that for each pair $w \in W$ for all its routes $r \in P_w$ from $x_r^w > 0$ we have $F_r^*(x) = \min_{q \in P_w} F_q^*(x) = \alpha_w$, and from $x_r^w = 0$ we have $F_r^*(x) \geq \min_{q \in P_w} F_q^*(x)$. Let also $y \in \chi$ to be an arbitrary feasible strategy profile.

The social costs function $SC(x)$ is convex, therefore

$$SC(y) - SC(x) \geq \sum_{w \in W} \sum_{r \in P_w} \frac{\partial SC(x)}{\partial x_r^w}(y_r^w - x_r^w)$$
$$= \sum_{w \in W} \sum_{r \in P_w} F_r^*(x)(y_r^w - x_r^w) \geq \sum_{w \in W} \alpha_w \sum_{r \in P_w}(y_r^w - x_r^w) = 0.$$

So, the profile x is a minimum of social costs.   □

Let x is an equilibrium in the game with a graph G and corresponding matrix $Corr$. We consider a new game on the same graph G but with corresponding matrix $\frac{1}{2}Corr$. The set of feasible profiles for the new game is $\frac{1}{2}\chi = \bigotimes_{w=(i,j) \in W} \{y_r^w \geq 0 | \sum_{r \in P_w} y_r^w = \frac{1}{2}Corr_{ij}\} = \{\frac{1}{2}x | x \in \chi\}$. The marginal costs for a profile $y = \frac{1}{2}x$ are $F_r^*(y) = \sum_{e \in r} t_e(1 + 2a_e\delta_e(y) + 2\sum_{l \in E} b_{el}\delta_l(y)) = \sum_{e \in r} t_e(1 + a_e\delta_e(x) + \sum_{l \in E} b_{el}\delta_l(x)) = F_r(x)$. They coincide with corresponding latencies on routes for the profile x. This result is formulates as the following lemma.

Lemma 1. *If a profile $x \in \chi$ is an equilibrium in the game with a graph G and a corresponding matrix $Corr$, then the profile $y = \frac{1}{2}x$ is an optimal in the system with the same graph G and corresponding matrix $\frac{1}{2}Corr$.*

Now we consider a socially optimal profile x^*, which provides a minimum for the social costs $SC(x^*) = \min_{x \in \chi} SC(x)$ in the game with a corresponding matrix $Corr$. We explore the social costs increasing if we substitute the corresponding matrix $Corr$ by new matrix $2Corr$.

Lemma 2. *Let x^* to be a socially optimal profile in the system with a corresponding matrix $Corr$ with minimal marginal costs α_w for each pair $w \in W$. If the corresponding matrix in the same system becomes $2Corr$, then social costs become not less than $SC(x^*) + \sum_{w=(i,j) \in W} \alpha_w Corr_{ij}$.*

Proof. Consider an arbitrary profile y from $2\chi = \underset{w=(i,j)\in W}{\otimes} \{y_r^w \geq 0 | \sum_{r\in P_w} y_r^w = 2Corr_{ij}\}$ Since the social costs function is convex, we obtain

$$
\begin{aligned}
SC(y) - SC(x^*) &\geq \sum_{w\in W}\sum_{r\in P_w} \frac{\partial SC(x^*)}{\partial x^*{}_r^w}(y_r^w - x^*{}_r^w) \\
&= \sum_{w\in W}\sum_{r\in P_w} F_r^*(x)(y_r^w - x^*{}_r^w) \geq \sum_{w\in W} \alpha_w \sum_{r\in P_w}(y_r^w - x^*{}_r^w) \\
&= \sum_{w\in W} \alpha_w(2Corr_{ij|w=(i,j)} - Corr_{ij|w=(i,j)}) \\
&= \sum_{w\in W} \alpha_w Corr_{ij|w=(i,j)}.
\end{aligned}
$$

□

6   Price of Anarchy

We consider a game with a correspondence matrix $Corr$. Suppose that a profile $x \in \chi$ is a Wardrop equilibrium with minimal latencies λ_w for each pair $w \in W$. That is in x for each pair $w \in W$ latencies are equal to $F_r(x) = \lambda_w$ on those its routes where $x_r^w > 0$. Then by Lemma 1 a profile $\frac{1}{2}x \in \frac{1}{2}\chi$ is an optimal for the game with the correspondence matrix $\frac{1}{2}Corr$ on the same graph. Minimal marginal costs for each $w \in W$ in the optimum $\frac{1}{2}x$ are $\alpha_w = \lambda_w$ and $F *_r(\frac{1}{2}x) = \lambda_w$ for the same routes with positive flows as in the profile x. Let's consider an arbitrary profile $y \in \chi$ in the game with the corresponding matrix $Corr$.

By Lemma 2

$$
\begin{aligned}
SC(y) &\geq SC(\tfrac{1}{2}x) + \sum_{w\in W} \alpha_w \tfrac{1}{2}Corr_{ij|w=(i,j)} \\
&= SC(\tfrac{1}{2}x) + \tfrac{1}{2}\sum_{w\in W} \lambda_w Corr_{ij|w=(i,j)} \\
&= SC(\tfrac{1}{2}x) + \tfrac{1}{2}\sum_{w\in W}\sum_{r=P_w} F_r(x)x_r^w \\
&= SC(\tfrac{1}{2}x) + \tfrac{1}{2}\sum_{w\in W}\sum_{r=P_w} x_r^w \sum_{e\in r} f_e(x) \\
&= SC(\tfrac{1}{2}x) + \tfrac{1}{2}\sum_{e\in E}\sum_{w\in W}\sum_{r=P_w|e\in r} x_r^w f_e(x) \\
&= SC(\tfrac{1}{2}x) + \tfrac{1}{2}\sum_{e\in E} \delta_e(x)f_e(x) \\
&= SC(\tfrac{1}{2}x) + \tfrac{1}{2}SC(x).
\end{aligned}
$$

At the same time

$$
\begin{aligned}
SC(\tfrac{1}{2}x) &= \sum_{e\in E} \delta_e(\tfrac{1}{2}x)f_e(\tfrac{1}{2}x) \\
&= \sum_{e\in E}\sum_{w\in W}\sum_{r=P_w|e\in r} \tfrac{1}{2}x_r^w f_e(\tfrac{1}{2}x) \\
&= \sum_{w\in W}\sum_{r=P_w} \tfrac{1}{2}x_r^w \sum_{e\in r} f_e(\tfrac{1}{2}x) \\
&= \sum_{w\in W}\sum_{r=P_w} \tfrac{1}{2}x_r^w F_r(\tfrac{1}{2}x) \\
&= \sum_{w\in W}\sum_{r=P_w}\sum_{e\in r} t_e \tfrac{1}{2}x_r^w(1 + \tfrac{1}{2}a_e\delta_e(x) + \sum_{l\in E} \tfrac{1}{2}b_{el}\delta_l(x)) \\
&= \tfrac{1}{4}\sum_{w\in W}\sum_{r=P_w}\sum_{e\in r} t_e x_r^w(2 + a_e\delta_e(x) + \sum_{l\in E} b_{el}\delta_l(x)) \\
&\geq \tfrac{1}{4}\sum_{w\in W}\sum_{r=P_w}\sum_{e\in r} t_e x_r^w(1 + a_e\delta_e(x) + \sum_{l\in E} b_{el}\delta_l(x)) \\
&= \tfrac{1}{4}SC(x).
\end{aligned}
$$

Then $SC(y) \geq \frac{3}{4}SC(x)$ for any profile $y \in \chi$ including the socially optimal profile. This result if formulated as an upper bound for the price of anarchy in the following theorem.

Theorem 5. *The price of anarchy in the game with general network and BPR linear delay fucntions with symmetric externalities does not exceed a value $\frac{4}{3}$.*

7 Conclusion

We considered a model of a general transport network and BPR linear delay functions with externalities. We explored the case where the impact of channel loads to the delay is pairwise symmetric. For this case we have shown that the game of traffic allocation among channels is potential. Therefore it always has a unique Wardrop equilibrium. Also it has allowed to estimate the upper bound for the price of anarchy. It is limited by the value $\frac{4}{3}$.

Acknowledgments. This research was supported by the Russian Science Foundation (No. 22-11-20015, https://rscf.ru/project/22-11-20015/), jointly with support of the authorities of the Republic of Karelia with funding from the Venture Investment Foundation of the Republic of Karelia.

References

1. Acemoglu, D., Ozdaglar, A.: Flow control, routing, and performance from service provider viewpoint. LIDS report, 74 (2004)
2. Braess, D.: Uber ein Paradoxon der Verkehrsplanung. Unternehmensforschung **12**, 258–268 (1968)
3. Chirkova, Y.V.: Potential game in parallel transport network with symmetric externalities. Matematicheskaya Teoriya Igr i Ee Prilozheniya **15**(4), 94–105 (2023) (in Russian)
4. Chirkova, J.V., Mazalov, V.V.: Optimal externalities in a parallel transportation network. Optim. Lett. **16**, 1971–1989 (2022)
5. Correa, J.R., Stier-Moses, N.E.: Wardrop Equilibria. John Wiley & Sons Inc. (2011)
6. Easley, D., Kleinberg, J.: Networks, Crowds, and Markets: Reasoning about Highly Connected World. Cambridge University Press, Cambridge (2010)
7. Gairing, M., Monien, B., Tiemann, K.: Routing (Un-) splittable flow in games with player-specific linear latency functions. In: Proceedings of the 33rd International Colloquium on Automata, Languages and Programming (ICALP 2006), LNCS, vol. 4051, pp. 501–512 (2006)
8. Gantmacher, F.R.: The Theory of Matrices. Chelsea Pub Co., New York (1959)
9. Gao, H., Mazalov, V.V., Xue, J.: Optimal Parameters of Service in a Public Transportation Market with Pricing. Journal of Advanced Transportation 2020 Safety, Behavior, and Sustainability under the Mixed Traffic Flow Environment (2020). https://doi.org/10.1155/2020/6326953
10. Gasnikov, A.V. (ed.): Introduction to Mathematical Modeling of Traffic Flows. MCCME Publishing House, Moscow (2013). (in Russian)
11. Holzman, R., Monderer, D.: Strong equilibrium in network congestion games: Increasing versus decreasing costs. Internat. J. Game Theory **44**, 647–666 (2015)

12. Jacobs, J.: The economy of cities. Random House, New York (1969)
13. Kuang, Z., Lian, Z., Lien, J.W., Zheng, J.: Serial and parallel duopoly competition in multi-segment transportation routes. Transp. Res. Part E Logist. Transp. Rev. **133**, 101821 (2020)
14. Kuang, Z., Mazalov, V.V., Tang, X., Zheng, J.: Transportation network with externalities. J. Comput. Appl. Math. **382**, 113091 (2021)
15. Mak, V., Seale, D.A., Gishces, E.J., et al.: The braess paradox and coordination failure in directed networks with mixed externalities. Prod. Oper. Manag. **27**(4), 717–733 (2018)
16. Milchtaich, I.: Congestion games with player-specific payoff functions. Games Econom. Behav. **13**, 111–124 (1996)
17. Milchtaich, I.: Network topology and the efficiency of equilibrium. Games Econom. Behav. **57**(2), 321–346 (2006)
18. Monderer, D., Shapley, L.: Potential games. Games Econom. Behav. **14**, 124–143 (1996)
19. Mtoi, E., Moses, R.: Calibration and evaluation of link congestion functions: applying intrinsic sensitivity of link speed as a practical consideration to heterogeneous facility types within urban network. J. Transp. Technol. **4**, 141–149 (2014). https://doi.org/10.4236/jtts.2014.42014
20. Papadimitriou, C.H., Koutsoupias, E.: Worst-Case Equilibria. LNSC **1563**, 404–413 (1999)
21. Rosenthal, R.W.: A class of games possessing pure-strategy Nash equilibria. Int. J. Game Theory. **2**, 65–67 (1973)
22. Roughgarden, T.: The price of anarchy is independent of the network topology. J. Comput. Syst. Sci. **67**, 341–364 (2003)
23. Roughgarden, T., Tardos, E.: How bad is selfish routing? J. ACM (JACM) **49**(2), 236–259 (2002)
24. Sheffi, Y.: Urban Transportation Networks: Equilibrium Analysis with Mathematical Programming Methods. Prentice-Hall (1984)
25. Sukharev, A.G., Timokhov, A.V., Fedorov, V.V.: Optimization Methods Course. Publ, Fizmatlit (1986)
26. U.S. Bureau of Public Roads. Traffic Assignment Manual. U.S. Department of Commerce, Washington, D.C. (1964)
27. Wardrop, J.G.: Some theoretical aspects of road traffic research. ICE Proc. Eng. Divisions **1**, 325–362 (1952)

Decision Analysis of Military Supply Chain Based on Stackelberg Game Model

Kuankuan Huang⑩, Yu Yue⑩, Yueyu Liu⑩, and Xuedong Liang⁽⊠⁾⑩

Business School, Sichuan University, Chengdu 610065, China
17709098080@163.com

Abstract. The military supply chain assumes a paramount role in national security and defense infrastructure development, entailing substantial economic implications. Scientifically informed managerial decisions serve as indispensable mechanisms for bolstering the operational efficiency and efficacy of the military supply chain, thereby posing a pragmatic challenge to contemporary decision-makers. This paper introduces the training task volume of military units as an indicator for measuring both the demand of the military and the efficiency of the supply chain for the first time. By constructing a Stackelberg game model, the decision-making behaviors of various members in the military supply chain under different leaderships are studied to explore the influence of leadership on optimal decision-making by each member in the supply chain. The research indicates that leadership dominance significantly affects the profits of members in the military supply chain. A centralized decision-making model is more effective than a decentralized decision-making model in maximizing the overall profits of the military supply chain, providing valuable insights for optimizing management and making scientifically informed decisions in the military supply chain.

Keywords: Stackelberg game · Military supply chain · Decision analysis

1 Introduction

In the contemporary geopolitical landscape, the significance of the military supply chain has garnered increasing attention, owing to its pivotal role not only in safeguarding national security interests but also in fostering the development of the defense industry and facilitating economic growth. Nonetheless, the confluence of heightened market competition and the diversification of military requirements have rendered decision-making processes within the military supply chain increasingly intricate and multifaceted. Therefore, it is of great significance to improve the scientific level of the management of the military supply chain and optimize the decision-making management to ensure the national security and the long-term stable development of the military industry. Introducing principles of supply chain management and dynamic game theory into the framework of the military supply chain represents an effective way towards optimizing decision-making processes. By means of simulated analyses and comparative optimizations of gaming strategies, the collaborative efficiency and holistic benefits of the military supply chain can be significantly augmented.

A. Eremeev et al. (Eds.): MOTOR 2024, LNCS 14766, pp. 243–256, 2024.
https://doi.org/10.1007/978-3-031-62792-7_17

Currently, the academic research around supply chain management is quite advanced. However, the determination of demand volumes within supply chains remains an inexorable focus and a difficult challenge [1, 2]. This is primarily due to the profound impact of demand uncertainties on the decision-making dynamics of supply chain stakeholders [3]. In contrast, demand in the military supply chain is more difficult to identify and predict. While many scholars have used price as a measure of demand [4–6], within military supply chains, demand volumes are contingent upon a plethora of factors, encompassing consumption rates of military supplies [7], training requisites, and operational imperatives, among others. Based on these, this paper innovatively introduces training task volume into the classical linear function of demand and price [8], thereby elucidating the nuanced fluctuations in demand within the military supply chain and associated benefits.

Existing literatures have meticulously delineated diverse decision-making models pertinent to supply chains. Notably, the Stackelberg game model, renowned for its prowess in capturing the intricacies of strategic interaction amidst static economic environments, has found widespread application in supply chain analyses [9–13]. Within the framework of this model, the presence of a leader and a follower underscores the dynamics of decision-making [14]. Employing the Stackelberg game model in supply chain management entails a comprehensive consideration of multiple decision-making entities and factors. In terms of decision subjects, existing studies mostly consider manufacturers, retailers [15–17] and third parties [18, 19], etc. In terms of decision factors, price [20–22], sales effort [23], green energy saving [24, 25], inventory management [26], market competition [27], etc. are generally considered. The differential decision-making modalities adopted by distinct stakeholders, in conjunction with variances in decision-making dominance within the supply chain hierarchy, yield disparate ramifications on overall supply chain profitability [28–30]. Within military supply chains, decision-making entities comprise Military Manufacturing Enterprises (MMEs) and Military Units (MUs) [31, 32], each governed by distinct imperatives. While MMEs predominantly factor pricing considerations into their decision-making calculus, MUs prioritize military efficacy parameters. In practice, MUs typically wield predominant decision-making authority within military supply chains [7]. However, since the army is often subordinate to and serves the overall interests of the country, the army also considers the supply chain as a whole when making decisions, and even sometimes the situation is dominated by the military industry.

Notwithstanding the Stackelberg game model is effective in addressing supply chain coordination quandaries, extant research remains relatively nascent in delving into equilibrium dynamics within military supply chains. This lacuna is attributable, in part, to a dearth of analyses and elucidating the dynamic interplay of decision-making information under disparate decision-making dominances among military supply chain stakeholders. Moreover, there exists a paucity of comprehensive assessments accounting for military efficacy considerations in decision-making processes. Consequently, this study pioneers the integration of military training task volumes and the utility of military product usage as indices for troop demands and supply chain benefits, respectively. Through the construction of a Stackelberg game model, this study investigates the decision-making comportment of military supply chain stakeholders under varying leadership scenarios, thereby furnishing invaluable insights to inform optimized decision-making management within military supply chains.

2 Problem Definition and Model Setup

2.1 Problem Definition

The military supply chain constituents under investigation in this study encompass two tiers: MMEs and MUs. MMEs are tasked with the production, packaging, and transportation of military products, thereby accruing revenue through product sales. Conversely, MUs procure military products and expend them during routine training tasks, thereby yielding military efficacy. The parameters and variables pertinent to this investigation are detailed in Table 1.

Table 1. Parameters and variables

Parameters & Variables	Clarification
p	The price of per unit military product
Q	The demand quantity for military products
q	The training task volume of MUs
ω	The utility generated per unit of military product
β	The coefficient of the utility per unit of military product on demand quantity
μ	The usage efficiency per unit of military product
$C(q)$	The additional costs associated with training tasks
λ	The coefficient of the impact of training task volume on additional costs
c_{MMEs}	The costs per unit of military product for MMEs
c_{MUs}	The costs incurred by MUs for receiving and storing each unit of military product
c	The total costs per unit product across the entire military supply chain
π_{MMEs}	The profits of MMEs
π_{MUs}	The profits of MUs
π_{MC}	The overall profits of the military supply chain

2.2 Model Assumptions

Assumption 1: Given the typical confidential nature of MMEs and their involvement in the production of military products, it is commonly observed that MMEs assume responsibility for the internal logistics and transportation of these goods, thereby eschewing external outsourcing. Consequently, the military supply chain primarily comprises two hierarchical levels: MMEs and MUs. Within this framework, both MMEs and MUs are presumed to be rational agents engaged in game-theoretic decision-making under conditions of complete and perfect information.

Assumption 2: Recognizing the specialized nature of MMEs, which often concentrate on the production of singular or complementary military goods characterized by uniform pricing denoted as p.

Assumption 3: The training task volume of MUs is deterministically established through internal planning processes. During the organization and execution of training exercises or missions, the consumption of military products yields utility, which can be quantitatively assessed in terms of value. This value is the sum of the purchase price of per unit military product p and its usage efficiency μ, where μ varies depending on the diverse types of training or tasks in which the MUs employ the military products. Therefore, the utility generated per unit of military product can be denoted as ω. It is calculated as Eq. (1).

$$\omega = p + \mu \tag{1}$$

The demand quantity Q for military products by MUs is influenced by the utility generated per unit of military product and the training volume q of the MUs. Drawing upon the methodology outlined by Mahdizadeh et al. [8], it is hypothesized that Q exhibits a linear inverse relationship with ω, while demonstrating a linear positive correlation with q. Specifically, this relationship can be articulated as Eq. (2).

$$Q(\omega, q) = \alpha - \beta\omega + q \tag{2}$$

In this context, β represents the coefficient indicating the impact of the utility per unit of military product on demand quantity. α is a constant, and its value equals the maximum demand quantity when MUs do not consider changes in military product prices or training task volume.

Assumption 4: It is posited that as MUs consume military products, they concurrently expend supplementary materials and manpower, denoted as additional training task costs, denoted by parameter $C(q)$. Drawing upon the methodology outlined by Tsay A. A. et al. [33], it is assumed that the function describing these costs follows a quadratic form, wherein parameter λ represents a constant signifying the coefficient of the impact of training task volume on additional costs. Thus, it can be expressed as Eq. (3).

$$C(q) = \lambda q^2 \tag{3}$$

Assumption 5: Assuming that the costs per unit of military product for MMEs is represented by c_{MMEs}, and the costs incurred by MUs for receiving and storing each unit of military product is denoted as c_{MUs}, the total costs per unit product across the entire military supply chain can be expressed as Eq. (4).

$$c = c_{MMEs} + c_{MUs} \tag{4}$$

The function representing the profits of MMEs π_{MMEs} can be expressed as Eq. (5).

$$\pi_{MMEs} = (p - c_{MMEs})Q = (p - c_{MMEs})(\alpha - \beta\omega + q) \tag{5}$$

The function representing the profits of MUs π_{MUs} can be expressed as Eq. (6).

$$\pi_{MUs} = (\omega - p - c_{MUs})Q - C(q) = (\omega - p - c_{MUs})(\alpha - \beta\omega + q) - \lambda q^2 \tag{6}$$

The function representing the profits of MUs π_{MC} can be expressed as Eq. (7).

$$\pi_{MC} = \pi_{MMEs} + \pi_{MUs} = (\omega - c_{MMEs} - c_{MUs})(\alpha - \beta\omega + q) - \lambda q^2 \tag{7}$$

In order to ensure the long-term stable operation of the entire military supply chain, the profits of all members within the supply chain should be greater than 0. Otherwise, the supply chain cannot function properly. It can be expressed as St. (8).

$$p > c_{MMEs} > 0, \omega > c_{MMEs} + c_{MUs} > 0, \alpha > 0, \beta > 0, \lambda > 0, \pi_{MMEs} > 0, \pi_{MUs} > 0 \tag{8}$$

3 Methodology and Model

In the military supply chain, the essence of the Stackelberg game between the MMEs and MUs is to select their strategies based on the potential strategies of the other party. This is to ensure the maximization of one's own interests under the opponent's strategy, ultimately leading to a Nash equilibrium. Within this military supply chain, there exist three decision-making models: the centralized decision-making model (Type MC), the decentralized decision-making model led by MMEs (Type MMEs), and the decentralized decision-making model led by MUs (Type MUs).

3.1 Type MC

In the centralized decision-making model, MMEs and MUs are integrated into a single entity for decision-making. In this mode, both parties make decisions with the aim of maximizing the overall profitability of the military supply chain. Deriving from Eq. (7):

$$\pi_{MC} = (\omega - c_{MMEs} - c_{MUs})(\alpha - \beta\omega + q) - \lambda q^2 \tag{9}$$

Solving for the partial derivative gives:

$$\frac{\partial \pi_{MC}}{\partial \omega} = -2\beta\omega + \alpha + q + \beta(c_{MMEs} + c_{MUs}), \frac{\partial^2 \pi_{MC}}{\partial \omega^2} = -2\beta$$

$$\frac{\partial \pi_{MC}}{\partial q} = -2\lambda q + \omega - c_{MMEs} - c_{MUs}, \frac{\partial^2 \pi_{MC}}{\partial q^2} = -2\lambda$$

The Hessian matrix of the overall profits the military supply chain π_{MC} with respect to ω and q is:

$$H_{MC} = \begin{bmatrix} -2\beta & 1 \\ 1 & -2\lambda \end{bmatrix}$$

As H_{MC} is a negative definite matrix, the maximum value point of function π_{MC} is at the stationary point.

Let $\frac{\partial \pi_{MC}}{\partial \omega} = 0$, $\frac{\partial \pi_{MC}}{\partial q} = 0$, if $\frac{\alpha - \beta(c_{MMEs} + c_{MUs})}{4\lambda\beta - 1} > 0$, the optimal solution can be obtained as follows:

- The optimal utility generated per unit of military product:

$$\omega = \frac{(2\lambda\beta - 1)(c_{MMEs} + c_{MUs}) + 2\lambda\alpha}{4\lambda\beta - 1}$$

- The optimal price of per unit military product: $p = \frac{(2\lambda\beta-1)(c_{MMEs}+c_{MUs})+2\lambda\alpha}{4\lambda\beta-1} - \mu$
- The optimal training task volume of MUs: $q = \frac{\alpha - \beta(c_{MMEs}+c_{MUs})}{4\lambda\beta-1}$
- The optimal demand quantity for military products: $Q = \frac{2\lambda\beta[\alpha - \beta(c_{MMEs}+c_{MUs})]}{4\lambda\beta-1}$
- The optimal overall profits of the military supply chain:

$$\pi_{MC} = \frac{\lambda[\alpha - \beta(c_{MMEs} + c_{MUs})]^2}{4\lambda\beta - 1}$$

3.2 Type MMEs

In this model, both MMEs and MUs make independent decisions with the aim of maximizing their own profits. Within the Type MMEs decentralized decision-making model, MMEs hold a dominant position. To obtain the perfect equilibrium value of this game model, an iterative induction method is employed, where MMEs first determine the optimal price of their per unit military product based on the reaction function of the MUs to their product prices. Subsequently, the MUs decide on the types and volumes of their training tasks. Both parties engage in a cyclic game under this model until they both achieve optimal decision-making, thus realizing equilibrium. Deriving from Eq. (6):

$$\pi_{MUs} = (\mu - c_{MUs})(\alpha - \beta p - \beta\mu + q) - \lambda q^2 \tag{10}$$

Taking the partial derivatives of μ and q in Eq. (10), respectively, it can obtain:

$$\frac{\partial \pi_{MUs}}{\partial \mu} = -2\beta\mu + \alpha - \beta p + q + \beta c_{MUs}$$

$$\frac{\partial \pi_{MUs}}{\partial q} = -2\lambda q + \mu - c_{MUs}$$

Let $\frac{\partial \pi_{MUs}}{\partial \mu} = 0$, $\frac{\partial \pi_{MUs}}{\partial q} = 0$, it can obtain:

$$q = \frac{\beta(c_{MUs} + p) - \alpha}{4\lambda\beta - 1}$$

$$\mu = \frac{2\lambda\beta(3c_{MUs} + p) - 2\lambda\alpha - c_{MUs}}{4\lambda\beta - 1}$$

Substituting them into Eq. (5) and taking the second partial derivative with respect to p, we get:

$$\frac{\partial^2 \pi_{MMEs}}{\partial p^2} = \frac{4\lambda\beta^2}{1 - 4\lambda\beta}$$

When $1 - 4\lambda\beta < 0$, $\alpha - \beta(c_{MMEs} + c_{MUs}) > 0$, we can derive $\frac{\partial^2 \pi_{MMEs}}{\partial p^2} < 0$.

The profit function of π_{MMEs} is a convex function, hence, it possesses a maximum value.

Let $\frac{\partial \pi_{MMEs}}{\partial p} = 0$, the maximum price per unit of product for MMEs can be obtained:

$$p = \frac{\alpha + \beta c_{MMEs} - \beta c_{MUs}}{2\beta}$$

Subsequently, the optimal solution can be computed based on p as follows:

- The optimal usage efficiency per unit of military product:

$$\mu = \frac{3\lambda\beta c_{MUs} + \lambda\alpha - c_{MUs} - \lambda\beta c_{MMEs}}{4\lambda\beta - 1}$$

- The optimal training task volume of MUs: $q = \frac{\alpha - \beta(c_{MMEs} + c_{MUs})}{8\lambda\beta - 2}$
- The optimal demand quantity for military products: $Q = \frac{\lambda\beta[\alpha - \beta(c_{MMEs} + c_{MUs})]}{4\lambda\beta - 1}$
- The optimal profits of MMEs: $\pi_{MMEs} = \frac{\lambda[\alpha - \beta(c_{MMEs} + c_{MUs})]^2}{8\lambda\beta - 2}$
- The optimal profits of MUs: $\pi_{MUs} = \frac{\lambda[\alpha - \beta(c_{MMEs} + c_{MUs})]^2}{16\lambda\beta - 4}$
- The optimal overall profits of the military supply chain:

$$\pi_{MC} = \frac{3\lambda[\alpha - \beta(c_{MMEs} + c_{MUs})]^2}{16\lambda\beta - 4}$$

3.3 Type MUs

In the Type MUs decentralized decision-making model, MUs hold the dominant position. In this model, the decision-making sequence of both parties is reversed compared to Type MMEs. Specifically, MUs first determine the types and volumes of their training tasks, after which MMEs determine the price of per unit military product based on this information. Similarly, the solution is obtained using the method of backward induction. Deriving from Eq. (5):

$$\pi_{MMEs} = (p - c_{MMEs})(\alpha - \beta p - \beta\mu + q) \tag{11}$$

Taking the first-order partial derivative of Eq. (8) with respect to p, it can obtain:

$$\frac{\partial \pi_{MMEs}}{\partial p} = -2\beta p + \alpha - \beta\mu + q + \beta c_{MMEs} = 0$$

$$p = \frac{\alpha - \beta\mu + q + \beta c_{MMEs}}{2\beta} \tag{12}$$

Substituting Eq. (12) into Eq. (6), it can obtain:

$$\pi_{MUs} = \frac{1}{2}(\mu - c_B)(\alpha - \beta\mu + q - \beta c_{MMEs}) - \lambda q^2 \qquad (13)$$

Solving for the partial derivative gives:

$$\frac{\partial \pi_{MUs}}{\partial \mu} = \frac{1}{2}(\alpha - 2\beta\mu + q - \beta c_{MMEs} + \beta c_{\pi_{MUs}}), \frac{\partial^2 \pi_{MUs}}{\partial \mu^2} = -\beta$$

$$\frac{\partial \pi_{MUs}}{\partial q} = \frac{1}{2}(\mu - c_{\pi_{MUs}}) - 2\lambda q, \frac{\partial^2 \pi_{MUs}}{\partial q^2} = -2\lambda$$

The Hessian matrix of the profits of MUs π_{MUs} with respect to μ and q is:

$$H_{MUs} = \begin{bmatrix} -\beta & \frac{1}{2} \\ \frac{1}{2} & -2\lambda \end{bmatrix}$$

As H_{MUs} is a negative definite matrix, the maximum value point of function π_{MUs} is at the stationary point.

Let $\frac{\partial \pi_{MUs}}{\partial \mu} = 0$, $\frac{\partial \pi_{MUs}}{\partial q} = 0$, the maximum profit of MUs, the optimal usage efficiency per unit of military product, and the optimal training task volume of MUs can be derived as follows:

- The maximum profits of MUs: $\pi_{MUs} = \frac{\lambda[\alpha - \beta(c_{MMEs} + c_{MUs})]^2}{8\lambda\beta - 1}$
- The optimal usage efficiency per unit of military product:

$$\mu = \frac{4\lambda[\alpha - \beta(c_{MMEs} - c_{MUs})] - c_{MUs}}{8\lambda\beta - 1}$$

- The optimal training task volume of MUs: $q = \frac{\alpha - \beta(c_{MMEs} + c_{MUs})}{8\lambda\beta - 1}$

Substituting μ and q into the function of the MMEs yields:

- The optimal price of per unit military product:

$$p = \frac{2\lambda[\alpha + \beta(3c_{MMEs} - c_{MUs})] - c_{MMEs}}{8\lambda\beta - 1}$$

- The optimal utility generated per unit of military product:

$$\omega = \frac{2\lambda[3\alpha - \beta(c_{MMEs} + c_{MUs})] - c_{MMEs} - c_{MUs}}{8\lambda\beta - 1}$$

- The optimal demand quantity for military products: $Q = \frac{2\lambda\beta[\alpha - \beta(c_{MMEs} + c_{MUs})]}{8\lambda\beta - 1}$
- The optimal profits of MMEs: $\pi_{MMEs} = \frac{4\lambda^2\beta[\alpha - \beta(c_{MMEs} + c_{MUs})]^2}{(8\lambda\beta - 1)^2}$
- The optimal overall profits of the military supply chain:

$$\pi_{MC} = \frac{\lambda(12\lambda\beta - 1)[\alpha - \beta(c_{MMEs} + c_{MUs})]^2}{(8\lambda\beta - 1)^2}$$

The equilibrium results of the Stackelberg game in the military supply chain under different modes are summarized in Table 2.

Table 2. Equilibrium results of different decision-making modes in the military supply chain

	Type MC	Type MMEs	Type MUs
p	$\frac{(2\lambda\beta-1)(c_{MMEs}+c_{MUs})+2\lambda\alpha}{4\lambda\beta-1}-\mu$	$\frac{\alpha+\beta c_{MMEs}-\beta c_{MUs}}{2\beta}$	$\frac{2\lambda[\alpha+\beta(3c_{MMEs}-c_{MUs})]-c_{MMEs}}{8\lambda\beta-1}$
μ	$\frac{(2\lambda\beta-1)(c_{MMEs}+c_{MUs})+2\lambda\alpha}{4\lambda\beta-1}-p$	$\frac{3\lambda\beta c_{MUs}+\lambda\alpha-c_{MUs}-\lambda\beta c_{MMEs}}{4\lambda\beta-1}$	$\frac{4\lambda[\alpha-\beta(c_{MMEs}-c_{MUs})]-c_{MUs}}{8\lambda\beta-1}$
ω	$\frac{(2\lambda\beta-1)(c_{MMEs}+c_{MUs})+2\lambda\alpha}{4\lambda\beta-1}$	$\frac{\alpha+\beta c_{MMEs}-\beta c_{MUs}}{2\beta}+\frac{3\lambda\beta c_{MUs}+\lambda\alpha-c_{MUs}-\lambda\beta c_{MMEs}}{4\lambda\beta-1}$	$\frac{2\lambda[3\alpha-\beta(c_{MMEs}+c_{MUs})]-c_{MMEs}-c_{MUs}}{8\lambda\beta-1}$
q	$\frac{\alpha-\beta(c_{MMEs}+c_{MUs})}{4\lambda\beta-1}$	$\frac{\alpha-\beta(c_{MMEs}+c_{MUs})}{8\lambda\beta-2}$	$\frac{\alpha-\beta(c_{MMEs}+c_{MUs})}{8\lambda\beta-1}$
Q	$\frac{2\lambda\beta[\alpha-\beta(c_{MMEs}+c_{MUs})]}{4\lambda\beta-1}$	$\frac{\lambda\beta[\alpha-\beta(c_{MMEs}+c_{MUs})]}{4\lambda\beta-1}$	$\frac{2\lambda\beta[\alpha-\beta(c_{MMEs}+c_{MUs})]}{8\lambda\beta-1}$
π_{MMEs}	-	$\frac{\lambda[\alpha-\beta(c_{MMEs}+c_{MUs})]^2}{8\lambda\beta-2}$	$\frac{4\lambda^2\beta[\alpha-\beta(c_{MMEs}+c_{MUs})]^2}{(8\lambda\beta-1)^2}$
π_{MUs}	-	$\frac{\lambda[\alpha-\beta(c_{MMEs}+c_{MUs})]^2}{16\lambda\beta-4}$	$\frac{\lambda[\alpha-\beta(c_{MMEs}+c_{MUs})]^2}{8\lambda\beta-1}$
π_{MC}	$\frac{\lambda[\alpha-\beta(c_{MMEs}+c_{MUs})]^2}{4\lambda\beta-1}$	$\frac{3\lambda[\alpha-\beta(c_{MMEs}+c_{MUs})]^2}{16\lambda\beta-4}$	$\frac{\lambda(12\lambda\beta-1)[\alpha-\beta(c_{MMEs}+c_{MUs})]^2}{(8\lambda\beta-1)^2}$

4   Comparative Analysis

For the purpose of comparison, Table 2 is simplified, with the numerical values of variables in the Type MC serving as standard units. This facilitated the conversion of corresponding data from the Type MMEs and Type MUs decision-making models, resulting in Table 3.

Through Table 3, it can be observed that:

(1) Among the three decision-making models, the centralized decision-making model (Type MC) maximizes the overall revenue of the military supply chain, followed by the decentralized decision-making model led by MMEs (Type MMEs), while the decentralized decision-making model led by MUs (Type MUs) performs the poorest, with its revenue being less than 75% of Type MC. The analysis suggests that in the Type MUs decision-making model, the MUs' training task volume q is the smallest, resulting in the least demand Q for military products and consequently lower profits from military product consumption.

(2) In the decentralized decision-making models of the military supply chain, the party with decision dominance has a positive impact on its own revenue. Comparing the revenues of MMEs and MUs in Type MMEs and Type MUs from Table 3, each party achieves maximum revenue in the decision-making model where it has decision dominance, and this maximum revenue is more than twice the revenue under the other decision-making model.

(3) In the decentralized decision-making models of the military supply chain, the party with decision dominance can relatively achieve higher profits. In the Type MMEs

decision-making model, where $\pi_{MMEs} = \frac{1}{2}\pi_{MC} = 2\pi_{MUs}$, the revenue of MMEs with decision dominance is twice that of the MUs. Similarly, in the Type MUs decision-making model, the revenue of MUs with decision dominance is more than twice that of MMEs.

Whether in Type MMEs or Type MUs decentralized decision-making models, when $\lambda > \frac{1}{2\beta}$, the revenue generated by a unit of military product is higher than that in the centralized decision-making model. This may be due to the fact that in the decentralized decision-making model, the force coordinates the development of training programs, scientifically and rationally arranges training plans, combines multiple batches of the same type of training tasks, and does a good job of supporting the single training task. Its additional cost impact factor will be higher than that of each single task, so the revenue that can be generated by the unit of military products is enhanced.

Table 3. Simplified equilibrium results of different decision-making models

	Type MC	Type MMEs	Type MUs
ω	ω_{MC}	*if* $\lambda > \frac{1}{2\beta}$, $\omega > \omega_{MC}$ *Otherwise* , $\omega \leq \omega_{MC}$	*if* $\lambda > \frac{1}{2\beta}$, $\omega > \omega_{MC}$ *Otherwise* , $\omega \leq \omega_{MC}$
q	q_{MC}	$\frac{1}{2}q_{MC}$	$\frac{1}{4}q_{MC} < q < \frac{1}{2}q_{MC}$
Q	Q_{MC}	$\frac{1}{2}Q_{MC}$	$< \frac{1}{2}Q_{MC}$
π_{MMEs}	-	$\frac{1}{2}\pi_{MC}$	$< \frac{1}{4}\pi_{MC}$
π_{MUs}	-	$\frac{1}{4}\pi_{MC}$	$\frac{1}{2}\pi_{MC} < \pi_{MUs} < \frac{3}{4}\pi_{MC}$
π_{MC}	π_{MC}	$\frac{3}{4}\pi_{MC}$	$< \frac{3}{4}\pi_{MC}$

Moreover, to verify the feasibility and effectiveness of the above models, numerical validation is conducted. Actual data from a certain military unit participating in peacekeeping operations is taken as an example. In this example, $Q = 2128, \alpha = 4980, p = 14998, q = 208, C(q) = 112920, c_A = 8200, c_B = 200$. The mission can be regarded as a decentralized decision-making model led by the MUs. Assuming $\mu = 2000$, we can derive $\lambda = 2.61, \beta = 0.18$. Substituting these values into Table 2, the numerical verification results for equilibrium decisions under different decision-making models can be obtained, as shown in Table 4. It can be observed that the above conclusion is validated.

Table 4. The numerical verification results for equilibrium decisions

	Type MC	Type MMEs	Type MUs
ω	28990.26	28328.47	16998
q	3944.49	1972.25	208
Q	3706.25	1853.12	2128
π_{MMEs}	-	17851758.78	14466144
π_{MUs}	-	8925879.39	3717480
π_{MC}	35703517.56	26777638.17	18183624

5 Conclusion

This paper constructs a Stackberg game model concerning MMEs and MUs, pioneering the introduction of military training task volume and the utility of military product usage as indices for troop demands and supply chain benefits, respectively. It investigates the optimal strategies and profit maximization challenges faced by MMEs and MUs within the military supply chain, considering scenarios of centralized decision-making and decentralized decision-making under varying degrees of leadership authority. The research findings indicate the following: a) Decision-making dominance significantly influences the revenue of members in the military supply chain. Members with decision-making dominance in the supply chain attain higher revenue compared to those in follower positions. Moreover, within the same decision-making model, members with decision-making dominance achieve higher revenue than those in follower positions. b) The centralized decision-making model yields the optimal total revenue, outperforming decentralized decision-making models in maximizing the overall revenue of the military supply chain. Compared to decentralized decision-making models, the centralized decision-making model results in military product demand and MUs' training intensity being more, thereby benefiting all members of military supply chain.

In practical scenarios, it's quite common for military supply chains to be led by MUs in decision-making. However, in the context of this study, MUs' dominance in decentralized decision-making leads to the lowest overall revenue of the supply chain. The results of this paper can be used as a reference for MUs' training programmers to develop more rational training programs, as a basis for pricing and production in MMEs' products, and as a basis for promoting deeper civil-military integration. To enhance the overall revenue of the military supply chain, MMEs and MUs should establish an in-depth cooperative mechanism, breaking the dichotomy between military and civilian sectors, and vigorously promoting deep integration of military and civilian sectors. This involves:

(1) Establishing a centralized decision-making, military-civilian coordination, and smooth and efficient organizational system to unify planning and decision-making, thereby maximizing the overall revenue of the military supply chain.

(2) Forming a unified pattern of national leadership, demand-driven, and market operation, encouraging supply chain members to break their respective interests barriers, overcome organizational obstacles, and build integration mechanisms.

(3) Establishing a workflow of resource aggregation, consensus-building, and collaborative division of labor, where MMEs leverage technical advantages, refine cost control systems, and minimize process losses, while MUs scientifically formulate training plans and adjust training content to maximize the benefits of military products, thereby enhancing the overall operational efficiency of military supply chain.

In order to simplify the model, this paper only studied the Stackelberg game model of a class of MMEs and MUs. And we only consider how the members of the military supply chain make optimal decisions under the influence of three factors: the price of per unit military product, the usage efficiency per unit of military product, and the training task volume of MUs. Given the actual complexity of the situation, future research could involve constructing game models involving multiple MUs and different MMEs, introducing other variables, then studying and comparing the overall revenue and optimal decisions of the military supply chain.

References

1. Dong J. F., Sun S.F., Gao G.C., Yang R.Y.: Pricing and strategy selection in a closed-loop supply chain under demand and return rate uncertainty. 4OR-A Quarterly J. Oper. Res. **19**, 501–530 (2021). https://doi.org/10.1007/s10288-020-00458-7

2. Ke H., Wu Y., Huang H.: Competitive pricing and remanufacturing problem in an uncertain closed-loop supply chain with risk-sensitive retailers. In: Editor, F., Editor, S. (eds.) CONFERENCE 2018, LNCS, vol. 35(1), p. 1850003. Springer, Heidelberg (2018). https://doi.org/10.1142/S0217595918500033

3. Lau, A.H.L., Lau, H.S., Wang, J.C.: Usefulness of resale price maintenance under different levels of sales-effort cost and system-parameter uncertainties. Eur. J. Oper. Res. **203**(2), 513–525 (2010). https://doi.org/10.1016/j.ejor.2009.08.011

4. Giri B.C., Glock C.H.: A closed-loop supply chain with stochastic product returns and worker experience under learning and forgetting. In: 9th International Proceedings on Proceedings, pp. 6760–6778. Publisher, Location (2017). https://doi.org/10.1080/00207543.2017.1347301

5. Tang, C.H., Yang, H.L., Cao, E.B., Lai, K.K.: Channel competition and coordination of a dual-channel supply chain with demand and cost disruptions. Appl. Econ. **50**(46), 4999–5016 (2018). https://doi.org/10.1080/00036846.2018.1466989

6. Dong, J., Sun, S., Gao, G., Yang, R.: Pricing and strategy selection in a closed-loop supply chain under demand and return rate uncertainty. 4OR **19**(4), 501–530 (2020). https://doi.org/10.1007/s10288-020-00458-7

7. Huang, G.L.: Game analysis of military supply chain based on Stackelberg Model (in Chinese). China Storage Transport Mag. **9**, 125–128 (2017). https://doi.org/10.16301/j.cnki.cn12-1204/f.2017.09.042

8. Mahdizadeh, R., Pourbaba, I., Fozooni, N., Abraham, A.: Addressing sustainable supply chain network using Stackelberg game. In: Proceedings of the 13th International Conference on Soft Computing and Pattern Recognition (2021). https://doi.org/10.1007/978-3-030-96302-6_31

9. Jaggi, C.K., Khanna, A., Mittal, M.: Credit financing for deteriorating imperfect-quality items under inflationary conditions. Int. J. Serv. Oper. Inf. **6**, 292 (2011). https://doi.org/10.1504/IJSOI.2011.045560

10. Yadav, R., Pareek, S., Mittal, M.: Supply chain models with imperfect quality items when end demand is sensitive to price and marketing expenditure. RAIRO Oper. Res. **52**, 725–742 (2018). http://www.numdam.org/articles/https://doi.org/10.1051/ro/2018011/
11. Yadav, R., Mittal, M., Lamba, N.K., Jayaswal, M.K.: A Stackelberg game approach in supply chain for imperfect quality items with learning effect in fuzzy environment. Soft Computing in Inventory Management (2021)
12. Yadav, R., Pareek, S., Mittal, M., Jayaswal, M.K.: Two-level supply chain models with imperfect quality items when demand influences price and marketing promotion. J. Manage. Analytics (2021). https://doi.org/10.1007/978-981-16-2156-7_7
13. Li, W., He, J., Shi, Y.: Firms' shareholding behavior in green supply chains: Carbon emissions reduction, power structures, and technology spillovers. Heliyon **10** (2024). https://doi.org/10.1016/j.heliyon.2024.e25086
14. Xia, L., Li, K., Wang, J., Xia, Y., Qin, J.: Carbon emission reduction and precision marketing decisions of a platform supply chain. Int. J. Prod. Econ. (2023). https://doi.org/10.1016/j.ijpe.2023.109104
15. Esmaeili, M., Hamedani, S.G.: A competitive facility location problem using distributor Stackelberg game approach in multiple three-level supply chains. Int. J. Appl. Manage. Sci. **14**(3), 205–220 (2022). https://doi.org/10.1504/IJAMS.2022.125125
16. Ren, X., Zhao, N.: Optimization of coordination configuration between product service system and supply chain based on Stackelberg game. J. Ind. Manage. Optim. (2024). https://doi.org/10.3934/jimo.2024009
17. Yu, Y., Qiu, R., Sun, M., Li, Z.: Supply chain quality and pricing decisions with retailer fairness concerns and consumer dynamic reference quality effects. J. Ind. Manage. Optim. (2024). https://doi.org/10.3934/jimo.2023163
18. Liu, W., et al.: Optimal pricing for a multi-echelon closed loop supply chain with different power structures and product dual differences. J. Clean. Prod. **257**, 120281 (2020). https://doi.org/10.1016/j.jclepro.2020.120281
19. Shan, L., Duan, C., Qiao, J.: A closed-loop supply chain operation decision under life cycle: ecological design, service design and recycling effort perspectives. RAIRO-Oper. Res. **58**(1), 341–371 (2024). https://doi.org/10.1051/ro/2023106
20. Khanlarzade, N., Zegordi, S.H., Nakhai Kamalabadi, I.: Pricing in two competing supply chains based on market power and market size under centralized and decentralized structures. Scientia Iranica **28**(1), 424–445 (2021). DOI:https://doi.org/10.24200/sci.2019.50740.1845
21. Wang, Z., Wang, H., Chen, X.: The impact of delayed fixed-price payment in the decentralized project supply chain. Int. J. Syst. Sci. Oper. Logistics **11**(1), 2308584 (2024). https://doi.org/10.1080/23302674.2024.2308584
22. Lu, L., Menezes, M.B.: Supply chain vertical competition and product proliferation under different power structures. Int. J. Prod. Econ. (2023). https://doi.org/10.1016/j.ijpe.2023.109097
23. Meng, L., Shi, Y.R., Zhang, X.X., Hong, M.X.: Dual-channel supply chain game considering the retailer's sales effort. Eng. Lett. **29**(4) (2021)
24. Qiao, S., Wang, P.: Research on emission reduction strategy of energy-saving service providers' participation in power supply chain cooperation based on matlab. In: 2022 International Conference on Artificial Intelligence, Internet and Digital Economy (ICAID 2022), pp. 942–949. Atlantis Press (2022). https://doi.org/10.2991/978-94-6463-010-7_94
25. Li, X., He, J.: Mechanism of the green supply chain profit of building materials considering the duopoly competition model and consumer green preference. Int. J. Syst. Sci. Oper. Logistics **11**(1), 2311283 (2024). https://doi.org/10.1080/23302674.2024.2311283
26. Khanlarzade, N., Farughi, H.: Modeling the Stackelberg game with a boundedly rational follower in deterioration supply chain-based interaction with the leader's hybrid pricing strategy. Expert Syst. Appl. **237**, 121302 (2023). https://doi.org/10.1016/j.eswa.2023.121302

27. He, Z., Ni, S., Jiang, X., Feng, C.: The influence of demand fluctuation and competition intensity on advantages of supply chain dominance. Mathematics (2023). Doi:https://doi.org/10.3390/math11244931
28. Shi, R., Zhang, J., Ru, J.: Impacts of power structure on supply chains with uncertain demand. Prod. Oper. Manag. **22**, 1232–1249 (2013). https://doi.org/10.1111/poms.12002
29. Wang, J., Wang, Y., Lai, F.: Impact of power structure on supply chain performance and consumer surplus. Int. Trans. Oper. Res. **26**, 1752–1785 (2019). https://doi.org/10.1111/itor.12466
30. Chen, X., Wang, X., Jiang, X.: The impact of power structure on the retail service supply chain with an O2O mixed channel. J. Oper. Res. Soc. **67**, 294–301 (2016). https://doi.org/10.1057/jors.2015.6
31. Nazeri, A., Soofifard, R., Asili, G.R.: Supplier selection and evaluation in military supply chain and Order allocation. Int. J. Procurement Manage. (2019). https://doi.org/10.1504/IJPM.2019.101226
32. Minie, S. M., Gendreau, M., Potvin, J., Berger, J., Boukhtouta, A., Thomson, D.: Military three-echelon disaster relief supply chain management. In: 2017 4th International Conference on Information and Communication Technologies for Disaster Management (ICT-DM), pp. 1–8 (2017). https://doi.org/10.1109/ICT-DM.2017.8275674
33. Tsay, A.A., Agrawal, N.: Channel dynamics under price and service competition. Manuf. Serv. Oper. Manag. **2**(4), 372–391 (2000). https://doi.org/10.1287/msom.2.4.372.12342

UCB Strategies in a Gaussian Two-Armed Bandit Problem

Alexander Kolnogorov[✉][iD]

Yaroslav-the-Wise Novgorod State University, Veliky Novgorod 173003, Russia
Alexander.Kolnogorov@novsu.ru

Abstract. We consider the two-armed bandit problem in the application to batch data processing if there are two alternative processing methods with different a priori unknown efficiencies, and income is understood as successfully processed data. It is necessary to determine a more
effective method and ensure its preferential use. Batch processing means
that the incomes in batches have Gaussian distributions with a priori
unknown one-step mathematical expectations and variances. This corresponds to a situation when the number of processed data batches and
their volumes are of moderate size. We use UCB strategies for control.
To calculate the regret, a recursive dynamic programming equation is
obtained. Using the properties of UCB strategies, this equation was presented in a more computationally convenient form and then presented
in an invariant form with a control horizon equal to one. The invariant
equation does not depend on the total number of processed data, but
only on the number of batches into which the data is divided and on the
number of internal packets for which the variance is estimated.

Keywords: two-armed bandit problem · batch processing · UCB
strategies · Bayesian approach · invariant description

1 Introduction

We consider the two-armed bandit problem [1,2]. A two-armed bandit is a slot
machine with two arms hereinafter referred to as actions. The choice of each
action is accompanied by a random income, the distribution of which depends
only on the selected action, is fixed during the game but is unknown to the player.
It is required, observing the statistics of the game, to determine the action that
corresponds to a larger value of the mathematical expectation of one-step income
and ensure its preferential use. The problem has applications in behavior modeling, adaptive control in a random environment, medicine, Internet technologies,
data processing, etc. [1–5].

We consider the problem in an application to batch data processing. Let
$N = KM$ data be given, for which two alternative methods (actions) can be
used, and N is large enough. The data is divided into K batches of M data
per batch, all data of each batch is processed by the same method. The total

number of successfully processed data in the batch can be considered as income. If the size of the batch is large enough, then the distribution of income in the batch is approximately Gaussian; further, for convenience, we consider it to be exactly Gaussian. Formally, a controlled random process ξ_k, $k = 1, 2, \ldots, K$, is considered, which values are interpreted as incomes, depend only on the actions y_k selected at time points k (processing moments of the kth batch) and have Gaussian distributions with densities $f_{MD_\ell}(x|Mm_\ell)$ if $y_k = \ell$, where

$$f_D(x|m) = (2\pi D)^{-1/2} \exp\left\{-(x - m)^2/(2D)\right\} \tag{1.1}$$

is the Gaussian distribution density, m_ℓ, D_ℓ are mathematical expectation and variance of one-step income for applying the ℓth action to a unit of data ($\ell = 1, 2$). Such a two-armed bandit is described by the parameter $\theta = (m_1, D_1, m_2, D_2)$. The value of the parameter is assumed to be a priori unknown but a set of its possible values $\Theta = \{\theta : |m_1 - m_2| \leq 2C < \infty, 0 < \underline{D} \leq D_\ell \leq \overline{D} < \infty, \ell = 1, 2\}$ is known.

A control strategy σ at time point $k + 1$ determines, generally speaking, the random choice of action y_{k+1} depending on the known current statistics $(X_1, S_1, k_1; X_2, S_2, k_2)$, where k_1, k_2 are current cumulative numbers of both actions applications ($k_1 + k_2 = k$), X_1, X_2 and S_1, S_2 are corresponding cumulative incomes and s^2-statistics. Let's denote $T_\ell = (X_\ell, S_\ell, k_\ell)$. Thus, $\sigma = \{\sigma_\ell(T_1; T_2)\}$, where $\sigma_\ell(T_1; T_2) = \Pr\{y_{k+1} = \ell|T_1; T_2\}$, $\ell = 1, 2$.

The goal is to maximize (in some sense) the total expected income. If the parameter θ were known then the optimal strategy would be always to choose an action corresponding to the larger of the values m_1, m_2. In this case, the total expected income would be $N \max(m_1, m_2)$. If the parameter is unknown then the regret

$$L_N(\sigma, \theta) = N \max(m_1, m_2) - \mathbf{E}_{\sigma,\theta}\left(\sum_{k=1}^{K} \xi_k\right) \tag{1.2}$$

describes the mathematical expectation of the loss of total income relative to the maximum possible value due to incomplete information. Here $\mathbf{E}_{\sigma,\theta}$ denotes the mathematical expectation calculated with respect to the measure generated by the strategy σ and the parameter θ. Note that considered problem has an interpretation as a game with nature. In this case, the set $\{\sigma\}$ contains the strategies of the person performing a control, the set Θ is the set of strategies of nature, and $L_N(\sigma, \theta)$ is the payment function. This game is antagonistic although only the person performing a control is interested in its outcome.

Note that previously the Gaussian two-armed bandit was considered with a priori known variances [6]. This situation occurs in the case of big data when the variances can be estimated during processing the first data batches. Since the regret changes little with a small change in variances, these estimates of variances can be used for control. But if the amount of data is moderate, then the variance estimation should be carried out in the control process. The following analogy from mathematical statistics may be useful here. When obtaining interval estimates of an unknown mathematical expectation, in the case of a large sample, a normal distribution is used, in which a point estimate replaces the variance as its approximate value. But in the case of a moderate sample size, the Student's distribution should be used.

Let us denote by $\lambda(\theta)$ a prior distribution density of the parameter θ on the set Θ and by

$$L_N^B(\sigma, \lambda) = \int_\Theta L_N(\sigma, \theta)\lambda(\theta)d\theta \tag{1.3}$$

the regret averaged with respect to $\lambda(\theta)$. Here $d\theta = dm_1 dm_2 dD_1 dD_2$. For computing a regret (1.3), one can apply a Bayesian approach that allows to find it using dynamic programming technique (for finding the regret (1.2), one should choose a degenerate density $\lambda(\theta)$). In Sect. 2, a dynamic programming equation is obtained for finding the regret (1.3) in general form.

Next, in the Sect. 3 we discuss UCB (Upper Confidence Bounds) strategies (see, e.g., [7,8]), as well as their properties that allow one to simplify the dynamic programming equation. The simplified equation and its invariant form with a control horizon equal to one are presented in Sect. 4. The invariant form of the equation does not depend on the total number of processed data but only on the number of batches into which it is divided and on the number of internal packets for which the variance of the batch is estimated. The results of numerical experiments are presented in Sect. 5. Section 6 contains the conclusion.

2 Recursive Equation for Computing a Regret

Let $x_{\ell i}$, $i = 1, \ldots, k_\ell$, $\ell = 1, 2$, denote incomes received in response to the ith application of the ℓth action to the data batches. We additionally assume that each batch contains $M_2 \geq 2$ small packets each of which contains M_1 data, so that $M = M_1 M_2$. Thus mathematical expectation and variance of income for small packet processing by ℓth action are $m'_\ell = M_1 m_\ell$, $D'_\ell = M_1 D_\ell$, where m_ℓ, D_ℓ are mathematical expectation and variance of income for data unit processing. Let us denote by $x'_{\ell i j}$ the income for processing the jth small packet in the ith large batch, so that $x_{\ell i} = \sum_{j=1}^{M_2} x'_{\ell i j}$. When processing small packets, it is also possible to calculate s^2-statistics for a large batch using the formula $s_{\ell i} = \sum_{j=1}^{M_2}(x'_{\ell i j} - x_{\ell i}/M_2)^2 = \sum_{j=1}^{M_2}(x'_{\ell i j})^2 - x_{\ell i}^2/M_2$. Suppose that after processing k batches, the first and second actions were applied k_1 and k_2 times. Then sufficient statistics are the total income X_ℓ and s^2-statistics S_ℓ:

$$X_\ell = \sum_{i=1}^{k_\ell} x_{\ell i}, \quad S_\ell = \sum_{i=1}^{k_\ell}\sum_{j=1}^{M_2}(x'_{\ell i j} - X_\ell/(M_2 k_\ell))^2,$$

$\ell = 1, 2$. Note that $X_\ell = 0$, $S_\ell = 0$ if $k_\ell = 0$ and consider how to update X_ℓ and S_ℓ after processing the next batch by ℓth action. For $k_\ell \geq 0$, let $x_{\ell,k_\ell+1} = Y$ be a new income and $s_{\ell,k_\ell+1} = U$ be a new value of s^2-statistics. Then $X_{\ell,new} =$

$\sum_{i=1}^{k_\ell+1} x_{\ell i} = X_\ell + Y$. Similarly, $S_{\ell,new} = U$ for $k_\ell = 0$. And for $k_\ell \geq 1$ we have

$$S_{\ell,new} = \sum_{i=1}^{k_\ell+1} \sum_{j=1}^{M_2} (x'_{\ell ij})^2 - \frac{X_{\ell,new}^2}{M_2(k_\ell+1)} = S_\ell + U$$

$$+\frac{Y^2}{M_2} + \frac{X_\ell^2}{k_\ell M_2} - \frac{(X_\ell+Y)^2}{M_2(k_\ell+1)} = S_\ell + U + \frac{(X_\ell - k_\ell Y)^2}{M_2 k_\ell(k_\ell+1)}.$$

Therefore, $X_\ell = 0$, $S_\ell = 0$ if $k_\ell = 0$ and then for $k_\ell \geq 1$ after processing the next large batch by ℓth action and obtaining the income Y and s^2-statistics U, they are updated according to the rule

$$X_\ell \leftarrow X_\ell + Y, \quad S_\ell \leftarrow S_\ell + U + \Delta(X_\ell, k_\ell, Y), \quad \ell = 1, 2, \qquad (2.1)$$

where $\Delta(0,0,Y) = 0$ and

$$\Delta(X_\ell, k_\ell, Y) = \frac{(X_\ell - k_\ell Y)^2}{M_2 k_\ell(k_\ell+1)}, \quad \text{if } k_\ell \geq 1. \qquad (2.2)$$

Let a prior distribution density $\lambda(\theta) = \lambda(m_1, D_2, m_2, D_2)$ be given and describe a posterior density. Denote by $\chi_k^2(x)$ the chi-squared distribution density with k degrees of freedom

$$\chi_k^2(x) = \frac{1}{2^{\frac{k}{2}}\Gamma\left(\frac{k}{2}\right)} x^{\frac{k}{2}-1} e^{-\frac{x}{2}}, \quad k \geq 1.$$

Note that random variables X_ℓ and S_ℓ after processing $k_\ell \geq 1$ large batches have exactly distribution densities $f_{k_\ell M_2 D'_\ell}(X_\ell | k_\ell M_2 m'_\ell)$ with $f(\cdot)$ defined in (1.1), and $\psi_{M_2 k_\ell - 1}(S_\ell / D'_\ell) = (D'_\ell)^{-1} \chi^2_{M_2 k_\ell - 1}(S_\ell / D'_\ell)$, where $m'_\ell = M_1 m_\ell$, $D'_\ell = M_1 D_\ell$, and X_ℓ, S_ℓ are independent random variables. In particular,

$$\mathbf{F}(X_\ell, S_\ell | m_\ell, D_\ell) = f_{M_2 D'_\ell}(X_\ell | M_2 m'_\ell)\, \psi_{M_2-1}(S_\ell / D'_\ell).$$

is the joint distribution density of X_ℓ and S_ℓ for one large batch. Distribution densities $f_{k_\ell M_2 D'_\ell}(X_\ell | k_\ell M_2 m'_\ell)$ and $\psi_{M_2 k_\ell - 1}(S_\ell / D'_\ell)$ could be used for finding a posterior density $\lambda(\theta | T_1, T_2)$. However, calculations are more simple if the posterior density is determined in a following equivalent way. Let us introduce functions $\tilde{\mathbf{F}}(T_\ell | m_\ell, D_\ell)$ satisfying conditions $\tilde{\mathbf{F}}(T_\ell | m_\ell, D_\ell) = 1$ if $k_\ell = 0$ and

$$\tilde{\mathbf{F}}(T_\ell | m_\ell, D_\ell) = (D'_\ell)^{-3/2} \tilde{f}_{M_2 k_\ell D'_\ell}(X_\ell | k_\ell M_2 m'_\ell)\, \tilde{\psi}_{M_2 k_\ell - 1}(S_\ell / D'_\ell) \qquad (2.3)$$

if $k_\ell \geq 1$, where

$$\tilde{f}_D(x|m) = \exp(-(x-m)^2/(2D)), \quad \tilde{\psi}_{M_2 k-1}(s) = s^{(M_2 k-1)/2-1} e^{-s/2}.$$

One can see that the posterior distribution density is

$$\lambda(\theta | T_1; T_2) = \frac{\tilde{\mathbf{F}}(T_1 | m_1, D_1)\tilde{\mathbf{F}}(T_2 | m_2, D_2)\lambda(\theta)}{\tilde{P}(T_1; T_2)} \qquad (2.4)$$

where

$$\tilde{P}(T_1; T_2) = \int_\Theta \tilde{\mathbf{F}}(T_1|m_1, D_1)\tilde{\mathbf{F}}(T_2|m_2, D_2)\lambda(\theta)d\theta, \tag{2.5}$$

and this formula remains valid if $k_1 = 0$ and/or $k_2 = 0$. Let us denote by

$$L^B(T_1; T_2) = \int_\Theta L_{M(K-k)}(\sigma, \theta)\lambda(\theta|T_1; T_2)d\theta$$

a regret (1.3) for application of the strategy σ to the latter $(K - k)$ batches, computed with respect to $\lambda(\theta|T_1, T_2)$. Let $x^+ = \max(x, 0)$. Then the following standard recursive equation holds true

$$L^B(T_1; T_2) = \sum_{\ell=1}^2 \sigma_\ell(T_1; T_2)L_\ell^B(T_1; T_2), \tag{2.6}$$

where $L_1^B(T_1; T_2) = L_2^B(T_1; T_2) = 0$ if $k = K$ and

$$L_1^B(X_1, S_1, k_1; T_2) = \int_\Theta \lambda(\theta|X_1, S_1, k_1; T_2)\big(M(m_2 - m_1)^+$$
$$+ \mathbf{E}_{Y,U}^{(1)} L^B(X_1 + Y, S_1 + U + \Delta(X_1, k_1, Y), k_1 + 1; T_2)\big)d\theta,$$

$$L_1^B(T_1; X_2, S_2, k_2) = \int_\Theta \lambda(\theta|T_1; X_2, S_2, k_2)\big(M(m_1 - m_2)^+ \tag{2.7}$$
$$+ \mathbf{E}_{Y,U}^{(2)} L^B(T_1; X_2 + Y, S_2 + U + \Delta(X_2, k_2, Y), k_2 + 1)\big)d\theta,$$

if $0 \leq k \leq K - 1$ with

$$\mathbf{E}_{Y,U}^{(\ell)} L(Y, U) = \iint_B L(Y, U)\mathbf{F}(Y, U|m_\ell, D_\ell)dY dU,$$

and $B = \{-\infty < Y < \infty, 0 \leq U < \infty\}$. Here $L_\ell^B(T_1; T_2)$ is a mathematical expectation of the regret at the remaining control horizon $(k + 1, K)$ if at first the ℓth action is applied and then a control is performed according to the strategy σ. A regret (1.3) is

$$L_N(\sigma, \lambda) = L^B(0, 0, 0; 0, 0, 0). \tag{2.8}$$

To find the regret (1.2), one should choose a degenerate prior distribution density concentrated at a single parameter θ. Let us give another form of the recursive equation, which is more convenient for calculations. Let us put

$$L_\ell(T_1; T_2) = L_\ell^B(T_1; T_2) \times \tilde{P}(T_1; T_2), \quad \ell = 1, 2, \tag{2.9}$$

where $\tilde{P}(T_1; T_2)$ is defined in (2.5). Note that
$L(T_1; T_2) = \sum_{\ell=1}^2 \sigma_\ell(T_1; T_2)L_\ell(T_1; T_2) = L^B(T_1; T_2) \times \tilde{P}(T_1; T_2)$.

Theorem 1. *Consider the recursive equation*

$$L(T_1; T_2) = \sum_{\ell=1}^{2} \sigma_\ell(T_1; T_2) L_\ell(T_1; T_2), \qquad (2.10)$$

where $L_1(T_1; T_2) = L_2(T_1; T_2) = 0$ *if* $k = K$ *and*

$$L_1(X_1, S_1, k_1; T_2) = MG_1(X_1, S_1, k_1; T_2)$$

$$+ \iint_B L(X_1 + Y, S_1 + U + \Delta(X_1, k_1, Y), k_1 + 1; T_2) H(X_1, S_1, k_1, Y, U) dY dU,$$

$$L_2(T_1; X_2, S_2, k_2) = MG_2(T_1; X_2, S_2, k_2) \qquad (2.11)$$

$$+ \iint_B L(T_1; X_2 + Y, S_2 + U + \Delta(X_2, k_2, Y), k_2 + 1) H(X_2, S_2, k_2, Y, U) dY dU,$$

if $0 \le k \le K - 1$. *Here* $\Delta(X_\ell, k_\ell, Y)$ *is defined in* (2.2),

$$G_1(T_1; T_2) = \int_\Theta (m_2 - m_1)^+ \tilde{\mathbf{F}}(T_1 | m_1, D_1) \tilde{\mathbf{F}}(T_2 | m_2, D_2) \lambda(\theta) d\theta,$$
$$G_2(T_1; T_2) = \int_\Theta (m_1 - m_2)^+ \tilde{\mathbf{F}}(T_1 | m_1, D_1) \tilde{\mathbf{F}}(T_2 | m_2, D_2) \lambda(\theta) d\theta, \qquad (2.12)$$

function $H(0, 0, 0, Y, U) = C(M_2)$ *if* $k = 0$ *and*

$$H(X, S, k, Y, U) = C(M_2) \times \frac{S^{(kM_2-1)/2-1} U^{(M_2-1)/2-1}}{(S + U + \Delta(X, k, Y))^{((k+1)M_2-1)/2-1}}, \qquad (2.13)$$

if $k > 0$. *Here* $C(M_2) = \left((2^{M_2} M_2 \pi)^{1/2} \Gamma((M_2 - 1)/2) \right)^{-1}$. *A regret* (1.3) *is*

$$L_N(\sigma, \lambda) = L(0, 0, 0; 0, 0, 0). \qquad (2.14)$$

Proof. Let's multiply Eqs. (2.6)–(2.7) by $\tilde{P}(T_1; T_2)$ defined in (2.5). As a result, we obtain (2.10)–(2.11), where $G_1(T_1; T_2)$, $G_2(T_1; T_2)$ are defined in (2.12). Let Δ below be defined in (2.2). For the first Eq. (2.11) function $H(X_1, S_1, k_1, Y, U)$ is

$$H(X_1, S_1, k_1, Y, U)$$

$$= \frac{\int_\Theta \tilde{\mathbf{F}}(X_1, S_1, k_1 | m_1, D_1) \tilde{\mathbf{F}}(T_2 | m_2, D_2) \mathbf{F}(Y, U | m_1, D_1) \lambda(\theta) d\theta}{\tilde{P}(X_1 + Y, S_1 + U + \Delta(X_1, S_1, k_1), k_1 + 1; T_2)} \qquad (2.15)$$

$$= \frac{\tilde{\mathbf{F}}(X_1, S_1, k_1 | m_1, D_1) \mathbf{F}(Y, U | m_1, D_1)}{\tilde{\mathbf{F}}(X_1 + Y, S_1 + U + \Delta(X_1, S_1, k_1), k_1 + 1 | m_1, D_1)}.$$

Therefore, for $k_1 \ge 1$ we have

$$H(X_1, S_1, k_1, Y, U)$$

$$= \frac{\tilde{f}_{k_1 M_2 D_1'}(X_1 | k_1 M_2 m_1') f_{M_2 D_1'}(Y | M_2 m_1')}{\tilde{f}_{(k_1+1)M_2 D_1'}(X_1 + Y | (k_1 + 1)M_2 m_1')} \frac{\tilde{\psi}_{k_1 M_2 - 1}(S_1/D_1') \psi_{M_2 - 1}(U/D_1')}{\tilde{\psi}_{(k_1+1)M_2 - 1}((S_1 + U + \Delta)/D_1')}.$$

Here

$$\frac{\tilde{f}_{k_1 M_2 D_1'}(X_1|k_1 M_2 m_1') f_{M_2 D_1'}(Y|M_2 m_1')}{\tilde{f}_{(k_1+1)M_2 D_1'}(X_1 + Y|(k_1+1)M_2 m_1')} = \left(\frac{1}{2\pi M_2 D_1'}\right)^{1/2} \exp\left(-\frac{\Delta}{2D_1'}\right)$$

and

$$\frac{\tilde{\psi}_{k_1 M_2 - 1}(S_1/D_1')\psi_{M_2-1}(U/D_1')}{\tilde{\psi}_{(k_1+1)M_2-1}((S_1 + U + \Delta)/D_1')} = \frac{1}{D_1' \times 2^{(M_2-1)/2}\Gamma((M_2-1)/2)}$$

$$\times \frac{(S_1/D_1')^{(k_1 M_2 - 1)/2-1}(U/D_1')^{(M_2-1)/2-1}}{((S_1 + U + \Delta)/D_1')^{((k_1+1)M_2-1)/2-1}} \times \frac{\exp\left(-S_1/(2D_1')\right)\exp\left(-U/(2D_1')\right)}{\exp\left(-(S_1 + U + \Delta)/(2D_1')\right)}$$

$$= \frac{(D_1')^{1/2}\exp\left(\Delta/(2D_1')\right)}{2^{(M_2-1)/2}\Gamma((M_2-1)/2)} \times \frac{S_1^{(k_1 M_2-1)/2-1}U^{(M_2-1)/2-1}}{(S_1 + U + \Delta)^{((k_1+1)M_2-1)/2-1}}.$$

Hence, $H(X_1, S_1, k_1, Y, U)$ satisfies (2.13) if $k_1 \geq 1$. If $k_1 = 0$ then $X_1 = 0$, $S_1 = 0$ and (2.15) takes the form

$$H(0, 0, 0, Y, U) = \frac{f_{M_2 D_1'}(Y|M_2 m_1')\psi_{M_2-1}(U/D_1')}{(D_1')^{-3/2}\tilde{f}_{M_2 D_1'}(Y|M_2 m_1')\tilde{\psi}_{M_2-1}(U/D_1')} = C(M_2).$$

Formula (2.14) follows from (2.8), (2.9) and equality $\tilde{P}(0, 0, 0; 0, 0, 0) = 1$.

3   UCB Strategies and Properties of a Prior Distribution

According to sufficient statistics X_ℓ, S_ℓ, the current estimates of mathematical expectation and variance of income for processing one data batch are done as

$$\hat{m}_\ell = \frac{X_\ell}{k_\ell}, \quad \hat{D}_\ell = \frac{M_2 S_\ell}{k_\ell M_2 - 1}, \quad \ell = 1, 2. \tag{3.1}$$

UCB strategy, at the beginning of control, each action applies an equal number of $k_0 \geq 1$ times. Next, at step number $k+1$, where $k = k_1 + k_2$ ($k_1 \geq k_0$, $k_2 \geq k_0$), an action is selected that corresponds to the largest of the values

$$Q_\ell(k_\ell) = \hat{m}_\ell + B_\ell(\hat{D}_1, \hat{D}_2, k_1, k_2)\left(\frac{\hat{D}_\ell}{k_\ell}\right)^{1/2}, \quad \ell = 1, 2. \tag{3.2}$$

If they are equal, we choose the first action for certainty. It is easy to see that $Q_1(k_1)$, $Q_2(k_2)$ are the upper bounds of the confidence intervals for estimates $\hat{m}_1$, $\hat{m}_2$. Functions $B_\ell(\hat{D}_1, \hat{D}_2, k_1, k_2)$ depend on the strategy. For the batch analogue of the Auer-Ceza-Bianchi-Fischer strategy [7], $B_\ell(\hat{D}_1, \hat{D}_2, k_1, k_2) = a_\ell(\hat{D}_1, \hat{D}_2)\ln^{1/2}(k)$. The positive functions $a_\ell(\hat{D}_1, \hat{D}_2)$ are chosen so as to minimize the maximum values of the regret (1.2). Such $Q_1(k_1)$, $Q_2(k_2)$ not only stimulate the choice of an action corresponding to the larger of the values $\hat{m}_1$, $\hat{m}_2$ but also ensure a sufficiently large number of applications of both actions.

Let us denote $n_\ell = Mk_\ell$, $\ell = 1, 2$, and consider strategies satisfying the following property

$$\sigma_\ell(X_1, S_1, k_1, X_2, S_2, k_2) = \sigma_\ell(X_1 + n_1 c, S_1, k_1, X_2 + n_2 c, S_2, k_2) \qquad (3.3)$$

for any fixed c and each history $X_1, S_1, k_1, X_2, S_2, k_2$ ($\ell = 1, 2$). One can see that UCB strategies (3.2) depend on X_1, X_2 in the form of dependence on $X_1/k_1 - X_2/k_2$ and, consequently, they satisfy the property (3.3).

Lemma 1. *Let a strategy σ satisfy the property (3.3). Denote $\tilde{\theta}_c = (m_1 + c, D_1, m_2 + c, D_2)$. The shift-transform $\tilde{\lambda}_c(\theta) = \tilde{\lambda}_c(m_1, D_1, m_2, D_2) = \lambda(\tilde{\theta}_c) = \lambda(m_1 + c, D_1, m_2 + c, D_2)$, performed for all m_1, D_1, m_2, D_2 and arbitrary fixed c, does not change a regret, i.e., $L_N^B(\sigma, \lambda) = L_N^B(\sigma, \tilde{\lambda}_c)$.*

Proof. Instead of incomes $x_{\ell i j}$, let's consider incomes $x_{\ell i j} + M_1 c$. Then statistics are transformed according to the rules $S_\ell \leftarrow S_\ell$ and $X_\ell \leftarrow X_\ell + n_\ell c$. Noting that $\tilde{\lambda}_c(\theta|X_1 + n_1 c, S_1, k_1, X_2 + n_2 c, S_2, k_2) = \lambda(\tilde{\theta}_c|X_1 + n_1 c, S_1, k_1, X_2 + n_2 c, S_2, k_2) = \lambda(m_1 + c, D_1, m_2 + c, D_2|X_1 + n_1 c, S_1, k_1, X_2 + n_2 c, S_2, k_2) = \lambda(m_1, D_1, m_2, D_2|X_1, S_1, k_1, X_2, S_2, k_2)$, we obtain by induction and using (2.6), (2.7) that the equalities $L_\ell^B(\tilde{\lambda}_c; X_1 + n_1 c, S_1, k_1, X_2 + n_2 c, S_2, k_2) = L_\ell^B(\lambda; X_1, S_1, k_1, X_2, S_2, k_2)$ are valid for $\ell = 1, 2$. Here, the dependence on a prior distribution density is explicitly indicated in the regret. Therefore, $L^B(\tilde{\lambda}_c; X_1 + n_1 c, S_1, k_1, X_2 + n_2 c, S_2, k_2) = L^B(\lambda; X_1, S_1, k_1, X_2, S_2, k_2)$ for all $X_1, S_1, k_1, X_2, S_2, k_2$ and arbitrary fixed c. We obtain a required result at $k_1 = k_2 = 0$. ∎

It follows from Lemma 1, that for the strategy, satisfying the property (3.3), a regret on the mixture of prior distribution densities $\tilde{\lambda}_{c_1}$, $\tilde{\lambda}_{c_2}$ is equal to original, i.e., $L_N^B(\sigma, \alpha_1 \tilde{\lambda}_{c_1} + \alpha_2 \tilde{\lambda}_{c_2}) = L_N^B(\sigma, \lambda)$, if $\alpha_1 + \alpha_2 = 1$, $\alpha_1 \geq 0$, $\alpha_2 \geq 0$.

In what follows, it is convenient to change the parametrization. Let us put $m_1 = m + v$, $m_2 = m - v$, then $\theta = (m + v, D_1, m - v, D_2)$, $\Theta = \{\theta : |v| \leq C, 0 < \underline{D} \leq D_\ell \leq \overline{D} < \infty, \ell = 1, 2\}$. In new variables, let us consider a prior distribution density $\nu(m, v, D_1, D_2)$. According to Lemma 1, for strategies, satisfying the property (3.3), the value of the regret is the same for the density $\tilde{\nu}_c(m, v, D_1, D_2) = \nu(m + c, v, D_1, D_2)$, where c is an arbitrary constant. This makes it possible to simplify a prior distribution density and the recursive equation for computing the regret. We do this like it was previously made in [6].

Consider a set of parameters $\Theta_0 = \{\theta : |v| \leq C, |m| \leq C_0, 0 < \underline{D} \leq D_\ell \leq \overline{D} < \infty, \ell = 1, 2\}$. We assume that $\theta \in \Theta_0$. Given some $b \gg C_0$, let us define a distribution density

$$\nu^{(1)}(m, v, D_1, D_2) = (2b)^{-1} \int_{-b}^{b} \nu(m + x, v, D_1, D_2)dx.$$

Obviously, $L_N^B(\sigma, \nu^{(1)}) = L_N^B(\sigma, \nu)$. Let us put $\rho(v, D_1, D_2) = \nu^{(1)}(0, v, D_1, D_2)$. One can see that $\nu^{(1)}(m, v, D_1, D_2) = \rho(v, D_1, D_2)$ if $|m| < b - C_0$,

$\nu^{(1)}(m, v, D_1, D_2) \leq \rho(v, D_1, D_2)$ if $b - C_0 < |m| < b + C_0$ and $\nu^{(1)}(m, v, D_1, D_2) = 0$ if $|m| > b + C_0$. Hence, there is such a, satisfying the condition $b - C_0 < a < b + C_0$, that a prior distribution density has the form

$$\nu(a, m, v, D_1, D_2) = \kappa_a(m)\rho(v, D_1, D_2), \qquad (3.4)$$

where $\kappa_a(m) = (2a)^{-1}$ is a uniform distribution density on the segment $|m| \leq a$. Note that $\nu(a, m, v, D_1, D_2) = \nu^{(1)}(m, v, D_1, D_2)$ if $|m| < b - C_0$. Therefore, $\iiiint_\Theta |\nu^{(1)}(m, v, D_1, D_2) - \nu(a, m, v, D_1, D_2)| dm dv dD_1 dD_2 \to 0$ as $b \to \infty$ and, hence, $L_N^B(\sigma, \nu(a, v, D_1, D_2)) - L_N^B(\sigma, \nu^{(1)}(m, v, D_1, D_2)) \to 0$ as $b \to \infty$.

Consider a prior distribution density (3.4). As in [6], it can be proved that there is a limit

$$\lim_{a \to \infty} L_N^B(\nu(a, m, v, D_1, D_2)) = L_N^B(\rho(v, D_1, D_2)). \qquad (3.5)$$

To obtain the formula for computing $L_N^B(\rho(v, D_1, D_2))$, note first that it follows from (2.4), (2.5) that a posterior density $\lambda(\theta|T_1; T_2)$ does not change if a prior density $\lambda(\theta)$ is multiplied by an arbitrary value. Therefore, we can formally consider the "limiting" prior distribution density with $\kappa(m) \equiv 1$ for all $m \in (-\infty, \infty)$. Let us use notations $n_\ell = k_\ell M$, $\tilde{n}_\ell = n_\ell/D_\ell$, $n_\ell^* = D_\ell n_\ell$, $\overline{X}_\ell = X_\ell/n_\ell$ $(\ell = 1, 2)$ and $\tilde{n} = \tilde{n}_1 + \tilde{n}_2$, $f_D(x) = f_D(x|0)$. In what follows, we need the following equality, which is checked straightforwardly

$$\tilde{f}_{n_1^*}(X_1|n_1(m + v))\tilde{f}_{n_2^*}(X_2|n_2(m - v))$$
$$= \tilde{f}_{\tilde{n}^{-1}}\left(Y + \tilde{n}^{-1}(\tilde{n}_2 - \tilde{n}_1)v - m\right)\tilde{f}_{n_1^* n_2^* \tilde{n}}\left(Z - 2vn_1 n_2\right), \qquad (3.6)$$

where $Y = \tilde{n}^{-1}(\tilde{n}_1\overline{X}_1 + \tilde{n}_2\overline{X}_2)$, $Z = X_1 n_2 - X_2 n_1$.

Consider the strategy which processes the first two batches with both actions in turn. Taking into account (2.4), (2.5), the "limiting" posterior distribution density, corresponding to $\kappa(m) \equiv 1$, for $n_1 = n_2 = M$, $n = 2M$ has the form

$$\nu(m, v, D_1, D_2|Y, Z, S_1, S_2) = K^{-1}(Z, S_1, S_2)\rho(v, D_1, D_2)$$
$$\times (D_1' D_2')^{-3/2} \tilde{f}_{\tilde{n}^{-1}}\left(Y + \tilde{n}^{-1}(\tilde{n}_2 - \tilde{n}_1)v - m\right)\tilde{f}_{n_1^* n_2^* \tilde{n}}\left(Z - 2vn_1 n_2\right) \qquad (3.7)$$
$$\times \tilde{\psi}_{M_2-1}(S_1/D_1')\tilde{\psi}_{M_2-1}(S_2/D_2'),$$

where $K(Z, S_1, S_2)$ corresponds to $\tilde{P}(T_1; T_2)$ computed for the "limiting" prior distribution density.

Note that for any c the density $\nu(m, v, D_1, D_2|Y - c, Z, S_1, S_2)$ can be obtained by shifting the argument m in the density (3.7) to c. This follows from the equalities $\nu(m, v|Y - c, Z, S_1, S_2) = \nu(m + c, v|Y, Z, S_1, S_2) = \tilde{\nu}_c(m, v|Y, Z, S_1, S_2)$. So, it follows from Lemma 1 that the regret $L_{M(K-2)}^B(\nu(m, v, D_1, D_2|Y, Z, S_1, S_2))$ does not depend on Y and can be calculated by the formula

$$L_N^B(\rho(v, D_1, D_2)) = L(\rho(v, D_1, D_2)) \qquad (3.8)$$
$$+ \mathbf{E}_{Z, S_1, S_2} L_{M(K-2)}^B(\nu(m, v, D_1, D_2|Y, Z, S_1, S_2)),$$

where

$$L(\rho(v, D_1, D_2)) = M \int_{\Theta'} 2|v|\rho(v, D_1, D_2) d\theta' \qquad (3.9)$$

describes the losses for choosing the first and second actions in the first two steps with $\Theta' = \{\theta' = (v, D_1, D_2) : |v| \le C, 0 < \underline{D} \le D_\ell \le \overline{D} < \infty, \ell = 1, 2\}$, $d\theta' = dv dD_1 dD_2$, and $\mathbf{E}_{Z,S_1,S_2}$ denotes the mathematical expectation relative to Z, S_1, S_2, i.e.,

$$\mathbf{E}_{Z,S_1,S_2} L(Z, S_1, S_2) = \iiint_{B_1} L(Z, S_1, S_2) P(Z; S_1, 1; S_2, 1) dZ dS_1 dS_2, \qquad (3.10)$$

where $B_1 = \{-\infty < Z < \infty, 0 \le S_1 < \infty, 0 \le S_2 < \infty\}$, and $P(Z, S_1, S_2)$ corresponds to $P(X_1, S_1, 1; X_2, S_2, 1) = \int_{\Theta} \mathbf{F}(X_1, S_1 | m_1, D_1) \mathbf{F}(X_2, S_2 | m_2, D_2) \lambda(\theta) d\theta$, computed with respect to the "limiting" prior distribution density.

Let us summarize the results. Consider a strategy that satisfies (3.3) and at the beginning of the control applies both actions in turn. Given a prior distribution density $\nu(m, v, D_1, D_2)$ there exists asymptotically unform prior distribution density of the form (3.4) so that corresponding regret is asymptotically equal to the original as $a \to \infty$. Moreover, there is a limit (3.5) equal to (3.8). Density $\nu(m, v, D_1, D_2 | Y, Z, S_1, S_2)$ is selected from the condition (3.7).

4   Computing the Regret with Respect to Asymptotically Uniform Prior Distribution

The property of asymptotic uniformity of a prior distribution density makes it possible to significantly simplify the recursive Eq. (2.10)–(2.11). Let us denote $\tilde{T}_1 = (S_1, k_1)$ and $\tilde{T}_2 = (S_2, k_2)$. Let us put $n_1 = Mk_1$, $n_2 = Mk_2$, $Z = X_1 n_2 - X_2 n_1$, and

$$L(Z; \tilde{T}_1; \tilde{T}_2) = L(T_1; T_2), \quad L_\ell(Z; \tilde{T}_1; \tilde{T}_2) = L_\ell(T_1; T_2), \quad \ell = 1, 2, \qquad (4.1)$$

where $L_1(T_1; T_2)$, $L_2(T_1; T_2)$, $L(T_1; T_2)$ are included in Eq. (2.10)–(2.11). The following theorem holds true.

Theorem 2. *Let $\nu(a, m, v, D_1, D_2)$ be selected from (3.4) and $a \to \infty$. Consider a strategy satisfying (3.3), which applies the first and second actions in turn at the beginning of the control. Then the regrets (4.1) satisfy the recursive equation*

$$L(Z; \tilde{T}_1; \tilde{T}_2) = \sum_{\ell=1}^{2} \sigma_\ell(Z; \tilde{T}_1; \tilde{T}_2) L_\ell(Z; \tilde{T}_1; \tilde{T}_2), \qquad (4.2)$$

where $L_1(Z; \tilde{T}_1; \tilde{T}_2) = L_2(Z; \tilde{T}_1; \tilde{T}_2) = 0$ if $k_1 + k_2 = K$ and

$$L_1(Z; S_1, k_1; \tilde{T}_2) = MG_1(Z; S_1, k_1; \tilde{T}_2) + n_2^{-1}$$

$$\times \iint_B L(Z + \tilde{Z}, S_1 + U + n_2^{-2}\Delta(Z, k_1, \tilde{Z}), k_1 + 1, \tilde{T}_2)H(Z, S_1, k_1, \tilde{Z}, U)d\tilde{Z}dU,$$

$$L_2(Z, \tilde{T}_1, S_2, k_2) = MG_2(Z, \tilde{T}_1; S_2, k_2) + n_1^{-1} \qquad (4.3)$$

$$\times \iint_B L(Z + \tilde{Z}, \tilde{T}_1, S_2 + U + n_1^{-2}\Delta(Z, k_2, \tilde{Z}), k_2 + 1)H(Z, S_2, k_2, \tilde{Z}, U)d\tilde{Z}dU,$$

if $2 \le k_1 + k_2 < K$, $k_1 \ge 1$, $k_2 \ge 1$. Here $B = \{-\infty < \tilde{Z} < \infty, 0 \le U < \infty\}$,

$$G_1(Z; \tilde{T}_1; \tilde{T}_2) = \int_{\Theta'} 2v^- G(\theta'; Z; \tilde{T}_1; \tilde{T}_2)\rho(v, D_1, D_2)d\theta',$$

$$G_2(Z; \tilde{T}_1; \tilde{T}_2) = \int_{\Theta'} 2v^+ G(\theta'; Z; \tilde{T}_1; \tilde{T}_2)\rho(v, D_1, D_2)d\theta',$$

$$G(\theta'; Z; \tilde{T}_1; \tilde{T}_2) = (2\pi/\tilde{n})^{1/2}(D_1'D_2')^{-3/2} \qquad (4.4)$$

$$\times \tilde{f}_{n_1^* n_2^* \tilde{n}}(Z - 2vn_1n_2)\tilde{\psi}_{M_2k_1-1}(S_1/D_1')\tilde{\psi}_{M_2k_2-1}(S_2/D_2')$$

with $v^+ = \max(v, 0)$, $v^- = \max(-v, 0)$, and

$$H(Z, S_\ell, k_\ell, \tilde{Z}, U) = \frac{C(M_2) \times S_\ell^{(k_\ell M_2-1)/2-1}U^{(M_2-1)/2-1}}{(S_\ell + U + n_{\bar{\ell}}^{-2}\Delta(Z, k_\ell, \tilde{Z}))^{((k_\ell+1)M_2-1)/2-1}} \qquad (4.5)$$

with $\bar{\ell} = 3 - \ell$ and $C(M_2) = \left((\pi M_2 2^{M_2})^{1/2}\Gamma((M_2-1)/2)\right)^{-1}$. A regret (1.3) is

$$L_N^B(\rho(v, D_1, D_2)) = L(\rho(v, D_1, D_2)) \qquad (4.6)$$

$$+H \iiint_{B_1} L(Z; S_1, 1; S_2, 1)dZdS_1dS_2,$$

where $L(\rho(v, D_1, D_2))$ is defined in (3.9), $H = C^2(M_2)$ and $B_1 = \{-\infty < Z < \infty, 0 \le S_1 < \infty, 0 \le S_2 < \infty\}$. To compute a regret (1.2), one should take a prior distribution density concentrated at a single parameter $\theta' = (v, D_1, D_2)$.

Proof. Let's check the validity of the first Eq. (4.3). According to (3.7)–(3.8), one should choose $\lambda(\theta) = \rho(v, D_1, D_2)\kappa(m)$ with $\kappa(m) \equiv 1$ in the first Eq. (2.11). Taking into account (3.6), we obtain that

$$\int_{-\infty}^{\infty} \tilde{f}_{n_1^*}(X_1|n_1(m+v))\tilde{f}_{n_2^*}(X_2|n_2(m-v))dm = (2\pi/\tilde{n})^{1/2}\tilde{f}_{n_1^* n_2^* \tilde{n}}(Z - 2vn_1n_2),$$

therefore, (4.4) is valid. Consider the function (2.13). Note that for $n_1 + M$, n_2, the value Z is recalculated according to the rule $Z \leftarrow (X_1+Y)n_2 - X_2(n_1+M) = Z + \tilde{Z}$ with $\tilde{Z} = Yn_2 - X_2M$. Noting that $MX_1 - n_1Y = n_2^{-1}(MZ - n_1\tilde{Z})$, we obtain

$$\Delta(X_1, k_1, Y) = \frac{(MX_1 - n_1Y)^2}{M^2 M_2 k_1(k_1+1)} = \frac{(MZ - n_1\tilde{Z})^2}{n_2^2 M^2 M_2 k_1(k_1+1)}$$

$$= \frac{(Z - k_1\tilde{Z})^2}{n_2^2 M_2 k_1(k_1+1)} = n_2^{-2}\Delta(Z, k_1, \tilde{Z}),$$

which corresponds to expressions in the first Eq. (4.3) and in (4.5). Next, replacing the integration variable Y by $\tilde{Z} = Y n_2 - X_2 M$ in the first Eq. (2.11), we obtain the first Eq. (4.3). The validity of the second Eq. (4.3) is similarly verified. Let us find the value H in (4.6). Let $\tilde{P}(Z; \tilde{T}_1; \tilde{T}_2)$ and $P(Z; \tilde{T}_1; \tilde{T}_2)$ correspond to $\tilde{P}(T_1; T_2)$ and $P(T_1; T_2)$ computed for the "limiting" distribution density, where $P(T_1; T_2) = \int_\Theta \mathbf{F}(T_1 | m_1, D_1) \mathbf{F}(T_2 | m_2, D_2) \lambda(\theta) d\theta$. Taking into account (3.8), (3.10) with $n_1 = n_2 = M$, $n = 2M$, we have

$$
\begin{aligned}
H &= \frac{P(T_1; T_2)}{\tilde{P}(T_1; T_2)} = \frac{(2\pi n_1^*)^{-1/2} \tilde{f}_{n_1^*}(X_1 | m_1 n_1)(2\pi n_2^*)^{-1/2} \tilde{f}_{n_2^*}(X_2 | m_1 2 n_2)}{(D_1')^{-1/2} \tilde{f}_{n_1^*}(X_1 | m_1 n_1)(D_2')^{-1/2} \tilde{f}_{n_2^*}(X_2 | m_1 2 n_2)} \\
&\quad \times \frac{(D_1')^{-1} \psi_{M_2-1}(S_1/D_1')(D_2')^{-1} \psi_{M_2-1}(S_2/D_2')}{(D_1' D_2')^{-1} \tilde{\psi}_{M_2-1}(S_1/D_1') \tilde{\psi}_{M_2-1}(S_2/D_2')} \\
&= \frac{1}{2\pi M_2 2^{M_2-1} \Gamma^2((M_2-1)/2)} = \frac{1}{\pi M_2 2^{M_2} \Gamma^2((M_2-1)/2)} = C^2(M_2).
\end{aligned}
$$

Formula (4.6) follows from (3.8).

Finally, let us obtain an invariant form of formulas (4.2)–(4.6) with a control horizon equal to one. We take the set of parameters $\Theta'_N = \{(v, D_1, D_2) : \underline{D} \leq D_\ell \leq \overline{D}, \ell = 1, 2; |v| \leq c(\overline{D}/N)^{1/2}\}$, where $c > 0$. If one puts $D_\ell = \beta_\ell \overline{D}$, $v = \alpha(\overline{D}/N)^{1/2}$ then the set of parameters takes the form $\Theta'_N = \{(\alpha, \beta_1, \beta_2) : |\alpha| \leq c; \underline{D}/\overline{D} = \beta_0 \leq \beta_\ell \leq 1; \ell = 1, 2\}$.

Consider the change of variables $Z = z\overline{D}^{1/2} N^{3/2}$, $\tilde{Z} = \tilde{z}\overline{D}^{1/2} N^{3/2}$, $U = u\overline{D} M_1$, $S_1 = s_1 \overline{D} M_1$, $S_2 = s_2 \overline{D} M_1$, $k_1 = t_1 K$, $k_2 = t_2 K$, $M/N = K^{-1} = \varepsilon$, $v = \alpha(\overline{D}/N)^{1/2}$, $D_\ell = \beta_\ell \overline{D}$, $\tilde{\tau}_1 = (s_1, t_1)$; $\tilde{\tau}_2 = (s_2, t_2)$.

Let us describe UCB strategies. According to (3.1), $\hat{D}_\ell = s_\ell M \overline{D}/(k_\ell M_2 - 1)$, $\ell = 1, 2$. In new variables, for UCB strategies (3.2) we have

$$
\begin{aligned}
Q_1(k_1) - Q_2(k_2) &= \frac{Z}{M k_1 k_2} + B_1(\cdot) \left(\frac{\hat{D}_1}{k_1}\right)^{1/2} - B_2(\cdot) \left(\frac{\hat{D}_2}{k_2}\right)^{1/2} \\
&= \frac{(\overline{D}N)^{1/2}}{K} \left(\frac{z}{t_1 t_2} + B_1(\cdot) \left(\frac{s_1}{(k_1 M_2 - 1) t_1}\right)^{1/2} - B_1(\cdot) \left(\frac{s_2}{(k_2 M_2 - 1) t_2}\right)^{1/2}\right).
\end{aligned}
$$

(4.7)

In what follows, we assume that $B_1(\cdot)$, $B_2(\cdot)$ depend on $\hat{D}_1$, $\hat{D}_2$ in the form of dependence on $\hat{D}_1/\hat{D}_2$. In this case, the sign of $Q_1(k_1) - Q_2(k_2)$ in (4) and, hence, the strategy itself does not depend on $\overline{D}$. Let's put

$$
L_\ell(Z; \tilde{T}_1; \tilde{T}_2) = (\overline{D})^{-2} M_1^{-3} l_\ell(z; \tilde{\tau}_1; \tilde{\tau}_2), \quad \text{if } k_1 \geq 1, \ k_2 \geq 1.
$$

Theorem 3. *Let $B_1(\cdot)$, $B_2(\cdot)$ in (3.2) depend on $\hat{D}_1$, $\hat{D}_2$ in the form of dependence on $\hat{D}_1/\hat{D}_2$. Consider a recursive equation*

$$
l(z; \tilde{\tau}_1; \tilde{\tau}_2) = \sum_{\ell=1}^{2} \sigma_\ell(z; \tilde{\tau}_1; \tilde{\tau}_2) l_\ell(z; \tilde{\tau}_1; \tilde{\tau}_2),
$$

(4.8)

where $l_1(z; \tilde{\tau}_1; \tilde{\tau}_2) = l_2(z; \tilde{\tau}_1; \tilde{\tau}_2) = 0$ *if* $t_1 + t_2 = 1$ *and*

$$l_1(z; s_1, t_1; \tilde{\tau}_2) = \varepsilon g_1(z; s_1, t_1; \tilde{\tau}_2)$$

$$+ t_2^{-1} \iint_B l(z + \tilde{z}, s_1 + u + t_2^{-2}\delta(z, t_1, \tilde{z}), t_1 + \varepsilon; \tilde{\tau}_2) h(z, s_1, t_1, \tilde{z}, u) d\tilde{z} du,$$

$$l_2(z; \tilde{\tau}_1, s_2, t_2) = \varepsilon g_2(z; \tilde{\tau}_1, s_2, t_2) \qquad (4.9)$$

$$+ t_1^{-1} \iint_B l(z + \tilde{z}; \tilde{\tau}_1, s_2 + u + t_1^{-2}\delta(z, t_2, \tilde{z}), t_2 + \varepsilon) h(z, s_2, t_2, \tilde{z}, u) d\tilde{z} du,$$

if $2\varepsilon \leq t_1 + t_2 < 1$, $t_1 \geq \varepsilon$, $t_2 \geq \varepsilon$. *Here* $B = \{-\infty < \tilde{z} < \infty, 0 \leq u < \infty\}$,

$$g_1(z; \tilde{\tau}_1; \tilde{\tau}_2) = 2\alpha^- g(\alpha, \beta_1, \beta_2; z; \tilde{\tau}_1; \tilde{\tau}_2),$$
$$g_2(z; \tilde{\tau}_1; \tilde{\tau}_2) = 2\alpha^+ g(\alpha, \beta_1, \beta_2; z; \tilde{\tau}_1; \tilde{\tau}_2),$$
$$g(\alpha, \beta_1, \beta_2; z; \tilde{\tau}_1; \tilde{\tau}_2) = (2\pi/\tilde{t})^{1/2}(\beta_1\beta_2)^{-3/2} \qquad (4.10)$$
$$\times \tilde{f}_{t_1^* t_2^* \tilde{t}}(z - 2\alpha t_1 t_2) \tilde{\psi}_{M_2 k_1 - 1}(s_1/\beta_1) \tilde{\psi}_{M_2 k_2 - 1}(s_2/\beta_2),$$

with $t_\ell^* = t_\ell \beta_\ell$, $\tilde{t}_\ell = t_\ell/\beta_\ell$, $\tilde{t} = \tilde{t}_1 + \tilde{t}_2$, $\alpha^+ = \max(\alpha, 0)$, $\alpha^- = \max(-\alpha, 0)$, *and*

$$h(z, s_\ell, t_\ell, \tilde{z}, u) = \frac{c(M_2) s_\ell^{(M_2 k_\ell - 1)/2 - 1} u^{(M_2 - 1)/2 - 1}}{(s_\ell + u + t_{\bar{\ell}}^{-2}\delta(z, t_\ell, \tilde{z}))^{((k_\ell + 1)M_2 - 1)/2 - 1}}, \qquad (4.11)$$

$$\delta(z, t_\ell, \tilde{z}) = \frac{(\varepsilon z - t_\ell \tilde{z})^2}{\varepsilon t_\ell(t_\ell + \varepsilon)} \qquad (4.12)$$

with $\bar{\ell} = 3 - \ell$, $c(M_2) = \left((\pi\varepsilon 2^{M_2})^{1/2}\Gamma((M_2 - 1)/2)\right)^{-1}$. *A regret* (1.2) *is*

$$L_N(v, D_1, D_2) = (\overline{D}N)^{1/2} l(\alpha, \beta_1, \beta_2), \qquad (4.13)$$

where

$$l(\alpha, \beta_1, \beta_2) = 2|\alpha|\varepsilon + h \iiint_{B_1} l(z, s_1, 1, s_2, 1) dz ds_1 ds_2,$$

and $h = c^2(M_2)$. *This description is invariant in the sense that it does not depend on the total number of processed data* N *but only on the number of batches* K *into which the data is divided and on the number of internal packets* M_2 *for which the variance is estimated.*

Proof. One should perform the above change of variables in (4.2)–(4.6). Clearly, for degenerate $\rho(v, D_1, D_2)$, which is concentrated at a single parameter $\theta' = (v, D_1, D_2)$ we have

$$G_\ell(Z; S_1, k_1; S_2, k_2) = (\overline{D}/N)^{1/2}(\overline{D}/N)^{1/2}(\overline{D}M_1)^{-3} g_\ell(z; s_1, t_1; s_2, t_2). \quad (4.14)$$

Next, $k_\ell + 1$ should be replaced with $K^{-1}(k_\ell + 1) = t_\ell + \varepsilon$, and $S_\ell + U + n_{\bar{\ell}}^{-2}\Delta(Z, k_\ell, \tilde{Z})$ should be replaced with $(S_\ell + U + n_{\bar{\ell}}^{-2}\Delta(Z, k_\ell, \tilde{Z}))(\overline{D}M_1)^{-1}$, i.e.,

$$s_\ell + u + n_{\bar{\ell}}^{-2} \frac{(Z - k_\ell \tilde{Z})^2}{\overline{D}Mk_\ell(k_\ell + 1)} = s_\ell + u + t_{\bar{\ell}}^{-2}\delta(z, t_\ell, \tilde{z}),$$

where $\delta(z, t_\ell, \tilde{z})$ is defined in (4.12). In addition,

$$n_\ell^{-1} H(Z, S_\ell, k_\ell, \tilde{Z}, U) d\tilde{Z} dU = t_\ell^{-1} h(z, s_\ell, t_\ell, \tilde{z}, u) d\tilde{z} du, \qquad (4.15)$$

where $H(Z, S_\ell, k_\ell, \tilde{Z}, U)$, $h(z, s_\ell, t_\ell, \tilde{z}, u)$ are given in (4.5), (4.11).

To check the validity of the first Eq. (4.9), one should make replacements (4.14)–(4.15) in the first Eq. (4.3). Next, $H \times L(Z; \tilde{T}_1; \tilde{T}_2) dZ dS_1 dS_2 = (\overline{D}N)^{1/2} h \times l(z; \tilde{\tau}_1; \tilde{\tau}_2) dz ds_1 ds_2$ and $L(\rho(v, D_1, D_2)) = M(\overline{D}/N)^{1/2} 2|\alpha|$ for degenerate $\rho(v, D_1, D_2)$. This implies a fulfillment of (4.13).

Corollary 1. *Let* $B_1(\cdot)$, $B_2(\cdot)$ *in (3.2) depend on* $\hat{D}_1$, $\hat{D}_2$ *in the form of dependence on* $\hat{D}_1/\hat{D}_2$. *In this case UCB strategy does not depend on* $\overline{D}$ *and one can choose* $\overline{D} = \max(D_1, D_2)$.

5   Numerical Results

Consider the batch version of the UCB strategy proposed in [7] and customized for batch processing in [9]. According to a condition in Theorem 3, assume that $B_1(\cdot)$, $B_2(\cdot)$ depend only on $\hat{D}_1/\hat{D}_2$. For upper confidence bounds (3.2), we choose $B_\ell(\hat{\beta}_1, \hat{\beta}_2, k_1, k_2) = a_\ell(\hat{\beta}_1, \hat{\beta}_2) \ln^{1/2}(k)$, $\ell = 1, 2$, where $\hat{\beta}_1 = \hat{D}_1/\hat{D}$, $\hat{\beta}_2 = \hat{D}_2/\hat{D}$ with $\hat{D} = \max(\hat{D}_1, \hat{D}_2)$. Denote $\overline{D} = D_1 = \max(D_1, D_2)$, $\beta_1 = 1$,

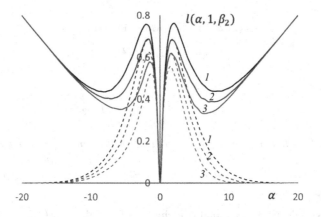

Fig. 1. Normalized regrets

$\beta_2 = D_2/\overline{D}$. For $0.5 \leq \beta_2^{1/2} \leq 1$, using Monte-Carlo simulations, we have found close to optimal functions $a_\ell(1, \beta_2)$ in the case of a priori known variances D_1, D_2 if $m_1 = m + \alpha(\overline{D}/N)^{1/2}$, $m_2 = m - \alpha(\overline{D}/N)^{1/2}$ (here m is arbitrary value). These functions are $a_1(1, \beta) = 0.93 - 0.5(1 - \beta^{1/2})$, $a_2(1, \beta) = 1.112 - 0.182/\beta^{1/2}$, and $a_1(\beta, 1) = a_2(1, \beta)$, $a_2(\beta, 1) = a_1(1, \beta)$. Then determined functions were used as $a_1(\hat{\beta}_1, \hat{\beta}_2)$, $a_2(\hat{\beta}_1, \hat{\beta}_2)$ in the case of a priori unknown variances.

The results of Monte-Carlo simulations for processing $K = 50$ data batches containing $M_2 = 4$ small packets are presented in Fig. 1. The set of parameters is $\Theta = \{m_1 - m_2 = 2\alpha(\overline{D}/N)^{1/2}; \beta_1 = 1; 0.5 \leq \beta_2^{1/2} \leq 1; \overline{D} = 1\}$, where $|\alpha| \leq 20$. In Fig. 1, we present normalized regrets $l(\alpha, 1, \beta_2)$ corresponding to a degenerate prior distribution density concentrated at the parameter $(\alpha, 1, \beta_2)$. Lines 1, 2 and 3 correspond to $\beta_2 = 1$, $\beta_2^{1/2} = 0.8$ and $\beta_2^{1/2} = 0.5$. Thick solid lines describe normalized total regrets, and thin dashed lines describe normalized total regrets without those at two initial stages when data batches are processed by two methods in turn. One can see that for large α regrets are mostly determined by the application of both actions at the initial stages. To reduce them, one should either increase the number of stages K or reduce the batch sizes at the initial stages. The maximum values of $l(\alpha, 1, \beta_2)$ are achieved with equal variances (i.e., with $\beta_2 = 1$) and are approximately equal to 0.76. Since the principal lower bound for the normalized maximum regret in the case of $K = 50$ batches and a priori known variances is achieved for equal variances and is approximately 0.65 (see, e.g., [5]), we can say that UCB batch strategies ensure high quality control.

6 Conclusion

The proposed strategies are easy to implement and provide values of maximum normalized regret close to the principal lower bounds in the case of a priori known variances. So, these strategies ensure high quality control.

Acknowledgement. This study was funded by Russian Science Foundation, project number 23-21-00447, https://rscf.ru/en/project/23-21-00447/.

References

1. Lattimore, T., Szepesvari, C.: Bandit Algorithms. Cambridge University Press, Cambridge (2020)
2. Cesa-Bianchi, N., Lugosi, G.: Prediction, Learning, and Games. Cambridge University Press, Cambridge (2006)
3. Sragovich, V.-G.: Mathematical theory of adaptive control. World Scientific. Interdisciplinary Mathematical Sciences, New Jersey, London, vol. 4 (2006)
4. Perchet, V., Rigollet, P., Chassang, S., Snowberg, E.: Batched bandit problems. Ann. Statist. **44**(2), 660–681 (2016)
5. Kolnogorov, A.-V.: Robust parallel control in a random environment and data processing optimization. Autom. Remote. Control. **75**(12), 2124–2134 (2014)
6. Kolnogorov, A.-V.: Gaussian two-armed bandit and optimization of batch data processing. Probl. Inf. Transm. **54**(1), 84–100 (2018)
7. Auer, P., Cesa-Bianchi, N., Fisher, P.: Finite-time analysis of the multi-armed bandit problem. Mach. Learn. **47**(2–3), 235–256 (2002)
8. Lai, T.-L.: Adaptive treatment allocation and the multi-armed bandit problem. Annals of Statist. **25**, 1091–1114 (1987)
9. Garbar, S.-V., Kolnogorov, A.-V., Lazutchenko, A.-N. UCB strategies and optimization of batch processing in a one-armed bandit problem. Mat. Teor. Igr Prilozh. **15** (4), 3–27 (2023) (Russian. English summary)

On a Global Search in Bilevel Optimization Problems with a Bimatrix Game at the Lower Level

Andrei V. Orlov[(✉)] [iD]

Matrosov Institute for System Dynamics and Control Theory of SB of RAS, Irkutsk, Russia
anor@icc.ru

Abstract. This paper addresses one class of bilevel optimization problems (BOPs) with an equilibrium at the lower level (in optimistic statement). Namely, we study BOPs with a convex quadratic optimization problem under linear constraints at the upper level and with a parametric non-normalized bimatrix game at the lower one, where we need to find a Nash equilibrium. In order to construct numerical methods for the problem in question, first, we transform the original bilevel problem into a single-level nonconvex optimization problem by replacing the lower level with its optimality conditions. Then we apply the Exact Penalization Theory and Global Search Theory (GST) to the resulting problem. According to the standard research of nonconvex problems by the GST, we construct the d.c. representations of all nonconvex functions from the original statement (into the differences of two convex functions), formulate the Global Optimality Conditions in terms of reduced penalized problem, and develop local and global search method taking into account the specific of problem in question. The main feature of the developed methods consists in the possibility of varying the penalty parameter within the methods themselves.

Keywords: Bilevel optimization · Bilevel problems with an equilibrium at the lower level · Optimistic solution · Bimatrix game · Nash equilibrium · Exact Penalization Theory · Global Search Theory · Local search · Global Search Scheme

1 Introduction

It is well known that the study of optimization problems with hierarchical structure is an urgent issue in Operations Research [1,2]. Especially, the study of bilevel problems with several players at the lower (Single-Leader-Multi-Follower-Problems (SLMFP)), at the upper level (Multi-Leader-Single-Follower-Problems (MLSFP)), or even Multi-Leader-Follower-Problems (MLFPs) is rather important (see [3–9] and Chapter 3 in [2]), since such models can describe complex hierarchical systems in ecology [3], economy [8], and other fields of the real life [7,9] (see, also, [1,10], as well as Chapters 5, 6, and 20 in [2]).

© The Author(s), under exclusive license to Springer Nature Switzerland AG 2024
A. Eremeev et al. (Eds.): MOTOR 2024, LNCS 14766, pp. 272–287, 2024.
https://doi.org/10.1007/978-3-031-62792-7_19

In such bilevel problems, besides coordination the interests of the upper and lower levels, say, by finding the Stackelberg equilibrium (optimistic solution) [1,11], it is also required to find, as a rule, a (generalized) Nash equilibrium among players at the same level of the hierarchy. The study of similar models is currently performed along the path of studying specific applied problems using their features, since it is not possible to develop any general approach to bilevel problems with equilibrium at the upper and/or lower levels in the contemporary state of optimization theory (see e.g. [2]).

It is primarily because the problem of finding an equilibrium usually has a nonconvex structure from the optimization point of view whenever it is possible to make a corresponding transformation. Wherein the majority of existing solution algorithms require a preliminary reduction of the bilevel problem in question to a standard single-level optimization problem (e.g. by the Karush—Kuhn—Tucker (KKT) conditions) [1,2,11]. Achieving this is simple only when if the lower level of the original bilevel problem is convex concerning the follower's variable. So the solving of discussed bilevel problems requires the involvement of a special apparatus, including methods of nonconvex optimization.

Currently, there is a few number of publications on the study of bilevel problems with a nonconvex lower level [12–17]. This scarcity could be explained by the immense complexity associated with such problems. Researchers often resort to employing diverse strategies and simplifications to make progress towards obtaining solutions for bilevel problems with a nonconvex lower level. At the same time, research is often limited to the construction of local methods, where authors seek only stationary points in bilevel problems (see e.g. [17]). So it is prudent to categorize problems into classes based on their structure and create specific methods for each class that leverage this structure and the unique characteristics of the problems being investigated.

In recent works of the author, the approach to the simplest classes of the bilevel problems with so-called non-normalized parametric matrix and bimatrix games [18,19] at the lower level [20,21] was developed. Here we continue this research, focusing on the developing for the problem in question, some special local search methods as well as a global search scheme, which allow to get out from the points providing by the local search.

The developed approach to BOPs with declared equilibrium is based on the transformation of the original problem to a single-level global optimization problem with d.c. constraints (see, e.g., [22,23]). Classical convex optimization techniques are recognized to be incapable of globally solving nonconvex optimization problems [23–25]. Therefore, to find global solutions to the resulting single-level problem, we employ the so-called Global Search Theory (GST) created by A.S. Strekalovsky [23,27–30]. Recently, GST has emerged as a potent tool for developing numerical methods to tackle various nonconvex Operations Research problems, including those with hierarchical and equilibrium structure, see [11,19,23,31–33].

The paper is organized in the following way Sect. 2 discusses the formulation of the problem and its transformation into a single-level one. Section 3 provides the Exact Penalization Theory and the Global Search Theory in terms of the constructed nonconvex problem. In Sect. 4, the variants of the Special Penalty Local Search Method are built. Section 5 is devoted to the Global Search Scheme, and Sect. 6 concludes the paper.

2   Problem Formulation and Its Transformation

Let us formulate the following BOPs with equilibrium at the lower level [21]:

$$\left.\begin{array}{c} \langle x, Cx \rangle + \langle c, x \rangle + \langle y, D_1 y \rangle + \langle d_1, y \rangle + \langle z, D_2 z \rangle + \langle d_2, z \rangle \uparrow \max_{x,y,z}, \\ x \in X, \ (y, z) \in NE(\Gamma B(x)), \end{array}\right\} \quad (\mathcal{BP}_{\Gamma B})$$

where $X = \{x \in \mathbb{R}^m \mid Ax \le a, \ x \ge 0, \ \langle b_1, x \rangle + \langle b_2, x \rangle = 1\}$, $NE(\Gamma B(x))$ is a set of Nash equilibria of the game (y is the variable of Player 1, z is the variable of Player 2)

$$\left.\begin{array}{ll} \langle y, B_1 z \rangle \uparrow \max_y, & y \in Y(x) = \{y \mid y \ge 0, \ \langle e_{n_1}, y \rangle = \langle b_1, x \rangle\}, \\ \langle y, B_2 z \rangle \uparrow \max_z, & z \in Z(x) = \{z \mid z \ge 0, \ \langle e_{n_2}, z \rangle = \langle b_2, x \rangle\}; \end{array}\right\} \quad (\Gamma B(x))$$

$c, b_1, b_2 \in \mathbb{R}^m$; $y, d_1 \in \mathbb{R}^{n_1}$; $z, d_2 \in \mathbb{R}^{n_2}$; $a \in \mathbb{R}^{m_1}$; $b_1 \ge 0$, $b_1 \ne 0$, $b_2 \ge 0$, $b_2 \ne 0$; A, B_1, B_2, C, D_1, D_2 are matrices of appropriate dimension, $e_{n_1} = (1, ..., 1)$, $e_{n_2} = (1, ..., 1)$ are vectors of appropriate dimension. $C = C^T$, $D_1 = D_1^T$, $D_2 = D_2^T$ are negative semidefinite matrices, so, the objective function of the leader is concave. Note, if a term like yB_1 is used, it means that y is a row vector, whereas the expression $B_1 z$ implies that z is a column vector.

It can be readily seen that, at the lower level, we formulate a so-called non-normalized bimatrix game with mixed strategies [21], which differs from a classical bimatrix game (see, e.g. [18,19]) in scalar parameters $\xi_1 := \langle b_1, x \rangle$, $\xi_2 := \langle b_2, x \rangle$ (x is fixed at the lower level). The equality $\langle b_1, x \rangle + \langle b_2, x \rangle = 1$ represents a resource that the leader is responsible for distributing among the followers.

The problem $(\mathcal{BP}_{\Gamma B})$ is written in the so-called optimistic formulation when the interests of the leader can be agreed with the actions of the followers [1,2]. In order to study the conditions guaranteeing the existence of a global solution in such formulation, one can use the corresponding theoretical results of bilevel optimization [1,2].

As for developing numerical methods for finding solutions to the bilevel problem $(\mathcal{BP}_{\Gamma B})$, where a standard Nash equilibrium [18,19,21] is sought at the lower level, we need to reformulate it as a single-level problem. For this purpose, we can employ the following theorem.

Theorem 1 [21]. *The pair $(y^*, z^*) \in NE(\Gamma B(x))$ (where x is fixed), if and only if there exist numbers α_* and β_*, such that the following system takes place:*

$$\left.\begin{array}{c} \xi_1(B_1 z^*) \le \alpha_* e_{n_1}, \ z^* \in Z(x), \quad \xi_2(y^* B_2) \le \beta_* e_{n_2}, \ y^* \in Y(x), \\ \langle y^*, (B_1 + B_2) z^* \rangle = \alpha_* + \beta_*. \end{array}\right\} \quad (1)$$

So, it is possible to replace the lower level of $(\mathcal{BP}_{\Gamma B})$ with the system (1) where $\xi_1 = \langle b_1, x \rangle$, $\xi_2 = \langle b_2, x \rangle$ (x is fixed). Note that the last equality with respect to a couple of variables y and z is nonconvex.

Hence, we can write down the following single-level mathematical optimization problem, which, from a global solutions perspective, is equivalent to the bilevel problem $(\mathcal{BP}_{\Gamma B})$:

$$
\left.
\begin{aligned}
-f_0(x, y, z) &:= \langle x, Cx \rangle + \langle c, x \rangle + \langle y, D_1 y \rangle + \langle d_1, y \rangle + \\
&\quad + \langle z, D_2 z \rangle + \langle d_2, z \rangle \uparrow \max_{x,y,z,\alpha,\beta}, \\
(x, y, z) &\in S := \{x, y, z \mid Ax \le a, \; x \ge 0, \; \langle b_1, x \rangle + \langle b_2, x \rangle = 1, \\
&\quad y \ge 0, \; \langle e_{n_1}, y \rangle = \langle b_1, x \rangle, \quad z \ge 0, \; \langle e_{n_2}, z \rangle = \langle b_2, x \rangle \}, \\
&\quad \langle b_1, x \rangle (B_1 z) \le \alpha e_{n_1}, \quad \langle b_2, x \rangle (y B_2) \le \beta e_{n_2}, \\
&\quad \langle y, (B_1 + B_2) z \rangle = \alpha + \beta.
\end{aligned}
\right\}
\quad (\mathcal{PB})
$$

Theorem 2 [21]. *The 3-tuple (x^*, y^*, z^*) is a global (optimistic) solution to the bilevel problem $(\mathcal{BP}_{\Gamma B})$ $((x^*, y^*, z^*) \in \mathrm{Sol}(\mathcal{BP}_{\Gamma B}))$, if and only if there exist numbers α_* and β_* such that the 5-tuple $(x^*, y^*, z^*, \alpha_*, \beta_*)$ is a global solution to the problem $(\mathcal{PB})$.* □

It is evident that the optimization problem $(\mathcal{PB})$ has a nonconvex feasible set (refer to references such as [22,23]). The nonconvex nature in problem $(\mathcal{PB})$ emanates from the $(n_1 + n_2)$ bilinear inequality constraints and the single bilinear equality constraint. It is well known that a bilinear function can be represented as a difference of two convex functions [19]. To address problem $(\mathcal{PB})$ with d.c. constraints [23,27,28], we will use Global Search Theory (GST) previously mentioned.

3 D.C. Approach to the Problem

Our next step involves constructing explicit representations of all nonconvex functions from the constraints of problem $(\mathcal{PB})$ as differences of two convex functions.

First, it is straightforward to see that the d.c. decomposition of the i-th scalar constraint from the first group of n_1 inequality constraints can be constructed based on the well-established property of an inner product [21]:

$$
f_i(x, z, \alpha) := \langle b_1, x \rangle \langle (B_1)_i, z \rangle - \alpha = g_i(x, z, \alpha) - h_i(x, z), \quad i = 1, \ldots, n_1, \quad (2)
$$

where $(B_1)_i$ is an i-th row of the matrix B_1, $g_i(x, z, \alpha) = \dfrac{1}{4} \|x Q_i + z\|^2 - \alpha$, $h_i(x, z) = \dfrac{1}{4} \|x Q_i - z\|^2$. Here $Q_i^T = (b_1^{(1)} (B_1)_i; \; b_1^{(2)} (B_1)_i; \; \ldots; \; b_1^{(m)} (B_1)_i)$, where $b_1^{(1)}, b_1^{(2)} \ldots, b_1^{(m)}$ are components of the vector b_1 (Q_i is a $(m \times n_2)$-matrix).

Similarly, we can write down the d.c. decompositions of n_2 inequality constraints from the second group [21]:

$$f_j(x, y, \beta) := \langle b_2, x \rangle \langle y, (B_2)_j \rangle - \beta = g_j(x, y, \beta) - h_j(x, y), \ j = n_1+1, ..., n_1+n_2, \tag{3}$$

where $(B_2)_j$ is a $(j-n_1)$-th column of the matrix B_2, $g_j(x, y, \beta) = \frac{1}{4}\|xR_j+y\|^2 - \beta$,

$h_j(x, y) = \frac{1}{4}\|xR_j - y\|^2$, $R_j^T = (b_2^{(1)}(B_2)_j^T;\ b_2^{(2)}(B_2)_j^T; ...; b_2^{(m)}(B_2)_j^T)$ (R_j is a $(m \times n_1)$-matrix). Finally, the d.c. representation of the last bilinear equality constraint has the following form:

$$f_{n_1+n_2+1}(y, z, \alpha, \beta) = g_{n_1+n_2+1}(y, z, \alpha, \beta) - h_{n_1+n_2+1}(y, z), \tag{4}$$

where $g_{n_1+n_2+1}(y, z, \alpha, \beta) = \frac{1}{4}\|y + B_1 z\|^2 + \frac{1}{4}\|yB_2 + z\|^2 - \alpha - \beta$,

$h_{n_1+n_2+1}(y, z) = \frac{1}{4}\|y - B_1 z\|^2 + \frac{1}{4}\|yB_2 - z\|^2$.

The problem $(\mathcal{PB})$ can now be formulated as a minimization problem featuring a convex quadratic objective function, convex set S, and $(n_1 + n_2 + 1)$ d.c. constraints:

$$\left. \begin{aligned} f_0(x, y, z) &\downarrow \min_{x,y,z,\alpha,\beta}, \quad (x, y, z) \in S, \\ f_i(x, z, \alpha) &:= g_i(x, z, \alpha) - h_i(x, z) \le 0, \quad i \in \{1, ..., n_1\} =: \mathcal{I}, \\ f_j(x, y, \beta) &:= g_j(x, y, \beta) - h_j(x, y) \le 0, \quad j \in \{n_1 + 1, ..., n_1 + n_2\} =: \mathcal{J}, \\ f_{n_1+n_2+1}(y, z, \alpha, \beta) &:= g_{n_1+n_2+1}(y, z, \alpha, \beta) - h_{n_1+n_2+1}(y, z) = 0, \end{aligned} \right\}$$
$$(\mathcal{DCC})$$

where the functions f_0; g_i, h_i $\forall i \in \mathcal{I} = \{1, ..., n_1\}$; g_j, h_j $\forall j \in \mathcal{J} = \{n_1+1, ..., n_1+n_2\}$; $g_{n_1+n_2+1}$, and $h_{n_1+n_2+1}$, are convex.

Consider $\mathcal{F}$ as a feasible set of the problem $(\mathcal{DCC})$ ($N := n_1 + n_2 + 1$):

$$\mathcal{F} := \{(x, y, z, \alpha, \beta) \mid (x, y, z) \in S; \ f_i(x, z, \alpha) \le 0, \ i \in \mathcal{I};$$
$$f_j(x, y, \beta) \le 0, \ j \in \mathcal{J}; \ f_N(y, z, \alpha, \beta) = 0\}.$$

The so-called basic nonconvexity in the problem $(\mathcal{DCC})$ arises from the functions $h_i(x, z)$, $i \in \mathcal{I}$, $h_j(x, y)$, $j \in \mathcal{J}$, and $h_N(y, z)$ (refer to [21] for detailed information). To address this problem, we will develop a Global Search Algorithm (GSA) founded on the Global Search Theory (GST) [23, 26–28, 30] using the previously established d.c. decomposition.

The GSA typically involves two key phases:

1) a specialized Local Search Method (LSM) tailored to the specific problem characteristics [11, 19, 23, 26, 30];

2) an improvement process for obtained points following the Local Search phase, based on the Global Optimality Conditions (GOCs) [23, 26–28, 30].

Our focus in this study is on developing the customized LSM for problem $(\mathcal{DCC})$. This involves further modifying the problem using the Exact Penalization Theory [21, 28].

Let us consider the penalized problem $(\theta := (x, y, z, \alpha, \beta))$.

$$\Phi_\sigma(\theta) := f_0(x, y, z) + \sigma W(\theta) \downarrow \min_\theta, \quad (x, y, z) \in S, \qquad (\mathcal{DC}(\sigma))$$

where, with $\sigma > 0$ as the penalty parameter, the function $W(\cdot)$ serves as the penalty function for problem $(\mathcal{DCC})$:

$$W(x, y, z, \alpha, \beta) := \max\{0, f_1(x, z, \alpha), \ldots, f_{n_1}(x, z, \alpha),$$
$$f_{n_1+1}(x, y, \beta), \ldots, f_{n_1+n_2}(x, y, \beta)\} + |f_N(y, z, \alpha, \beta)|.$$

It is easy to see that when σ is fixed, this problem belongs to the class of d.c. minimization problems with a convex feasible set [23, 26]. Indeed (see [21–23]),

$$\Phi_\sigma(\theta) \overset{\triangle}{=} f_0(x, y, z) + \sigma \max\{0, f_i(x, z, \alpha), i \in \mathcal{I}; f_j(x, y, \beta), j \in \mathcal{J}\} \qquad (5)$$
$$+ \sigma |f_N(y, z, \alpha, \beta)| =: G_\sigma(\theta) - H_\sigma(\theta),$$

where

$$G_\sigma(\theta) := f_0(x, y, z) + \sigma \max\left\{ \sum_{k \in \mathcal{I}} h_k(x, z) + \sum_{k \in \mathcal{J}} h_k(x, y); \right.$$
$$\left. \left[g_l(\cdot) + \sum_{\substack{k \in \mathcal{I} \cup \mathcal{J} \\ k \neq l}} h_k(\cdot) \right], l \in \mathcal{I} \bigcup \mathcal{J} \right\} + 2\sigma \max\{g_N(y, z, \alpha, \beta); h_N(y, z)\}, \qquad (6)$$

$$H_\sigma(\theta) := \sigma \left[\sum_{i \in \mathcal{I}} h_i(x, z) + \sum_{j \in \mathcal{J}} h_j(x, y) + g_N(y, z, \alpha, \beta) + h_N(y, z) \right] \qquad (7)$$

are both convex functions.

It is known from the Classical Penalty Theory [24, 25] that if for some value of the parameter σ the 5-tuple $(x(\sigma), y(\sigma), z(\sigma), \alpha(\sigma), \beta(\sigma)) =: \theta(\sigma)$ is a solution to the problem $(\mathcal{DC}(\sigma))$ $(\theta(\sigma) \in \mathrm{Sol}(\mathcal{DC}(\sigma)))$, and $\theta(\sigma)$ is feasible in the problem $(\mathcal{DCC})$ $(\theta(\sigma) \in \mathcal{F})$, i.e. $W(\theta(\sigma)) = 0$, then $\theta(\sigma)$ is a global solution to the problem $(\mathcal{DCC})$ [24, 25, 27, 28]. In addition, we know [24, 25] that if the equality $W(\theta(\sigma)) = 0$ holds for some $\sigma := \hat{\sigma}$ at a solution $\theta(\sigma)$, then this solution to the problem $(\mathcal{DC}(\sigma))$ is a solution to the problem $(\mathcal{DCC})$ for all $\sigma \geq \hat{\sigma}$. It means that

$$\mathrm{Sol}(\mathcal{DCC}) = \mathrm{Sol}(\mathcal{DC}(\sigma)) \quad \forall \sigma > \hat{\sigma}, \qquad (8)$$

so that the problems $(\mathcal{DCC})$ and $(\mathcal{DC}(\sigma))$ are equivalent (in the sense of (8)) [24, 25, 27, 28].

The core moment of the Exact Penalization Theory [24, 25] is the existence of a threshold value $\hat{\sigma} > 0$ for the penalty parameter σ, ensuring that $W(\theta(\sigma)) = 0$ for all $\sigma \geq \hat{\sigma}$. The validity of this fact can be established under specific regularity conditions for the problem at hand, detailed in [21, 27, 28].

Consequently, by leveraging Theorem 2 in tandem with the aforementioned result, we can assert that the relationship between the problems $(\mathcal{DC}(\sigma))$ and $(\mathcal{DCC})$ allows us to find an optimistic solution to the problem $(\mathcal{BP}_{\Gamma B})$ through solving the single problem $(\mathcal{DC}(\sigma))$ (with a fixed $\sigma > \hat{\sigma}$) instead of dealing with problem $(\mathcal{DCC})$.

Let $S' := \{(x, y, z, \alpha, \beta) \in \mathbb{R}^{m+n_1+n_2+2} \mid (x, y, z) \in S\}$, and formulate the necessary Global Optimality Conditions (GOCs) in relation to the problem $(\mathcal{DC}(\sigma))$. These conditions serve as the foundation for developing global search algorithms.

Theorem 3 [21,27,28]. *Suppose, a feasible point $\theta_* \in \mathcal{F} \subset \mathbb{R}^{m+n_1+n_2+2}$, $\zeta := f_0(x_*, y_*, z_*)$ is a global solution to the problem $(\mathcal{DCC})$, and a number $\sigma : \sigma \geq \hat{\sigma} > 0$ is fixed, where $\hat{\sigma}$ is a threshold value of the penalty parameter (i.e. equality (8) is fulfilled).*
Then $\forall (\eta, \gamma) \in \mathbb{R}^{m+n_1+n_2+2} \times \mathbb{R}$, such that

$$H_\sigma(\eta) = \gamma - \zeta, \tag{9}$$

and

$$G_\sigma(\eta) \leq \xi \leq \sup(G_\sigma, S') \tag{10}$$

the inequality

$$G_\sigma(\theta) - \gamma \geq \langle \nabla H_\sigma(\eta), \theta - \eta \rangle \qquad \forall \theta \in S' \tag{11}$$

holds.   □

The conditions (9)–(11) exhibit the so-called constructive (algorithmic) property, as discussed in [23,26–28]. This property implies that if the main inequality (11) of the GOCs is violated, it is possible to find a feasible point that will be better than the current solution.

Furthermore, Theorem 3 gives rise to convex (linearized) problems defined as

$$\Psi_{\sigma\eta}(\theta) := G_\sigma(\theta) - \langle \nabla H_\sigma(\eta), \theta \rangle \downarrow \min_\theta, \quad \theta \in S', \tag{$\mathcal{P}_\sigma\mathcal{L}(\eta)$}$$

where $\eta := (\bar{x}, \bar{y}, \bar{z}, \bar{\alpha}, \bar{\beta}) \in \mathbb{R}^{m+n_1+n_2+2}$ satisfies the equality (9). The linearization is performed with respect to the function $H_\sigma(\cdot)$, which accumulated all the nonconvexities of both problems $(\mathcal{DC}(\sigma))$ and $(\mathcal{DCC})$.

By considering the decompositions (2)–(4) and (5)–(7), we can compute the gradient of the function $H_\sigma(\cdot)$ used in formulating the linearized problem:

$$\nabla H_\sigma(\eta) = \sigma \begin{pmatrix} \frac{1}{2}\sum_{i \in \mathcal{I}}[Q_i(\bar{x}Q_i - \bar{z})] + \frac{1}{2}\sum_{j \in \mathcal{J}}[R_j(\bar{x}R_j - \bar{y})]; \\ -\frac{1}{2}\sum_{j \in \mathcal{J}}(\bar{x}R_j - \bar{y}) + \frac{1}{2}[\bar{y} + B_1\bar{z} + B_2(\bar{y}B_2 + \bar{z})] + \\ \frac{1}{2}[\bar{y} - B_1\bar{z} + B_2(\bar{y}B_2 - \bar{z})]; \\ -\frac{1}{2}\sum_{i \in \mathcal{I}}(\bar{x}Q_i - \bar{z}) + \frac{1}{2}[(\bar{y} + B_1\bar{z})B_1 + \bar{y}B_2 + \bar{z}] - \\ \frac{1}{2}[(\bar{y} - B_1\bar{z})B_1 + \bar{y}B_2 - \bar{z}]; \\ -1; \\ -1 \end{pmatrix}.$$

Our objective is to vary the parameters (η, γ) to potentially violate the inequality (11). It has been noted in literature [23, 26–28] that combining this process with a local search is convenient. Hence, the central focus lies in devising and analyzing a specialized Local Search Method tailored to the unique characteristics of the problem at hand. This task precedes the development of a Global Search Algorithm [11, 19, 23, 31].

4   Local Search Methods and Their Specificity

To develop the LSM, we draw upon the concepts outlined in [35]. The Local Search Scheme proposed in [35] initially adheres to the established practice of linearizing with respect to the basic nonconvexity of the current problem (also referenced in [23, 34]). Additionally, it incorporates procedures for dynamically adjusting the penalty parameter. This method is termed the Special Penalty Local Search Method (SPLSM).

Consider a starting point $(x_0, y_0, z_0) \in S$ and an initial penalty parameter value $\sigma^0 > 0$. Assume that at iteration s of the SPLSM, we have the triplet $(x^s, y^s, z^s) \in S$ and a penalty parameter value $\sigma_s \geq \sigma^0$. Compute $\alpha_s := \langle y^s, B_1 z^s \rangle$ and $\beta_s := \langle y^s, B_2 z^s \rangle$ based on the optimality conditions in the non-normalized bimatrix game (as outlined in Theorem 1 in [21]). And introduce the following notations: $G_s(\cdot) := G_{\sigma_s}(\cdot)$, $H_s(\cdot) := H_{\sigma_s}(\cdot)$, and $\theta^s := (x^s, y^s, z^s, \alpha_s, \beta_s)$.

The key aspect of the Special Penalty Local Search Method (SPLSM) lies in generating a sequence $\{\theta^s\}$ where each element is an approximate solution to the linearized problem $(\mathcal{P}_s \mathcal{L}_s) = (\mathcal{P}_{\sigma_s} \mathcal{L}(\theta^s))$. At iteration s, the formulation of the linearized problem reads in the following way:

$$\Psi_s(\theta) := G_s(\theta) - \langle \nabla H_s(\theta^s), \theta \rangle \downarrow \min_{\theta}, \quad \theta \in S'. \qquad (\mathcal{P}_s \mathcal{L}_s)$$

The point θ^{s+1} satisfies the inequality $\Psi_s(\theta^{s+1}) \overset{\triangle}{=} G_s(\theta^{s+1}) - \langle \nabla H_s(\theta^{s+1}), \theta^{s+1} \rangle \leq \mathcal{V}(\mathcal{P}_s \mathcal{L}_s) + \delta_s$, where $\delta_s \geq 0$, $s = 0, 1, 2, ...$, $\sum_{s=0}^{\infty} \delta_s < +\infty$.

While all components of the problem $(\mathcal{DCC})$ are differentiable, it is evident that the problem $(\mathcal{P}_s \mathcal{L}_s)$, more precisely the function $G_s(\cdot)$, lacks smoothness.

To address this challenge, we employ the well-known lemma (see, e.g. Sect. 5 of [29]), which transforms the problem $(\mathcal{P}_s \mathcal{L}_s)$ into a new formulation:

$$
\left.
\begin{aligned}
& f_0(x, y, z) - \langle \nabla H_s(\theta^s), \theta \rangle + \sigma_s \omega + 2\sigma_s t \downarrow \min_{\theta, \omega, t}, \\
& (\theta, \omega, t) \in \mathcal{D} := \{\theta \in S', \ \omega \in \mathbb{R}, \ t \in \mathbb{R} \mid g_l(\cdot) + \sum_{\substack{k \in \mathcal{I} \cup \mathcal{J} \\ k \neq l}} h_k(\cdot) \leq \omega, \\
& l \in \mathcal{I} \cup \mathcal{J}; \quad \sum_{k \in \mathcal{I}} h_k(x, z) + \sum_{k \in \mathcal{J}} h_k(x, y) \leq \omega; \\
& g_N(y, z, \alpha, \beta) \leq t, \quad h_N(y, z) \leq t\}.
\end{aligned}
\right\} \quad (\mathcal{AP}_s \mathcal{L}_s)
$$

The problem $(\mathcal{AP}_s\mathcal{L}_s)$ presents a convex optimization problem with solely inequality constraints (excluding the linear constraints from the set S'), unlike from the structure of the original problem $(\mathcal{DCC})$. Given the smoothness of the components in $(\mathcal{DCC})$, the auxiliary problem $(\mathcal{AP}_s\mathcal{L}_s)$ retains its smooth nature. Moreover, the adding of only two supplementary scalar variables poses minimal hardship in numerically solving this problem.

As part of the SPLSM scheme [35, 36], we need to use an additional auxiliary convex problem focused on minimizing the penalty function $W(\theta)$:

$$\Psi_W(\theta) := G_W(\theta) - \langle \nabla H_W(\theta^s), \theta \rangle \downarrow \min_{\theta}, \quad \theta \in S', \qquad (\mathcal{P}_W\mathcal{L}_s)$$

where $G_W(\theta) := \dfrac{1}{\sigma_s}[G_s(\theta) - f_0(x, y, z)]$ and $H_W(\theta) := \dfrac{1}{\sigma_s}[H_s(\theta)]$ represent the components within the d.c. decomposition of the penalty function: $W(\theta) = G_W(\theta) - H_W(\theta)$. It can be readily seen that $\nabla H_W(\theta^s) := \dfrac{1}{\sigma}\nabla[H_\sigma(\theta^s)]$. Similar to the problem $(\mathcal{P}_s\mathcal{L}_s)$, the problem $(\mathcal{P}_W\mathcal{L}_s)$ lacks smoothness due to the properties of the function $G_W(\cdot)$.

The smooth auxiliary problem derived from the problem $(\mathcal{P}_W\mathcal{L}_s)$ and formulated following the aforementioned lemma [29] can be represented as:

$$-\langle \nabla H_W(\theta^s), \theta \rangle + \omega + 2t \downarrow \min_{\theta,\omega,t}, \quad (\theta, \omega, t) \in \mathcal{D}. \qquad (\mathcal{AP}_W\mathcal{L}_s)$$

Considering the possibility of approximately solving these linearized problems, the scheme of the SPLSM-1 for the problem $(\mathcal{DC}(\sigma))$ looks in the following way.

Let there be given a starting point $(x_0, y_0, z_0) \in S$, an initial penalty parameter value $\sigma^0 > 0$, two scalar parameters $\mu_1, \mu_2 \in]0, 1[$ for the method, a numerical sequence $\{\delta_s\}$ such that $\delta_s > 0$, $s = 0, 1, 2, ...$; $\delta_s \downarrow 0$ $(s \to \infty)$, as well as small tolerances $\varepsilon_0 > 0$ and $\tau > 0$.

Step 0. Set $s := 0$, $\sigma_s := \sigma^0$; $(x^s, y^s, z^s) := (x_0, y_0, z_0)$, $\alpha_s := \langle y^s, B_1 z^s \rangle$, $\beta_s := \langle y^s, B_2 z^s \rangle$, and $\theta^s := (x^s, y^s, z^s, \alpha_s, \beta_s)$.

Step 1. By solving the problem $(\mathcal{AP}_s\mathcal{L}_s)$ find an approximate solution $\theta(\sigma_s)$ to the linearized problem $(\mathcal{P}_s\mathcal{L}_s)$, so that $\theta(\sigma_s) \in \delta_s$-Sol$(\mathcal{P}_s\mathcal{L}_s)$.

Step 2. If $W(\theta(\sigma_s)) \leq \varepsilon$ then define $\sigma_+ := \sigma_s$, $\theta(\sigma_+) := \theta(\sigma_s)$ and proceed to **Step 7.**

Step 3. Otherwise (if $W(\theta(\sigma_s)) > \varepsilon$), by addressing the subproblem $(\mathcal{P}_W\mathcal{L}_s)$ (through the problem $(\mathcal{AP}_W\mathcal{L}_s)$) find $\theta_W^s \in \delta_s$-Sol$(\mathcal{P}_W\mathcal{L}_s)$.

Step 4. If $W(\theta_W^s) \leq \varepsilon$ then solve multiple problems $(\mathcal{P}_{\sigma_s}\mathcal{L}(\theta_W^s))$ (by potentially increasing σ_s), attempting to obtain $\sigma_+ > \sigma_s$ and the vector $\theta(\sigma_+) \in \delta_s$-Sol$(\mathcal{P}_{\sigma_+}\mathcal{L}(\theta_W^s))$, ensuring $W(\theta(\sigma_+)) \leq \varepsilon$ and move to **Step 7.**

Step 5. Else, if $W(\theta_W^s) > \varepsilon$, and no $\sigma_+ > \sigma_s$ satisfying $W(\theta(\sigma_+)) \le \varepsilon$ is found during Step 4, then find $\sigma_+ > \sigma_s$ that fulfills the inequality

$$W(\theta^s) - W(\theta(\sigma_+)) \ge \mu_1[W(\theta^s) - W(\theta_W^s)]. \tag{12}$$

Step 6. Increase σ_+, if necessary, to satisfy the inequality

$$\Psi_s(\theta^s) - \Psi_{\sigma_+}(\theta(\sigma_+)) \ge \mu_2\sigma_+[W(\theta^s) - W(\theta(\sigma_+))]. \tag{13}$$

Step 7. If the following system of inequalities is satisfied:

$$\text{a) } W(\theta(\sigma_+)) \le \varepsilon; \quad \text{b) } \Psi_{s+1}(\theta^s) - \Psi_{s+1}(\theta(\sigma_+)) \le \frac{\tau}{2}, \tag{14}$$

then stop. $\theta(\sigma_+)$ represents the output of the SPSLM-1, else $\sigma_{s+1} := \sigma_+$, $\theta^{s+1} := \theta(\sigma_+)$, $s := s + 1$ and return to **Step 1.**   □
The ideas of additional minimization of the penalty function $W(\cdot)$ and using the parameters μ_1 and μ_2 (they estimate the progress made at the current iteration with respect to the penalty and the linearized functions) were proposed in [37,38]. In these works also, one can find the practical rules on how to select the parameters of the scheme and how to increase the value of a penalty parameter σ.

The concept of additional minimizing the penalty function $W(\cdot)$ and using the parameters μ_1 and μ_2 (to gauge progress regarding the penalty and linearized functions) was introduced in [37,38]. These works also offer practical rules for selecting scheme parameters and increasing the penalty parameter σ effectively.

Detailed theoretical results on the convergence of the presented SPLSM are elaborated in [35]. Specifically, the stopping criterion (14) is introduced and substantiated in [35]: the inequality (14a) ensures that the point $\theta(\sigma_+)$ is an approximately feasible point of the problem $(\mathcal{DCC})$; if inequality (14b) holds with $\delta_s \le \tau/2$, as per Remark 7.5 in [35], the point θ^s derived by the SPLSM presents a τ-critical point and a τ-solution to the linearized problem $(\mathcal{P}_{\sigma_+}\mathcal{L}_s)$. Hence, the criterion (14) is applicable for computational experiments.

Besides the SPLSM-1, we also examine its simplified version the SPLSM-2 (somewhat more specialized), which relies on the local search method proposed in [30].

Given a starting point $(x_0, y_0, z_0) \in S$, an initial value of the penalty parameter $\sigma^0 > 0$, two scalar parameters $\mu_2 \in]0,1[$ and $\bar{\mu} > 0$, a numerical sequence $\{\delta_s\}: \delta_s > 0$, $s = 0,1,2,...$; $\delta_s \downarrow 0$ $(s \to \infty)$, and tolerances $\varepsilon_0 > 0$ and $\tau > 0$.

Step 0. Set $s := 0$, $\sigma_s := \sigma^0$; $(x^s, y^s, z^s) := (x_0, y_0, z_0)$, $\alpha_s := \langle y^s, B_1 z^s \rangle$, $\beta_s := \langle y^s, B_2 z^s \rangle$, and $\theta^s := (x^s, y^s, z^s, \alpha_s, \beta_s)$.

Step 1. By solving the problem $(\mathcal{AP}_s\mathcal{L}_s)$ find an approximate solution $\theta(\sigma_s)$ to the linearized problem $(\mathcal{P}_s\mathcal{L}_s)$, so that $\theta(\sigma_s) \in \delta_s\text{-Sol}(\mathcal{P}_s\mathcal{L}_s)$.

Step 2. If $W(\theta(\sigma_s)) \le \varepsilon_0$, then define $\sigma_+ := \sigma_s$, $\theta(\sigma_+) := \theta(\sigma_s)$ and proceed to **Step 6.**

Step 3. Otherwise (if $W(\theta(\sigma_s)) > \varepsilon_0$), if the inequality

$$\Psi_s(\theta^s) - \Psi_s(\theta(\sigma_s)) \ge \mu_2\sigma_s[W(\theta^s) - W(\theta(\sigma_s))] \tag{15}$$

holds, then set $\sigma_+ := \sigma_s$, $\theta(\sigma_+) := \theta(\sigma_s)$ and go to **Step 6**.

Step 4. Increase the number σ_s so that $\sigma_+ := \bar{\mu}\sigma_s$ and solve the linearized problem $(\mathcal{P}_{\sigma_+}\mathcal{L}(\theta(\sigma_s)))$, therefore $\theta(\sigma_+) \in \delta_s\text{-Sol}(\mathcal{P}_{\sigma_+}\mathcal{L}(\theta(\sigma_s)))$.

Step 5. Set $\theta(\sigma_s) := \theta(\sigma_+)$, $\sigma_s := \sigma_+$ and move to **Step 2**.

Step 6. If the subsequent conditions takes place:

$$\text{a) } W(\theta(\sigma_+)) \leq \varepsilon_0; \quad \text{b) } \Psi_{s+1}(\theta^s) - \Psi_{s+1}(\theta(\sigma_+)) \leq \frac{\tau}{2}, \tag{16}$$

then stop. $\theta(\sigma_+)$ is the result of the SPLSM-2, else $\sigma_{s+1} := \sigma_+$, $\theta^{s+1} := \theta(\sigma_+)$, $s := s + 1$ and return to **Step 1**.   □

To begin, it is worth noting that within the SPLSM-2, there are no steps directly involving the solution of a penalty linearized problem $(\mathcal{P}_W\mathcal{L}_s)$. It seems that since the stopping criterion of a local search necessitates obtaining points that are approximately feasible in problem $(\mathcal{DCC})$ (within an accuracy of ε_0), the need for additional minimization of the penalty function and the monitoring of its decreasing progress may be unnecessary. Ultimately, the question of the practical effectiveness comparison of these methods can only be conclusively addressed through computational experiments.

Furthermore, it is important to highlight that the search for an increased penalty parameter value ("hardwired" into Step 4 of SPLSM-1) is explicitly ordered in the second version of the method. The rationale behind this scheme and its corresponding stopping criteria can be referenced in [30].

Let us now move on to the description of the Global Search Scheme (GSS) in the problem under study. The main module of the GSS is the local search methods presented above.

5   Global Search Scheme

In this section, we introduce a global search scheme in the problem $(\mathcal{DC}(\sigma))$. Its key distinguishing aspect, akin to the local search method discussed earlier, lies in the adaptive adjustment of the penalty parameter $\sigma > 0$ to align the algorithm more precisely with the characteristics of the specific problem at hand. This scheme was initially proposed in [30] for general d.c. optimization problems. We now present its adapted version tailored specifically for the current issue in a more structured algorithmic form.

Let us denote k as the current iteration number within the global search process. Through the operation of the SPLSMs, an approximate critical point $\theta^k := \theta(\sigma_+) \in S'$ is derived, alongside the determination of the corresponding penalty parameter value, denoted as $\bar{\sigma} := \sigma_+$. Then one of the possible schemes for attaining a global solution to the problem $(\mathcal{DC}(\sigma))$, where the penalty parameter $\sigma_k > 0$ dynamically changes, with the goal of improving the objective function value $\Phi_k(\theta^k) := \Phi_{\sigma_k}(\theta^k)$, can be outlined in the following way.

Given a starting point $(x^0, y^0, z^0) \in S$, an initial penalty parameter value $\sigma^0 > 0$, numerical sequences $\{\tau_k\}$, $\{\delta_k\}$ (τ_k, $\delta_k > 0$, $k = 1, 2, ...$; $\tau_k \downarrow 0$, $\delta_k \downarrow 0$ ($k \to \infty$)), a set of directions $Dir = \{(\bar{u}^1, \bar{v}^1, \bar{w}^1), ..., (\bar{u}^N, \bar{v}^N, \bar{w}^N) \in$

$I\!\!R^{m+n_1+n_2} \mid (\bar{u}^r, \bar{v}^r, \bar{w}^r) \neq 0, \ r = 1, ..., N\}$, numbers $\xi_- \overset{\triangle}{=} \inf(G_{\sigma^0}, S')$ and $\xi_+ \overset{\triangle}{=} \sup(G_{\sigma^0}, S')$ $(S' \overset{\triangle}{=} \{(x, y, z, \alpha, \beta) \in I\!\!R^{m+n_1+n_2+2} \mid (x, y, z) \in S\})$, an accuracy of feasibility ε_0, and the algorithm parameters $M \in \mathbb{N}$ and $\nu \in I\!\!R_+$.

Step 0. Define $k := 1$, $(\bar{x}^k, \bar{y}^k, \bar{z}^k) := (x^0, y^0, z^0)$, $\bar{\alpha}_k := \langle \bar{y}^k, B_1 \bar{z}^k \rangle$, $\bar{\beta}_k := \langle \bar{y}^k, B_2 \bar{z}^k \rangle$ and $\bar{\theta}^k := (\bar{x}^k, \bar{y}^k, \bar{z}^k, \bar{\alpha}_k, \bar{\beta}_k)$; $r := 1$, $\xi := \xi_-$, $\triangle\xi = (\xi_+ - \xi_-)/M$; $\sigma_k := \sigma^0$, $G_k(\cdot) := G_{\sigma_k}(\cdot)$, $H_k(\cdot) := H_{\sigma_k}(\cdot)$.

Step 1. Utilizing the SPLSM from the preceding section, beginning from the point $(\bar{x}^k, \bar{y}^k, \bar{z}^k)$ with penalty parameter $\sigma = \sigma^0$, build a τ_k-critical point $\theta^k := (x^k, y^k, v^k, \alpha_k, \beta_k) \in S'$ in the problem $(\mathcal{DC}(\sigma))$ with an updated, generally different, penalty parameter $\sigma_k \geq \sigma^0$ ($\sigma_k := \bar{\sigma}_k$, where $\bar{\sigma}_k$ denotes the final penalty parameter value in the local search). So that $\zeta_k := \Phi_k(\theta^k) \leq \Phi_k(\bar{\theta}^k)$, satisfying $W(\theta^k) \leq \varepsilon_0$.

Step 2. Using the triple $(\bar{u}^r, \bar{v}^r, \bar{w}^r) \in Dir$ construct the point (u^r, v^r, w^r) from the approximation $\mathcal{A}_k = \{(u^r, v^r, w^r) \mid H_k(u^r, v^r, w^r) = \xi - \zeta_k, r = 1, ..., N\}$ of the level surface $\mathcal{U}(\zeta_k) = \{(x, y, z) \mid H_k(x, y, z) = \xi - \zeta_k\}$ pertaining to the convex function $H_k(\cdot)$. Compute $\alpha_r := \langle \bar{y}^r, B_1 \bar{z}^r \rangle$, $\beta_r := \langle \bar{y}^r, B_2 \bar{z}^r \rangle$, $\theta^r := (u^r, v^r, w^r, \alpha_r, \beta_r)$.

Step 3. If $G_k(\theta^r) > \xi + \nu\xi$, $r < N$ and $\xi < \xi_+$, then proceed to **Step 2** by incrementing $r := r + 1$.

Step 4. If $G_k(\theta^r) > \xi + \nu\xi$, $r = N$ and $\xi < \xi_+$, then set $r := 1$, $\xi := \xi + \triangle\xi$ and return to **Step 2**.

Step 5. If $G_k(\theta^r) > \xi + \nu\xi$, $r = N$ and $\xi = \xi_+$, then stop; $(x^k, y^k, z^k, \alpha_k, \beta_k)$ is the resultant solution to the problem $(\mathcal{DC}(\sigma))$.

Step 6. Solve the problem $(\mathcal{AP}_k\mathcal{L}_r) = (\mathcal{AP}_{\sigma_k}\mathcal{L}(\theta^r))$ to obtain an approximate δ_k-solution $(\bar{x}^r, \bar{y}^r, \bar{z}^r, \bar{\alpha}_r, \bar{\beta}_r)$ for the linearized problem

$$G_k(\theta) - \langle \nabla H_k(\theta^r), \theta \rangle \downarrow \min_{\theta}, \quad \theta \in S'. \qquad (\mathcal{P}_{\sigma_k}\mathcal{L}(\theta^r))$$

Step 7. Starting from the point $(\bar{x}^r, \bar{y}^r, \bar{z}^r)$ with the penalty parameter value $\sigma = \sigma^0$, employ the SPLSM to construct the τ_k-critical point in problem $(\mathcal{DC}(\sigma))$ as $\hat{\theta}^r := (\hat{x}^r, \hat{y}^r, \hat{z}^r, \hat{\alpha}_r, \hat{\beta}_r) \in S'$ with a new penalty parameter value $\sigma_r \geq \sigma^0$ ($\sigma_r := \bar{\sigma}_r$, where $\bar{\sigma}_r$ is the final penalty parameter value from the local search).

Step 8. If $\Phi_k(\hat{\theta}^r) < \Phi(\theta^k)$, then set $(\bar{x}^{k+1}, \bar{y}^{k+1}, \bar{z}^{k+1}) := (\hat{x}^r, \hat{y}^r, \hat{z}^r)$, $\xi := \xi_-$, $\sigma^0 := \max\{\sigma_k, \sigma_r\}$, $k := k + 1$, $r := 1$ and move to **Step 1**.

Step 9. If $\Phi_k(\hat{\theta}^r) \geq \Phi(\theta^k), r < N$, then increase $r := r + 1$ and go to **Step 2**.

Step 10. If $\Phi_k(\hat{\theta}^r) \geq \Phi(\theta^k)$, $r = N$ and $\xi < \xi_+$, then set $\xi := \xi + \triangle\xi$, $r := 1$ and return to **Step 2**.

Step 11. If $\Phi_k(\hat{\theta}^r) \geq \Phi(\theta^k)$, $r = N$ and $\xi = \xi_+$, then stop. $\theta^k = (x^k, y^k, z^k, \alpha_k, \beta_k)$ is the resulting solution to the problem $(\mathcal{DC}(\sigma))$.   □

It is easy to see that the main work on increasing the penalty parameter in the presented GSS mainly lies in the local search. At the same time, within each iteration of the global search ($k = 1, 2, ...$), the SPLSM not only generates a critical vector but also an ε_0-feasible one. The SPLSM then constructs the penalty parameter σ_k at each GSS iteration by forming a "local" sequence $\{\sigma_s\}$

of penalty parameters, where $\sigma_k := \bar{\sigma}_k$, with $\bar{\sigma}_k = \sigma_+$ being the maximum penalty parameter value in the local search at step 1. Similarly, $\sigma_r := \bar{\sigma}_r$ at step 7. These values, σ_k and σ_r, are used for finding points that violate the Global Optimality Conditions for problem $(\mathcal{DC}(\sigma))$ and improving the objective function value. The convergence of this global search scheme is substantiated in [30].

Also, it's crucial to note that the Global Search Scheme does not follow a traditional algorithmic structure, as certain steps lack specific instructions. For instance, the processes of selecting algorithm parameters, constructing a starting point and an approximation of the function's level surface $H_k(\cdot)$, implementing a local search, and solving linearized problems are not explicitly defined. The resolution of these uncertainties depends on actual numerical data and draws from previous computational experiences in handling nonconvex problems [11, 19, 23, 26, 31]. Therefore, the ultimate decisions in these aspects must be confirmed and justified through successful computational experiments.

6   Concluding Remarks

This paper introduced the novel Special Penalty Local Search Methods and Global Search Scheme designed for one class of bilevel optimization problems (BOPs) in an optimistic statement with equilibrium at the lower level. These methods are grounded in A.S. Strekalovsky's Global Search Theory for general d.c. optimization problems and the Exact Penalization Theory.

To operationalize these theories, first, the initial problem is transformed into a nonconvex single-level equivalent by leveraging the optimality conditions for non-normalized games [21]. Subsequently, necessary theoretical frameworks are built to develop the aforementioned methods. Key emphasis is placed on constructing d.c. decompositions of nonconvex functions in the problem formulations, along with the local search. An inherent highlight of the proposed local and global search schemes lies in their possibility to vary the penalty parameter within the methods themselves.

The primary target of future research endeavors in this domain is constructing test cases falling within the class of these bilevel problems, alongside implementing and evaluating the computational efficacy of the local and global search techniques. Drawing on our recent computational experiences (evidenced by our results in solving other bilevel problems [11, 32, 33]), we anticipate that the developed methodology holds promise for efficiently addressing bilevel problems with an equilibrium problem at the lower level.

Acknowledgement. The research was funded by the Ministry of Education and Science of the Russian Federation within the framework of the project "Theoretical foundations, methods and high-performance algorithms for continuous and discrete optimization to support interdisciplinary research" (No. of state registration: 121041300065-9, code FWEW-2021-0003).

References

1. Dempe, S.: Foundations of Bilevel Programming. Kluwer Academic Publishers, Dordrecht (2002)
2. Dempe, S., Zemkoho, A. (eds.): Bilevel Optimization: Advances and Next Challenges. Springer, Cham (2020)
3. Ramos, M., Boix, M., Aussel, D., Montastruc, L., Domenech, S.: Water integration in eco-industrial parks using a multi-leader-follower approach. Comput. Chem. Eng. **87**, 190–207 (2016). https://doi.org/10.1016/j.compchemeng.2016.01.005
4. Yang, Z., Ju, Y.: Existence and generic stability of cooperative equilibria for multi-leader-multi-follower games. J. Glob. Optim. **65**, 563–573 (2016). https://doi.org/10.1007/s10898-015-0393-1
5. Hu, M., Fukushima M.: Multi-leader-follower games: models, methods and applications. J. Oper. Res. Soc. Japan **58**, 1–23 (2015). https://doi.org/10.15807/jorsj.58.1
6. Zewde, A.B., Kassa, S.M.: Multi-parametric approach for multilevel multi-leader-multi-follower games using equivalent reformulations. J. Math. Comput. Sci. **11**(3), 2955–2980 (2021). https://doi.org/10.28919/jmcs/5641
7. Jiang, S., Li, X., Wu, J.: Multi-leader multi-follower Stackelberg game in mobile blockchain mining. IEEE Trans. Mob. Comput. **21**(6), 2058–2071 (2022). https://doi.org/10.1109/TMC.2020.3035990
8. Aussel, D., Lepaul, S., von Niederhausern, L.: A multi-leader-follower game for energy demand-side management. Optim. **72**(2), 351–381 (2023). https://doi.org/10.1080/02331934.2021.1954179
9. Ramos, M.A., Boix, M., Aussel, D., Montastruc, L.: Development of a multi-leader multi-follower game to design industrial symbioses. Comput. Chem. Eng. **183**, 108598 (2024). https://doi.org/10.1016/j.compchemeng.2024.108598
10. Dempe, S., Kalashnikov, V.V., Perez-Valdes, G.A., Kalashnykova, N.: Bilevel Programming Problems: Theory, Algorithms and Applications to Energy Networks. Springer, Heidelberg (2015)
11. Strekalovsky, A.S., Orlov, A.V.: Linear and Quadratic-Linear Problems of Bilevel Optimization. SB RAS publishing, Novosibirsk (2019). (in Russian)
12. Mitsos, A., Lemonidis, P., Barton, P.I.: Global solution of bilevel programs with a nonconvex inner program. J. Glob. Optim. **42**, 475–513 (2008). https://doi.org/10.1007/s10898-007-9260-z
13. Lin, G.-H., Xu, M., Ye, J.J.: On solving simple bilevel programs with a nonconvex lower level program. Math. Program. Ser. A **144**, 277–305 (2014). https://doi.org/10.1007/s10107-013-0633-4
14. Zhu, X., Guo, P.: Approaches to four types of bilevel programming problems with nonconvex nonsmooth lower level programs and their applications to newsvendor problems. Math. Methods Oper. Res. **86**, 255–275 (2017). https://doi.org/10.1007/s00186-017-0592-2
15. Liu, R., Liu, Y., Zeng, Sh., Zhang, J.: Towards gradient-based bilevel optimization with non-convex followers and beyond. In: Ranzato M. et al. (eds.) Advances in Neural Information Processing Systems, vol. 34, pp. 8662–8675. Curran Associates, Inc. (2021)
16. Arbel, M., Mairal, J.: Non-convex bilevel games with critical point selection maps. In: Koyejo S. et al. (eds.) Advances in Neural Information Processing Systems, vol. 35, pp. 8013–8026. Curran Associates, Inc. (2022)

17. Huang, F.: On Momentum-Based Gradient Methods for Bilevel Optimization with Nonconvex Lower-Level. arXiv:2303.03944v4 (2023). https://doi.org/10.48550/arXiv.2303.03944

18. Mazalov, V.: Mathematical Game Theory and Applications. John Wiley & Sons, New York (2014)

19. Strekalovsky, A.S., Orlov, A.V.: Bimatrix games and bilinear programming. Fiz-MatLit, Moscow (2007). (in Russian)

20. Orlov, A.V., Gruzdeva, T.V.: The Local and Global Searches in Bilevel Problems with a Matrix Game at the Lower Level. In: Khachay M. et al. (eds.) MOTOR 2019. LNCS, vol. 11548, pp. 172–186. Springer, Cham (2019). https://doi.org/10.1007/978-3-030-22629-9_13

21. Orlov A.V.: On solving bilevel optimization problems with a nonconvex lower level: the case of a bimatrix game. In: Pardalos P., Khachay M., Kazakov A. (Eds.) MOTOR 2021. LNCS, vol. 12755, pp. 235–249. Springer, Cham (2021). https://doi.org/10.1007/978-3-030-77876-7_16

22. Horst, R., Tuy, H.: Global Optimization. Deterministic approaches. Springer, Berlin (1993)

23. Strekalovsky, A.S.: Elements of nonconvex optimization. Nauka, Novosibirsk (2003). (in Russian)

24. Nocedal, J., Wright, S.J.: Numerical Optimization. Springer, Heidelberg (2000)

25. Bonnans, J.-F., Gilbert, J.C., Lemarechal, C., Sagastizabal, C.A.: Numerical Optimization: Theoretical and Practical Aspects. Springer, Heidelberg (2006)

26. Strekalovsky, A.S.: On Solving Optimization Problems with Hidden Nonconvex Structures. In: Rassias, T.M., Floudas, C.A., Butenko, S. (Eds.) Optimization in Science and Engineering, pp. 465–502, Springer, N.Y. (2014). https://doi.org/10.1007/978-1-4939-0808-0_23

27. Strekalovsky, A.S.: Global optimality conditions and exact penalization. Optim. Lett. **13**, 597–615 (2019). https://doi.org/10.1007/s11590-017-1214-x

28. Strekalovsky, A.S.: On a global search in D.C. optimization problems. In: Jacimovic, M., Khachay, M., Malkova, V., Posypkin, M. (Eds.) Optimization and Applications. OPTIMA 2019. CCIS, vol. 1145, pp. 222–236. Springer, Cham (2020). https://doi.org/10.1007/978-3-030-38603-0_17

29. Strekalovsky, A.S.: On global optimality conditions for d.c. minimization problems with d.c. constraints. J. Appl. Numer. Optim. **3**(1), 175–196 (2021). https://doi.org/10.23952/jano.3.2021.1.10

30. Strekalovsky, A.S.: Minimizing Sequences in a Constrained DC Optimization Problem. Proc. Steklov Inst. Math. **323**(suppl. 1), S255–S278 (2023)

31. Orlov, A.V., Strekalovsky, A.S., Batbileg, S.: On computational search for Nash equilibrium in hexamatrix games. Optim. Lett. **10**(2), 369–381 (2016). https://doi.org/10.1007/s11590-014-0833-8

32. Orlov, A.V.: The global search theory approach to the bilevel pricing problem in telecommunication networks. In: Kalyagin, V.A., et al. (eds.) Computational Aspects and Applications in Large Scale Networks, pp. 57–73, Springer International Publishing AG (2018). https://doi.org/10.1007/978-3-319-96247-4_5

33. Strekalovsky, A.S., Orlov, A.V.: Global search for bilevel optimization with quadratic data. In: Dempe, S., Zemkoho A. (Eds.) Bilevel Optimization: Advances and Next Challenges, pp. 313–334, Springer, Cham (2020). https://doi.org/10.1007/978-3-030-52119-6_11

34. Tao, P.D., Souad, L.B.: Algorithms for solving a class of non convex optimization. Methods of subgradients. In: Hiriart-Urruty J.-B. (ed.) Fermat days 85, pp. 249–271. Elservier Sience Publishers B.V., North Holland (1986)

35. Strekalovsky, A.S.: Local search for nonsmooth DC optimization with DC equality and inequality constraints. In: Bagirov, A.M., et al. (eds.) Constraints Numerical Nonsmooth Optimization: State of the Art Algorithms, pp. 229–261, Springer, Cham (2020). https://doi.org/10.1007/978-3-030-34910-3_7
36. Barkova, M.V., Strekalovsky, A.S.: Computational study of local search methods for a d.c. optimization problem with inequality constraints. In: Olenev, N.N. et al. (eds.) Optimization and Applications, LNCS 13078, pp. 94–109, Springer, Cham (2021). https://doi.org/10.1016/j.apm.2017.07.031
37. Byrd, R.H., Nocedal, J., Waltz, R.A.: Steering exact penalty methods for nonlinear programming. Optim. Methods Softw. **23**, 197–213 (2008). https://doi.org/10.1080/10556780701394169
38. Byrd, R.H., Lopez-Calva, G., Nocedal, J.: A line search exact penalty method using steering rules. Math. Programming, Ser. A **133**, 39–73 (2012). https://doi.org/10.1007/s10107-010-0408-0

Differential Network Games with Different Types of Players Behavior

Leon Petrosyan[ID] and Yaroslavna Pankratova[(✉)][ID]

St Petersburg State University, Saint-Petersburg, Russia
{l.petrosyan,y.pankratova}@spbu.ru

Abstract. In the paper, a cooperative differential network game with infinite duration in which players follow different types of behavior (to cooperate or to act individually in their own interests) is considered. As solutions the core and the Shapley value are proposed, and non-emptiness of the core is proved. The results are illustrated on an example.

Keywords: Dynamic network game · the Shapley value · core

Differential n-person network games (see [2,3,5–7,9,11,15,17,20]) with infinite duration and different types of players behaviour (see [1]) are considered. It is supposed that the set of players N ($|N| = n$) is represented as union of two disjoint subsets $N = S \cup M$. The set S contains players who wish to cooperate to achieve the maximal joint payoff and after that redistribute this payoff among members of this set S using optimally principles from the classical cooperative game theory. The set M consists of players who play individually trying to maximize each one his own payoff. Since players are connected in the network the payoff of each player depends upon behaviour of his neighbours.

It is assumed that each player can cut connections with his neighbours at any time instant. In the special case, when payoff functions take positive values this assumption greatly simplifies the construction of characteristic function of the game and, as result, the calculation of the corresponding optimality principles.

1 Differential Network Games

Consider a class of n-person differential games on network. Players are connected in a network system. We use $N = \{1, 2, \ldots, n\}$ to denote the set of players in the network. The nodes of the network are used to represent players from the set N. We also denote the set of nodes by N and denote the set of all arcs in network N by L. The arcs in L are the arc $(i, j) \in L$ for players $i, j \in N$. For notational convenience, we denote the set of players connected to player i as $K(i) = \{j : arc(i, j) \in L, j \neq i\}$, for $i \in N$.

Supported by the Russian Science Foundation grant No. 22-11-00051, https://rscf.ru/en/project/22-11-00051/..

Let $x^i(\tau) \in R^m$ be the state variable of player $i \in N$ at time τ, and $u^i(\tau) \in U^i \subset R^k$ the control variable of player $i \in N$, U_i is a compact set, and $u_i(\tau)$ is piecewise continuous function with a finite number of discontinuity points, $i \in N$.

Every player $i \in N$ can cut the connection with any other player from the set $K(i)$ at any instant of time.

The state dynamics of the game is

$$\dot{x}^i(\tau) = f^i(x^i(\tau), u^i(\tau)), \ x^i(t_0) = x_0^i, \ \text{for } \tau \in [t_0, \infty) \ \text{and } i \in S \cup M. \quad (1)$$

Functions $f^i(x^i, u^i)$ are continuously differentiable in x^i and u^i and satisfy the conditions of existence, uniqueness and continuability of the solution on the interval (t_0, ∞) for all admissible piecewise continuous controls with a finite number of discontinuity points. For notational convenience, we use $x(t)$ to denote the vector $(x^1(t), x^2(t), \cdots, x^n(t))$.

As usual in classical differential game theory we suppose that at each time instant $t \in [t_0, \infty)$ players $i \in N$ have information about this time instant and state variable $x(t)$ and based on this information choose their controls and the action to cut or to keep the connection with players from $K(i)$.

We consider a special case, when the payoff of player i depends upon his state variable and the state variables of players from the set $K(i)$. Thus, if the connections remain valued, the payoff of player i is given as

$$H_i(x_0^1, \ldots, x_0^n, u^1, \ldots, u^n) = \sum_{j \in K(i)} \int_{t_0}^{\infty} e^{-\rho(\tau - t_0)} h_i^j(x^i(\tau), x^j(\tau)) d\tau, \ i \in N, \quad (2)$$

provided that the players do not interrupt communication. Since the game has infinite duration, the payoffs are discounted with discount factor $e^{-\rho(\tau - t_0)}$, $\rho > 0$. Players have the ability to interrupt connection with neighboring players at any time moment t, $t \in [t_0, +\infty)$. In case, the player i interrupts the connection with player j at some time instant t functions $h_i^j(x^i(\tau), x^j(\tau))$ and $h_j^i(x^j(\tau), x^i(\tau))$ will be set 0 for all $\tau \geq t$.

If $K(i) = \emptyset$ (if player i has no connection with other players) then

$$H_i(x_0^1, \ldots, x_0^n, u^1, \ldots, u^n) = 0.$$

The term $h_i^j(x^i(\tau), x^j(\tau))$ is the instantaneous gain that player i can obtain through network links with player $j \in K(i)$.

The functions $h_i^j(x^i(\tau), x^j(\tau))$, for $j \in K(i)$, $i \in N$ are positive, continuous and uniformly bounded.

From (2) we can see that the payoff of player i is computed as sum of payoffs which he gets interacting with players $j \in N\backslash\{i\}$.

Let $x_0 = (x_0^1, \ldots, x_0^n)$ and denote this game as $\Gamma(x_0)$.

1.1   Cooperation and Characteristic Function

Consider the cooperative version of the game. As noted earlier, it is assumed that the set of players is divided into two disjoint subsets: the coalition S (player S) consists of players striving for cooperation in terms of maximizing the total joint payoff and its subsequent redistribution between cooperating players in accordance with certain optimality principle, the set M consists of players acting individually in order to maximize their own payoffs.

Definition 1. *Strategy profile* $((u^1)^*, \ldots, (u^i)^*, \ldots, (u^n)^*)$ *is called Nash equilibrium in game* $\Gamma(x_0)$ *if inequalities*

$$H_i(x_0^1, \ldots, x_0^n; (u^1)^*, \ldots, (u^i)^*, \ldots, (u^m)^*) \geq$$

$$H_i(x_0^1, \ldots, x_0^n; (u^1)^*, \ldots, (u^{i-1})^*, (u^i)(u^{i+1})^*, \ldots, (u^n)^*)$$

are satisfied for every (u^i), $i \in N$.

Suppose that in the game $\Gamma(x_0)$ there exists Nash equilibrium in feedback strategies and denote the corresponding strategy profile by

$$((u^1)^*, \ldots, (u^i)^*, \ldots, (u^n)^*).$$

As usual in n-person differential game theory literature, we suppose that we consider only those strategy profiles in feedback strategies for which there exist unique solution of motion equations for any initial conditions.

Denote by $V(x_0, t_0; S)$ the joint payoff of the players from the coalition S under this strategy profile

$$V(x_0, t_0; S) = \sum_{i \in S} \sum_{j \in S \cap K(i)} \int_{t_0}^{\infty} e^{-\rho(\tau - t_0)} h_i^j(x^{i*}(\tau), x^{j*}(\tau)) d\tau +$$

$$\sum_{i \in S} \sum_{j \in M \cap K(i)} \int_{t_0}^{\infty} e^{-\rho(\tau - t_0)} h_i^j(x^{i*}(\tau), x^{j*}(\tau)) d\tau =$$

$$\sum_{i \in S} \sum_{j \in (S \cup M) \cap K(i)} \int_{t_0}^{\infty} e^{-\rho(\tau - t_0)} h_i^j(x^{i*}(\tau), x^{j*}(\tau)) d\tau \qquad (3)$$

subject to dynamics (1), where $x^{i*}(\tau)$ is the corresponding Nash trajectory.

Note that there can be many Nash equilibria in this games $\Gamma(x_0)$ [18]. Consider now one special case when the maximal joint (the sum) payoff of players $j \in S \cup M$ is achieved in Nash equilibrium. The following proposition shows that this is possible.

Proposition 1. *The cooperative outcome (maximal sum of players' payoffs) in the game* $\Gamma(x_0)$ *can be achieved in Nash equilibrium.*

Proof. Consider the following behaviour of players in $\Gamma(x_0)$. Suppose that all players starting from time instant t_0 choose cooperative controls (controls which maximise the sum of players payoffs in $\Gamma(x_0)$)

$$\bar{u}(t) = (\bar{u}_1(t), \dots, \bar{u}_i(t), \dots, \bar{u}_n(t)),$$

then the game will evolve along the corresponding cooperative trajectory $\bar{x}(t)$. If in some time instant $\bar{t}$ the state $x(\bar{t})$ of the game will be seen outside the cooperative trajectory $x(\bar{t}) \neq \bar{x}(t)$, then players strategy from time instant $\bar{t}$ is to cut connections with all neighbours. Since the motion trajectory of the game is continuous function the time instant $\bar{t}$ can be choosen very close to the first moment of the deviation of player $i_0 \in N$ from cooperative behaviour $\bar{u}_{i_0}(t)$.

Now suppose that one player $i_0 \in N$ deviates at some time instant t' from the cooperative behaviour $\bar{u}_{i_0}(t)$ (choosing $u_{i_0}(t') = \bar{u}_{i_0}(t')$). Then if

$$f^{i_0}(x^{i_0}(t'), u^{i_0}(t')) \neq f^{i_0}(\bar{x}^{i_0}(t'), \bar{u}^{i_0}(t'))$$

the motion of player $i_0 \in N$ will change and as result $x(t) \neq \bar{x}(t)$ for any t, $t > t'$ sufficiently close to t'. Since players have information about the state variable x at each time instant of the game there will exist such $\bar{t}$ close to t', $(\bar{t} > t')$ that they will be informed about $x(\bar{t}) \neq \bar{x}(\bar{t})$. Then according to the prescribed behaviour each of players $i \in N \setminus \{i_0\}$ will cut connections with all his neighbours, and the player i_0 deviating from cooperative trajectory will get payoff equal to 0 till the end of the game. On the time interval $[t', \bar{t}]$, the deviating player i_0 can get not more than $V(x_0, t_0; N)$ (here $V(x_0, t_0; N)$ is the total maximal payoff of all players in the game).

Since the function $h_i^j > 0$ for $j \in K(i)$ and the infinite duration of the game it is clear that the player $i_0 \in N$ by not deviating will get some positive amount $A > 0$. And if we take

$$\bar{t} - t' < \frac{A}{V(x_0, t_0; N)}$$

player i_0 will lose by deviating from cooperative trajectory. This means that the proposed behaviour constitutes a Nash equilibrium in $\Gamma(x_0)$ with cooperative outcome.

In this case, the set of Nash equilibrium trajectories $x^{i*}(\tau)$, $i \in S \cup M = N$ (here we consider the game $\Gamma(x_0)$) will coincide with the cooperative trajectories maximizing the joint payoff of players from the sets $S \cup M$. Then the whole construction of finding a solution is significantly simplified, and instead of finding Nash equilibria in a game with players $\{1, \dots, i, \dots, n\}$ it is sufficient to conduct one maximization operation

$$V(x_0, t_0; N) = \max_{u_i,\ i \in S \cup M} \sum_{i \in N} \sum_{j \in (M \cup S) \cap K(i)} \int_{t_0}^{\infty} e^{-\rho(t-t_0)} h_i^j(x^i(\tau), x^j(\tau)) d\tau =$$

$$\sum_{i \in N} \sum_{j \in (M \cup S) \cap K(i)} \int_{t_0}^{\infty} e^{-\rho(t-t_0)} h_i^j(\bar{x}^i(\tau), \bar{x}^j(\tau)) \tag{4}$$

subject to dynamics (1). We denote by $\bar{x}(t) = (\bar{x}^1(t), \bar{x}^2(t), \cdots, \bar{x}^n(t))$ and by $\bar{u}(t) = (\bar{u}^1(t), \bar{u}^2(t), \cdots, \bar{u}^n(t))$ the optimal cooperative trajectory and the optimal cooperative control in the problem of maximization (4) subject to (1).

Note that in (4), instead of Nash equilibrium trajectories (see (3)), we have cooperative trajectories (Nash equilibrium trajectories which correspond to considered above Nash eqlibrium coincide with cooperative trajectories) obtained as a result of one maximization procedure defined above. Thus, in this special case, it turns out that the implementation of the optimality principles (Core or the Shapley value), based on the construction of characteristic functions does not require finding Nash equilibria or minimax or maxmin solutions, which in case of deferential games is an almost unsolvable problem.

We will be interested in the values of characteristic function defined on subsets A of the set $S \subset N$. Define these values as

$$V(x_0, t_0; A) = \sum_{i \in A} \sum_{j \in (A \cup M) \cap K(i)} \int_{t_0}^{\infty} e^{-\rho(\tau - t_0)} h_i^j(\bar{x}^i(\tau), \bar{x}^j(\tau)) d\tau. \qquad (5)$$

From (5) we see that the values of characteristic function consist of two terms. The first term corresponds to the joint payoff of players from A who have connections with each other when using strategies corresponding to the above-described Nash equilibrium strategy profile. The second term is the joint payoff of players from A, obtained when interacting with players from the coalition M. Note that the expression (5) does not include the payoff of players from A interacting with players from $S \setminus A$, since by the definition $V(x_0, t_0; A)$ is a guaranteed payoff of coalition A under the worst behavior of players from $S \setminus A$. And since the worst behavior is to cut off all connections with players from A, the corresponding payoffs are zero (note that instantaneous payoffs are positive).

For simplicity in notations, we denote the gain that player i can obtain through the network link with player $j \in K(i)$ as

$$\alpha_{ij}(\bar{x}(t), t) = \int_t^{\infty} e^{-\rho(\tau - t_0)} h_i^j(\bar{x}^i(\tau), \bar{x}^j(\tau)) d\tau, \qquad (6)$$

for $t \in [t_0, \infty]$.

Since (6) we have

$$V(x_0, t_0; A) = \sum_{i \in A} \sum_{j \in (A \cup M) \cap K(i)} \alpha_{ij}(x_0, t_0). \qquad (7)$$

Definition 2. *The characteristic function $V(S)$ is called* convex *if*

$$V(x_0, t_0; S_1 \cup S_2) \geq V(x_0, T; S_1) + V(x_0, t_0; S_2),$$
$$-V(x_0, t_0; S_1 \cap S_2).$$

The following statements can be proved.

Proposition 2. *Characteristic function $V(x_0, t_0; A)$ is convex.*

Proof of the proposition 2 is similar to corresponding proposition from [14], where the differential games with finite time horizon were considered.

The imputation of the cooperatives game is a vector $b = (\beta_1, \ldots, \beta_n)$ such that

$$\sum_{i \in N} \beta_i = V(x_0, t_0; N), \ \beta_i \geq V(x_0, t_0; \{i\}), \ i \in N.$$

The subset C of imputation set is called core if for each $b \in C$ the following condition holds

$$\sum_{i \in S} \beta_i \geq V(x_0, t_0; S), \ S \subset N. \tag{8}$$

The cooperative game is called convex if its characteristic function is convex. It is well known that in convex game the core is not empty, and the Shapley value belongs to the core [8].

Similarly, the convexity of the above defined characteristic function holds in any subgame along the cooperative trajectory $\bar{x}(t)$

$$V(\bar{x}(t), t; S_1 \cup S_2) \geq V(\bar{x}(t), t; S_1) + V(\bar{x}(t), t; S_2),$$
$$-V(\bar{x}(t), t; S_1 \cap S_2), \text{ for } t \in [t_0, \infty).$$

Proposition 3. *Vector* $(\bar{H}_1, \ldots, \bar{H}_s)$ *of players' payoffs under cooperation*

$$\bar{H}_i = \sum_{j \in (S \cup M) \cap K(i)} \alpha_{ij}(t_0, x_0), \ i = 1, \ldots, s.$$

belongs to the Core in the game with the characteristic function $V(x_0, t_0; A)$ *(see (5)).*

Proof. We have

$$\bar{H}(A) = \sum_{i \in A} \bar{H}_i = \sum_{i \in A} \sum_{j \in (S \cup M) \cap K(i)} \alpha_{ij}(t_0, x_0) \geq$$

$$\sum_{i \in A} \sum_{j \in (A \cup M) \cap K(i)} \alpha_{ij}(x_0, t_0) = V(x_0, t_0; A).$$

The above inequality is true for $A \subset S$ since $A \cap K(i) \subset S \cap K(i)$ and $h_i^j > 0$, and shows that vector $(\bar{H}_1, \ldots, \bar{H}_s)$ belongs to the core.

2   The Shapley Value

In this section, we develop a dynamic Shapley value [12,13,16] imputation using the defined characteristic function.

$$Sh_i(x_0, t_0) = \sum_{\substack{A \subset S, \\ A \ni i}} \frac{(|A| - 1)!(s - |A|)!}{s!} \times [V(x_0, t_0, A) - V(x_0, t_0, A \backslash \{i\})], \tag{9}$$

for $i \in S$.

Using (6), we can rewrite the formula for the Shapley value

$$Sh_i(x_0, t_0) = \sum_{\substack{A \subseteq S \\ A \ni i}} \frac{(|A| - 1)!(s - |A|)!}{s!} \times \left\{ \sum_{l \in A} \sum_{j \in K(l) \cap (A \cup M)} \alpha_{lj}(x_0, t_0) - \right.$$

$$\left. \sum_{l \in A \setminus \{i\}} \sum_{j \in K(l) \cap ((A \setminus \{i\}) \cup M)} \alpha_{lj}(x_0, t_0) \right\} . =$$

$$= \sum_{\substack{A \subseteq S \\ A \ni i}} \frac{(|A| - 1)!(s - |A|)!}{s!} \times$$

$$\left\{ \sum_{l \in A} \sum_{j \in K(l) \cap (A \cup M)} \int_{t_0}^{\infty} e^{-\rho(\tau - t_0)} h_i^j(\bar{x}^i(\tau), \bar{x}^j(\tau)) d\tau - \right.$$

$$\left. \sum_{l \in A \setminus \{i\}} \sum_{j \in K(l) \cap ((A \setminus \{i\}) \cup M)} \int_{t_0}^{\infty} e^{-\rho(\tau - t_0)} h_i^j(\bar{x}^i(\tau), \bar{x}^j(\tau)) d\tau \right\} .$$

Proposition 4. *The Shapley value imputation in (9) satisfies the time consistency property (see [12]).*

Proof. By direct computation we get

$$Sh_i(x_0, t_0) = \sum_{\substack{A \subseteq S \\ A \ni i}} \frac{(|A| - 1)!(s - |A|)!}{s!} \times$$

$$\left\{ \sum_{l \in A} \sum_{j \in K(l) \cap (A \cup M)} \int_{t_0}^{t} e^{-\rho(\tau - t_0)} h_i^j(\bar{x}^i(\tau), \bar{x}^j(\tau)) d\tau - \right. \tag{10}$$

$$\left. \sum_{l \in A \setminus \{i\}} \sum_{j \in K(l) \cap ((A \setminus \{i\}) \cup M)} \int_{t_0}^{t} e^{-\rho(\tau - t_0)} h_i^j(\bar{x}^i(\tau), \bar{x}^j(\tau)) d\tau \right\} +$$

$$e^{-\rho(t - t_0)} Sh_i(\bar{x}(t), t), \ t \in [t_0, +\infty]$$

which exhibits the time consistency property of the Shapley value imputation.

Time-consistency property for optimal solutions in differential cooperative games was first introduced in [10] and for games with infinite horizon in [12]. In differential cooperative games the time-consistency of solutions (in our case the Shapley Value) means that it is possible to redistribute the instantaneous payments players get according to the Shapley value along the cooperative trajectory in such a way that the amount they expect to get in any subgame with initial conditions on cooperative trajectory will coincide with the Shapley value computed for this subgame. As we see from (10) this takes place in our case without even any necessary redistribution if instantaneous payments.

This is a very rare event that the Shapley value measure itself in a dynamic framework fulfils the property of time consistency (see existing dynamic Shapley value measures which do not share this property in [4, 19].

Example. Consider the following $|N| = 5$ player network game (see Fig. 1).

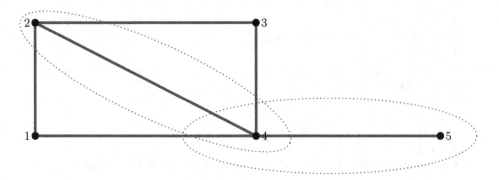

Fig. 1. 5 player network game

Suppose that the set of players which wish to cooperate consists of 2, 4 and 5 players, i.e. $S = \{2, 4, 5\}$ (in Fig. 1 nodes 2, 4 and 5 are inside dotted lines), and $M = \{1, 3\}$.

The values of characteristic function for $A \subset S$ are

$V(\{2\}) = \alpha_{23} + \alpha_{21}$,
$V(\{4\}) = \alpha_{43} + \alpha_{41}$,
$V(\{5\}) = 0$,
$V(\{2, 4\}) = \alpha_{24} + \alpha_{42} + \alpha_{21} + \alpha_{23} + \alpha_{41} + \alpha_{43}$,
$V(\{2, 5\}) = \alpha_{21} + \alpha_{23}$,
$V(\{4, 5\}) = \alpha_{45} + \alpha_{54} + \alpha_{41} + \alpha_{43}$,
$V(\{2, 4, 5\}) = \alpha_{24} + \alpha_{42} + \alpha_{45} + \alpha_{54} + \alpha_{21} + \alpha_{23} + \alpha_{41} + \alpha_{43}$.

Computing the Shapley Value for players $i \in S = \{2, 4, 5\}$ we get

$$Sh_2 = \alpha_{21} + \alpha_{23} + \frac{\alpha_{24} + \alpha_{42}}{2},$$

$$Sh_4 = \alpha_{41} + \alpha_{43} + \frac{\alpha_{45} + \alpha_{54} + \alpha_{24} + \alpha_{42}}{2},$$

$$Sh_5 = \frac{\alpha_{54} + \alpha_{45}}{2}.$$

For payoffs of players $i \in M = \{1, 3\}$ we get following expressions

$$H_1 = \alpha_{12} + \alpha_{14}, \quad H_3 = \alpha_{32} + \alpha_{34}.$$

Suppose that we change the set of players which wish to cooperate increasing it. Now we add player 1 to the coalition S and calculate the value of characteristic function and the corresponding Shapley value.

$V(\{1\}) = 0,$
$V(\{2\}) = \alpha_{23},$
$V(\{4\}) = \alpha_{43},$
$V(\{5\}) = 0,$
$V(\{1,2\}) = \alpha_{12} + \alpha_{21} + \alpha_{23},$
$V(\{1,4\}) = \alpha_{14} + \alpha_{41} + \alpha_{43},$
$V(\{1,5\}) = 0,$
$V(\{2,4\}) = \alpha_{24} + \alpha_{42} + \alpha_{23} + \alpha_{43},$
$V(\{2,5\}) = \alpha_{23},$
$V(\{4,5\}) = \alpha_{45} + \alpha_{54} + \alpha_{43},$
$V(\{1,2,4\}) = \alpha_{12} + \alpha_{21} + \alpha_{41} + \alpha_{41} + \alpha_{23} + \alpha_{43}.$
$V(\{1,4,5\}) = \alpha_{14} + \alpha_{41} + \alpha_{45} + \alpha_{54} + \alpha_{43}.$
$V(\{1,2,5\}) = \alpha_{12} + \alpha_{21} + \alpha_{23}.$
$V(\{2,4,5\}) = \alpha_{24} + \alpha_{42} + \alpha_{45} + \alpha_{54} + \alpha_{23} + \alpha_{43}.$
$V(\{1,2,4,5\}) = \alpha_{21} + \alpha_{12} + \alpha_{24} + \alpha_{42} + \alpha_{45} + \alpha_{54} + \alpha_{23} + \alpha_{43}.$

Computing the Shapley Value for players $i \in S = \{1,2,4,5\}$ we get

$$Sh_1 = \frac{\alpha_{12} + \alpha_{21} + \alpha_{14} + \alpha_{41}}{2},$$

$$Sh_2 = \alpha_{23} + \frac{\alpha_{21} + \alpha_{12} + \alpha_{24} + \alpha_{42}}{2},$$

$$Sh_4 = \alpha_{43} + \frac{\alpha_{14} + \alpha_{41} + \alpha_{45} + \alpha_{54} + \alpha_{24} + \alpha_{42}}{2},$$

$$Sh_5 = \frac{\alpha_{54} + \alpha_{45}}{2}.$$

For player $i \in M = \{3\}$ the payoff will be $H_3 = \alpha_{32} + \alpha_{34}$. In the case $N = S$ (all players cooperate) we get

$$Sh_1 = \frac{\alpha_{12} + \alpha_{21} + \alpha_{14} + \alpha_{41}}{2},$$

$$Sh_2 = \frac{\alpha_{12} + \alpha_{21} + \alpha_{23} + \alpha_{32} + \alpha_{24} + \alpha_{42}}{2},$$

$$Sh_3 = \frac{\alpha_{23} + \alpha_{32} + \alpha_{34} + \alpha_{43}}{2},$$

$$Sh_4 = \frac{\alpha_{14} + \alpha_{41} + \alpha_{43} + \alpha_{34} + \alpha_{24} + \alpha_{42} + \alpha_{45} + \alpha_{54}}{2},$$

$$Sh_5 = \frac{\alpha_{45} + \alpha_{54}}{2}.$$

Differential Network Games with Different Type of Players Behavior 297

Table 1. Players' payoffs for two different cases when $S = \{2,4,5\}$ and $S = \{1,2,4,5\}$

Coalition Player	$\{2,4,5\}$	$\{1,2,4,5\}$
1	$\alpha_{12} + \alpha_{14}$	$\dfrac{\alpha_{12} + \alpha_{21} + \alpha_{14} + \alpha_{41}}{2}$
2	$\alpha_{21} + \alpha_{23} + \dfrac{\alpha_{24} + \alpha_{42}}{2}$	$\alpha_{23} + \dfrac{\alpha_{21} + \alpha_{12} + \alpha_{24} + \alpha_{42}}{2}$
3	$\alpha_{32} + \alpha_{34}$	$\alpha_{32} + \alpha_{34}$
4	$\alpha_{41} + \alpha_{43} + \dfrac{\alpha_{45} + \alpha_{54} + \alpha_{24} + \alpha_{42}}{2}$	$\alpha_{43} + \dfrac{\alpha_{14} + \alpha_{41} + \alpha_{45} + \alpha_{54} + \alpha_{24} + \alpha_{42}}{2}$
5	$\dfrac{\alpha_{54} + \alpha_{45}}{2}$	$\dfrac{\alpha_{54} + \alpha_{45}}{2}$

In the Table 1, it is shown how the payoffs of the players changed if player 1 from the set M changed the behavior and decided to joint to coalition S. If $\alpha_{ij} \neq \alpha_{ji}$, in order to stimulate, for example, the first player for cooperation, it is necessary that the following condition be fulfilled

$$H_1 = \alpha_{12} + \alpha_{14} < Sh_1 = \frac{\alpha_{12} + \alpha_{21} + \alpha_{14} + \alpha_{41}}{2},$$

or

$$\alpha_{12} + \alpha_{14} < \alpha_{21} + \alpha_{41}.$$

The next three inequalities show that the payoff of player 1 can increase if payoffs of players 2 or 4 or both decrease after player 1 joints to coalition S.

$$\alpha_{12} + \alpha_{14} < \alpha_{21} + \alpha_{41}, \quad \alpha_{21} < \alpha_{12}, \quad \alpha_{41} < \alpha_{14}.$$

It is obvious, that in case $\alpha_{ij} = \alpha_{ji}$, the payoffs will coincide for both types of players behaviour.

Table 2. Players' payoffs for two different cases when $S = \{2,3,4,5\}$ and $S = \{1,2,3,4,5\}$

Coalition Player	$\{2,3,4,5\}$	$\{1,2,3,4,5\}$
1	$\alpha_{12} + \alpha_{14}$	$\dfrac{\alpha_{12} + \alpha_{21} + \alpha_{14} + \alpha_{41}}{2}$
2	$\alpha_{21} + \dfrac{\alpha_{23} + \alpha_{32} + \alpha_{24} + \alpha_{42}}{2}$	$\dfrac{\alpha_{21} + \alpha_{12} + \alpha_{23} + \alpha_{32} + \alpha_{24} + \alpha_{42}}{2}$
3	$\dfrac{\alpha_{23} + \alpha_{32} + \alpha_{34} + \alpha_{43}}{2}$	$\dfrac{\alpha_{23} + \alpha_{32} + \alpha_{34} + \alpha_{43}}{2}$
4	$\alpha_{41} + \dfrac{\alpha_{43} + \alpha_{34} + \alpha_{24} + \alpha_{42} + \alpha_{45} + \alpha_{54}}{2}$	$\dfrac{\alpha_{14} + \alpha_{41} + \alpha_{43} + \alpha_{34} + \alpha_{24} + \alpha_{42} + \alpha_{45} + \alpha_{54}}{2}$
5	$\dfrac{\alpha_{54} + \alpha_{45}}{2}$	$\dfrac{\alpha_{45} + \alpha_{54}}{2}$

From Table 2 we can also see that the full cooperation can increase the payoff of one or more players by redistributing the payoffs of other players since the next system is not compatible

$$\alpha_{12} + \alpha_{14} < \alpha_{21} + \alpha_{41}, \quad \alpha_{21} + \alpha_{23} < \alpha_{12} + \alpha_{32}, \quad \alpha_{41} + \alpha_{43} < \alpha_{14} + \alpha_{34}.$$

3   Conclusion

In the paper, a differential network game with infinite duration in which players follow different types of behavior (to cooperate or to act individually in their own interests) is considered. In such a game, a new type of characteristic function is proposed, and the Shapley value is computed. Non-emptiness of the core is proved. Some properties of mentioned solutions are investigated.

References

1. Bilbao J.M. et al.: Bicooperative games Cooperative games on combinatorial structures. Kluwer Acad. 131–295 (2000)
2. Bulgakova, M., Petrosyan, L.: About one multistage non-antagonistic network game (in Russian), Vestnik S.-Petersburg Univ. Ser. 10. Prikl. Mat. Inform. Prots. Upr. **5**(4), 603–615 (2019). https://doi.org/10.21638/11702/spbu10.2019.415
3. Cao, H., Ertin, E. and Arora, A.: MiniMax equilibrium of networked differential games. ACM TAAS. **3**(4) (1963). https://doi.org/10.1145/1452001.1452004
4. Gromova, E.: The Shapley Value as a Sustainable Cooperative Solution in Differential Games of Three Players. In: Petrosyan, L., Mazalov, V. (eds.) Recent Advances in Game Theory and Applications. Static & Dynamic Game Theory: Foundations & Applications, pp. 67–89. Birkhäuser, Cham (2016). https://doi.org/10.1007/978-3-319-43838-2_4
5. Isaacs, R.: Differential Games. Wiley, New York (1965)
6. Meza, M.A.G. and Lopez-Barrientos, J.D.: A differential game of a duopoly with network externalities. In: Petrosyan, L., Mazalov, V. (eds.) Recent Advances in Game Theory and Applications. Static & Dynamic Game Theory: Foundations & Applications, pp. 49–66. Birkhäuser, Cham (2016). https://doi.org/10.1007/978-3-319-43838-2
7. Pai, H.M.: A differential game formulation of a controlled network. Queueing SY. **64**(4), 325–358 (2010)
8. Peleg, B., Sudhölter, P.: Introduction to the Theory of Cooperative Games. Springer, Heidelberg (2007). https://doi.org/10.1007/978-3-540-72945-7
9. Petrosyan, L.A.: Cooperative Differential Games on Networks (in Russian), Trudy Inst. Mat. i Mekh. UrO RAN **16**(5), 143–150 (2010)
10. Petrosyan, L.A., Danilov, N.N.: Stability of solutions in non-zero sum differential games with transferable payoffs. Vestnik of Leningrad Universtiy **1**, 52–59 (1979)
11. Petrosyan, L., Yeung, D., Pankratova, Y.: Dynamic cooperative games on networks. In: Strekalovsky, A., Kochetov, Y., Gruzdeva, T., Orlov, A. (eds.) MOTOR 2021. CCIS, vol. 1476, pp. 403–416. Springer, Cham (2021). https://doi.org/10.1007/978-3-030-86433-0_28
12. Petrosyan, L., Zaccour, G.: Time-consistent shapley value allocation of pollution cost reduction. J. Econ. Dyn. Control **27**, 381–398 (2003). https://doi.org/10.1016/S0165-1889(01)00053-7
13. Petrosyan, L.A., Yeung, D.W.K.: Shapley value for differential network games: theory and application. JDG **8**(2), 151–166 (2020). https://doi.org/10.3934/jdg.2020021
14. Petrosyan, L., Yeung, D., Pankratova, Y.: Differential Network Games with Infinite Duration. In: Petrosyan, L.A., Mazalov, V.V., Zenkevich, N.A. (eds.) Frontiers of Dynamic Games. Trends in Mathematics. Birkhauser, Cham (2021). https://doi.org/10.1007/978-3-030-93616-7_15

15. Petrosyan, L., Yeung, D., Pankratova, Y.: Characteristic functions in cooperative differential games on networks. J. Dyn. Games **11**(2), 115–130 (2024). https://doi.org/10.3934/jdg.2023017
16. Shapley, L.S.: A Value for N-person Games. In: Kuhn, H., Tucker, A. (eds.) Contributions to the Theory of Games, pp. 307–317. Princeton University Press, Princeton (1953)
17. Tur, A. and Petrosyan, L.: The core of cooperative differential games on networks. In: Pardalos, P., Khachay, M., Mazalov, V. (eds.) MOTOR 2022. LNCS, vol. 13367. Springer, Cham (2022). https://doi.org/10.1007/978-3-031-09607-5_21
18. Wie, B.W.: A differential game model of nash equilibrium on a congested traffic network. Networks **23**, 557–565 (1993)
19. Yeung, D.W.K.: Time consistent shapley value imputation for cost-saving joint ventures. Mat. Teor. Igr Pril. **2**(3), 137–149 (2010)
20. Zhang, H., Jiang, L.V., Huang, S., Wang, J., Zhang, Y.: Attack-defense differential game model for network defense strategy selection. IEEE Access (2018). https://doi.org/10.1109/ACCESS.2018.2880214

Network Structure Properties and Opinion Dynamics in Two-Layer Networks with Hypocrisy

Chi Zhao[✉][iD] and Elena M. Parilina[iD]

Saint Petersburg State University, St. Petersburg, Russia
st081292@student.spbu.ru, e.parilina@spbu.ru

Abstract. We extended a social Zachary's karate club network by adding the second layer or internal layer of communication. Different models of opinion dynamics are represented on such two-layer network. The presence of internal layer is motivated by the fact that individuals can share their real opinions with their close friends. In external layer, individuals express their opinions publicly. Starting from a real structure of the external layer defined as Zachary's karate club, we observe how opinion dynamics is affected by internal layer, i.e. how consensus time and winning rate change depending on an internal layer structure. We find significantly strong correlation between internal graph density and consensus time, as well as group degree centrality and consensus time.

Keywords: Opinion Dynamics · Voter Model · Concealed Voter Model · General Concealed Voter Model · Zachary's karate club

1 Introduction

There are macroscopic and microscopic opinion dynamics models. Macroscopic models including Ising model [23], Sznajd model [31], voter model [18], concealed voter model (CVM) [13,14], and macroscopic version of general concealed voter model (GCVM) [36] examine social networks using statistical-physical and probability-theoretical methods to analyze distribution of opinions.

Within GCVM [36], it is supposed that the individuals communicate in two layers (internal and external) and can interact in internal or private layer. The latter assumption is different from CVM, where individuals do not express their true opinions in internal layer.

Examples of microscopic models of opinion dynamics are DeGroot model [7], Friedkin-Johnsen (F-J) model [12], and bounded confidence models [6,17]. In the F-J model, actors can also factor their initial prejudices into every iteration of opinion [25]. The models of opinion dynamics based on DeGroot and F-J models, and Markov chains with possibility to control agents' opinions are proposed in [5,19,22,27]. The levels of influence and opinion dynamics with the presence of agents with different levels of influence is examined in [15,16].

A. Eremeev et al. (Eds.): MOTOR 2024, LNCS 14766, pp. 300–314, 2024.
https://doi.org/10.1007/978-3-031-62792-7_21

A bounded confidence model (BCM) is a model, in which agents ignore opinions that are very far from their own opinions [4,24]. The BCM includes two essential models: the Deffuant-Weisbuch model (D-W) proposed in paper [6], and the Hegselman-Krause (H-K) model introduced in the work [17]. In the D-W model, two individuals are randomly chosen, and they determine whether to interact according to the bounded confidence [34].

The micro version of GCVM is introduced in [35], and the difference between macro and micro versions is that in the micro version we do not need to adjust the simulation program according to different network structures. As long as network structure is given, the program automatically produces simulations. Therefore, we can use this program to simulate real networks. But for the macro version, we should adjust the corresponding state transition formulae for different network structures.

Since network structure in GCVM is two-layer, it is interesting to examine how not only this structure in general, but also network characteristics, e.g. different centrality measures [21], affect opinion dynamics and resulting opinion in consensus if it is reached. We consider two key performance indicators of opinion dynamics, namely, winning rate and consensus time.

In this paper, we examine Zachary's karate club and opinion dynamics in this network by adding internal layer of communication among network members. Therefore, in our model external structure is always Zachary's karate club and internal structures are star, two-star, complete, empty, and finally, Zachary's karate club. We examine if consensus time and winning rate of an opinion depend on internal layer and network characteristics in general. By the latter, we mean different centrality measures. The primary conclusion drawn from this research indicates that correlation between internal graph density and consensus time, as well as group degree centrality and consensus time, is significantly strong.

The rest of this paper is organized as follows. Section 2 introduces a model. Network properties are discussed in Sect. 3. Section 4 presents simulation experiments and results. We briefly conclude and discuss our future work in Sect. 5.

2   Model

We examine Zachary's Karate Club network which is a well-known social network representing friendship relations among 34 members of a karate club at the US university in the 1970 s. The data was collected by Wayne Zachary in 1977 [33]. The study became famous in data and network analytical literature since it highlighted a conflict between manager (Node 0) and director (Node 33), which eventually led to the split of the club into two different groups. We present this one-layer network in Fig. 1. Below we examine opinion dynamics on this network using BVM.

Figure 2 shows how one-layer Zachary's Karate Club network can be extended for two-layer network if we assume internal layer exists within the model of opinion dynamics in this network. As discussed above, in CVM the nodes in internal layer are not connected, i.e. internal layer is represented by an empty

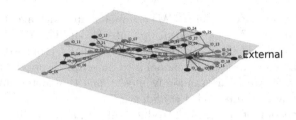

Fig. 1. One-layer Zachary's Karate Club network

graph (Fig. 2a). But in GCVM there may be nonempty network representing internal communication of individuals. In Fig. 2b we represent a star internal structure. The colors in Fig. 2 represent individuals' opinions. The color, blue or red, is randomly initialized according to the given parameters ($\rho = 0.75$ for BVM and $\rho_{r_e} = 0.75, \rho_{r_i} = 0.25, \rho_r = 0.2$ for CVM and GCVM). The particular difference between CVM and GCVM will be introduced in details in Sects. 2.1–2.3.

2.1 Opinion Dynamics in One-Layer Networks

Basic Voter Model. We briefly describe a basic voter model (BVM). BVM assumes that anyone in a network can express his opinion publicly. There is a predefined network G and we use following notations:

- N: number of individuals/agents in network;
- $a_i, i = 0, \ldots, N - 1$: individual/agent;
- t: current time;
- $\omega(a_i, t) \in \{0, 1\}$: opinion of individual a_i as time t (0 is represented by blue and 1 is marked by red color).
- π_c: copying rate, that is probability that an individual adapts his/her neighbor's opinion;
- r: number of individuals with red opinion;
- $\rho = r/N$: proportion of individuals with red opinion.

Algorithm of BVM:

Step 1. Initialize $t = 0$.
Step 2. Choose an individual a_i, uniformly random from N individuals in predefined network G.
Step 3. Pick up a neighbor of a_i randomly among all a_i's neighbors. Let it be individual a_j.
Step 4. Generate a random number $x \sim U(0, 1)$, a_i adapts a_j's opinion if $x < \pi_c$.
Step 5. Increase t by 1. If the group has reached a consensus[1], stop the iteration. Otherwise, go back to step 2.

---

[1] By consensus in BVM model it is meant that all individuals in a network hold the same opinion, i.e. we say that consensus is reached.

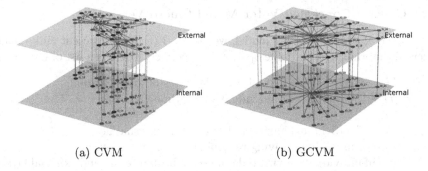

(a) CVM   (b) GCVM

Fig. 2. Two-layer networks used in CVM and GCVM: (a) CVM: two-layer network with external Zachary's karate club and empty internal layer, (b) GCVM: two-layer network with external Zachary's karate club and star internal layer.

2.2 Opinion Dynamics in Two-Layer Networks

A two-layer network with N individuals/agents is defined by

- N: number of individuals/agents in the network;
- $a_i = (a_i^E, a_i^I)$: individual/agent i, where $i = 1, \ldots, N$, a_i^E (a_i^I) is a representation of agent i in the external (internal) layer (i.e. a set of individuals/agents is the same for both layers);
- $G_E(\mathcal{V}_E, \mathcal{E}_E)$: predefined external network, where $\mathcal{V}_E = \{a_i^E\}, i = 1, \ldots, N$ represents a set of individuals and $\mathcal{E}_E$ is a set of edges between individuals in the external layer;
- $G_I(\mathcal{V}_I, \mathcal{E}_I)$: predefined internal network, where $\mathcal{V}_I = \{a_i^I\}, i = 1, \ldots, N$ represents a set of individuals and $\mathcal{E}_I$ is a set of edges between individuals in the internal layer;
- $\mathcal{E}_C = \{(a_i^E, a_i^I) | i = 1, \ldots, N\}$: set of edges connecting individuals in external and internal layers.

We define a two-layer network with N individuals/agents as

$$G(\mathcal{V}, \mathcal{E}), \qquad (1)$$

where $\mathcal{V} = \mathcal{V}_E \cup \mathcal{V}_I$, $|\mathcal{V}_E| = |\mathcal{V}_I| = N$, and $\mathcal{E} = \mathcal{E}_E \cup \mathcal{E}_I \cup \mathcal{E}_C$. This definition is independent of a specific network structure, i.e. external/internal networks can be different.

2.3 General Concealed Voter Model (macro Version)

Zhao and Parilina proposed a general concealed voter model (GCVM) in [36] based on a concealed voter model (CVM) introduced in [13].

2.4   General Concealed Voter Model (micro Version)

In the general concealed voter model, we use R, B (r, b) to represent individuals' external (internal) red and blue opinions respectively. There is a list of notations:

- $S = \{Rr, Rb, Br, Bb\}$: set of all possible states of an individual;
- $\omega(a_i, t) \in S$: opinion of individual a_i at time t, where $i = 1, \ldots, N$ and $t = 0, 1, \ldots$;
- ρ_{r_e}: ratio of individuals having red opinion in external layer;
- ρ_{r_i}: ratio of individuals having red opinion in internal layer;
- ρ_r: ratio of individuals having red opinion in both internal and external layers, $\rho_r \leq \rho_{r_e}$ and $\rho_r \leq \rho_{r_i}$;
- r_e: number of individuals having red opinion in external layer;
- r_i: number of individuals having red opinion in internal layer;
- r: number of individuals having red opinion in both internal and external layers, $r \leq r_e$ and $r \leq r_i$;
- π_{c_e}: external copy rate, that is a probability of an individual to copy opinion of his/her external neighbor;
- π_{c_i}: internal copy rate, that is a probability of an individual to copy opinion of his/her internal neighbor;
- π_e: externalization rate, that is a probability of hypocrisy[2] choosing to publicly express his/her internal opinion;
- π_i: internalization rate, that is a probability of hypocrisy accepting his/her external opinion.

We describe GCVM in a two-layer network.

Two-Layer Network Initialization. We start by setting two networks G_E and G_I (we read these networks from the file and add the edges between external and internal representations of individuals). This results in a two-layer network G we store as an adjacency list.

Initialization of Individuals' Initial States. Denote a number of individuals in state $s \in S$ by $\#s$. We have the following relations:

$$N = \#Rr + \#Rb + \#Br + \#Bb,$$
$$r_e = \#Rr + \#Rb,$$
$$r_i = \#Rr + \#Br,$$
$$r = \#Rr,$$
$$\#Bb = N - r_e - r_i + r,$$
$$\#Rb = r_e - r,$$
$$\#Br = r_i - r.$$

(2)

---

[2] By hypocrisy we mean a node having different opinions in external and internal layers, i.e., the nodes in states Rb and Br.

Assuming a uniform distribution for each agent to belong to any state $s \in S$ at the initial time, we adopt the following rule of setting the initial state $\omega(a_i, 0)$ for any agent a_i at time $t = 0$:

$$\omega(a_i, 0) := f(x) = \begin{cases} Rr, & 0 \le x < \rho_r, \\ Rb, & \rho_r \le x < \rho_{r_e}, \\ Br, & \rho_{r_e} \le x < \rho_{r_e} + \rho_{r_i} - \rho_r, \\ Bb, & \rho_{r_e} + \rho_{r_i} - \rho_r \le x \le 1, \end{cases} \tag{3}$$

where $x \sim U(0, 1)$.

Opinion Transmission Process. We divide individuals into hypocrites and nonhypocrites based on consistency of their external and internal opinions. Hypocrites are individuals who have different opinions in internal and external layers, while nonhypocrites have the same opinions in both layers.

In our analysis we focus on two measurements (KPIs[3]):

- **Consensus time:** T_{cons} is consensus time in (G)CVM, that is, the time required for all individuals to form the same opinion in internal and external layers (i.e., $\rho_{r_e} = \rho_{r_i} = \rho_r = 0$ or 1 for T_{cons}).
- **Winning rate:** ρ is a winning rate of red opinion in a series of simulations. For the opinion, to win means that there is no other opinion that agents have in the whole network (i.e. in a series of simulations, the number of simulations, in which red opinion wins blue opinion divided by the number of simulations).

We define the actions available for a randomly chosen individual a_i:

- **Picking up a_i's neighbor:** Randomly choose a neighbor among all a_i's neighbors. Let it be individual a_j (this is a prerequisite action for external/internal copying);
- **External copying:** a_i copies a_j's external opinion with probability π_{c_e};
- **Internal copying:** a_i copies a_j's internal opinion with probability π_{c_i};
- **Externalization:** a_i expresses his/her internal opinion with probability π_e (this action is available only for hypocrite);
- **Internalization:** a_i accepts his/her external opinion with probability π_i (this action is available only for hypocrite).

Externalization and internalization are meaningless for nonhypocrites, so they have only two possible actions (external and internal copying).

Algorithm of GCVM:

Step 1. Initialize $t = 0$.
Step 2. Choose an individual a_i, uniformly random from N individuals in two-layer network G;

---

[3] Key Performance Indicators.

Step 3. Check all valid actions of individual a_i (depending on his/her state) and randomly choose one of the valid actions with equal probabilities:

  I. a_i is a hypocrite, then he/she has four possible actions: (i) external copying, (ii) internal copying, (iii) externalization, and (iv) internalization. Any action is chosen with a probability of 0.25.

  II. a_i is a nonhypocrite, then he/she can perform only external or internal copying. Any action is chosen with a probability of 0.5.

Step 4. Generate random number $x \sim U(0,1)$. Perform the action chosen in Step 3:

  a. If external copying is chosen in Step 3 and $x < \pi_{c_e}$, then a_i copies a_j's external opinion;

  b. If internal copying is chosen in Step 3 and $x < \pi_{c_i}$, then a_i copies a_j's internal opinion;

  c. If externalization is chosen in Step 3 and $x < \pi_e$, then a_i expresses his/her internal opinion;

  d. If internalization is chosen in Step 3 and $x < \pi_i$, then a_i accepts his/her external opinion.

Step 5. Increase t by 1. If consensus is reached[4], stop iteration. Otherwise, go back to Step 2.

3   Network Properties

Network structure has huge impact on KPIs. We define some characteristics of a network which, in our opinion, have the most significant correlation with opinion dynamics KPIs.

3.1   Pairwise Average Shortest Path

Define d_E as the pairwise average shortest path for external layer:

$$d_E = \sum_{s,t \in \mathcal{V}_E} \frac{d_E(s,t)}{n_E(n_E - 1)}, \tag{4}$$

where $d_E(s,t)$ is a length of the shortest path between s and t in external layer, $\mathcal{V}_E$ is a set of nodes in external layer, $n = |\mathcal{V}_E|$ is a number of nodes in external layer. Similarly, we can define d_I as a pairwise average shortest path for internal layer.

3.2   Density

A graph density is defined as a ratio of the number of edges $|\mathcal{E}|$ with respect to the maximal number of edges. Since internal layer is represented by an undirected graph, we define an internal graph density as in [1]:

$$D_I = \frac{2|\mathcal{E}_I|}{|\mathcal{V}_I|(|\mathcal{V}_I| - 1)}. \tag{5}$$

---

[4] The algorithm will be stopped when all individuals in both layers hold the same opinion, i.e. consensus is reached.

3.3   Centrality Measures

Betweenness Centrality. Betweenness centrality is a basic concept in a network analysis, which was suggested in [10]. Betweenness centrality of a node gives the number of geodesics between all nodes that contain this node. It reflects the level of node participation in the dissemination of information between other nodes in a graph. It is calculated by the formula:

$$C_b(v) = \frac{1}{n_b} \sum_{s,t \in V} \frac{\sigma_{s,t}(v)}{\sigma_{s,t}}, \tag{6}$$

where $\sigma_{s,t}$ indicates the number of shortest paths between nodes s and t, and $\sigma_{s,t}(v)$ is the number of shortest paths between nodes s and t containing node v. Normalization coefficient is $n_b = (|V| - 1)(|V| - 2)$ for $v \notin \{s,t\}$, otherwise $n_b = |V|(|V| - 1)$, where $|V|$ is the number of nodes in a one-layer network [21]. If $s = t, \sigma_{s,t} = 1$ and if $v \in \{s,t\}$, then $\sigma_{s,t}(v) = 0$.

Group Betweenness Centrality. Group betweenness centrality measure indicates a proportion of shortest paths connecting pairs of nongroup members that pass through the group [8], and it is defined by formula:

$$C_{gb}(X) = \frac{1}{n_{gb}} \sum_{s,t \in V \setminus X} \frac{\sigma_{s,t}(X)}{\sigma_{s,t}}, \tag{7}$$

where $\sigma_{s,t}(X)$ is the number of shortest paths between nodes s and t passing through some nodes in group X. Normalization coefficient is $n_{gb} = (|V| - |X|)(|V| - |X| - 1)$, where $|X|$ is the number of nodes in group X.

Closeness Centrality. In a connected graph, closeness centrality of node u is the reciprocal of a sum of lengths of the shortest paths between u and all other nodes in the graph [3,11,28]. When calculating closeness centrality, its normalized form is usually referred to as the one representing the average length of the shortest path instead of their sum, and it is calculated like this:

$$C_c(u) = \frac{n_c}{\sum_{v \in V \setminus \{u\}} d(v, u)}, \tag{8}$$

where normalization coefficient is $n_c = |V| - 1$.

Group Closeness Centrality. Group closeness centrality is the reciprocal of the sum of the shortest distances from the group to all nodes outside the group [8,9,37]. It is defined by the formula:

$$C_{gc}(X) = \frac{n_{gc}}{\sum_{v \in V \setminus X} d(v, X)}, \tag{9}$$

where $d(v, X)$ is the shortest distance between group X and v. Normalization coefficient is $n_{gc} = |V - X|$.

Degree Centrality. Degree centrality of node v [26] is defined as

$$C_d(v) = \frac{v_d}{n_d},\tag{10}$$

where v_d is a degree of node v, and normalization coefficient is $n_d = |V| - 1$.

Group Degree Centrality. Group degree centrality is the number of nodes outside the group connected with the nodes from this group [8,9]. Normalized group degree centrality for group X is given by the formula:

$$C_{gd}(X) = \frac{|\{v_i \in V \setminus X | v_i \text{ is connected to } v_j \in X\}|}{n_{gd}},\tag{11}$$

where normalization coefficient is $n_{gd} = |V| - |X|$.

4 Experiments

4.1 General Description

We have done simulations of opinion dynamics in one-layer Zachary's karate club network and two-layer network with Zachary's karate club network being external layer and different internal structures. There are internal layers we use in our analysis:

1. **karate**: Zachary's karate club network;
2. **star**: star structure with node 0 being the center;
3. **two-star**: two central nodes 0 and 17, nodes 1–16 are linked with node 0, nodes 18–33 are linked with node 17. Moreover, nodes 0 and 17 are linked;
4. **cycle**: node 0 is linked with node 1, node 1 is linked with node 2, and so on. Finally, node 33 is linked with node 0;
5. **two-clique**: nodes 0–16 belong to the first clique, nodes 17–33 belong to the second clique, and these two cliques are connected through link between nodes 0 and 17;
6. **complete**: all nodes are linked with each other.

We start with simulations of opinion dynamics of BVM, and then CVM and GCVM implementing different internal structures and their affect on consensus time and winning rate. Network properties (in this paper, centrality measurements) will also change with changes in the network structure.

4.2 Main Results and Observations

Figure 3a shows how internal average shortest path d_I varies with the network structure.[5] Figure 3b shows how internal density varies with the network structure.

---

[5] "Karate-34" and "karate-empty-34" refer to a one-layer Zachary's karate club network and to a two-layer network with Zachary's karate club network in external layer and empty internal layer respectively (i.e. d_I does not exist for these two structures, in particular, it is equal to infinity. But in Fig. 3a, we use a value of 99 instead of infinity).

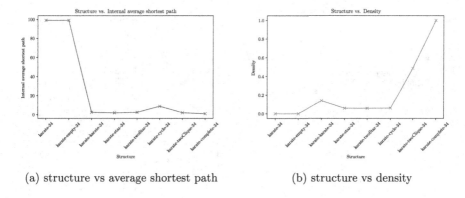

(a) structure vs average shortest path   (b) structure vs density

Fig. 3. Internal average shortest path and density for different structures

Figures 4, 5 and 6 show betweenness centrality, closeness centrality, and degree centrality for different network structures. Each figure consists of three parts representing external centrality, internal centrality, and group centrality from left to right respectively with fixed external Zachary's karate club structure. For external/internal centrality, we start from a one-layer Zachary's Karate Club network (BVM) and two-layer network with empty internal layer (CVM), then there are two-layer GCVMs with different internal structures described above. When calculating centrality in a two-layer network, we define centralities of external/internal nodes separately. Group centrality is calculated only for two-layer networks since each group is determined by a pair of nodes representing an individual in both layers, i.e. group X_i is $X_i = a_i = (a_i^E, a_i^I)$.

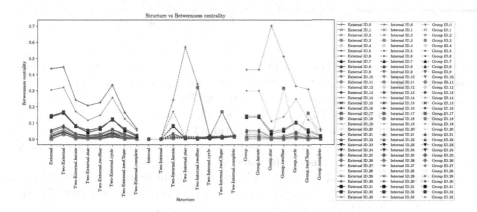

Fig. 4. Betweenness centrality for different structures

Figure 7 shows how KPIs vary with different network structures. Looking at Fig. 4, we may note that empty internal structure increases external betweenness

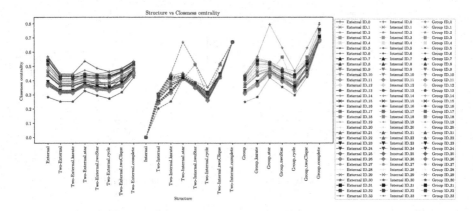

Fig. 5. Closeness centrality for different structures

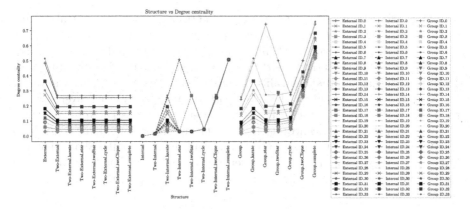

Fig. 6. Degree centrality for different structures

centrality of any node (the left-most part of any graph), and then when we add internal layer, the trend is exactly the same as in Fig. 3a. Figure 3b also demonstrates exactly the same trend as in Fig. 7b.

Figure 7a shows that the structure has a great impact on winning rate, but we have not yet found reasonable explanations for this.

If we exclude nodes 0 and 17[6] from Fig. 6, a trend of group degree centrality (the right-most part) is exactly the same as in Fig. 7b.

We can also make an interesting observation about group closeness centrality (the right-most part) from Fig. 5, that is, group closeness centrality of some nodes (2, 31, 32, 33) do not increase when we change internal structure from "Zachary's karate club" to "star". The reasons of these observations deserve further discussion.

---

[6] Nodes 0 and 17 are the centers of a two-star structure, therefore, they have a much higher degree in comparison with other nodes.

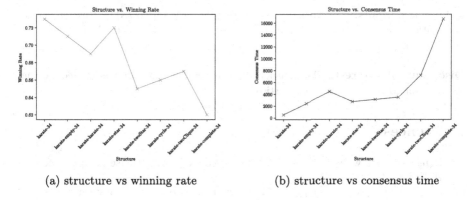

(a) structure vs winning rate   (b) structure vs consensus time

Fig. 7. Winning rate and consensus time for different structures

We conducted correlation tests on the above observation results by SciPy [2,32]. Correlation coefficients are presented in Table 1.[7]

Since all correlation coefficients in Table 1 are greater than 0.8 and they are all statistically significant, we can conclude that correlation between external betweenness centrality and d_I, D_I and T_{cons}, as well as group degree centrality and T_{cons}, is significantly strong.

Table 1. Correlation coefficients

	Pearson [30]	Kendall [20]	Spearman [29]
External betweenness centrality vs d_I	0.830	0.982	0.994
D_I vs T_{cons}	0.983	0.964	0.988
Group degree centrality vs T_{cons}	0.924	0.878	0.955

5 Conclusions and Future Work

We examined opinion dynamics models including BVM, CVM, and GCVM in Zachary's karate club. In CVM and GCVM, we assume different structures of internal layer and conducted simulations for all these internal layers. We examined how internal network structure affects consensus time and winning rate and if these KPIs correlate with network centrality measures. We think the following developments of our work are interesting: (i) incorporating stubbornness and redefining consensus conditions to observe the impact of stubbornness on KPIs, (ii) formalizing consensus time and winning rate, (iii) exploring the reasons of winning rate variation when network structure changes.

---

[7] We choose node 33 as the input for centrality. Actually, if we choose another node as the input, the conclusion is still valid.

Acknowledgments. The work of the second author was supported by Russian Science Foundation grant, grant no. 22-11-00051. https://rscf.ru/en/project/22-11-00051/.

Disclosure of Interests. The authors have no competing interests to declare that are relevant to the content of this article.

References

1. Dense graph - wikipedia. https://en.wikipedia.org/wiki/Dense_graph. Accessed 20 Feb 2024
2. scipy.stats.pearsonr - scipy v1.12.0 manual. https://docs.scipy.org/doc/scipy/reference/generated/scipy.stats.pearsonr.html. Accessed 11 Feb 2024
3. Bavelas, A.: Communication patterns in task-oriented groups. J. Acoust. Soc. Am. **22**(6), 725–730 (1950)
4. Bernardo, C., Altafini, C., Proskurnikov, A., Vasca, F.: Bounded confidence opinion dynamics: a survey. Automatica **159**, 111302 (2024). https://doi.org/10.1016/j.automatica.2023.111302, https://www.sciencedirect.com/science/article/pii/S0005109823004661
5. Bolouki, S., Malhame, R.P., Siami, M., Motee, N.: Éminence grise coalitions: on the shaping of public opinion. IEEE Trans. Control Netw. Syst. **4**(2), 133–145 (2017). https://doi.org/10.1109/TCNS.2015.2482218
6. Deffuant, G., Neau, D., Amblard, F., Weisbuch, G.: Mixing beliefs among interacting agents. Adv. Complex Syst. **3**(01n04), 87–98 (2000)
7. DeGroot, M.H.: Reaching a consensus. J. Am. Stat. Assoc. **69**(345), 118–121 (1974)
8. Everett, M.G., Borgatti, S.P.: The centrality of groups and classes. J. Math. Sociol. **23**(3), 181–201 (1999)
9. Everett, M.G., Borgatti, S.P.: Extending centrality. Models Methods Soc. Netw. Anal. **35**(1), 57–76 (2005)
10. Freeman, L.C.: A set of measures of centrality based on betweenness. Sociometry, 35–41 (1977)
11. Freeman, L.C.: Centrality in social networks conceptual clarification. Soc. Netw. **1**(3), 215–239 (1978). https://doi.org/10.1016/0378-8733(78)90021-7, https://www.sciencedirect.com/science/article/pii/0378873378900217
12. Friedkin, N.E., Johnsen, E.C.: Social influence and opinions. J. Math. Sociol. **15**(3–4), 193–206 (1990)
13. Gastner, M.T., Oborny, B., Gulyás, M.: Consensus time in a voter model with concealed and publicly expressed opinions. J. Stat. Mech: Theory Exp. **2018**(6), 063401 (2018)
14. Gastner, M.T., Takács, K., Gulyás, M., Szvetelszky, Z., Oborny, B.: The impact of hypocrisy on opinion formation: a dynamic model. PLoS ONE **14**(6), e0218729 (2019)
15. Gubanov, D.A.: Methods for analysis of information influence in active network structures. Autom. Remote. Control. **83**(5), 743–754 (2022). https://doi.org/10.1134/S0005117922050071
16. Gubanov, D., Chkhartishvili, A.: Influence levels of users and meta-users of a social network. Autom. Remote. Control. **79**(3), 545–553 (2018)
17. Hegselmann, R., Krause, U.: Opinion dynamics and bounded confidence models, analysis and simulation. J. Artif. Soc. Soc. Simul. **5** (2002)

Network Structure Properties and Opinion Dynamics in Two-Layer Networks   313

18. Holley, R.A., Liggett, T.M.: Ergodic theorems for weakly interacting infinite systems and the voter model. Ann. Probab., 643–663 (1975)
19. Kareeva, Y., Sedakov, A., Zhen, M.: Influence in social networks with stubborn agents: from competition to bargaining. Appl. Math. Comput. **444**, 127790 (2023). https://doi.org/10.1016/j.amc.2022.127790, https://www.sciencedirect.com/science/article/pii/S009630032200858X
20. Kendall, M.G.: The treatment of ties in ranking problems. Biometrika **33**(3), 239–251 (1945)
21. Mazalov, V., Chirkova, J.V.: Networking Games: Network Forming Games and Games on Networks. Academic Press, Cambridge (2019)
22. Mazalov, V., Parilina, E.: The Euler-equation approach in average-oriented opinion dynamics. Mathematics **8**(3) (2020). https://doi.org/10.3390/math8030355, https://www.mdpi.com/2227-7390/8/3/355
23. McKeehan, L.: A contribution to the theory of ferromagnetism. Phys. Rev. **26**(2), 274 (1925)
24. Noorazar, H.: Recent advances in opinion propagation dynamics: a 2020 survey. Eur. Phys. J. Plus **135**, 1–20 (2020)
25. Parsegov, S.E., Proskurnikov, A.V., Tempo, R., Friedkin, N.E.: Novel multidimensional models of opinion dynamics in social networks. IEEE Trans. Autom. Control **62**(5), 2270–2285 (2016)
26. Powell, J., Hopkins, M.: 9 - library networks-coauthorship, citation, and usage graphs. In: Powell, J., Hopkins, M. (eds.) A Librarian's Guide to Graphs, Data and the Semantic Web, pp. 75–81. Chandos Information Professional Series, Chandos Publishing (2015). https://doi.org/10.1016/B978-1-84334-753-8.00009-9, https://www.sciencedirect.com/science/article/pii/B9781843347538000099
27. Rogov, M., Sedakov, A.: Coordinated influence on the opinions of social network members. Autom. Remote. Control. **81**, 528–547 (2020)
28. Sabidussi, G.: The centrality index of a graph. Psychometrika **31**(4), 581–603 (1966)
29. Spearman, C.: The proof and measurement of association between two things. Am. J. Psychol. **100**(3/4), 441–471 (1987)
30. STUDENT: PROBABLE ERROR OF A CORRELATION COEFFICIENT. Biometrika **6**(2-3), 302–310 (1908). https://doi.org/10.1093/biomet/6.2-3.302
31. Sznajd-Weron, K., Sznajd, J.: Opinion evolution in closed community. Int. J. Mod. Phys. C **11**(06), 1157–1165 (2000)
32. Virtanen, P., et al.: SciPy 1.0: fundamental algorithms for scientific computing in Python. Nature Methods **17**, 261–272 (2020) https://doi.org/10.1038/s41592-019-0686-2
33. Zachary, W.W.: An information flow model for conflict and fission in small groups. J. Anthropol. Res. **33**(4), 452–473 (1977)
34. Zha, Q., et al.: Opinion dynamics in finance and business: a literature review and research opportunities. Financ. Innov. **6**, 1–22 (2020)
35. Zhao, C., Parilina, E.: Consensus time and winning rate based on simulations in two-layer networks with hypocrisy. In: 2023 7th Scientific School Dynamics of Complex Networks and their Applications (DCNA). pp. 68–71 (2023). https://doi.org/10.1109/DCNA59899.2023.10290478

36. Zhao, C., Parilina, E.: Opinion dynamics in two-layer networks with hypocrisy. J. Oper. Res. Soci. China **12**(1), 109–132 (2024). https://doi.org/10.1007/s40305-023-00503-2
37. Zhao, J., Lui, J.C., Towsley, D., Guan, X.: Measuring and maximizing group closeness centrality over disk-resident graphs. In: Proceedings of the 23rd International Conference on World Wide Web, pp. 689–694 (2014)

Dynamic Stability of Coalition Structures in Network-Based Pollution Control Games

Jiangjing Zhou[1]([✉]) and Vladimir Mazalov[1,2,3]

[1] Applied Mathematics and Control Processes, St.Petersburg State University,
Universitetskiy Prospekt, 35, Petergof, St. Petersburg 198504, Russia
st092028@student.spbu.ru
[2] Institute of Applied Mathematical Research, Karelian Research Center of the
Russian Academy of Sciences, Pushkinskaya Street, 11, Petrozavodsk,
Karelia 185610, Russia
vmazalov@krc.karelia.ru
[3] School of Mathematics and Statistics, Qingdao University, No. 308 Ningxia Road,
Qingdao 266071, China

Abstract. This paper investigates the dynamics of coalition stability in pollution control games that are built on networks. It specifically focuses on the concept of dynamically stable coalition partitions involving three asymmetric players. At first, the research presents the model and clearly identifies optimal strategies for players in various coalition forms. In order to evaluate the dynamic stability of these coalitions, the system additionally calculates the time-consistent Imputation Distribution Procedure (IDP) for players belonging to different coalitions. This work presents a novel idea of dynamically Nash stable coalition partitions, which is defined using the IDP, in contrast to the traditional definition. By accurately identifying the IDP as specified, we may attain dynamically Nash stable coalition partitions in the simulation results.

Keywords: Dynamically Nash stable coalition partitions ·
Network-based pollution control games · Asymmetric players

1 Introduction

Dynamic games, a subset of game theory, examine the process of choosing decisions in circumstances that change over time, frequently employing difference equations. These games play a crucial role in comprehending strategic interactions in dynamic environments, such as pollution control. The model proposed by [2] depicts a network in which nodes symbolize agents that act as pollution sources, and the connections between them reflect the transmission of pollution. In contrast to the river pollution game analyzed in [19], which considers network externalities and the influence of both a participant's own decisions and those of their predecessors in the network on their state variables, our model extends this concept by asserting that a player's strategy is influenced by the actions of their immediate neighbors, rather than just the preceding nodes.

A. Eremeev et al. (Eds.): MOTOR 2024, LNCS 14766, pp. 315–333, 2024.
https://doi.org/10.1007/978-3-031-62792-7_22

Cooperative games play a significant role in the theory of dynamic games. The coalition's benefit is derived from the characteristic function. There are various methods to construct the characteristic function. The outcome is dependent upon the conduct of players who are not part of this coalition. Members of an opposing coalition have the ability to take actions that work against the coalition being discussed. They can employ strategies that maintain a state of balance, among other tactics. In the classical version, the characteristic function exhibits superadditivity, hence incentivizing participants to form coalitions of maximum size. Upon reaching a mutually beneficial agreement in a game with changing circumstances, a collaborative path is determined that all players are required to adhere to, along with the allocation of rewards among the participants. In order to ensure consistency over time, the reward for each participant is spread over time, as described in [15]. Simultaneously, in order to incentivize each player to continue cooperating and not withdraw from the agreement, the division must meet the requirements of rational behavior [20].

In this research, we recognize that the superadditivity condition of the characteristic function in cooperative games is not universally maintained. This divergence allows for the analysis of a wider range of coalition structures, including situations where coalitions of different sizes compete. This is in contrast to the typical model where one large coalition competes against individual participants, as mentioned in many prior studies [14, 21]. Hence, the act of creating larger coalitions may not always be beneficial for players. The theory of cooperative games with a coalition structure is relevant when considering the intricacies of payoff distributions [3]. An essential element of this theory involves identifying the most advantageous and enduring coalition divisions for players. In this context, stability refers to different types, including internal, external, Nash, individual, and core stability [1,5,13], as well as other less traditional types mentioned in the literature [10].

When assessing a coalition partition, it is essential to take into account not only the stability from external and internal factors, but also the dynamics of participants transitioning between different coalitions. The idea of inter-coalition stability is discussed in detail in [7]. The majority of studies on coalition stability, particularly in the context of agreements to reduce emissions, places significant emphasis on this particular component. The use of these notions has been examined in several research, including those focused on fisheries, as evidenced by the references [6,8,12,17,18,21].

Coalition partition stability refers to a situation where players lack any motivation to abandon their existing coalition in favor of joining another one. In static games, where the reward of a coalition is not affected by how the coalition is divided, stability is typically guaranteed for any characteristic function [9]. Nevertheless, within the domain of dynamic games, this particular element has not been well studied, with only a limited number of research papers identifying stable divisions in specific scenarios [18,21]. The objective of our research is to provide a comprehensive analysis of this topic, with a specific emphasis on creating a model that demonstrates this situation in the context of pollution control games.

This study focuses on dynamic games on fixed network structures, which is different from traditional network game models that incorporate strategies related to link generation or dissolution. Specifically, we explore the theories proposed by [11] on social network development and [4] on network dynamics. We are primarily interested in examining how a specific network structure influences the strategies adopted by players, where the rewards are determined by the fixed structure of the network. This approach enables a thorough examination of strategic interactions inside static network structures, introducing a novel aspect to the field of network game theory.

2   Theoretical Background

2.1   Model Formulation

The pollutant stock for country i in a classical model of pollution control is represented by the following difference equation:

$$x_i(t+1) = \epsilon x_i(t) + r u_i(t), \quad i \in N,$$

where $x_i(t+1)$ denotes the pollution stock of country i at time $t+1$. The term $\epsilon x_i(t)$ represents the residual pollution at time t, considering natural degradation, with $1 > \epsilon > 0$ being the degradation rate constant. The expression $r u_i(t)$ represents the incremental pollution produced at time t as a result of the activities of country i, where $1 > r > 0$ is a fixed value. In this context, N denotes the set comprising all players, and $n = |N|$ represents the total number of players.

The profit for player i in this dynamic game is given by:

$$J_i = \sum_{t=0}^{\infty} \delta^t \left[(p - u_i(t)) u_i(t) - c_i x_i(t) \right], i \in N,$$

where J_i is the cumulative discounted profit over time for player i. The discount factor δ (with $0 < \delta < 1$) accounts for the time value of money. The term $(p - u_i(t)) u_i(t)$ represents the revenue generated by player i, and $c_i x_i(t)$ is the cost associated with the pollution level $x_i(t)$ for player i.

In this model, a coalition partition is denoted by π, representing the division of the entire set of players N into smaller, separate coalitions. Mathematically, we represent this as $\pi = \{K_1, K_2, \ldots, K_l\}$, where each K_j represents a coalition, and together, these coalitions cover the complete set N ($\bigcup_{j=1}^{l} K_j = N$). It is important to emphasize that these coalitions are mutually exclusive, meaning that no two different coalitions have any players in common ($\forall i, j \in \{1, 2, \ldots, l\}, i \neq j : K_i \cap K_j = \emptyset$).

The notation k_i is used to denote the number of agents or participants in a specific coalition K_i, where i ranges from 1 to l. Here, l represents the total number of coalitions in the partition.

Furthermore, for any player i, $K_\pi(i)$ represents the particular coalition to which player i belongs within the partition π. This implies that player i belongs to the coalition $K_\pi(i)$, which is a component of the coalition partition π ($i \in K_\pi(i) \in \pi$).

Generally speaking, a player i who is part of a coalition $K_\pi(i)$ from a partition π can take into account the coalition structure π. Therefore, we will assume that the control vector $u(t, \pi) = (u_1(t, K_\pi(1)), ..., u_n(t, K_\pi(n)))$ and the corresponding trajectory $x(t, \pi) = (x_1(t, K_\pi(1)), ..., x_n(t, K_\pi(n)))$ correspond to the dynamics

$$x_i(t+1, K) = \epsilon x_i(t, K) + r u_i(t, K), \quad i \in K, \ K \in \pi. \tag{1}$$

The value function of a coalition $K \in \pi$ at time τ is defined under the assumption that players not in the coalition have the power to cut off links with members of the coalition. The definition is as follows:

$$V_{\pi,\tau}(K, x_K) = \max_{u_i, i \in K} \sum_{t=\tau}^{\infty} \delta^t \left[\sum_{i \in K} \left(p - \sum_{j \in S(i) \cap K} u_j(\cdot) \right) u_i(t, K) - c_K x_K(t, K) \right],$$

where c_K represents the minimum cost among all members' costs within a coalition K. This parameter is structured to ensure the cost for any coalition remains below or equal to the minimum cost that any single player within it would face, thereby encouraging coalition formation. Suppose that $S_0(i)$ is the set of neighbors of player i in the network, $S(i) = S_0(i) \cup \{i\}$. The term x_K under the coalition partition π is defined by $x_K(t, K) = \sum_{i \in K} x_i(t, K)$, representing the aggregated state of players within K at time t under partition π.

In this framework, the concept of a coalition partition, represented by π, categorizes players into various coalitions. The term $\phi_i(\pi, V_{\pi,t}, K_\pi(i), t)$ quantifies the individual payoff or utility of player i within their respective coalition at a given time t. This calculation is influenced by the specific coalition partition π and the payoff function V, which assigns values to the coalitions in π.

We define a coalition partition, π, as a specific arrangement of players into groups. For clarity, we use $D_i(\pi)$ to represent the set of all possible coalition partitions that can arise from π when a player i either joins an existing group or forms a new one. This set is defined as:

$$D_i(\pi) = \{\pi\} \cup \{\{K_\pi(i) \backslash \{i\}, A \cup \{i\}, \pi \backslash \{K_\pi(i) \cup A\}\} \mid A = \emptyset \vee A \in \pi \backslash K_\pi(i)\}.$$

To illustrate, consider a player set $N = \{1, 2, 3\}$ and a partition $\pi = \{\{1, 2\}, \{3\}\}$. The set $D_1(\pi)$ can be expressed as:

$$\{\{\{1, 2\}, \{3\}\}, \{\{1\}, \{2\}, \{3\}\}, \{\{2\}, \{1, 3\}\}\}.$$

Suppose that at time t a coalition partition is π and a player $i \in K_\pi(i)$ decides to change the coalition $K_\pi(i)$ to a coalition $K_\rho(i), K_\rho(i) \neq K_\pi(i)$. So, the new coalition partition is $\rho \in D_i(\pi)$.

The dynamics described by equation (1) at time $t+1$ should be changed to

$$x'(t+1, \rho) = T_{i,\rho} x(t, \pi),$$

where shift operator $T_{i,\rho}x(t,\pi)$ is determined by the equations

$$
x'_j(t+1, K_\rho(j)) = \begin{cases} \epsilon x_j(t, K_\pi(j)) + r u_j(t, K_\pi(j)), & j \in K \in \pi \cap \rho \\ \epsilon x_j(t, K_\pi(j)) + r u_j(t, K_\rho(j)), & otherwise. \end{cases} \tag{2}
$$

The operator $T_{i,\rho}x(t,\pi)$ is designed to capture the dynamic evolution of a system's trajectory as it transitions between two distinct coalition structures, π and $\rho \in D_i(\pi)$, at a specific time t. At time t, there is a pivotal transition to a new coalition structure ρ, marking a shift in the system's dynamics.

We will provide a graphical example to illustrate the meaning of this operator:

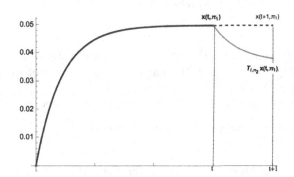

Fig. 1. State trajectory across the transition from π_1 to π_2 at t_1

Figure 1 illustrates the state trajectory of player 1 under the transition from the coalition partition π_1 to π_2. The thick blue line represents player 1's state over time at each moment t under the coalition partition π_1, where player 1 acts alone following the optimal strategy. At time t_1, player 1 realizes that forming a coalition with player 2 offers greater benefits, thus deviating from the initial coalition at t_1. Henceforth, the state trajectory follows the previously defined operator as Eq. (2) from t_1 onward. The dashed line, which does not occur, represents the hypothetical continuation of player 1's trajectory under π_1 if they had not deviated at $t \geq t_1$.

2.2 Imputation Distribution Procedure

In this framework, the function $\phi_j(\pi, V_{\pi,t}, K_\pi(j), t)$ indicates the payoff for player j in coalition $K_\pi(j)$ from time t, based on the collective payoff $V_{\pi,t}(K_\pi(j), x)$. For imputation, one may choose from cooperative optimality principles like proportional solutions, C-core, n-core, or the Shapley value. One of the important problems in dynamic game is the time-consistency. An imputation distribution procedure (IDP) might be suggested [16].

Definition 1. *The vector* $\beta(\pi, t) = (\beta_1(\pi, V_{\pi,0}, K_\pi(1), t), \ldots, \beta_n(\pi, V_{\pi,0},$ $K_\pi(n), t))$ *is an imputation distribution procedure (IDP) if*

$$\phi_i(\pi, V_{\pi,0}, K_\pi(i), 0) = \sum_{t=0}^{\infty} \delta^t \beta_i(\pi, V_{\pi,t}, K_\pi(i), t), \quad i = 1, 2, \ldots, n.$$

The term $\beta_i(\pi, V_{\pi,t}, K_\pi(i), t)$ signifies the payment to player i in coalition $K_\pi(i)$ at time t.

Definition 2.   *The vector* $\beta(\pi, t) = (\beta_1(\pi, V_{\pi,t}, K_\pi(1), t), \ldots, \beta_n(\pi, V_{\pi,t}, K_\pi(n), t))$ *is a time-consistent IDP if for every* $t \geq 0$

$$\phi_i(\pi, V_{\pi,0}, K_\pi(i), 0) = \sum_{\tau=0}^{t} \delta^\tau \beta_i(\pi, V_{\pi,\tau}, K_\pi(i), \tau) + \delta^{t+1} \phi_i(\pi, V_{\pi,t+1}, K_\pi(i), t+1),$$

$$i = 1, 2, \ldots, n.$$

This strategy guarantees that participants constantly adhere to the same allocation principle throughout the game, without any motivation to depart from the initial agreement.

The time-consistency of the IDP $\beta(\pi, t)$ is demonstrated by

$$\beta_i(\pi, V_{\pi,t}, K_\pi(i), t) = \phi_i(\pi, V_{\pi,t}, K_\pi(i), t) - \delta\phi_i(\pi, V_{\pi,t+1}, K_\pi(i), t+1), \tag{3}$$

$$i = 1, 2, \ldots, n.$$

Definition 3. *A coalition* π *is dynamically Nash stable, if for* $\forall i \in N$, $\forall t \geq 1$,

$$\phi_i(\pi, V_{\pi,t}(\cdot, x(t, \pi)), K_\pi(i), t) \geq \phi_i(\rho, V_{\rho,t}(\cdot, T_{i,\rho}x(t-1, \pi)), K_\rho(i), t), \quad \forall \rho \in D_i(\pi).$$

This approach ensures that participants consistently adhere to the same allocation principle throughout the game, without any incentive to deviate from the initial agreement.

Definition 4. *The IDP* $\beta(\pi, t)$ *supports the stability of the coalition partition* π *if the following conditions satisfied for all* $t \geq 0$,

$$\phi_i(\pi, V_{\pi,t}(\cdot, x(t, \pi)), K_\pi(i), t)$$
$$\geq \beta_i(\pi, V_{\pi,t}(\cdot, x(\pi, t)), K_\pi(i), t) + \delta\phi_i(\rho, V_{\rho,t+1}(\cdot, T_{i,\rho}x(t, \pi)), K_\rho(i), t+1), \tag{4}$$
$$\forall \rho \in D_i(\pi), i \in N,$$

where $\beta_i(\pi, V_{\pi,t}(\cdot, x(\pi, t)), K_\pi(i), t)$ *is time-consistent IDP.*

Evidently, that (3) and (4) yield

$$\phi_i(\pi, V_{\pi,t}(\cdot, x(\pi, t)), K_\pi(i), t) \geq \phi_i(\rho, V_{\rho,t}(\cdot, T_{i,\rho}x(t-1, \pi)), K_\rho(i), t),$$
$$\forall \rho \in D_i(\pi), i \in N.$$

Inequalities (4) indicate that when a player chooses a coalition, they do so strategically by considering both immediate rewards and future benefits. This approach aims to maintain stability and prevent unnecessary changes. Essentially, it guarantees players that remaining in their existing coalition is equally or more advantageous than any potential alternative they might consider.

3 Asymmetric Three-Player Model in a Star Network Structure

In this section, we explore a scenario involving three neighboring industries or countries, denoted as the set of players $N = \{1, 2, 3\}$. This model is characterized by its asymmetry, both in terms of the network structure, which is not a complete network as shown in Fig. 2, and in the distinct pollution control costs for each player, represented by c_i for each player $i \in N$. The assumption regarding the transformation of costs is strategically crafted to encourage cooperation among the players. As the size of the coalition increases, the cost incurred by each participating player is reduced, thus enhancing the appeal of forming coalitions. Specifically, in asymmetric contexts, we posit the subsequent assumptions on the cost parameters to further incentivize cooperative behavior among the players:

$$c_{12} \leq \min\{c_1, c_2\},$$
$$c_{13} \leq \min\{c_1, c_3\},$$
$$c_{23} \leq \min\{c_2, c_3\},$$
$$c_{123} \leq \min\{c_{12}, c_{23}, c_{13}\},$$

where c_K represents the coalition cost for coalition K. These cost parameters are designed such that the cost to any coalition does not exceed the minimum cost faced by any individual member within it, thereby lowering barriers to coalition formation.

Fig. 2. Star network structure for the three-player model

Our objective is to investigate the coalition partition $\pi_1 = \{\{1\}, \{2\}, \{3\}\}$ and determine the conditions that make it dynamically stable, considering specific parameter restrictions.

3.1 Cooperative Case

In the case of cooperative network partitioning, we specifically consider the scenario where the grand coalition $\pi_0 = \{\{1, 2, 3\}\}$. This involves examining the strategic actions of the participants who work together to maximize their overall gain. We analyze the most efficient combination of strategies, the progression of the state trajectory, and the value function $V_{\pi_0, t}(N, x_{123})$, which represents the collective reward to the coalition over time. In addition, we evaluate how the payoffs are distributed among the members in this coalition split, together with the IDP at each time step. These aspects are essential for comprehending the dynamics and results of collaboration among the participants in the networked game.

Proposition 1. *In the grand coalition partition* $\pi_0 = \{\{1,2,3\}\}$, *the optimal control strategies and corresponding state trajectories for the players are as follows:*

$$u_1^*(t, \{1,2,3\}) = u_3^*(t, \{1,2,3\}) = 0, \quad u_2^*(t, \{1,2,3\}) = \frac{p(1-\delta\epsilon) - \delta c_{123} r}{2(1-\delta\epsilon)},$$

$$x_i^*(t, \{1,2,3\}) = \epsilon^t x_i^0, i = 1, 3, \quad x_2^*(t, \{1,2,3\}) = \epsilon^t x_2^0 + r\frac{(p(1-\delta\epsilon) - \delta r c_{123})(1-\epsilon^t)}{2(1-\delta\epsilon)(1-\epsilon)}.$$

Proof. Let $W_{\pi_0,t}(N, x_{123})$ be the value function of the grand coalition, which is aimed to solve the Hamilton-Jacobi-Bellman (HJB) equation:

$$W_{\pi_0,t}(N, x_{123}) =$$

$$\max_{u_i \geq 0} \left\{ p\Big(\sum_{i=1}^{3} u_i\Big) - \sum_{i=1}^{3} u_i^2 - 2(u_1 u_2 + u_2 u_3) - c_{123} x_{123} + \delta W_{\pi_0,t+1}(N, x) \right\}. \tag{5}$$

Assuming a linear-state structure for the model, the value function is hypothesized as:

$$W_{\pi_0,t}(N, x_{123}) = A_{123} x_{123} + B_{123},$$

where $x_{123} = x_1 + x_2 + x_3$.

Differentiating the right-hand side of Eq.(5) with respect to u_i and setting it to zero, we obtain:

$$p - 2u_i - 2u_2^* + \delta A_{123} r = 0, \quad i = 1, 3,$$
$$p - 2u_2 - 2(u_1^* + u_3^*) + \delta A_{123} r = 0. \tag{6}$$

Solving Eq.(6) gives us the optimal strategies

$$u_i^*(t, \{1,2,3\}) = 0, i = 1, 3,$$

$$u_2^*(t, \{1,2,3\}) = \frac{p + \delta A_{123} r}{2}.$$

By substituting these strategies back into Eq.(5) and equating coefficients, we derive A_{123} and B_{123}:

$$A_{123} = -\frac{c_{123}}{1 - \delta\epsilon}, \quad B_{123} = \left(\frac{p(1 - \delta\epsilon) - \delta r c_{123}}{2(1 - \delta\epsilon)\sqrt{1 - \delta}}\right)^2. \tag{7}$$

This results in the optimal strategies and trajectories for the players under the grand coalition.

This statement and its proof determine the most effective control tactics and state trajectories for all members in the grand coalition, taking into account the structure of the network and the collective goal of maximizing the total payoff.

Therefore, the characteristic function for the grand coalition {1,2,3} under the coalition partition π_0 along the trajectory x is given by:

$$V_{\pi_0,t}(N, x_{123}) = A_{123}x_{123}(t, \cdot) + B_{123},$$

where A_{123} and B_{123} are defined as per Eq.(7), and the term $x_{123}(t, \cdot)$ represents the aggregate state of all players in the coalition at time t, specifically $x_{123}(t, \cdot) = \sum_{i=1}^{3} x_i(t, \cdot)$. This expression reflects the collective dynamics of the grand coalition's strategy and its overall impact on the game's outcomes.

3.2 Non-cooperative Case

In the non-cooperative case with network partitioning, we focus on the scenario under the coalition partition π_1. This entails analyzing the strategic behavior of the players who individually aim to optimize their own payoff. We investigate the Nash equilibrium strategy profile, the evolution of the state trajectory, and the value function $V_{\pi_1,t}(\{i\}, x_i)$, $i = 1, 2, 3$, which represents the payoff to the player i over time.

Proposition 2. *Given the condition $\frac{p(1-\delta\epsilon)}{\delta r} \geq \max\{c_1, c_2, c_3\}$, the equilibrium strategies in an asymmetric scenario for players' pollution policies are given by:*

$$u_i^*(t, \{i\}) = \frac{p(1 - \delta\epsilon) - \delta r c_i}{2(1 - \delta\epsilon)}, \quad i = 1, 2, 3,$$

with the corresponding pollution stock trajectories:

$$x_i^*(t, \{i\}) = \epsilon^t x_i^0 + r\frac{(p(1 - \delta\epsilon) - \delta r c_i)(1 - \epsilon^t)}{2(1 - \delta\epsilon)(1 - \epsilon)}, \quad i = 1, 2, 3. \tag{8}$$

These trajectories indicate that as $t \to \infty$, the pollution stock of player i converges to $\frac{r(p(1-\delta\epsilon)-\delta r c_i)}{2(1-\delta\epsilon)(1-\epsilon)}$.

Proof. Assuming the value function $W_i(x_i)$ for each player is linear, i.e., $W_i(x_i) = A_i x_i(t) + B_i$, we derive the first-order conditions for optimality:

$$p - 2u_i + \delta A_i r = 0.$$

From which the Nash equilibrium strategies for each player can be defined:

$$u_i^*(t) = \frac{p + \delta A_i r}{2}.$$

Substituting the Nash strategies into the value function, we obtain a system of equations to solve for the coefficients A_i and B_i. Simplifying this system yields:

$$A_i = -\frac{c_i}{1 - \delta\epsilon}, \quad B_i = \left(\frac{p(1 - \delta\epsilon) - \delta r c_i}{2(1 - \delta\epsilon)\sqrt{1 - \delta}}\right)^2. \tag{9}$$

The characteristic function for player i under the coalition partition π_1 is:

$$V_{\pi_1,t}(\{i\}, x_i) = A_i x_i + B_i,$$

which illustrates the payoff of a single player considering the time-dependent effects of the strategies and state variables. This function reflects the discounted benefits over time, considering both immediate control actions and future states.

3.3   Formation of Partial Coalitions

Optimal Strategies Under Partition $\pi_2 = \{\{1, 2\}, \{3\}\}$

Proposition 3. *Under the coalition partition $\pi_2 = \{\{1,2\},\{3\}\}$, the optimal control strategies for players 1 and 2 are denoted as $u_1^*(t, \{1,2\})$ and $u_2^*(t, \{1,2\})$, respectively. These strategies are determined by the maximization of the joint payoff for the coalition $\{1,2\}$, and are given by:*

$$u_1^*(t, \{1,2\}) = u_2^*(t, \{1,2\}) = \frac{p(1 - \delta\epsilon) - \delta c_{12} r}{4(1 - \delta\epsilon)}.$$

Correspondingly, the state trajectories for players under this partition are:

$$x_i^*(t, \{1,2\}) = \epsilon^t x_i^0 + r\frac{(p(1 - \delta\epsilon) - \delta r c_{12})(1 - \epsilon^t)}{4(1 - \delta\epsilon)(1 - \epsilon)}, \quad i = 1, 2. \qquad (10)$$

Proof. The value function for coalition {1,2} under partition π_2 is given by:

$$V_{\pi_2,t}(\{1,2\}, x_{12}) = \max_{u_1,u_2} \sum_{\tau=t}^{\infty} \delta^\tau \left[(p - (u_1 + u_2))(u_1 + u_2) - c_{12}(x_{12}(t, \cdot))\right].$$

To find the optimal strategies $u_1^*(t, \{1,2\})$ and $u_2^*(t, \{1,2\})$, we differentiate the above function with respect to u_1 and u_2 and set the derivatives to zero, leading to:

$$p - 2u_1 - 2u_2^*(t, \{1,2\}) + \delta A_{12} r = 0,$$

$$p - 2u_1^*(t, \{1,2\}) - 2u_2 + \delta A_{12} r = 0.$$

Upon substituting the optimal strategies u_1^* and u_2^*, respectively, and organizing the terms, we can derive the coefficients associated with x. The coefficient A_{12} is obtained as:

$$A_{12} = -\frac{c_{12}}{1 - \delta\epsilon}. \qquad (11)$$

Additionally, by rearranging the constant terms in the HJB equation, we can determine the value of B_{12} given by:

$$B_{12} = \frac{(\delta c_{12} r - (1 - \delta\epsilon)p)^2}{4(1 - \delta)(1 - \delta\epsilon)^2}. \qquad (12)$$

The characteristic function for the coalition $\{1,2\}$ under the coalition parti-tion π_2 is denoted as $V_{\pi_2,t}(\{1,2\}, x_{12})$. It is expressed as a linear combination of the states of players 1 and 2, given by:

$$V_{\pi_2,t}(\{1,2\}, x_{12}) = A_{12}x_{12}(t, \cdot) + B_{12},$$

where A_{12} and B_{12} are coefficients defined in Eq. (11) and Eq. (12), respectively. The term $x_{12}(t, \cdot)$ represents the aggregate state of players 1 and 2 at time t in the coalition, specifically $x_{12}(t, \cdot) = x_1(t, \cdot) + x_2(t, \cdot)$. This formulation reflects the interconnected nature of the players' decisions and their collective influence on the coalition's performance.

Optimal Strategies Under Partition $\pi_3 = \{\{1\}, \{2, 3\}\}$

Proposition 4. *Under the coalition partition $\pi_3 = \{\{1\}, \{2, 3\}\}$, the optimal control strategies for players 2 and 3 are given by $u_2^*(t, \{2, 3\})$ and $u_3^*(t, \{2, 3\})$, respectively, and are obtained by maximizing the joint payoff for the coalition $\{2,3\}$. The specific forms of these strategies are:*

$$u_2^*(t, \{2, 3\}) = u_3^*(t, \{2, 3\}) = \frac{p(1 - \delta\epsilon) - \delta c_{23}r}{4(1 - \delta\epsilon)}.$$

The corresponding state trajectories for players 2 and 3 are:

$$x_i^*(t, \{2, 3\}) = \epsilon^t x_i^0 + r\frac{(p(1 - \delta\epsilon) - \delta r c_{23})(1 - \epsilon^t)}{4(1 - \delta\epsilon)(1 - \epsilon)}, \quad i = 2, 3. \tag{13}$$

Proof. The proof follows a similar approach as for π_2. The value function for coalition $\{2,3\}$ under partition π_3 is formulated, and the optimal strategies are derived by setting the first-order conditions to zero and solving the resulting system of equations.

Similarly, the characteristic function for the coalition $\{2,3\}$ under the coali-tion partition π_3 is represented as $V_{\pi_3,t}(\{2, 3\}, x_{23})$. This function similarly cap-tures the combined effect of the strategies of players 2 and 3 who form this coalition. The characteristic function is formulated as:

$$V_{\pi_3,t}(\{2, 3\}, x_{23}) = A_{23}x_{23}(t, \cdot) + B_{23},$$

where the coefficients A_{23} and B_{23} are analogous to A_{12} and B_{12} from the coalition $\{1,2\}$, with c_{12} replaced by c_{23} in their respective formulas.

Optimal Strategies Under Partition $\pi_4 = \{\{1, 3\}, \{2\}\}$

Proposition 5. *Under the coalition partition $\pi_4 = \{\{1, 3\}, \{2\}\}$, each player's optimal strategy, denoted as $u^*(t, \pi_4)$, is equivalent to their strategy in the non-cooperative case π_1. Specifically, $u^*(t, \pi_4) = u^*(t, \pi_1)$.*

Proof. For players 1 and 3, this equivalence arises because players 1 and 3, though part of a coalition in π_4, are not directly connected. The value functions for each player are given by:

$$V_{\pi_4,t}(\{1\}, x_1) = V_{\pi_1,t}(\{1\}, x_1), \quad V_{\pi_4,t}(\{3\}, x_3) = V_{\pi_1,t}(\{3\}, x_3).$$

The optimal strategies $u_1^*(t, \{1\})$ and $u_3^*(t, \{3\})$ are then obtained by maximizing these individual value functions. Consequently, their strategic positioning and optimal choices in π_4 remain the same as if they were in single-player coalitions, akin to the non-cooperative scenario π_1.

For player 2, who is in a single-player coalition under π_4, the strategy follows the same rationale.

4 Analysis of Dynamically Stable Coalition Partition π_1

In this section, we analyze the coalition partitions and the distribution of payoffs in a three-player game. Our special focus is on the initial coalition partition $\pi_1 = \{\{1\}, \{2\}, \{3\}\}$ and how it evolves over time.

For the initial coalition partition π_1, each player forms a separate coalition. We consider the sets of potential coalitions for each player as follows:

$$D_1(\pi_1) = \{\{\{1\}, \{2\}, \{3\}\}, \{\{1, 2\}, \{3\}\}, \{\{1, 3\}, \{2\}\}\} = \{\pi_1, \pi_2, \pi_4\},$$
$$D_2(\pi_1) = \{\{\{1\}, \{2\}, \{3\}\}, \{\{1, 2\}, \{3\}\}, \{\{2, 3\}, \{1\}\}\} = \{\pi_1, \pi_2, \pi_3\},$$
$$D_3(\pi_1) = \{\{\{1\}, \{2\}, \{3\}\}, \{\{2, 3\}, \{1\}\}, \{\{1, 3\}, \{2\}\}\} = \{\pi_1, \pi_3, \pi_4\}.$$

In order to allocate the payoffs among these coalitions, we employ the Aumann-Dreze value methodology. This technique takes into account the payoffs that every player may obtain in various coalition arrangements. Our specific objective is to demonstrate the dynamic stability of the coalition partition π_1. We also ensure that the vector $\beta(\pi_1, t)$ is a time-consistent IDP.

Payoff distribution in coalition partition $\pi_1 = \{\{1\}, \{2\}, \{3\}\}$

We obtain the following expressions for each player along the trajectory $x^*(t, \pi_1)$:

$$\phi_i(\pi_1, V_{\pi_1,t}, K_{\pi_1}(i), t) = V_{\pi_1,t}(\{i\}, x^*(t, \pi_1)) = A_i x_i^*(t, \{i\}) + B_i, \quad i = 1, 2, 3,$$

where A_i and B_i are coefficients that have been previously defined and elaborated upon in Eq. (9). The term $x_i^*(t, \{i\})$ represents the state trajectory of player i under the coalition partition π_1, as outlined in Eq. (8).

Following the computation of the payoff functions under the coalition partition π_1, we derive the following expression for each player's payoff at any given time t. The calculation incorporates the dynamics of the players' state trajectories and their corresponding strategies:

$$\phi_i(\pi_1, V_{\pi_1,t}, K_{\pi_1}(i), t) =$$
$$\frac{(p(1 - \delta\epsilon) + \delta c_i r)(c_i r(\delta + \delta\epsilon + 2(1 - \delta)e^t - 2) - p(1 - \epsilon)(1 - \delta\epsilon))}{4(1 - \delta)(1 - \epsilon)(1 - \delta\epsilon)^2}, \quad i = 1, 2, 3.$$

The IDP as described in Equation (3) is formulated as follows:

$$\beta_i(\pi_1, V_{\pi_1,t}, K_{\pi_1}(i), t) =$$
$$\frac{(p(1-\delta\epsilon) + \delta c_i r)(c_i r(\delta + \delta\epsilon + 2(1-\delta\epsilon)\epsilon^t - 2) - p(1-\epsilon)(1-\delta\epsilon))}{4(1-\epsilon)(1-\delta\epsilon)^2}, \quad i = 1, 2, 3.$$

The IDP $\beta(\pi_1, t)$ specifically captures the immediate payoff allocated to a player at a particular moment, reflecting a time-consistent approach to distribution within the coalition.

Payoff distribution in coalition partition $\pi_2 = \{\{1,2\},\{3\}\}$

The payoff distributions for the players in the coalition partition π_2 are formulated as follows:

$$\phi_1(\pi_2, V_{\pi_2,t}, K_{\pi_2}(1), t)$$
$$= \frac{1}{2}\left(V_{\pi_2,t}(\{1,2\}, x^*(t, \pi_2)) - V_{\pi_1,t}(\{2\}, x^*(t, \pi_2)) + V_{\pi_1,t}(\{1\}, x^*(t, \pi_2))\right),$$
$$\phi_2(\pi_2, V_{\pi_2,t}, K_{\pi_2}(2), t)$$
$$= \frac{1}{2}\left(V_{\pi_2,t}(\{1,2\}, x^*(t, \pi_2)) - V_{\pi_1,t}(\{1\}, x^*(t, \pi_2)) + V_{\pi_1,t}(\{2\}, x^*(t, \pi_2))\right),$$
$$\phi_3(\pi_2, V_{\pi_2,t}, K_{\pi_2}(3), t) = V_{\pi_1,t}(\{3\}, x^*(t, \pi_1)),$$

where $x^*(t, \pi_2)$ represents the aggregate state of the coalition $\{1,2\}$ at time t.

The time-consistent IDP for each player at time t in relation to the coalition partition π_2 is calculated and explained in Appendix A. To determine the time-consistent IDP for each player at time t in the coalition partition π_3, we calculate it according to the instructions provided in Appendix B.

Our research focuses on the payoffs ϕ_i and β_i that each player obtains under various coalition arrangements (without coalition transition) at any given point t.

Our next objective is to prove that the coalition partition π_1 fulfills the requirement mentioned in Eq. (4). However, in this particular situation, we will particularly demonstrate the inequality for player 1, who has no intention of deviating from their decision to form a coalition with player 2 at any point in time t.

The following condition guarantees that, at any given moment t, player 1 lacks any motivation to deviate from cooperating with player 2:

$$\phi_1(\pi_1, V_{\pi_1,t}(\{1\}, x(t, \pi_1)), K_{\pi_1}(1), t)$$
$$\geq \beta_1(\pi_1, V_{\pi_1,t}(\{1\}, x(t, \pi_1)), K_{\pi_1}(1), t)$$
$$+ \delta\phi_1(\pi_2, V_{\pi_2,t+1}(\{1,2\}, T_{1,\pi_2}x(t, \pi_1)), K_\pi(1), t+1)$$
$$\geq \phi_1(\pi_2, V_{\pi_2,t}(\{1\}, x(t, \pi_2)), K_{\pi_2}(1), t).$$

In the following, we will present the allocation that player i receives from time $t+1$ until the end of the game if they choose to deviate to the coalition structure π_2 at time $t+1$.

Firstly, we calculate the state transition at time $t+1$ as specified by:

$$x_i'(t+1,\{1,2\}) = \epsilon x_i(t,\{i\}) + r\frac{(p(1-\delta\epsilon) - \delta r c_{12})}{4(1-\delta\epsilon)}, \quad i = 1,2, \qquad (14)$$

subsequently integrating (14) into the characteristic function for coalition π_2:

$$V_{\pi_2,t+1}(\{1,2\}, T_{1,\pi_2}x(t,\pi_1)) = A_{12}T_{1,\pi_2}x(t,\pi_1) + B_{12}. \qquad (15)$$

Therefore,

$$\phi_1(\pi_2, V_{\pi_2,t+1}(\{1,2\}, T_{1,\pi_2}x(t,\pi_1)), K_\pi(1), t+1)$$
$$= \frac{1}{2}(V_{\pi_2,t+1}(\{1,2\}, T_{1,\pi_2}x(t,\pi_1)) - V_{\pi_1,t+1}(\{1\}, T_{1,\pi_2}x(t,\pi_1))) \qquad (16)$$
$$+ \frac{1}{2}V_{\pi_1,t+1}(\{2\}, T_{1,\pi_2}x(t,\pi_1)).$$

This formulation computes the discounted distribution for player 1 deviating to coalition π_2 at time t.

The configuration of parameters for our example model is detailed as follows: a product price (p) of 0.5, a discount factor (δ) of 0.8, pollution control costs for individual players with $c_1 = 0.1$, $c_2 = 0.2$, $c_3 = 0.3$, pollution control costs for coalitions at $c_{12} = 0.1$, $c_{23} = 0.2$, $c_{123} = 0.1$, an initial pollution level (x_0) of 0, a rate of pollution removal (r) of 0.7, and a natural decay factor (ϵ) of 0.1.

We present the numerical simulation results for the IDPs under different coalition structures, using the aforementioned parameters:

Graphical Analysis

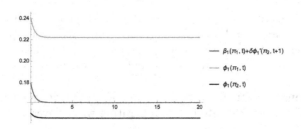

Fig. 3. Payoff analysis of coalition transition for player 1

Figure 3 depicts the simulation results of the distribution of payoffs in the setting of coalition development over time. The green line, denoted as $\phi_1(\pi_1, t)$, illustrates the payout that player 1 obtains by sustaining a coalition consisting of only themselves from time t until the end of the game. The red line represents the sum of player 1's immediate payment while in the single-person coalition at time t, and the discounted value of the payoffs that start at time t+1 when player 1 deviates to join a coalition with player 2. The blue line represents the payoffs that player 1 would earn if they were to create a coalition with player 2, starting from the initial time. Player 1 consistently chooses to remain in the single-player coalition, indicating a preference for this option over forming a coalition with player 2. This confirms the dynamic stability of the single-player strategy in the simulated environment.

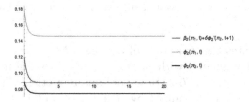

(a) Stability analysis of player 2's single coalition versus forming a coalition with player 1

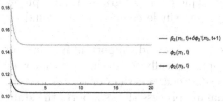

(b) Player 2's incentive to maintain single status against coalition formation with player 3

Fig. 4. Comparative distribution analysis for player 2: single versus coalition formation

In Fig. 4, player 2 autonomously establishes a coalition, with a pollution control cost of $c_2 = 0.2$. When player 1 joins the coalition, the combined cost of controlling pollution is reduced to $c_{12} = 0.1$. Although the expenses for pollution management have fallen, player 2's production quantity decreases inside the coalition. Therefore, the reduction in pollution costs does not offset the decline in output. As a result, player 2 is disinclined to establish an alliance with player 1. Another potential departure occurs when player 2 decides to establish a coalition $\{2, 3\}$ with player 3. This variation is evidently harmful to player 2. Although the coalition $\{2, 3\}$ has a cost parameter of 0.2, player 2's stake in the coalition is reduced due to player 3's higher cost parameter of 0.3. The allocation of resources to player 2 inside the coalition $\{2, 3\}$ is dependent upon player 3's cost parameter. Consequently, player 2 is unwilling to form a coalition with a partner who may diminish its share.

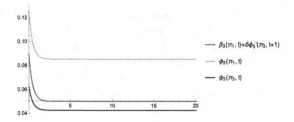

Fig. 5. Player 3's payoff: single vs. coalition with player 2

Figure 5 illustrates the trajectory of the reward for player 3. Player 3's decision to create a new alliance with player 2 results in a fall in pollution control cost from $c_3 = 0.3$ to $c_{23} = 0.2$. However, this reduction is not enough to off-set the decrease in revenue caused by reduced output in the coalition $\{2, 3\}$. This suggests that although the cost of pollution control is lower in the coalition consisting of players 2 and 3, player 3's decision to cut their production within this coalition leads to a decrease in their share of the allocation. This ultimately leads to the conclusion depicted in the graph. Hence, player 3 has no motivation to depart from the existing single coalition as it would not enhance his or her welfare.

To summarize, the graphs provide important insights into the process of coali-tion development. People typically resist changes that would raise the expenses of pollution management or decrease their production levels, which demonstrates the logical decision-making process that supports the stability of coalitions.

5   Conclusions

The results of this study make a substantial contribution to our comprehen-sion of strategic conduct in the field of environmental economics, specifically in the realm of pollution control games that are built on networks. This article presents a new method for defining dynamically stable coalition partitions using the IDP, providing a strong foundation for studying the stability of coalitions among members with different characteristics. The simulation findings confirm the accuracy of the theoretical model and also illustrate the actual usefulness of IDP in maintaining the dynamic stability of coalitions. This offers vital guidance to policymakers and stakeholders in developing more efficient pollution control techniques. This innovative method enhances the predictability and stability of coalition formations, underscoring the importance of strategic planning and cooperation in addressing environmental challenges.

Acknowledgments. This research was supported by the Russian Science Foundation (No. 22-11-20015, No. 22-11-00051).

Disclosure of Interests. The authors have no competing interests to declare that are relevant to the content of this article.

A   Time-consistent IDP calculations for π_2

The payoffs for players 1 and 2 in the coalition partition π_2 are computed in the following manner, taking into account their individual state trajectories and the cooperative interaction dynamics within the coalition:

$$\phi_1(\pi_2, V_{\pi_2,t}, K_{\pi_2}(1), t) = -\frac{1}{8(1-\delta)(1-\epsilon)(1-\delta\epsilon)^2}[(1-\epsilon)(1-\delta\epsilon)^2 p^2$$
$$+ (1-\delta\epsilon)pr(-c_1 - \delta c_1 - 2c_{12} + c_2 + \delta c_2 + 2\delta c_1\epsilon + 2\delta c_{12}\epsilon - 2\delta c_2\epsilon)$$
$$+ \delta c_{12}(c_1 + 2c_{12} - c_2)(1 - \epsilon^t) - (1-\delta\epsilon)pr(1-\delta)(c_1 + 2c_{12} - c_2)\epsilon^t$$
$$+ \delta^2 r^2(c_1^2(1-\epsilon) - c_2^2(1-\epsilon) + c_{12}^2(1+\epsilon - 2\epsilon^t) - c_1 c_{12}(1-\epsilon^t) + c_{12}c_2(1-\epsilon^t))],$$

$$\phi_2(\pi_2, V_{\pi_2,t}, K_{\pi_2}(2), t) = \frac{1}{8(1-\delta)(1-\epsilon)(1-\delta\epsilon)^2}[-((1-\epsilon)(1-\delta\epsilon)^2 p^2)$$
$$+ (1-\delta\epsilon)pr(2c_{12} + c_2 + (1-\delta)(2c_{12} + c_2)\epsilon^t + \delta(c_2 - 2(c_{12} + c_2)\epsilon))$$
$$+ \delta c_{12}(c_1 - 2c_{12} - c_2)(1 - \epsilon^t) - (1-\delta\epsilon)prc_1(1 + \delta - 2\delta\epsilon + (1-\delta)\epsilon^t)$$
$$+ \delta^2 r^2(c_1^2(1-\epsilon) - c_2^2(1-\epsilon) - c_1 c_{12}(1-\epsilon^t) + c_{12}c_2(1-\epsilon^t) + c_{12}^2(-1-\epsilon + 2\epsilon^t))].$$

The equations provide the rewards for players 1 and 2 when they establish a coalition {1,2} under partition π_2. These rewards are based on their joint contribution to the coalition's performance and the subsequent distribution of the total reward.

Under the coalition partition π_2, we compute the time-consistent IDP for each player at time t:

$$\beta_1(\pi_2, V_{\pi_2,t}, K_{\pi_2}(1), t) = \frac{1}{8(1-\epsilon)(1-\delta\epsilon)^2}[(1-\epsilon)(1-\delta\epsilon)^2 p^2$$
$$+ (1-\delta\epsilon)pr(-c_1 - \delta c_1 - 2c_{12} + c_2 + \delta c_2 + 2\delta c_1\epsilon + 2\delta c_{12}\epsilon - 2\delta c_2\epsilon)$$
$$+ \delta c_{12}(c_1 + 2c_{12} - c_2)(1 - \epsilon^t) - (1-\delta\epsilon)pr(c_1 + 2c_{12} - c_2)\epsilon^t(1-\delta\epsilon)$$
$$+ \delta^2 r^2(c_1^2(1-\epsilon) - c_1 c_{12}(1-\epsilon^t) + c_{12}^2(1+\epsilon - 2\epsilon^t) + c_{12}c_2(1-\epsilon^t) + c_2^2(1-\epsilon))],$$

$$\beta_2(\pi_2, V_{\pi_2,t}, K_{\pi_2}(2), t) = \frac{1}{8(1-\epsilon)(1-\delta\epsilon)^2}[(1-\epsilon)(1-\delta\epsilon)^2 p^2$$
$$+ (1-\delta\epsilon)pr(c_1 + \delta c_1 - 2c_{12} - c_2 - \delta c_2 - 2\delta c_1\epsilon + 2\delta c_{12}\epsilon + 2\delta c_2\epsilon)$$
$$+ \delta c_{12}(-c_1 + 2c_{12} + c_2)(1 - \epsilon^t) + (1-\delta\epsilon)pr(c_1 - 2c_{12} - c_2)\epsilon^t(1-\delta\epsilon)$$
$$+ \delta^2 r^2(c_1^2(1-\epsilon) - c_1 c_{12}(1-\epsilon^t) + c_{12}^2(-1-\epsilon + 2\epsilon^t) + c_{12}c_2(1-\epsilon^t) + c_2^2(1-\epsilon))].$$

These equations represent the time-consistent IDPs for players 1, 2 in the coalition partition π_2. The IDPs capture the individual payoffs at time t, adjusted for the dynamics of the coalition structure and the players' strategic interactions within it.

B   Time-consistent IDP calculations for π_3

In the coalition partition $\pi_3 = \{\{2,3\},\{1\}\}$, the time-consistent IDPs for players 2 and 3 are computed as follows:

$$\beta_2(\pi_3, V_{\pi_3,t}, K_{\pi_3}(2), t) = \frac{1}{8(1-\epsilon)(1-\delta\epsilon)^2}[(1-\epsilon)(1-\delta\epsilon)^2 p^2$$
$$+ (1-\delta\epsilon)pr(-c_2 - \delta c_2 - 2c_{23} + c_3 + \delta c_3 + 2\delta c_2\epsilon + 2\delta c_{23}\epsilon - 2\delta c_3\epsilon)$$
$$+ \delta c_{23}(c_2 + 2c_{23} - c_3)(1-\epsilon^t) - (1-\delta\epsilon)pr(c_2 + 2c_{23} - c_3)\epsilon^t(1-\delta\epsilon)$$
$$+ \delta^2 r^2(c_2^2(1-\epsilon) - c_2 c_{23}(1-\epsilon^t) + c_{23}^2(1+\epsilon-2\epsilon^t) + c_{23}c_3(1-\epsilon^t) + c_3^2(1-\epsilon))],$$

$$\beta_3(\pi_3, V_{\pi_3,t}, K_{\pi_3}(3), t) = \frac{1}{8(1-\epsilon)(1-\delta\epsilon)^2}[(1-\epsilon)(1-\delta\epsilon)^2 p^2$$
$$+ (1-\delta\epsilon)pr(c_2 + \delta c_2 - 2c_{23} - c_3 - \delta c_3 - 2\delta c_2\epsilon + 2\delta c_{23}\epsilon + 2\delta c_3\epsilon)$$
$$+ \delta c_{23}(-c_2 + 2c_{23} + c_3)(1-\epsilon^t) + (1-\delta\epsilon)pr(c_2 - 2c_{23} - c_3)\epsilon^t(1-\delta\epsilon)$$
$$+ \delta^2 r^2(c_2^2(1-\epsilon) - c_2 c_{23}(1-\epsilon^t) + c_{23}^2(-1-\epsilon+2\epsilon^t) + c_{23}c_3(1-\epsilon^t) + c_3^2(1-\epsilon))].$$

These equations capture the individual payoffs for players 2 and 3 within the coalition {2,3} at time t, considering the combined contributions of the coalition members and the dynamics of their cooperation.

References

1. Abe, T.: Stable coalition partitions in symmetric majority games: a coincidence between myopia and farsightedness. Theory Decis., 1–22 (2018)
2. Anastasiadis, E., Deng, X., Krysta, P., et al.: Network pollution games. Algorithmica **81**, 124–166 (2019)
3. Aumann, R.J., Dreze, J.H.: Cooperative games with coalition partitions. Internat. J. Game Theory **3**(4), 217–237 (1974)
4. Bala, V., Goyal, S.: A noncooperative model of network formation. Econometrica **68**(5), 1181–1229 (2000)
5. Bogomolnaia, A., Jackson, M.O.: The stability of hedonic coalition partitions. Games Econom. Behav. **38**(2), 201–230 (2002)
6. Breton, M., Keoula, M.Y.: A great fish war model with asymmetric players. Ecol. Econ. **97**, 209–223 (2014)
7. Carraro, C.: The structure of international environmental agreements. In: Paper Presented at the FEEM/IPCC/Stanford EMF Conference on International Environmental Agreements on Climate Change, pp. 309–328. Venice (1997)
8. Fischer, R., Mirman, L.: Strategic dynamic interaction: fish wars. J. Econ. Dyn. Control **16**(2), 267–287 (1992)
9. Gusev, V., Mazalov, V.: Potential functions for finding stable coalition partitions. Oper. Res. Lett. **47**(6), 478–482 (2019)
10. Hoefer, M., Vaz, D., Wagner, L.: Dynamics in matching and coalition formation games with structural constraints. Artif. Intell. **262**, 222–247 (2018)
11. Jackson, M.O., Wolinsky, A.: A strategic model of social and economic networks. J. Econ. Theory **71**(1), 44–74 (1996)

12. Lindroos, M.: Coalitions in international fisheries management. Nat. Resour. Model. **21**, 366–384 (2008)
13. Ohta, K., Barrot, N., Ismaili, A., Sakurai, Y., Yokoo, M.: Core stability in hedonic games among friends and enemies: impact of neutrals. IJCAI, 359–365 (2017)
14. Parilina, E., Sedakov, A.: Stable cooperation in a game with a major player. Int. Game Theory Rev. **18**(02), 1640005 (2016)
15. Petrosjan, L., Zaccour, G.: Time-consistent Shapley value allocation of pollution cost reduction. J. Econ. Dyn. Control **7**, 381–398 (2003)
16. Petrosyan, L., Danilov, N.: Stable solutions in non-zero sum differential games with transferable payoffs. Vestnik of Leningrad Univ. Ser. **1**(1), 46–54 (1979)
17. Pintassilgo, P., Lindroos, M.: Coalition formation in straddling stock fisheries: a partition function approach. Int. Game Theory Rev. **10**(2), 303–317 (2008)
18. Rettieva, A.N.: Stable coalition partition in bioresource management problem. Ecol. Model. **235**, 102–118 (2012)
19. Sedakov, A., Qiao, H., Wang, S.: A model of river pollution as a dynamic game with network externalities. Eur. J. Oper. Res. **290**(3), 1136–1153 (2021)
20. Yeung, D.: An irrational-behavior-proof condition in cooperative differential games. Int. Game Theory Rev. **8**(4), 739–744 (2006)
21. Zeeuw, A.D.: Dynamic effects on the stability of international environment agreements. J. Environ. Econ. Manag. **55**(2), 163–174 (2008)

Operations Research

Robustness of Graphical Lasso Optimization Algorithm for Learning a Graphical Model

Valeriy Kalyagin[1,2]([⊠]) [iD] and Ilya Kostylev[1] [iD]

[1] HSE University, Laboratory of Algorithms and Technologies for Network Analysis, Nizhny Novgorod, Russia
vkalyagin@hse.ru, idkostylev@edu.hse.ru
[2] Steklov Mathematical Institute, Russian Academy of Sciences, Moscow, Russia
https://nnov.hse.ru/en/latna/

Abstract. Problem of learning a graphical model (graphical model selection problem) consists of recovering a conditional dependence structure (concentration graph) from data given as a sample of observations from a random vector. Various algorithms to solve this problem are known. One class of algorithms is related with convex optimization problem with additional lasso regularization term. Such algorithms are called graphical lasso algorithms. Various properties and practical efficiency of graphical lasso algorithms were investigated in the literature. In the present paper we study sensitivity of uncertainty (level of error) of graphical lasso algorithms to the change of distribution of the random vector. This issue is not well studied yet. First, we show that uncertainty of the classical version of graphical lasso algorithm is very sensitive to the change of distribution. Next, we suggest simple modifications of this algorithm which are much more robust in the large class of distributions. Finally, we discuss a future development of the proposed approach.

Keywords: graphical model · concentration graph · uncertainty · graphical lasso optimization · robustness

1 Introduction

Let $X = (X_1, X_2, ..., X_N)$ be a random vector in R^N. The graphical model associated with X is a graph G, that represents the dependence structure of the components of X. Graphical models are widely used in various applications, such as in image processing [1], natural language processing [2], computational biology [3], and bioinformatics and genomics [4,5] and others. The problem of learning the graphical model, also known as the graphical model selection problem, is to recover the associated dependency graph G given a sample of observations drawn from the distribution of X.

Graphical model selection is an active field of research that has attracted growing attention in recent decades (see the recent reviews [6,7]). Various algorithms have been proposed and studied to identify the dependency graph G from

© The Author(s), under exclusive license to Springer Nature Switzerland AG 2024
A. Eremeev et al. (Eds.): MOTOR 2024, LNCS 14766, pp. 337–348, 2024.
https://doi.org/10.1007/978-3-031-62792-7_23

observations. One class of these algorithms involves solving a maximum likelihood optimization problem with an additional lasso regularization term, and these algorithms are known as graphical lasso methods. A popular version of this method was proposed in [3] for selecting Gaussian graphical models, where the vector X has a multivariate Gaussian distribution and the corresponding graph G represents pairwise conditional dependencies between X_i and X_j, given all other X_k, $i \neq j$, $i, j = 1, 2, \ldots, N$. Various modifications and variants of the graphical lasso method have been discussed in the literature (see for example recent advances in [9]).

The quality of any graphical model selection algorithm can be measured by its ability to accurately identify the graphical model given observations. One can consider several factors related to this error, including the measure of dependence between variables, the distribution of vector X, the size of the sample used, and the dimension of vector X. In this paper, we investigate how the identification error of a particular algorithm depends on the distribution of random vector X. A desirable property of an algorithm in this context is "robustness", or low sensitivity of identification error to changes in distribution. This is of practical importance because it allows algorithms to function well in situations where we do not have sufficient information about the distribution of X. As far as we know, this topic has not been well-studied yet. Another sense of "robustness" used in the literature is related with sensitivity of identification error to outliers in data. Robust in this sense algorithm were proposed in [10]. It is based on the minimum covariance determinant (MCD) estimator for the variance-covariance matrix. We will call this algorithm *MCD algorithm for graphical model selection* and we will use it to compare it's robustness with our versions of graphical lasso algorithms.

To investigate sensitivity of graphical lasso algorithms to changes in the distribution, we propose a new general framework for the graphical model selection problem and introduce a wide class of distributions with the same dependency graph structure. This allows us to make a fair comparison of the robustness of different versions of graphical model selection algorithms. First, we demonstrate that the traditional version of the graphical lasso method is highly sensitive to changes in the distribution. To handle this problem, we suggest two new modifications of the classical lasso algorithm and show their robustness. We compare these algorithms with MCD algorithm and show their advantages in robustness with respect to distribution. Finally, we discuss a future development of the proposed approach.

The paper is organized as follows: In Sect. 2, we describe our general setting for the graphical model selection problem. Section 3 focuses on a large class of distributions that have the same graphical model structure. In Sect. 4, we present classical and new versions of the graphical lasso algorithm. In Sect. 5, we compare the described algorithms in terms of their sensitivity to identification errors due to changes in the underlying distribution. Finally, in Sect. 6, we discuss the results and provide directions for future research.

2   Graphical Model

Let $X = (X_1, X_2, ..., X_N)$ be a random vector in R^N. We consider the following graphical model related with conditional dependence structure between random variables X_i, X_j, $i \neq j$, $i, j = 1, 2, ..., N$. Denote by $\rho^{i,j}$ partial correlation between X_i, X_j in the collection of random variables $\{X_1, X_2, ..., X_N\}$. Graphical model, considered in this paper, is undirected simple graph $G = (V, E)$ with N vertices, $|V| = N$, and the set of edges E defined by: $(i, j) \in E$, if and only if $\rho^{i,j} = 0$. If X has a Gaussian distribution, then the condition $\rho^{i,j} = 0$ is equivalent to the conditional independence of X_i, X_j all other X_k being fixed. In this case the vector X has so-called pairwise Markov property with respect to the graph G (in others words: X form a Markov random field with respect to G). When X is not Gaussian, the graphical model G reflects some conditional dependence structure of the vector X.

Let $\Sigma = (\sigma_{i,j})$ be the covariance matrix of the vector X. Partial correlations $\rho^{i,j}$ can be calculated as follows

$$\rho^{i,j} = -\frac{\sigma^{i,j}}{\sqrt{\sigma^{i,i}\sigma^{j,j}}},$$

where $\sigma^{i,j}$ are the elements of the matrix $\Sigma^{-1} = (\sigma^{i,j})$. Therefore, the edge (i, j) is included in the graphical model G if and only if $\sigma^{i,j} = 0$. It implies that defined graphical model is associated with zero elements of the matrix Σ^{-1}. Matrix Σ^{-1} is sometimes called concentration matrix of the vector X, and associated graph is called concentration graph of the random vector X.

The presented graphical model is a natural generalization of the well-studied Gaussian Graphical Model, allowing us to consider probabilistic distributions other than Gaussian. In the following section, we describe a broad class of such distributions that have the same underlying graph structure, for any given covariance matrix Σ.

3   Class of Distributions

We consider the class of elliptical distributions. Random vector X has elliptical distribution with parameters μ (vector), Λ (matrix), $g(y)$ (function) if its density function has the form [11]:

$$f(x; \mu, \Lambda, g) = |\Lambda|^{-\frac{1}{2}} g\{(x - \mu)\Lambda^{-1}(x - \mu)^T\}, \quad x \in R^{1 \times N}, \tag{1}$$

where $\Lambda = (\lambda_{i,j})_{i,j=1,2,...,N}$ is positive definite symmetric matrix, $g(x) \geq 0$, and $(y \in R^{1 \times N})$

$$\int_{-\infty}^{\infty} \cdots \int_{-\infty}^{\infty} g(yy^T) dy_1 dy_2 \cdots dy_N = 1.$$

Two classical multivariate distributions, Gaussian distribution and Student's t-distribution are elliptical. One has for the Gaussian distribution from $N(\mu, \Lambda)$:

$$f_{Gauss}(x) = \frac{1}{(2\pi)^{p/2}|\Lambda|^{\frac{1}{2}}} e^{-\frac{1}{2}(x-\mu)\Lambda^{-1}(x-\mu)^T},$$

and similarly for the multivariate Student's t-distribution with ν degrees of freedom:

$$f_{Student}(x) = \frac{\Gamma\left(\frac{\nu+N}{2}\right)}{\Gamma\left(\frac{\nu}{2}\right)\nu^{N/2}\pi^{N/2}}|\Lambda|^{-\frac{1}{2}}\left[1 + \frac{(x-\mu)\Lambda^{-1}(x-\mu)^T}{\nu}\right]^{-\frac{\nu+N}{2}}.$$

Various properties of elliptical distributions are similar to Gaussian distributions. In particular, if for an elliptical vector X its expectation $E(X)$ exists then $E(X) = \mu$. Moreover, if for an elliptical vector X its covariance matrix $Cov(X) = \Sigma = (\sigma_{i,j})$ exists then the following relation holds

$$\sigma_{i,j} = Cov(X_i, X_j) = C \cdot \lambda_{i,j}, \tag{2}$$

where C is a constant and

$$C = \frac{2\pi^{\frac{1}{2}N}}{\Gamma(\frac{1}{2}N)}\int_0^{+\infty} r^{N+1}g(r^2)dr.$$

For Gaussian distribution one has $Cov(X_i, X_j) = \lambda_{i,j}$. For multivariate Student's t-distribution with ν degrees of freedom ($\nu > 2$) one has $\sigma_{i,j} = \nu/(\nu-2)\lambda_{i,j}$. Detailed study of elliptical distributions can be found in [11].

Let EGM(Λ, μ) denote the class of elliptical distributions with expectation μ, and whose matrix parameter is proportional to the fixed matrix Λ. The EGM abbreviation stands for the Elliptical Graphical Model. The following lemma is the basis of our study of the sensitivity of graphical model selection algorithms to changes in the underlying distribution.

Lemma 1. *All distributions X from the class EGM(Λ, μ) have the same graphical model G.*

Proof: Let matrix parameter for the elliptical distribution $X \in$ EGM(Λ) be $K \cdot \Lambda$, for a constant $K \in R$, $K \neq 0$. The relation (2) implies for the covariance matrix $\Sigma(X) = C \cdot (K \cdot \Lambda)$, and $\Sigma^{-1} = \frac{1}{(C\cdot K)}\Lambda^{-1}$. Therefore

$$\rho^{i,j}(X) = -\frac{\sigma^{i,j}}{\sqrt{\sigma^{i,i}\sigma^{j,j}}} = -\frac{\lambda^{i,j}/(C\cdot K)}{\sqrt{(\lambda^{i,i}/(C\cdot K))(\lambda^{j,j}/(C\cdot K))}} = -\frac{\lambda^{i,j}}{\sqrt{\lambda^{i,i}\lambda^{j,j}}},$$

where $\Lambda^{-1} = (\lambda^{i,j})$. It means that all $X \in$ EGM(Λ) have the same partial correlations, and for two vectors $X^{(1)}, X^{(2)} \in$ EGM(Λ) one has $\rho^{i,j}(X^{(1)}) = 0$ if and only if $\rho^{i,j}(X^{(2)}) = 0$. The statement follows.

4   Graphical Lasso Algorithms

Learning a graphical model from observations (the graphical model selection problem) can be viewed as a binary classification problem for pairs (i,j), where $i \neq j$ and $i, j = 1, 2, ..., N$. The class "YES" is the set of pairs (i,j) such that there is an edge between nodes i and j in the graphical model G, and the class "NO" is the complement of this set. Any graphical model selection algorithm outputs an estimate $\hat{G}$ of the true graphical model G. The quality of the algorithm can be assessed by measuring the expected value of the error, which is the difference between G and $\hat{G}$. In this paper, we will focus on specific measures of accuracy, such as True Negative Rate (TNR), False Discovery Rate (FDR), False Omission Rate (FOR), True Positive Rate (TPR), Balanced Accuracy (BA), F1 Score, and Matthew's Correlation Coefficient. These measures will help us understand the robustness of different algorithms when the data distribution changes.

The graphical lasso algorithm takes as input a sample of observations from the random vector X, and as output it gives a graphical model associated with the vector X. This algorithm is based on the following optimization problem described in [8]:

$$Trace(S\Omega) - log \det(\Omega) + \gamma ||\Omega||_1 \to min,$$

over all positive semi-definite matrices $\Omega = (\omega_{i,j})$ with $||\Omega||_1 = \sum_{i=1}^{N} \sum_{j=1}^{N} |\omega_{i,j}|$, $\gamma > 0$ being a regularization parameter. In the classical version of the algorithm, the matrix S is the sample covariance matrix. The objective function of this optimization problem is convex, so many optimization algorithms can be applied to solve this problem. The output of this optimization procedure is an estimate of the inverse covariance matrix Σ^{-1}. To obtain the associated graphical model, one needs to identify the zero elements of this solution to the optimization problem, which give all absent edges in the graph.

We will show in the next section, that classical version of the graphical lasso algorithm is highly sensitive to changes in the distribution in the class $EGM(\Lambda, \mu)$ for a wide range of the matrix Λ. To handle this problem we suggest two new versions of the algorithm and show their robustness. The main idea comes form our previous investigation of uncertainty of graph structure in random variable networks [12]. Proposed modification of graphical lasso algorithm is not in optimization technique but in the more appropriate use of the data given by observations. Despite a simplicity of the proposed approach one get a new and efficient algorithms for graphical model selection.

Let X be a random vector from the class $EGM(\Lambda, \mu)$, $X \in EGM(\Lambda, \mu)$. It is known, that sample covariance matrix is not a robust (to the change of distribution) estimation of the covariance matrix Σ. To obtain more robust estimations of the covariance matrix one can use another type of correlations at the place of Pearson correlations. It was previously observed, that useful correlations in this setting are Fechner correlations [13]. In the present paper we suggest to use two type of correlations in the estimation of the covariance matrix and in the new versions of graphical lasso algorithm, Kendall and Fechner correlations. For a pair of random variables U, V Kendall correlation between U and V is formally defined as follows:

$$\rho^{Kendall}(U, V) = 2P\{(U_1 - U_2)(V_1 - V_2) > 0\} - 1, \tag{3}$$

where U_1, U_2 are independent random variables with the same distribution as U, and V_1, V_2 are independent random variables with the same distribution as V.

Usual Kenlall rank correlation $\hat{\rho}^{Kendall}$ is an unbiased and consistent estimation of the $\rho^{Kendall}$. For the Fechner correlation one has by definition:

$$\rho^{Fechner}(U, V) = 2P((U - E(U))(V - E(V)) > 0) - 1. \qquad (4)$$

Unbiased and consistent estimation of $\rho^{Fechner}$ can be easily obtained from calculation of the sign coincidence of the observed values of U and V. More details can be found in [12]. Our modifications of graphical lasso algorithm are based on the following statement, proved in [12]

Lemma 2. [12] Let random vector X has elliptical distribution. Then for any $i, j = 1, 2, \ldots, N$ the following relations hold

$$\rho^{Pearson}((X_i, X_j)) = \sin(\frac{\pi}{2}\rho^{Fechner}(X_i, X_j)) = \sin(\frac{\pi}{2}\rho^{Kendall}(X_i, X_j)). \quad (5)$$

Based on this fact we propose two new versions of graphical lasso algorithm:

Graphical Lasso via Kendall: In the classical version of the graphical lasso algorithm, replace the estimations of Pearson's correlations (covariances) with its estimations using the sample value of Kendall's correlations

$$\sin(\frac{\pi}{2}\hat{\rho}^{Kendall}(X_i, X_j)).$$

Graphical Lasso via Fechner: In the classical version of the graphical lasso algorithm, replace the estimations of Pearson's correlations (covariances) with its estimations using the sample value of Fechner's correlations

$$\sin(\frac{\pi}{2}\hat{\rho}^{Fechner}(X_i, X_j)).$$

In what follows, we will call the classical graphical lasso algorithm as *graphical lasso via Pearson*.

5   Robustness of Graphical Model Selection Algorithms

In this Section we study sensitivity of identification error of graphical model selection algorithms to the change in distribution. We use distributions from the class $EGM(\Lambda, \mu)$ and consider a wide class of the matrix Λ. We will compare robustness of the following algorithms: graphical lasso via Pearson, graphical lasso via Kendall, graphical lasso via Fechner, and "robust" MCD (Minimum Covariance Determinant estimator) algorithm from [10]. Our theoretical study in the Sect. 5.2 shows that the graphical lasso via Fechner algorithm guaranties the same level of identification error for any $X \in EGM(\Lambda, \mu)$, and therefore is robust with respect to the change in distribution in the class $EGM(\Lambda, \mu)$. Our numerical experiments in the Sect. 5.4 show that classical graphical lasso algorithm (graphical lasso via Pearson) is very sensitive to the change in distribution in the class $EGM(\Lambda, \mu)$. The graphical lasso via Kendall algorithm has some advantages with respect to other algorithms both in robustness and in accuracy of the graphical model identification. The MCD (Minimum Covariance Determinant estimator) algorithm is inferior to the graphical lasso via Kendall algorithm in efficiency and in robustness as well. Therefore, both new graphical model selection algorithms are very competitive with existing approaches. Another approaches to improve performance of the graphical lasso algoruthm was developed in [14].

5.1 Measures of Error for Graphical Model Selection Algorithms

It was already discussed above, that graphical model selection problem can be considered as a binary classification problem. To measure the quality of graphical model identification algorithms we propose to use the following characteristics:

- True Negative Rate: $TNR = \frac{TN}{TN+FP}$, proportion of correctly identified absences of edges in the graphical model.
- True Positive Rate: $TPR = \frac{TP}{TP+FN}$, proportion of correctly identified edges in the graphical model.
- False Discovery Rate: $FDR = \frac{FP}{TP+FP}$, proportion of incorrectly identified edges between all identifies edges.
- False Omission Rate: $FOR = \frac{FN}{TN+FN}$, proportion of incorrectly identified absences of edges between all identifies absences of edges.

This four characteristics give in our opinion a complete picture of the quality of any graphical model identification algorithm. Between various generalized criterion we will chose three:

- Balanced accuracy: $BA = \frac{1}{2}(TPR + TNR)$.
- F1 score: $F1 = 2\frac{(1-FDR)TPR}{(1-FDR)+TPR}$.
- Matthew Correlation Coefficient:

$$MCC = \frac{TN \times TP - FN \times FP}{\sqrt{(TP+FP)(TP+FN)(TN+FP)(TN+FN)}}.$$

Balanced accuracy and F1 score are standard in binary classification. Recently it was shown that MCC has some advantages with respect to BA and F1 in applications [15].

5.2 Robustness. Theoretical Study

The main result of this Section is the following

Theorem 1. *Let Λ be a positive definite matrix, $\Lambda \in R^{N \times N}$, $\mu = 0$. For any random vectors X from the class $EGM(\Lambda, 0)$ expected value of any measure of error of graphical model identification by the graphical lasso via Fechner algorithm is the same, and it is equal to the expected value of error of graphical model identification by the graphical lasso via Fechner algorithm for the Gaussian distribution from the class $N(0; \Lambda)$.*

Sketch of the proof: We model observations by a sequence

$$X(1), \; X(2), \; \ldots, \; X(n),$$

of independent identically distributed random variables with distribution X. Estimation of Fechner correlations $\hat{\rho}^{Fechner}(X_i, X_j)$ is based on the following statistics (remember that we suppose that $\mu = 0$)

$$T_{i,j} = \frac{1}{2} \sum_{t=1}^{n} (1 + \text{sign}(X_i(t))\text{sign}(X_j(t))).$$

These statistics give the number of the sign coincidence in two sequences $X_i(t)$ and $X_j(t)$, $t = 1, 2, \ldots, n$. Estimation of the Fechner correlation in this case can be calculated as follows

$$\hat{\rho}^{Fechner}(X_i, X_j) = 2\frac{T_{i,j}}{n} - 1.$$

It was proven in [12] (Chap. 5) that for any $X \in \text{EGM}(\Lambda, 0)$ distributions of statistics $T_{i,j}$ are the same. It means that estimations $\hat{\rho}^{Fechner}(X_i, X_j)$ have the same distributions for any $X \in \text{EGM}(\Lambda, 0)$. Graphical lasso via Fechner algorithm uses this estimations in calculation of the estimation $\hat{\Sigma}^{-1}$ of the concentration matrix Σ^{-1}. Therefore, distribution of the elements of the matrix $\hat{\Sigma}^{-1}$ is the same for any $X \in \text{EGM}(\Lambda, 0)$. It implies that probabilities of TP, TN, FP, FN are the same for any $X \in \text{EGM}(\Lambda, 0)$. The theorem follows.

5.3   Graphical Model Generator

To make our conclusions on numerical experiments more valuable we use a specific graphical model generator, known as Dominant Diagonal Generator (see [16]). This generator produces a random family of graphical models for a given dimension of the vector X and a given expected density of the graph G. The generator works as follows:

- Input: N – number of vertices, p – probability of an edge in a random graph.
- Generate random Erdos-Renyi graph $G(n, p) = (V, E)$ with given number of vertices N and probability of an edge equal to p.
- Generate matrix $A = (a_{i,j})$ with the following structure:

$$a_{i,j} = \text{Uniform}([-1, -0.5] \cap [0.5, 1]) \text{if } (i, j) \in E, \; a_{i,j} = 0, \text{ otherwise.}$$

- Divide each element in an i-th row of matrix A by $1.5(\sum_{k=1}^{N} |a_{i,k}|)$:

$$a_{i,j} = \frac{a_{i,j}}{1.5(\sum_{k=1}^{N} |a_{i,k}|)}.$$

- Calculate concentration matrix by symmetrizing matrix A and adding identity matrix: $\Sigma^{-1} = I + \frac{1}{2}(A + A^T)$.
- Calculate covariance matrix as inverse of the generated concentration matrix: $\Sigma = (\Sigma^{-1})^{-1}$.
- Output: covariance matrix Σ and associated graphical model.

Note, that the expected density of the obtained graphical models is p

5.4   Robustness. Numerical Experiments

To study robustness of the investigated algorithms we conducted a series of experiments. One experiment is conducted in 6 steps:

Step 1. Generate a graphical model with a given expected density and fix obtained covariance matrix Σ and graphical model G.

Step 2. Generate a sample of the size n from distribution of the random vector X.

Step 4. Apply a graphical model selection algorithms and identify the graph G.

Step 4. Calculate quality characteristics TPR, TNR, FDR, FOR, BA, F1, MCC.

Step 5. Repeat S_{obs} times the Step 2–4 and calculate average values of quality characteristics.

Step 6. Repeat S_{exp} times the Step 1–5 and calculate average values of quality characteristics.

We use the following parameters settings:

- Dimension of the random vector X (number of vertices in the graphical model), $N = 20$.
- Values of the concentration graph density: $d = 0.2$.
- Value of the regularization parameter in graphical lasso optimization problem $\gamma = 0, 1$.
- Distribution of the random vector X: mixture of the multivariate Gaussian and multivariate Student distributions with zero means and covariance matrix Σ. Density of the vector X is given by

$$f(x) = \epsilon f_{Student}(x) + (1 - \epsilon) f_{Gauss}(x),$$

where $\epsilon \in [0, 1]$, a mixture parameter.
- Degree of freedom for Student distribution: $\nu = 3$, $\nu = 2, 5$.
- Sample size (number of observations) $n = 100, 1000$.
- Number of replications for estimation of expected values $S_{obs} = 50$, $S_{exp} = 100$.

We evaluate robustness of an algorithms with respect to the change in distribution by evaluation of the quality characteristics TPR, TNR, FDR, FOR, BA, F1, MCC with evaluation of the mixture parameter ϵ from 0 to 1. The Fig. 1 shows dependence of the quality characteristics BA, F1, MCC on the value of the mixture parameter ϵ for the sample size $n = 1000$. One can see that quality of the classical graphical lasso algorithm (graphical lasso via Pearson algorithm) is highly sensitive to the change in distribution. Quality of the graphical lasso via Fechner algorithm is stable and does not change with distribution (Theorem 1). Interesting to note, that the quality of the graphical lasso via Kendall algorithm is superior to both other algorithms both in robustness and accuracy of the graphical model identification.

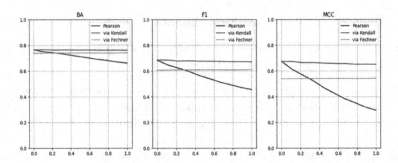

Fig. 1. Dependence of BA, F1, MCC on the mixture parameter ϵ for different versions of graphical lasso algorithm. Sample size $n = 1000$.

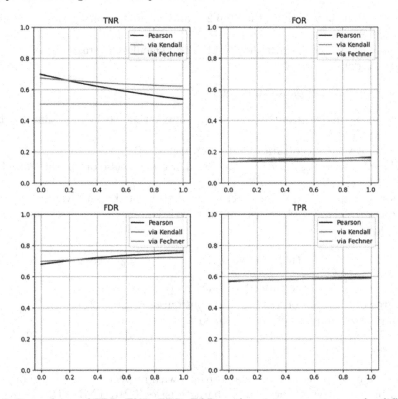

Fig. 2. Dependence of TPR, TNR, FDR, FOR on the mixture parameter ϵ for different versions of graphical lasso algorithm. Sample size $n = 100$.

Figure 2 presents the dependence of the quality characteristics TPR, TNR, FDR, FOR on the value of the mixture parameter ϵ for the sample size $n = 100$. The change in the sample size does not change the qualitative behavior of these characteristics. Quality of the classical graphical lasso algorithm (graphical lasso via Pearson algorithm) is still sensitive to the change in distribution. Quality of

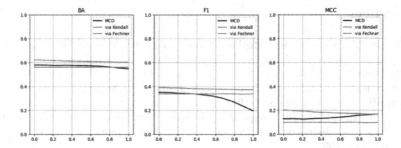

Fig. 3. Dependence of BA, F1, MCC on the mixture parameter ϵ for MCD algorithm in comparison with new versions of Graphical Lasso algorithm. Sample size $n = 100$.

the graphical lasso via Fechner algorithm is stable and does not change with distribution (Theorem 1). Quality of the graphical lasso via Kendall algorithm is still superior to other algorithms.

Figure 3 presents the results of comparison of the quality characteristics BA, F1, MCC of the robust MCD algorithm with our new versions of graphical lasso algorithms for n=100. One can see, that MCD algorithm is not robust with respect to the change in distribution (it was designed to be robust with respect to outliers in the data), and it's quality is inferior to both new algorithms.

Programming code and complete collection of results of numerical experiments are available on https://github.com/cofofprom/GGMS_framework/tree/master and https://github.com/cofofprom/GGMS_framework/blob/master/Results_corr_matr_observ_generator.pdf.

Note that the dependence of the accuracy of graphical model selection algorithms on graph density has recently been investigated in [17].

6 Conclusion

The robustness of the graphical lasso optimization algorithm for the graphical model selection problem is investigated in a wide range of distributions. It is demonstrated that the classical version of the algorithm is highly sensitive to changes in the distribution. New variants of the graphical lasso algorithms are proposed and analyzed. It is revealed that they outperform existing algorithms in terms of both robustness and accuracy in identifying graphical models. The proposed methodology holds promise for further exploration.

Acknowledgements. Sections 2 and 3 with general problem setting were prepared within the framework of the Basic Research Program at the National Research University Higher School of Economics (HSE University), results of the Sects. 4, 5 and 6 about graphical lasso algorithms are obtained with a support from RSF grant 22-11-00073. Numerical experiments were conducted using HSE HPC resources [18].

References

1. Geman, S., Geman, D.: Stochastic relaxation, Gibbs distributions, and the Bayesian restoration of images. IEEE Trans. Pattern Anal. Mach. Intel., (6), 721–741 (1984)
2. Manning, C.D., Schütze, H.: Foundations of Statistical Natural Language Processing. MIT Press, Cambridge (1999)
3. Durbin, R., Eddy, S.R., Krogh, A., Mitchison, G.: Biological Sequence Analysis: Probabilistic Models of Proteins and Nucleic Acids. Cambridge University Press, Cambridge (1998)
4. Liu, J., Peissig, P., Zhang, C., Burnside, E., McCarty, C., Page, D.: Graphical-model based multiple testing under dependence, with applications to genome-wide association studies. In: Uncertainty in artificial intelligence: proceedings of the Conference on Uncertainty in Artificial Intelligence, vol. 2012, p. 511, NIH Public Access (2012)
5. Zhou, L., Wang, L., Liu, L., Ogunbona, P., Dinggang, S.: Learning discriminative Bayesian networks from high-dimensional continuous neuroimaging data. IEEE Trans. Pattern Anal. Mach. Intell. **38**(11), 2269–2283 (2016)
6. Drton, M., Maathuis, M.H.: Structure learning in graphical modeling. Ann. Rev. Stat. Appl. **4**, 365–393 (2017)
7. Cordoba, I., Bielza, C., Larranaga, P.: A review of Gaussian Markov models for conditional independence. J. Stat. Plann. Infer. **206**, 127–144 (2020)
8. Friedman, J., Hastie, T., Tibshirani, R.: Sparse inverse covariance estimation with the graphical lasso. Biostatistics **9**(3), 432–441 (2008)
9. Seal, S., Li, Q., Basner, E.B., Saba, L.M., Kechris, K.: RCFGL: rapid condition adaptive fused graphical lasso and application to modeling brain region co-expression networks. PLoS Comput. Biol. **19**(1), e1010758 (2023)
10. Gottard, A., Pacillo, S.: Robust concentration graph model selection. Comput. Stat. Data Anal. **54**(12), 3070–3079 (2010)
11. Anderson, T.W.: An Introduction to Multivariate Statistical Analysis, 3rd edn. Wiley Interscience, New York (2003)
12. Kalyagin, V.A., Koldanov, A.P., Koldanov, P.A., Pardalos, P.M.: Statistical Analysis of Graph Structures in Random Variable Networks. SO, Springer, Cham (2020). https://doi.org/10.1007/978-3-030-60293-2
13. Kalyagin, V., Koldanov, A., Koldanov, P.: Robust identification in random variable networks. J. Stat. Plann. Infer. **181**, 30–40 (2017)
14. Cisneros-Velarde, P., Petersen, A., Oh, S.-Y.: Distributionally robust formulation and model selection for the graphical lasso. In: Proceedings of the Twenty Third International Conference on Artificial Intelligence and Statistics (AISTATS), PMLR, vol. 108, pp. 756–765 (2020)
15. Chicco, D., Jurman, G.: The advantages of the Matthews correlation coefficient (MCC) over F1 score and accuracy in binary classification evaluation. BMC Genomics **21**, 6 (2020)
16. Peng, J., Wang, P., Zhou, N., Zhu, J.: Partial correlation estimation by joint sparse regression models. J. Am. Stat. Assoc. **104**(486), 735–746 (2009)
17. Kalyagin, V., Kostylev, I.: Graph density and uncertainty of graphical model selection algorithms, Commun. Comput. Inf. Sci. **1913**, 188–201, Springer Cham (2023). https://doi.org/10.1007/978-3-031-48751-4_14
18. Kostenetskiy, P., Chulkevich, R., Kozyrev, V.: HPC resources of the higher school of economics. J. Phys. Conf. Ser. **1740**, 012050 (2021)

A Unified Framework of Multi-stage Multi-winner Voting: An Axiomatic Exploration

Shengjie Gong[1]([✉]), Lingxiao Huang[2], Shuangping Huang[1], Yuyi Wang[3], Zhiqi Wang[4], Tao Xiao[5], Xiang Yan[5], and Chunxue Yang[5]

[1] South China University of Technology, Guangzhou, China
jackeygong2002@gmail.com
[2] State Key Laboratory of Novel Software Technology, Nanjing University, Nanjing, China
[3] CRRC Zhuzhou Institute, Zhuzhou, China
[4] Shanghai University of Finance and Economics, Shanghai, China
[5] Huawei Taylor Lab, Shanghai, China

Abstract. Multi-winner voting plays a crucial role in selecting representative committees based on voter preferences. Previous research has predominantly focused on single-stage voting rules, which are susceptible to manipulation during preference collection. In order to mitigate manipulation and increase the cost associated with it, we propose the introduction of multiple stages in the voting procedure, leading to the development of a unified framework of multi-stage multi-winner voting rules. To shed light on this framework of voting methods, we conduct an axiomatic study, establishing provable conditions for achieving desired axioms within our model. Our theoretical findings can serve as a guide for the selection of appropriate multi-stage multi-winner voting rules.

Keywords: Multi-stage voting · Multi-winner voting · Axiomatic exploration

1 Introduction

The problem of multi-winner voting is to select a winning committee of size k from m candidates by n voters with individual preferences of candidates, which has various applications in political, social, or business settings [18,24], e.g., parliamentary elections, the selection of awards judging committee, and movie selection on the front page of Netflix. Due to the importance, there is a large body of study for multi-winner voting, including axiomatic study [11], computational complexity [13,22], and approximate optimal committee selection [18,25].

Given the wide-ranging applications of multi-winner voting, there has been a growing concern regarding potential manipulations aiming at influencing the election results in favor of certain individuals through the misrepresentation of their preferences. A promising approach to address this concern is the adoption of multi-stage voting protocols, which have proven effective in enhancing

A. Eremeev et al. (Eds.): MOTOR 2024, LNCS 14766, pp. 349–373, 2024.
https://doi.org/10.1007/978-3-031-62792-7_24

computational resistance to manipulation [5,9,19]. Recent studies have illuminated the advantageous characteristics of multi-stage voting rules, notably their efficacy in enhancing computational resistance to manipulation. Such manipulations typically aim at influencing the election results in favor of certain individuals through the misrepresentation of their preferences, as highlighted in seminal works by [5,9,19]. However, this paper shifts focus from manipulation-proofness to more desired properties, known as axioms in literature. Intuitively, if a single-stage voting rule satisfies an axiom, one may wonder whether the corresponding multi-stage voting rule also does.

In this paper, we develop the axiomatic study for the multi-stage multi-winner voting. It is the initial phase of a theoretical investigation into multi-stage multi-winner voting, aiming to enhance our understanding of its design principles. Notably, we place less emphasis on the computational aspects of multi-stage voting in this study, as the outcomes may either yield negative results as extensions of existing NP-hard proofs or positive results that aggregate established single-stage rules. Instead, we focus on determining whether and under what conditions the specific desired axioms or properties can be satisfied by multi-stage multi-winner voting mechanisms, which holds significant importance for the practical selection of voting rules.

1.1   Technique Contributions

We first introduce a unified framework of multi-stage multi-winner voting rules, denoted as $\mathcal{R} = (R_1, \ldots, R_t)$ $(t \geq 1)$, where different multi-winner voting rules R_r can be employed in distinct stages $r \in [t]$, and the winning committee of the r-th stage serves as the candidate pool for the $(r+1)$-th stage (Definition 2). Our approach allows for the capture of classic multi-winner voting rules, most importantly the score-based rules (Definition 3), which generalizes existing multi-stage voting rules.

We then establish a set of desired axioms for multi-winner voting and provide sufficient conditions for the satisfaction or violation of these axioms within our model. Specifically, we demonstrate that multi-stage score-based voting maintains solid coalition (Sect. 3.1) but breaches committee monotonicity, candidate monotonicity, and consistency (Sect. 3.2). These findings offer valuable insights for selecting appropriate rules in multi-stage voting, recommending the inclusion of rules that satisfy solid coalitions, such as SNTV, in each stage (Sect. 4).

1.2   Other Related Work

A large body of literature studies axioms of multi-winner voting. The closest work to ours is [11], which considered multiple axioms including consistency, adapted from the single-winner setting; committee monotonicity [4], which ensures that the selected committee can be extended without removing anyone from it when increasing the target committee size; solid coalitions, which are weaker than but reflect the same idea as Dummett's proportionality [10] or the Droop proportionality criterion [30]. Other axioms that are theoretically important but not discussed in [11] include Pareto efficiency [21] and justified representation [12]. Some axiomatic studies of multi-stage single-winner voting have been done in [19]. In this paper, we study the above axioms in multi-stage settings and extend to the multi-winner case.

The design of multi-winner voting rules has also attracted a lot of attention for decades. Elkind et al. [11] suggested two natural ways to identify the similar internal structure between many of the known multi-winner rules: best-k rules such as SNTV and k-Borda, and committee scoring rules such as Bloc and CC, which is a subclass of the score-based voting rules defined in Sect. 2.1. They picked ten single-round voting rules as examples of different ideas on (score-based) multi-winner elections, and studied whether these rules satisfy the axioms aforementioned. There are some commonly used rules that have not been discussed in [11], such as approval-based rules [3] and Condorcet committee-related rules [12]. Besides, majority rule is another well-established and extensively researched voting rule within modern theoretical computer science, intersecting with rapidly evolving fields such as machine learning and game theory [1,2,14,16]. In this work, we consider the unified framework for the voting rules, extending the analysis for score-based rules to the multi-stage case, and add a discussion of approval-based rules.

The axiomatic study for multi-stage voting is started by Smith [27], who considers rules of successive candidate elimination in a single-winner setting and showed that all scoring runoff rules fail monotonicity. Narodytska and Walsh [19] studied two-stage single-winner voting rules, corresponding axioms, and their computing complexity. They showed that a two-stage voting rule offers advantages by inheriting appealing properties from the two stages. For instance, the Black process inherits the Condorcet consistency from the first stage and the properties of monotonicity, participation, and relevance to Condorcet losers from the second stage. On the other hand, the vulnerability to manipulation and control can be seen as an undesirable characteristic of two-stage voting rules. Davies et al. [9] considered a model similar to ours except that they use scoring vectors and only allow a single winner. Their work shows that the process of sequential elimination of candidates is often considered a means to make manipulation computationally challenging. Borodin et al. [6] focused on the primary elections within political parties, followed by a general referendum. Their paper points out that, in the real world, electoral and decision-making processes are often more complex, involving multiple stages. In this paper, we propose a unified framework and provide a systematic analysis for the axioms in multi-stage voting.

2 Unified Framework for Multi-stage Multi-winner Voting

In this section, we establish the framework for multi-stage multi-winner voting. We denote an election by $E = (C, V)$, which consists of a set C of m candidates and a group V of n voters. Before the discussion on multi-stage rules, we revisit the definition of single-stage multi-winner voting rules.

Definition 1 (Multi-Winner Voting Rules). *A multi-winner voting rule, denoted as R, is a function that, given an election $E = (C, V)$ and a positive integer k $(1 \leq k < m)$, outputs a collection $R(E, k)$ of possible winning committees $S \subseteq C$ of size k.*

We now introduce a concept of multi-stage multi-winner voting rules, wherein each stage employs a multi-winner voting rule (which may vary across stages), and the winning committee from each stage serves as the candidate set for the subsequent stage.

Definition 2 (t-Stage Multi-Winner Voting Rules). *A t-stage multi-winner voting rule, denoted as $\mathcal{R} = (R_1, R_2, \ldots, R_t)$, is defined as a function that, given an election $E = (C, V)$ and an integer vector $\boldsymbol{v} = (k_1, k_2, \ldots, k_t)$ where $m > k_1 > k_2 > \ldots > k_t$, outputs a collection $\mathcal{R}(E, \boldsymbol{v})$ of possible winning committees $S \subseteq C$ of size k_t. This output is contingent on the existence of a sequence $(S_0, S_1, \ldots, S_t)$ that satisfies the following conditions: (i) $S_0 = C$ and $S_t = S$; and (ii) $S_r \in R_r(E_r, k_r)$ for each $r \in [t]$, where $E_r = (S_{r-1}, V)$.*

Notice that a multi-stage multi-winner voting rule can be considered a special case within the scope of Definition 1. This implies that multi-stage voting represents one among various methodologies for determining the final winning committees. The sequence $(S_0, S_1, \ldots, S_t)$ in a t-stage voting rule may be interpreted as a trajectory of successive shortlists, that certificates S_t as the ultimate winning committee.

A multi-winner voting rule determines its output according to the voters' preferences for candidates. We assume that a voter v's preference is represented by a ranking of the candidates. For each candidate c, we define $p_v(c) = l$ to indicate that candidate c is ranked as the l-th favorite candidate by voter v. Thus $(p_v(c_1), \ldots, p_v(c_m))$ forms a permutation of $[m]^1$. For multi-stage voting rules, we need to establish the ranking of voters specifically on the candidate set S_{r-1} during the r-th stage instead of the complete set C.

Assumption 1 (Rankings are preserved within any subset). *For any subset $S \subseteq C$, let p_v^S represent the ranking of v on S. We assume that for any pair of candidates $c, c' \in S$, $p_v^S(c) < p_v^S(c')$ holds if and only if $p_v(c) < p_v(c')$.*

2.1 Score-Based Rules

We introduce a unified framework of multi-stage multi-winner voting rules, known as *score-based rules* (Definition 3). It encompasses a wide range of important multi-winner voting procedures that has been well studied in the literature (e.g., [8,11]). We begin by defining single-stage score-based rules before extending them to the multi-stage context.

Two parameters characterize a single-stage score-based rule: first, how a voter assigns scores to individual candidates; and second, how a voter evaluates a committee comprised of candidates that she supports in different degree. Score-based

---

[1] In this paper, we assume that there is no tie in the order of preferences.

rules operate under the assumption that a voter v assigns a score $\gamma^{m,k}(p_v(c))$ to each individual candidate c, and this score is non-increasing based on c's position $p_v(c)$ in v's preference list. The score $\gamma^{m,k}(\cdot)$ may depend on the size m of the candidate set and the size k of the winning committees. This is because a voter's preference for candidates may vary with changes in m and k, influencing the determination of position scores. As an illustration, consider the approval score, wherein a uniform score is assigned to the candidates ranking within the top k positions. Another example is the Borda score, which assigns a score of $m - p$ to the candidate occupying the p-th position in a voter's preference list. To prepare for implementing a position score function in multi-stage voting, where the sizes of the candidate pool and the target committee vary in different stages, it is imperative to precisely define the score function for distinct cases of (m, k).

We allow two distinct forms for a voter's assigned score to a committee: it can be either the cumulative sum of scores assigned to individual candidates within the committee, or the score specifically designated for the voter's preferred candidate within the same committee. Let us consider the committee's score as a norm β applied to the score vector assigned to all candidates in the committee. When $\beta = \ell_1$, each candidate c in a committee S contributes a utility of $\gamma^{m,k}(p_v(c))$ to voter v. These rules with $\beta = \ell_1$ are also referred to as *weakly separable committee scoring rules* in [11]. In the case of $\beta = \ell_{\max}$, a voter v evaluates a committee S based solely on one representative candidate within S. With the parameters β and γ, we formally define score-based rules.

Definition 3 (Score-based rules; (β, γ)-rule). *A score-based rule is parameterized by a norm $\beta \in \{\ell_1, \ell_{\max}\}$ and an infinite-dimensional vector function*

$$\gamma = \left(\gamma^{2,1}, \gamma^{3,1}, \gamma^{3,2}, ..., \gamma^{m,k}, ...\right),$$

where $\gamma^{m,k} : [m] \to \mathbb{R}$ is a non-increasing position score function. Given an election $E = (C, V)$ and a target committee size k, the score $f_v(S)$ of a committee S given by a voter $v \in V$ is defined as

$$f_v(S) = \begin{cases} \displaystyle\sum_{c \in S} \gamma^{m,k}(p_v(c)) & \text{if } \beta = \ell_1; \\ \displaystyle\max_{c \in S} \gamma^{m,k}(p_v(c)) & \text{if } \beta = \ell_{\max}. \end{cases}$$

The (β, γ)-rule voting returns the committees with maximum scores of $\sum_{v \in V} f_v(S)$ over all possible S of size k.

Examples of Score-Based Rules. The class of score-based rules encompasses multiple commonly-used multi-winner voting rules; summarized as follows.

- Single Non-Transferable Vote (SNTV): the (ℓ_1, PLU)-rule, where PLU is the plurality score $\text{PLU}^{m,k}(p) = \mathbb{I}_{\{p=1\}}$.
- Bloc: the (ℓ_1, APP)-rule, where APP represents the approval score $\text{APP}^{m,k}(p) = \mathbb{I}_{\{p \leq k\}}$.

- Borda: the (ℓ_1, BORDA)-rule, where BORDA is the Borda score $\text{BORDA}^{m,k}(p) = m - p$.
- Chamberlin-Courant (CC): the $(\ell_{\max}, \text{BORDA})$-rule.

We can extend score-based rules introduced in Definition 3 to the multi-stage case (Definition 2). Specifically, we define $\mathcal{R}$ as a multi-stage (β, γ)-rule if it employs the (β, γ)-rule in each stage of the voting process. This framework for multi-stage multi-winner voting consists of a broad spectrum of previously defined voting procedures. For instance, *Single Transferable Vote (STV)* can be regarded as a $(m - 1)$-stage rule with a vector $\boldsymbol{v} = (m - 1, m - 2, \ldots, 1)$, where each stage employs a (ℓ_1, PLU)-rule. In practical terms, this represents that STV eliminates one candidate with the minimum plurality score at each stage. Another example within this framework is *Baldwin's rule* (refer to [19]). Baldwin's rule is also structured as a $(m - 1)$-stage rule with $\boldsymbol{v} = (m - 1, m - 2, \ldots, 1)$, but each stage utilizes a (ℓ_1, BORDA)-rule.

Notice that Definition 2 enables the utilization of different position score functions γ across various stages. However, for the sake of simplicity, we do not delve into the exploration of such general multi-stage voting rules within the scope of this paper. The investigation of employing distinct γ-scores in different stages is an intriguing direction for future research.

2.2 Axioms Related to Score-Based Rules

Next, we present some axioms that are desirable for score-based rules. The first axiom is *Solid Coalition* representing an implementation of proportionality, which is a notion that requires a voting rule to select a committee representing as precisely as possible the opinions of the society. A typical example of scenarios where proportionality is attached importance is parliamentary elections [23].

Definition 4 (Solid Coalition). *A score-based rule R satisfies Solid Coalition iff $\forall E = (C, V)$ and $k \in [m]$, if at least $\frac{n}{k}$ voters rank some candidate c first then c belongs to every committee in $R(E, k)$.*

Our next axiom, *Committee Monotonicity*, requires that a target committee size increase should not result in the elimination of any of the currently selected candidates. This requirement arises when we are looking for the "best" candidates. An example is a hiring process in which the number of candidates is not determined beforehand. A committee monotone rule actually produces a ranking of candidates, indicating which one should be hired for an extra position [17].

Definition 5 (Committee Monotonicity). *Given an integer $t \geq 1$, a t-stage multi-winner voting rule $\mathcal{R}$ is committee monotone if, for any election $E = (C, V)$ and t-dimensional vectors $\boldsymbol{v}_1 = (k_1^1, k_2^1, \ldots, k_t^1)$ and $\boldsymbol{v}_2 = (k_1^2, k_2^2, \ldots, k_t^2)$ with $k_t^1 + 1 = k_t^2$ in the last dimension, we have (i) if a committee $S \in \mathcal{R}(E, \boldsymbol{v}_1)$ then there exists a $S' \in \mathcal{R}(E, \boldsymbol{v}_2)$ such that $S \subset S'$; (ii) if $S \in \mathcal{R}(E, \boldsymbol{v}_2)$ then there exists a $S' \in \mathcal{R}(E, \boldsymbol{v}_1)$ such that $S' \subset S$.*

The following two axioms, *Candidate Monotonicity* and *Consistency*, are generally desirable if one is interested in rules that are fair to candidates. As Janson [15] points out, no perfect election exists, but particularly disturbing are cases when changing some votes in favor of an elected candidate may result in her losing the election. Candidate Monotonicity is essential when candidates are not inanimate objects.

Definition 6 (Candidate Monotonicity). *A score-based rule R is candidate monotone iff $\forall E = (C, V), k \in [m]$ and $c \in C$, if $c \in S$ for some $S \in R(E, k)$, then if we shift c one position forward in a voter v and obtain an election E', we have $c \in S'$ for some $S' \in R(E', k)$.*

Consistency requires that if two disjoint groups of voters have the same election result, a voting rule should also arrive at this outcome if the two groups are united.

Definition 7 (Consistency). *A multi-winner voting rule R is consistent iff $\forall E_1(C, V_1), E_2(C, V_2)$ and $k \in [m]$, if $R(E_1, k) \cap R(E_2, k) \neq \varnothing$, then we have $R((C, V_1 + V_2), k) = R(E_1, k) \cap R(E_2, k)$.*

Remark on Axiom Selection. The reason why we choose the aforementioned axioms for our study is based on the belief that a "good" multi-stage multi-winner rule has distinct meanings when applied in different scenarios. Firstly, the objective may be to select the "optimal" committee, necessitating the consideration of Committee Monotonicity (Definition 5). However, in other cases, fairness assumes greater importance. In such instances, we typically examine the axioms of Solid Coalitions (Definition 4), aiming for the elected members to represent the opinions of voters, thereby prioritizing fairness to voters. If our focus shifts towards fairness to the candidates and the need for the multi-stage rule to be easily understandable by them, we then take Candidate Monotonicity (Definition 6) and Consistency (Definition 7) into account.

3 Axiomatic Study for Multi-stage Score-Based Rules

In this section, we present the axiomatic study for score-based rules. We first demonstrate that if a set of single-stage rules satisfies Solid Coalition, this axiom can be maintained in multi-stage voting scenarios, where each stage employs one of these single-stage rules. Subsequently, we provide election examples to illustrate that the axioms of Committee Monotonicity, Candidate Monotonicity, and Consistency are violated in the multi-stage setting.

3.1 Preserving Axioms in Multi-stages

The following theorem suggests that when employing rules with the Solid Coalition property at each stage, the multi-stage voting rule also satisfies the Solid Coalition property. For example, we use SNTV rule in each stage of STV procedure, and SNTV satisfies the Solid Coalition [11]. Consequently, STV inherits this property throughout its multi-stage process.

Theorem 1 (Solid Coalition preserves in multi-stage voting). *Let $t \geq 1$ be an integer and $\mathcal{R} = (R_1, R_2, \ldots, R_t)$ be a t-stage multi-winner voting rule. If R_r satisfies Solid Coalition for each $r \in [t]$, $\mathcal{R}$ also satisfies Solid Coalition.*

Proof. Fix any $E = (C, V)$ and vector $\boldsymbol{v} = (k_1, k_2, \ldots, k_t)$. Suppose that there are at least $\frac{n}{k_t}$ voters ranking a candidate c first. Due to the fact that R_1 satisfies the Solid Coalition property and that at least $\frac{n}{k_t} \geq \frac{n}{k_1}$ voters rank c first, c must be one of the winners in R_1. We observe that the number of voters who assign the highest rank to v will not decrease in the successive stages of voting. Therefore, we can similarly show that c appears in each stage's winning committees, which completes the proof. □

3.2 Violating Axioms in Multi-stages

In this section, we discuss the axioms including Committee Monotonicity, Candidate Monotonicity, and Consistency.

Before presenting the main theorem in this section, we first clarify the non-triviality for multi-stage $(\beta, \boldsymbol{\gamma})$ rules. For a multi-stage $(\ell_1, \boldsymbol{\gamma})$-rule, any stage of voting that applies a position score function $\gamma^{m,k}$ such that $\gamma^{m,k}(1) = \gamma^{m,k}(m)$ has all the committees of size k as the winning committee. Moreover, in the case of $\beta = \ell_{\max}$, if $\gamma^{m,k}(1) = \gamma^{m,k}(m - k + 1)$, then all the committees of size k have the same score of $f_v(S) = \gamma^{m,k}(1)$ for each voter v. These cases are deemed to be of triviality and minimal merit consideration. Thus, we propose the following definition of rationality to specify the non-trivial rules.

Definition 8 (Rationality of $(\beta, \boldsymbol{\gamma})$-rules). *We define a multi-stage $(\beta, \boldsymbol{\gamma})$-rule as rational if, for any given pair (m, k), the following conditions hold:*

- *if $\beta = \ell_1$, $\gamma^{m,k}(1) > \gamma^{m,k}(m)$;*
- *if $\beta = \ell_{\max}$, $\gamma^{m,k}(1) > \gamma^{m,k}(m - k + 1)$.*

We now show that the axioms of Committee Monotonicity, Candidate Monotonicity, and Consistency do not preserve in any rational multi-stage $(\beta, \boldsymbol{\gamma})$-rule. Here we begin with the structure of our counter-example for Candidate Monotonicity. The proof ideas for Committee Monotonicity and Consistency are analogous.

Consider an election process that, in the $(t-1)$-th stage, selects two candidates from a pool of three, and in the final stage, chooses one candidate from the previously selected two. Let there be three candidates a, b, c before the $(t-1)$-th stage. The main structure that we use in the construction of voters' preferences is a cycle of (a, b, c) (Definition 9). The characteristic of this cycle is that each candidate experiences an equal number of occurrences in every possible position. Specifically, to form a cycle of (a, b, c), we can construct a Group of three voters whose preference order are $a \succ b \succ c$, $b \succ c \succ a$, and $c \succ a \succ b$ respectively. The voter set is formed by duplicating this Group of voters multiple copies. Due to this characteristic of cycles, each candidate can be the final winner. It can be observed that the final winning candidate will be a, if c is eliminated in the

$(t-1)$-th stage, while the final winner is c, if b is knocked out first. With rationality of γ, there are two possibilities: $\gamma^{3,2}(2) > \gamma^{3,2}(3)$, or $\gamma^{3,2}(1) > \gamma^{3,2}(2)$. In the former case, we transfer a voter with preference list $c \succ b \succ a$ to that with $c \succ a \succ b$. If the latter, we shift a one position forward in a voter with preference list $b \succ a \succ c$. Consequently, c becomes the sole winner in the last stage since b becomes less important and must be eliminated in the $(t-1)$-th stage. Therefore, the axiom of Candidate Monotonicity is violated.

The proofs for Committee Monotonicity and Consistency are based on similar constructions of such cycles. Formally, we define the cycle of a sequence and the permutation of a voter's preference list.

Definition 9 (The cycle of a sequence). *Given a sequence* $s = (s_1, s_2, \ldots, s_m)$ *of* m *elements and an integer* $i \in [m]$, *the cycle* $\rho_i(s)$ *is the sequence* s' *with:*

$$s'_j = \begin{cases} s_{j+i} & j \leq m - i, \\ s_{j+i-m} & otherwise. \end{cases}$$

In other words, $\rho_i(s)$ shifts all elements in the sequence s by i positions to the right. If an element is shifted past the end of the sequence, it wraps around to the front of the sequence.

Definition 10 (Full permutation of a sequence). *A permutation is a bijection of a sequence to itself. Let* $\Pi(s)$ *denote the set of all permutations of sequence* s.

We use the aforementioned concepts to simplify the description of preference lists. For example, consider a candidate set $C = \{a, b, c, d, e\}$. A ranking $a \succ \rho_1(b, c, d) \succ e$ can be interpreted as $a \succ c \succ d \succ b \succ e$.

Theorem 2 (Committee Monotonicity does not preserve in a rational multi-stage (β, γ)-rule). *Let* $t \geq 2$ *be an integer,* $\beta \in \{\ell_1, \ell_{\max}\}$, *and* $\mathcal{R}$ *be a rational* t-*stage* (β, γ)-*rule. Then* $\mathcal{R}$ *does not satisfy Committee Monotonicity.*

According to this theorem, any multi-stage voting rule that consists of scoring-based rules does not exhibit Committee Monotonicity, even if the individual rules applied in every stage satisfy this property. For instance, when combining several Borda rules to form a multi-stage voting rule, the resulting multi-stage rule no longer exhibit the Committee Monotonicity, though each single-stage rule individually satisfies it.

Proof (Proof of Theorem 2). We construct a two-stage voting procedure, which indicates that Committee Monotonicity does not preserve in a rational multi-stage (β, γ)-rule. Note that this example can be simply extended to arbitrary stages by adding dummy candidates.

For (ℓ_1, γ)-rules, let $C = \{a, b, c, d, e\}$, $\boldsymbol{v_1} = (4, 2)$, and $\boldsymbol{v_2} = (2, 1)$. Committee Monotonicity requires that for any $S \in \mathcal{R}(E, \boldsymbol{v_2})$, there exists a $S' \in \mathcal{R}(E, \boldsymbol{v_1})$ such that $S \subset S'$. However, we show that it is possible to construct a set V of voters such that $\{a\} \in \mathcal{R}(E, \boldsymbol{v_2})$ but $\mathcal{R}(E, \boldsymbol{v_1}) = \{\{b, c\}\}$, which does not include a possible winning committee S' such that $\{a\} \subset S'$. Due to the rationality assumption, either $\gamma^{4,2}(1) > \gamma^{4,2}(3)$ or $\gamma^{4,2}(2) > \gamma^{4,2}(4)$ holds.

– $\gamma^{4,2}(1) > \gamma^{4,2}(3)$. We construct V with 6 groups of voters, each having preference lists as follows:

$$
\begin{aligned}
&\text{Group 1:} && 1 \times \ \rho_i(b,c,a,e) \succ d \quad \forall i \in \{1,2,3,4\};\\
&\text{Group 2:} && 1 \times \ \rho_i(c,b,a,e) \succ d \quad \forall i \in \{1,2,3,4\};\\
&\text{Group 3:} && 1 \times \ \rho_i(a,b,c,d) \succ e \quad \forall i \in \{1,2,3,4\};\\
&\text{Group 4:} && 1 \times \ \rho_i(a,c,b,d) \succ e \quad \forall i \in \{1,2,3,4\};\\
&\text{Group 5:} && 200 \times \ \pi \succ e \quad \forall \pi \in \Pi(a,b,c,d);\\
&\text{Group 6:} && 100 \times \ \pi \succ d \quad \forall \pi \in \Pi(a,b,c,e).
\end{aligned}
$$

We begin our analysis with $v_1 = (4,2)$. In the first stage, e is eliminated as a result of preference of the voters in Group 5. Then in the second stage, d cannot win due to the presence of Group 6. The scores of $a, b,$ and c are initially equivalent. However, after e is knocked out, the score of a is less than that of b and c since $\gamma^{4,2}(1) + \gamma^{4,2}(2) > 2\gamma^{4,2}(3)$. Therefore, $\{b,c\}$ is the final winning committee.

On the other hand, consider the voting procedure that applies $\mathcal{R}$ on (E, v_2). In the first stage, d and e must be eliminated because of the existence of Group 5 and Group 6. Further, the scores of a, b and c are the same, and thus $\{a,b\}$ is a possible winning committee. In the second stage, the scores of both candidates are also the same, so a is a possible winner. The Committee Monotonicity is violated.

– $\gamma^{4,2}(2) > \gamma^{4,2}(4)$. We modify Group 1 in the voter set V constructed above as $1 \times \ d \succ \rho_i(b,c,a,e) \ \forall i \in \{1,2,3,4\}$ and Group 2 as $1 \times \ d \succ \rho_i(c,b,a,e)$ $\forall i \in \{1,2,3,4\}$. Then the difference between the score of a and b, c is $\gamma^{4,2}(2) + \gamma^{4,2}(3) - 2\gamma^{4,2}(4) > 0$. The previously established analysis remains applicable.

Next, we provide counter-examples for $(\ell_{\max}, \gamma)$-rules. We also construct a two-stage election. Let $C = \{a,b,c,d,e,f,g\}$ and $v_1 = (2,1), v_2 = (5,2)$. The voter set V consists of 5 groups of voters:

$$
\begin{aligned}
&\text{Group 1:} && 200 \times \ \pi \succ g \quad \forall \pi \in \Pi(a,b,c,d,e,f);\\
&\text{Group 2:} && 200 \times \ \pi \succ f \quad \forall \pi \in \Pi(a,b,c,d,e,g);\\
&\text{Group 3:} && 100 \times \ \pi \succ e \quad \forall \pi \in \Pi(a,b,c,d,f,g);\\
&\text{Group 4:} && 100 \times \ \pi \succ d \quad \forall \pi \in \Pi(a,b,c,e,f,g);
\end{aligned}
$$

$$
\begin{aligned}
\text{Group 5:} \quad
& f \succ b \succ d \succ e \succ c \succ g \succ a && f \succ c \succ d \succ e \succ b \succ g \succ a\\
& f \succ b \succ d \succ e \succ a \succ g \succ c && f \succ c \succ d \succ e \succ a \succ g \succ b\\
& d \succ a \succ f \succ g \succ b \succ e \succ c && d \succ a \succ f \succ g \succ c \succ e \succ b\\
& d \succ f \succ b \succ e \succ c \succ a \succ g && d \succ f \succ c \succ e \succ b \succ a \succ g\\
& d \succ e \succ a \succ f \succ b \succ c \succ g && d \succ e \succ a \succ f \succ c \succ b \succ g\\
& d \succ f \succ b \succ e \succ a \succ c \succ g && d \succ f \succ c \succ e \succ a \succ b \succ g.
\end{aligned}
$$

We next show that $\{a\} \in \mathcal{R}(E, v_1)$ but $\mathcal{R}(E, v_2) = \{b,c\}$. In $\mathcal{R}(E, v_1)$, it is guaranteed by Group 1–4 that candidates d, e, f, g are eliminated in the first stage. Additionally, the voters in Group 5 ensure that the score of the committee $\{a, b\}$ is greater than or equal to that of $\{b, c\}$ and $\{a, c\}$. This is because the score of $\{a, b\}$ is $4(\gamma^{7,2}(2) + \gamma^{7,2}(3) + \gamma^{7,2}(4))$, while both $\{a, c\}$ and $\{b, c\}$ score $4(\gamma^{7,2}(2) + \gamma^{7,2}(3) + \gamma^{7,2}(5))$. Thus $\{a, b\}$ is the winning committees in the first stage. In the second stage, $\{a\}$ and $\{b\}$ have identical scores, therefore $\{a\} \in \mathcal{R}(E, \boldsymbol{v}_1)$.

In $\mathcal{R}(E, \boldsymbol{v}_2)$, the voters from Groups 1–4 ensures that $\{a, b, c, d, e\}$ is the only winning committee in the first stage. Then in the second stage, Groups 3 and 4's voters guarantee the elimination of d, e. Through calculation of the scores given by the voters in Group 5, it can be verified that both $\{a, b\}$ and $\{a, c\}$ achieve a score of $2\gamma^{5,2}(1) + 4\gamma^{5,2}(2) + 2\gamma^{5,2}(3) + 4\gamma^{5,2}(4)$, while $\{b, c\}$ gets a score of $4\gamma^{5,2}(1) + 4\gamma^{5,2}(2) + 2\gamma^{5,2}(3) + 2\gamma^{5,2}(4)$. Consequently, we have $\mathcal{R}(E, \boldsymbol{v}_2) = \{\{b, c\}\}$.   □

Theorem 3 (Candidate Monotonicity does not preserve in a rational multi-stage (β, γ)-rule). *Let $t \geq 2$ be an integer, $\beta \in \{\ell_1, \ell_{\max}\}$, and $\mathcal{R}$ be a rational t-stage (β, γ)-rule. Then $\mathcal{R}$ does not satisfy Candidate Monotonicity.*

To provide an intuition for our analysis, consider a point runoff system (see e.g. [27]), which involves the successive use of the ℓ_1 scoring function to eliminate candidates with lower scores until a single winner is obtained. It can be considered as a specific instance of our framework with $\beta = \ell_1$. Smith [27] demonstrated that no point runoff system involving two or more stages exhibits Candidate Monotonicity. In Theorem 3, we extend our analysis beyond the ℓ_1 scoring function used in point runoff systems and include the $\ell_{\max}$ scoring function, such as Chamberlin-Courant's (CC) rules. We reach a more comprehensive conclusion that regardless of how we combine the ℓ_1 and $\ell_{\max}$ scoring functions, the resulting multi-stage voting rule fails to satisfy Candidate Monotonicity.

Proof (Proof of Theorem 3). We present below a counter-example which works for both (ℓ_1, γ)-rules and $(\ell_{\max}, \gamma)$-rules. Let there be three candidates a, b, and c. The target size of the winning committees in each stage is represented by $\boldsymbol{v} = (2, 1)$. There are two groups of voters in V with preferences as below.

$$\text{Group 1:} \quad 10 \times \rho_i(c, a, b) \quad \forall i \in \{1, 2, 3\};$$
$$\text{Group 2:} \quad 1 \times \pi \quad \forall \pi \in \Pi(a, b, c).$$

We run an election that selects a winning committee of size 2 in the first stage and a sole winner in the second stage. If a (β, γ)-rule is applied in each stage, $\{a, b\}$ is one of the winning committees for the first stage and a is the final winner between a and b.

If $\gamma^{3,2}(1) = \gamma^{3,2}(2) > \gamma^{3,2}(3)$ holds, we shift a one position forward in $c \succ b \succ a$. If $\gamma^{3,2}(1) > \gamma^{3,2}(2)$, we shift a one position forward in $b \succ a \succ c$. Then $\{a, c\}$ becomes the sole winning committee in both cases under the assumption that γ is rational. In the second stage, $\{c\}$ is the final winning committee, which does not include a.

This result can be generalized to any $m \geq 3$, $t \geq 2$, and $\boldsymbol{v} = (k_1, \ldots, k_{t-1}, k_t)$ with $k_t = k_{t-1} - 1$. Specifically, we can add $(k_t - 1)$ candidates to the front and

$(k_1 - k_{t-1})$ to the end of each voter's preference list presented above. Then the candidates at the end of the preference lists will be eliminated in the first $(t-2)$ stages, and the candidates who are positioned at the front will be all included in the final winning committee. All other procedures remain the same.   □

Theorem 4 (Consistency does not preserve in a rational multi-stage (β, γ)-rule). *Let $t \geq 2$ be an integer, $\beta \in \{\ell_1, \ell_{\max}\}$, and $\mathcal{R}$ be a rational t-stage (β, γ)-rule. Then $\mathcal{R}$ does not satisfy Consistency.*

According to Skowron's work [26], only the scoring rules that generate weak linear orders over the committees exhibit Consistency. However, the multi-stage rules examined in this study fail to produce linear orders. This is due to the greedy nature of the multi-stage process, where candidates are eliminated at each stage based on their current contribution to the committee rather than their overall value. As a result, it cannot be asserted that the final winning committee necessarily outperform a committee consisting of candidates eliminated at different stages. Therefore, non-linear order multi-stage rules are deemed inconsistent.

Proof (Proof of Theorem 4). We present below a two-stage election that works as a counter-example for both (ℓ_1, γ)-rules and $(\ell_{\max}, \gamma)$-rules. There are five candidates a, b, c, d, and e. The target size in each stage is represented by $\boldsymbol{v} = (4, 1)$. With rationality of γ, we have at least one of the following inequalities holds: $\gamma^{4,1}(1) > \gamma^{4,1}(2), \gamma^{4,1}(2) > \gamma^{4,1}(3)$, or $\gamma^{4,1}(3) > \gamma^{4,1}(4)$.

– $\gamma^{4,1}(1) > \gamma^{4,1}(2)$. We construct V_1 as below:

> Group 1: $1 \times \rho_i(a, b, e) \succ c \succ d$ $\forall i \in \{1, 2, 3\}$;
> Group 2: $1 \times \rho_i(b, a, c) \succ d \succ e$ $\forall i \in \{1, 2, 3\}$;
> Group 3: $300 \times \pi \succ c$ $\forall \pi \in \Pi(a, b, d, e)$;
> Group 4: $100 \times \pi \succ d$ $\forall \pi \in \Pi(a, b, c, e)$;
> Group 5: $400 \times \pi \succ e$ $\forall \pi \in \Pi(a, b, c, d)$.

We construct Groups 3–5 to ensure that e is knocked out in the first stage and the final winner cannot be c or d. It can be observed that the scores of a and b in Groups 2–5 are the same in both stages. In Group 1, after e is eliminated, the score of a is $2\gamma^{(4,1)}(1) + \gamma^{(4,1)}(2)$, and that of b is $\gamma^{(4,1)}(1) + \gamma^{(4,1)}(2)$. Therefore, a is the final winner. Further, we construct V_2 similarly as follows:

> Group 1: $1 \times \rho_i(a, b, d) \succ c \succ e$ $\forall i \in \{1, 2, 3\}$;
> Group 2: $1 \times \rho_i(b, a, c) \succ d \succ e$ $\forall i \in \{1, 2, 3\}$;
> Group 3: $300 \times \pi \succ c$ $\forall \pi \in \Pi(a, b, d, e)$;
> Group 4: $400 \times \pi \succ d$ $\forall \pi \in \Pi(a, b, c, e)$;
> Group 5: $100 \times \pi \succ e$ $\forall \pi \in \Pi(a, b, c, d)$.

A similar analysis to that of V_1 indicates that d is knocked out in the first stage, and the winner of the second stage is also a.

However, upon a combination of V_1 and V_2, the candidate eliminated in the first stage shall be transferred to c, as the total number of votes ranking c last is greater than that of d and e; specifically, $300 \times 2 > 400 + 100$. After c is eliminated, the scores of a and b are the same in Groups 1, 3, 4, and 5 in both V_1 and V_2. In Group 2, the score of b is $2\gamma^{4,1}(1) + \gamma^{4,1}(2)$ in the second stage, and that of a is $\gamma^{4,1}(1) + 2\gamma^{4,1}(2)$. Thus the final winner becomes b.

– $\gamma^{4,1}(2) > \gamma^{4,1}(3)$ and $\gamma^{4,1}(3) > \gamma^{4,1}(4)$. We modify our example election by letting Group 1 of V_1 be $1 \times c \succ \rho_i(a, b, e) \succ d \ \forall i \in \{1, 2, 3\}$. Through such modification, a gets a score of $2\gamma^{4,1}(2) + \gamma^{4,1}(3)$ and b gets $\gamma^{4,1}(2) + 2\gamma^{4,1}(3)$ in the second stage, so the winner is a. Note that although c performs better than a and b in Group 1, it cannot be the winner for the existence of Group 3. We can implement a similar modification in Group 2 in V_1 and Groups 1–2 in V_2 to obtain a counter-example for the case of $\gamma^{4,1}(2) > \gamma^{4,1}(3)$. The case of $\gamma^{4,1}(3) > \gamma^{4,1}(4)$ can also be solved with similar modification.   □

4   Discussion on Voting Rules Selection

This study's findings provide guidance for the selection of rules in multi-stage voting scenarios. When seeking to choose a multi-stage score-based rule, some criteria such as Committee Monotonicity, Candidate Monotonicity, and Consistency should not be considered due to the absence of rules that satisfy them. However, if the axiom of Solid Coalitions is considered important, a single-stage rule that satisfies Solid Coalitions (such as SNTV) can be employed to construct a multi-stage rule.

From the perspective of axioms, those of monotonicity, which requires that an increase in support for some elected candidates should not result in their knockout, are unlikely to preserve in multi-stage voting. The support for candidates may be increased by changing a voter's preference (Candidate Monotonicity) or by adding a set of voters in favor of them (Consistency). The reason why the axioms of monotonicity usually do not preserve is that increasing support for certain candidates in these ways can change the outcome of the election for other candidates. While the candidates receiving more support may still be elected by a single-stage rule, it is possible for the new members of this stage's winning committee to surpass them in the subsequent stages.

Notice that there is another commonly used family of multi-winner voting rules, called approval-based rules, under which voters are identified with their approval ballots[2]. In the case of approval-based rules, the axioms of Candidate Monotonicity and Consistency need not be taken into account. Basically, in multi-stage voting, approval-based rules are less prone to not preserving the desirable axioms as compared to score-based ones, specifically, Committee Monotonicity is preserved for approval-based rules but not for score-based rules.[3] This

---

[2] Due to the space limit, detailed discussion on approval-based rules and certain experimental results are relegated to the appendix. Interested readers are encouraged to see the full version accessible on arXiv: https://arxiv.org/abs/2402.02673.

is because, for score-based rules, a voter's satisfaction with regard to a committee (represented by the score of the committee given by the voter) exhibits less consistency across multiple stages. For example, it is likely that a voter assigns higher scores to the remaining candidates in the subsequent stages as a result of the elimination of the candidate that she initially favors. By contrast, in an approval-based voting, each voter's satisfaction with a committee solely depends on the selected number of her approved candidates, which does not change over different stages.

While our framework encompasses a wide range of multi-stage voting rules, it is crucial to recognize its limitations. Certain procedures, such as Cup and Black's method [19], fall outside the scope of our framework. The Cup rule entails representing candidates as leaf nodes in a binary tree and comparing them pairwise until a unique winner is identified. On the other hand, Black's procedure first checks for the presence of a Condorcet winner. If such a candidate exists, it was presented as the winner of this election. Otherwise, the winner is determined based on the Borda rule. Additionally, Nanson's rule [20] bears resemblance to our definition of multi-stage, expect that in each stage, candidates are eliminated based on whether their score falls below the average Borda score of all candidates. Therefore, it is not possible to pre-define the number of remaining candidates at each stage, i.e., the vector v. Although these methods are notable in the field of multi-stage voting, they are not covered by our framework.

To summarize, the research presented in this paper establishes a foundation for the theoretical investigation of multi-stage voting and opens up several avenues for future exploration. One possible direction involves conducting additional axiomatic studies for other rules in multi-stages, such as Moreno's rules, greedy CC, and similar approaches, which were not covered in this paper. Furthermore, an interesting line of inquiry is to explore how the introduction of multi-stages can mitigate various manipulation techniques. According to our experimental results in the appendix, multi-stage voting shows potential fairness advantages, warranting further theoretical analysis on whether multi-stage voting can effectively be employed for debiasing purposes.

A Axiomatic Study for Multi-stage Approval-Based Rules

A.1 Approval-Based Rules and Related Axioms

When voters are identified with their approval ballots, we refer to the voting method for selecting committees as the approval-based rule. This appendix

---

[3] Note that this observation may not hold if we take more general class of approval-based rules into consideration. For example, the "non-standard" rule of *satisfaction approval voting* defines that the score of a committee S given by a voter v not only depends on $|A_v \cap S|$ but also relates to the size $|A_v|$ itself. The score can change a lot across stages as the candidate pool shrinks, and therefore the axioms are unlikely to be preserved in multi-stages.

focuses on a quite general class of approval-based rules, known as *Thiele methods*, introduced by [29]. There has been extensive research on this class of rules and its special cases in the social choice community (e.g., [7,8,28]). The class of Thiele methods consists of all the rules that maximize the sum of voters' individual satisfaction, where each voter v's satisfaction with committee S solely depends on the number of v's approved candidates in S. For each voter $v \in V$, we denote by $A_v \subseteq C$ voter v's approval ballot, i.e., the set that consists of the first $|A_v|$ candidates in her preference list.

Definition 11 (Thiele Methods, ω-Thiele). *A Thiele method is parameterized by a nondecreasing function $\omega : \mathbb{N} \to \mathbb{R}$ with $\omega(0) = 0$. The score of a committee S given a set V of voters is defined as $score_\omega(V, S) = \sum_{v \in V} \omega(|S \cap A_v|)$. The ω-Thiele method returns committees with maximum scores.*

With different functions ω, ω-Thiele can cover a wide range of approval-based rules. An example is the most natural rule, *Approval Voting* (AV for short), which selects the k candidates that are approved by most voters. AV is the ω-Thiele with $\omega(x) = x$. Another example is *Proportional Approval Voting* (PAV for short), which is the ω-Thiele with $\omega(x) = \sum_{j=1}^{x} \frac{1}{j}$. The function ω depending on the sequence of harmonic numbers captures the property of diminishing returns. We refer to a rule $\mathcal{R}$ as a multi-stage ω-Thiele if it employs the same ω-Thiele method with the same approval ballots A_v's in each stage.

Axioms Related to Approval-Based Rules. We present below some axioms that may be desirable for approval-based rules. First, we consider *Pareto Efficiency* in which a dominated committee must never be output. When the goal of a multi-winner rule is to select the "best" committee, Pareto Efficiency is often considered to be a minimal requirement.[4]

Definition 12 (Pareto Efficiency). *A committee S_1 dominates a committee S_2 if (i) every voter has at least as many approved candidates in S_1 as in S_2 (for $v \in V$ it holds that $|A_v \cap S_1| \geq |A_v \cap S_2|$), and (ii) there is one voter with strictly more approved candidates (there exists $u \in V$ with $|A_u \cap S_1| > |A_u \cap S_2|$). An approval-based rule R satisfies Pareto Efficiency if R never outputs dominated committees.*

For approval-based rules, the definition of Committee Monotonicity is identical to that for score-based rules (Definition 5). Our next axiom is *Justified Representation*, which is a criterion for accessing whether an approval-based rule can be considered proportional. The intuition of Justified Representation is that if k candidates are to be selected, then a set of $\frac{n}{k}$ voters that are completely unrepresented can demand that at least one of their all-approved candidates be selected.

---

[4] Pareto Efficiency can also be defined for score-based rules. However, it is not easy to extend the notion of Pareto Efficiency to multi-stage. Hence, we only consider Pareto Efficiency for approval-based rules in this paper.

Definition 13 (Justified Representation). *An approval-based rule R satisfies Justified Representation iff* $\forall E = (C, V)$ *and* $k \in [m]$, *for any committee* $S \in R(E, k)$, *there does not exist a set of voters* $V^* \subseteq V$ *with* $\|V^*\| \geq \frac{n}{k}$ *such that* $\bigcap_{v \in V^*} A_v \neq \varnothing$ *and* $A_v \cap S = \varnothing$ *for all* $v \in V^*$.

Candidate Monotonicity and Consistency should also be considered in approval-based voting when the fair treatment of candidates is necessary.

Definition 14 (Candidate Monotonicity for Approval-Based Rules). *An approval-based rule R is candidate monotone iff* $\forall E = (C, V)$, $k \in [m]$ *and* $c \in C$, *if* $c \in S$ *for some* $S \in R(A, k)$, *then if a voter v additionally approves the candidate c and we obtain an election* E', *we have* $c \in S'$ *for some* $S' \in R(E', k)$.

In Definition 14, the support of a candidate increases by a voter approving the candidate additionally, instead of shifting her one position forward (in Definition 6). On the other hand, Consistency for approval-based rules is defined in the same way as in Definition 7.

Next, we provide an axiomatic study for approval-based rules. We focus on the case when single-stage rules satisfy a certain desirable property and investigate whether the property can be preserved in multi-stages that apply one of the single-stage rules in each stage.

A.2 Preserving Axioms in Multi-stages

We find the axioms of Committee Monotonicity (Theorem 5) and Justified Representation (Theorem 6) preserved in multi-stage approval-based voting.

Committee Monotonicity is a very demanding criterion for Thiele methods. We show that only the ω-Thiele methods with linear ω-functions satisfy Committee Monotonicity (Lemma 1). If a multi-stage rule $\mathcal{R}$ is composed of rules that maximize $\sum_{v \in V} |A_v \cap S|$ over possible committees S in each stage, it must finally select the candidates approved by most voters after multiple stages. In fact, the multi-stage rule $\mathcal{R}$ produces a ranking of candidates and thus satisfies Committee Monotonicity by definition.

Theorem 5 (Committee Monotonicity preserves in a multi-stage approval-based rule). *Let* $t \geq 2$ *be an integer and* $\mathcal{R} = (R_1, R_2, \ldots, R_t)$ *be a t-stage approval-based rule. If* R_r *is a single-stage* ω-*Thiele rule that satisfies Committee Monotonicity for each* $r \in [t]$, $\mathcal{R}$ *also satisfies Committee Monotonicity.*

Before proof of Theorem 5, we would like to first clarify what single-stage committee monotone rules are like. The following lemma indicates that only the ω-Thiele methods with linear ω-functions satisfy Committee Monotonicity.

Lemma 1. *If R is a single-stage* ω-*Thiele rule that satisfies Committee Monotonicity, then*

$$\omega(i) - \omega(i-1) = \omega(j) - \omega(j-1)$$

holds for any $i, j \in [1, \infty)$.

Proof. Let p_i denote the difference between $\omega(i)$ and $\omega(i-1)$ for each $i \geq 1$. Suppose for the sake of contradiction that there exists a nondecreasing function ω with $p_{i_0} \neq p_1$ for some $i_0 > 1$ that parameterizes a rule R satisfying Committee Monotonicity. Without loss of generality, we assume i_0 is the smallest integer such that $p_{i_0} \neq p_1$. There are two cases:

- $p_{i_0} < p_1$. There is a rational number $\frac{n_1}{n_2}$ such that $\frac{p_{i_0}}{p_1} < \frac{n_1}{n_2} < 1$, where n_1 and n_2 are both integers. Let us consider the voters with approval ballots as below:

$$n_1 \times \{a\} \quad n_2 \times \{a, c_1, c_2, \ldots, c_{i_0-1}\}$$
$$n_1 \times \{b\} \quad n_2 \times \{b, c_1, c_2, \ldots, c_{i_0-1}\}.$$

  R chooses $\{c_1, c_2, \ldots, c_{i_0-1}\}$ for $k = i_0 - 1$. However, for $k = i_0$, R chooses $\{a, b, c_{l1}, c_{l2}, \ldots, c_{l,i_0-2}\}$ as winning committee, where $c_{l1}, \ldots, c_{l,i_0-2}$ are (i_0-2) candidates chosen arbitrarily from $\{c_1, \ldots, c_{i_0-1}\}$. The only winning committee in $R(E, i_0-1)$ is not a subset of $\{a, b, c_{l1}, \ldots, c_{l,i_0-2}\}$, contradicting Committee Monotonicity of R.

- $p_{i_0} > p_1$. There is a rational number $\frac{n_1}{n_2}$ such that $\frac{p_{i_0}}{p_1} > \frac{n_1}{n_2} > 1$. Consider an election E with voters described below:

$$n_1 \times \{c, d_1, \ldots, d_{i_0-2}\} \quad n_2 \times \{a, b, d_1, \ldots, d_{i_0-2}\}.$$

  R chooses $\{c, d_1, \ldots, d_{i_0-2}\}$ for $k = i_0 - 1$ but chooses $\{a, b, d_1, \ldots, d_{i_0-2}\}$ for $k = i_0$. There is no committee S in $R(E, i_0)$ such that $\{c, d_1, \ldots, d_{i_0-2}\} \subseteq S$, contradicting Committee Monotonicity of R. $\square$

With Lemma 1, we can now complete proof of Theorem 5.

Proof (Proof of Theorem 5). For each $r \in [t]$, the score of a committee S in R_r is equivalently defined as $\sum_{v \in V} |A_v \cap S|$ by Lemma 1. In each stage, R_r selects the candidates that are approved by most voters. Then the t-stage rule $\mathcal{R} = (R_1, \ldots, R_t)$ finally selects k candidates approved by most voters. $\mathcal{R}$ actually produces a ranking of candidates and satisfies Committee Monotonicity by definition. $\square$

There are many single-stage rules that fail to satisfy the criterion of Justified Representation, though it is considered to be one of the less stringent definitions of "proportional representation" among several possible options. An example is AV, mentioned in Sect. A.1, which simply selects the candidates approved by most voters [3]. Despite this, the good news is that we can simply prove by contradiction that a multi-stage rule satisfies Justified Representation if it is composed of single-stage ω-Thiele methods that satisfy Justified Representation.

Theorem 6 (Justified Representation preserves in a multi-stage approval-based rule). *Let $t \geq 1$ be an integer and $\mathcal{R} = (R_1, R_2, \ldots, R_t)$ be a t-stage multi-winner approval-based voting rule. If R_r is a single-stage ω-Thiele rule that satisfies Justified Representation for each $r \in [t]$, $\mathcal{R}$ also satisfies Justified Representation.*

Proof. Fix an election $E = (C, V)$ and a $\boldsymbol{v} = (k_1, ..., k_t)$. Let S be one of the winning committees in the output of $\mathcal{R} = (R_1, \ldots, R_t)$ on $(E, \boldsymbol{v})$. Suppose for the sake of contradiction that R_r satisfies Justified Representation for each $r \in [t]$, but there exists a set $V^* \subseteq V$ with $\|V^*\| \geq \lceil \frac{n}{k} \rceil$ such that $\bigcap_{v \in V^*} A_v \neq \varnothing$ and $\left(\bigcup_{v \in V^*} A_v \right) \cap S = \varnothing$.

For each $r \in [t]$, let S_r be the output of R_r. Since $\left(\bigcup_{v \in V^*} A_v \right) \cap S = \varnothing$ and R_t satisfies Justified Representation by assumption, $\left(\bigcap_{v \in V^*} A_v \right) \cap S_{t-1}$ must be $\varnothing$. We have $S = S_t \subset \cdots \subset S_1 \subset S_0 = C$, $\left(\bigcap_{v \in V^*} A_v \right) \cap S_{t-1} = \varnothing$ and $\left(\bigcap_{v \in V^*} A_v \right) \neq \varnothing$. Let $r_0 \in [1, t]$ be the last stage such that $\left(\bigcap_{v \in V^*} A_v \right) \cap S_{r_0-1} \neq \varnothing$, i.e., $\left(\bigcap_{v \in V^*} A_v \right) \cap S_{r_0} = \varnothing$. This means the candidates in $\left(\bigcap_{v \in V^*} A_v \right) \cap S_{r_0-1}$ are not included in the winning set S_{r_0}. There are $\|V\| \geq \frac{n}{k_{r_0}}$ voters unrepresented in the r_0-th stage, which contradicts Justified Representation of R_{r_0}. □

A.3   Violating Axioms in Multi-stages

We discuss in this part the axioms of Candidate Monotonicity (Theorem 7), Consistency (Theorem 8), and Pareto Efficiency (Theorem 9). It is known that all of the single-stage increasing Thiele methods satisfy Candidate Monotonicity, Consistency, and Pareto Efficiency (e.g., see [17]). However, a multi-stage approval-based rule may not satisfy them even though it is composed of single-stage Thiele methods.

Theorem 7 (Candidate Monotonicity does not preserve in a multistage approval-based rule). *Let $t \geq 2$ be an integer and $\mathcal{R}$ be a t-stage ω-Thiele rule, where R is a single-stage ω-Thiele rule that satisfies Candidate Monotonicity. If for any integer $i \geq 0$, $\omega(i)$ is a rational number and*

$$0 \neq \omega(i_0 - 1) - \omega(i_0 - 2) \neq \omega(i_0) - \omega(i_0 - 1)$$

holds for some i_0 in the function ω, the multi-stage rule $\mathcal{R}$ is not candidate monotone.

Candidate Monotonicity requires that an increase in support for some elected candidate by changing a voter's preference should not result in their knock-out. The reason why this axiom does not preserve is that changing some votes in favor of an elected candidate can change the outcome of the election for other candidates. While the candidate receiving more support may still be elected by a single-stage rule, it is possible for the new members of this stage's winning committee to surpass them in the subsequent stages.

Proof (Proof of Theorem 7). For the function ω parameterizing R, let p_i denote the difference between $\omega(i)$ and $\omega(i-1)$ for each $i \geq 1$. There are two cases: $p_{i_0} < p_{i_0-1}$ and $p_{i_0} > p_{i_0-1}$. Our plan is to construct an election $E = (C, V)$ and a vector $\boldsymbol{v} = (k_1, k_2, \ldots, k_t)$ for each case to show that $\mathcal{R}$ does not satisfy Candidate Monotonicity.

$-$ $p_{i_0} < p_{i_0-1}$. We can always find a rational number $\frac{n_1}{n_2}$ such that n_1 and n_2 are both integers, and $\frac{n_1}{n_2} = \frac{p_{i_0-1}}{p_{i_0}}$. Without loss of generality, we assume that $n_1 - n_2 > 1$. Otherwise, we can use $c \cdot n_1$ and $c \cdot n_2$ for some integer c instead. Then consider the following set of voters V:

$$n_1 \times \{a, c, d_1, \ldots, d_{i_0-2}\} \quad n_1 \times \{b, c, d_1, \ldots, d_{i_0-2}\}$$
$$n_2 \times \{a, d_1, \ldots, d_{i_0-2}\} \quad n_2 \times \{b, d_1, \ldots, d_{i_0-2}\}$$
$$1 \times \{e_1\} \quad 1 \times \{e_2\} \quad \cdots \quad 1 \times \{e_{t-2}\}.$$

We eliminate one candidate in each stage. In particular, $\mathcal{R}$ selects i_0 candidates from $(i_0 + 1)$ ones in the $(t-1)$-th stage, and picks $(i_0 - 1)$ candidates from i_0 ones in the last stage.

It is straightforward that candidate $e_1, \ldots, e_{t-3}$ and e_{t-2} are eliminated in the first $(t-2)$ stages. In the $(t-1)$-th stage, $\{a, b, d_1, \ldots, d_{i_0-2}\}$ is one of the winning committees. Its score is equal to that of $\{a, c, d_1, \ldots, d_{i_0-2}\}$ or $\{b, c, d_1, \ldots, d_{i_0-2}\}$, as $n_1 p_{i_0} = n_2 p_{i_0-1}$. Then there must be a winning committee S in the last stage such that $b \in S$. However, if a voter who has approved $\{a, d_1, \ldots, d_{i_0-2}\}$ additionally approves b, then $\{b, c, d_1, \ldots, d_{i_0-2}\}$ becomes the only winning committee in the $(t-1)$-th stage, which outperforms $\{a, b, d_1, \ldots, d_{i_0-2}\}$. Hence the final winning committee is $\{c, d_1, \ldots, d_{i_0-2}\}$, which does not include b.

$-$ $p_{i_0} > p_{i_0-1}$. There exists a rational number $\frac{n_1}{n_2} < 1$ such that n_1 and n_2 are both integers and $\frac{n_1}{n_2} = \frac{p_{i_0-1}}{p_{i_0}}$. Without loss of generality, we assume that $n_2 - n_1 > 1$. Then consider voters with the following approval ballots:

$$n_1 \times \{a, b, d_1, \ldots, d_{i_0-2}\} \quad n_2 \times \{c, d_1, \ldots, d_{i_0-2}\}$$
$$1 \times \{e_1\} \quad 1 \times \{e_2\} \quad \cdots \quad 1 \times \{e_{t-2}\}.$$

We still consider an election that eliminates one candidate in each stage. Candidates $e_1, \ldots, e_{t-2}$ are eliminated in the first $(t-2)$ stages. In the $(t-1)$-th stage, $\{a, b, d_1, \ldots, d_{i_0-2}\}$ is one of the winning committees. Then there is a winning committee S in the last stage such that $b \in S$. However, if a voter who has approved $\{c, d_1, \ldots, d_{i_0-2}\}$ additionally approves b, then $\{b, c, d_1, \ldots, d_{i_0-2}\}$ becomes the only winning committee in the $(t-1)$th-stage, which outperforms $\{a, b, d_1, \ldots, d_{i_0-2}\}$. Hence the final winning committee is $\{c, d_1, \ldots, d_{i_0-2}\}$ as $n_2 > n_1+1$. The only winning committee does not include b. $\qquad\square$

Theorem 8 (Consistency does not preserve in a multi-stage approval-based rule). *Let $t \geq 2$ be an integer and $\mathcal{R}$ be a t-stage ω-Thiele rule, where $\mathcal{R}$ is a single-stage ω-Thiele rule that satisfies Consistency. If*

$$\omega(i_0 - 1) - \omega(i_0 - 2) \neq \omega(i_0) - \omega(i_0 - 1)$$

holds for some i_0 in the function ω, the multi-stage rule $\mathcal{R}$ does not satisfy Consistency.

The reason why Consistency does not inherit is similar to that of Candidate Monotonicity. Increasing support for certain candidates by adding a set of voters in favor of them not only affects the candidates themselves with support but also alters the outcome for other candidates. Although the candidates receiving additional support may still be elected at first, it is possible for the new members of the winning committee in this stage to knock out them in the subsequent stages. Consistency therefore cannot preserve in multi-stage voting.

Proof (Proof of Theorem 8). For the function ω parameterizing R, let p_i denote the difference between $\omega(i)$ and $\omega(i-1)$ for each $i \geq 1$. Without loss of generality, we assume i_0 is the smallest integer such that $p_{i_0} \neq p_1$. There are two cases: $p_{i_0} < p_1$ and $p_{i_0} > p_1$. We plan to construct an election $E = (C, V)$ and a vector $\boldsymbol{v} = (k_1, k_2, \ldots, k_t)$ for each case to show that $\mathcal{R}$ does not satisfy Consistency.

- $p_{i_0} < p_1$. Consider an election that eliminates one candidate in each stage, i.e., $\boldsymbol{v} = (i_0 + t - 2, i_0 + t - 3, \ldots, i_0, i_0 - 1)$. There is a rational number $\frac{n_1}{n_2}$ such that $\frac{p_{i_0}}{p_1} < \frac{n_1}{n_2} < 1$ and $n_1 > 2$. Let us consider the set of voters V_1 with approval ballots as below:

$$n_1 \times \{a\} \quad n_2 \times \{b, c_1, c_2, \ldots, c_{i_0-1}\}$$
$$1 \times \{e_1\} \quad 1 \times \{e_2\} \quad \cdots \quad 1 \times \{e_{t-2}\}.$$

It is obvious that candidates $e_1, \ldots, e_{t-2}$ are eliminated in the first $(t-2)$ stages. In the $(t-1)$-th stage, $\mathcal{R}$ may choose $\{a, c_1, \ldots, c_{i_0-1}\}$ or $\{a, b, c_{l1}, c_{l2}, \ldots, c_{l,i_0-2}\}$ as winning committee, where $c_{l1}, \ldots, c_{l,i_0-2}$ are arbitrary $(i_0 - 2)$ candidates chosen from $c_1, \ldots, c_{i_0-1}$. In the last stage, a will be eliminated due to that $n_2 > n_1$. Hence the final winning committee may be $\{c_1, \ldots, c_{i_0-1}\}$ or $\{b, c_{l1}, c_{l2}, \ldots, c_{l,i_0-2}\}$.
We construct the set of voters V_2 as below:

$$n_1 \times \{b\} \quad n_2 \times \{a, c_1, c_2, \ldots, c_{i_0-1}\}$$
$$1 \times \{e_1\} \quad 1 \times \{e_2\} \quad \cdots \quad 1 \times \{e_{t-2}\}.$$

Similarly, candidates $e_1, \ldots, e_{t-2}$ are eliminated in the first $(t-2)$ stages. In the $(t-1)$-th stage, $\mathcal{R}$ may choose $\{b, c_1, \ldots, c_{i_0-1}\}$ or $\{a, b, c_{l1}, c_{l2}, \ldots, c_{l,i_0-2}\}$ as winning committee, where $c_{l1}, \ldots, c_{l,i_0-2}$ are arbitrary $(i_0 - 2)$ candidates chosen from $c_1, \ldots, c_{i_0-1}$. In the last stage, b will be eliminated due to that $n_2 > n_1$. Hence the final winning committee may be $\{c_1, \ldots, c_{i_0-1}\}$ or $\{a, c_{l1}, c_{l2}, \ldots, c_{l,i_0-2}\}$. Therefore $\mathcal{R}((C, V_1), \boldsymbol{v}) \cap \mathcal{R}((C, V_2), \boldsymbol{v}) = \{\{c_1, \ldots, c_{i_0-1}\}\}$.
However, if we combine V_1 and V_2, $\mathcal{R}$ will choose $\{a, b, c_{l1}, \ldots, c_{l,i_0-2}\}$ as winning committee in the $(t-1)$-th stage since $p_1 n_1 > p_{i_0} n_2$. In the last stage, committee $\{c_1, \ldots, c_{i_0-1}\}$ must not be in $\mathcal{R}((C, V_1 + V_2), i_0 - 1)$, and thus $\mathcal{R}$ does not satisfy Consistency.
- $p_{i_0} > p_1$. Let $\boldsymbol{v} = (i_0 + t - 2, i_0 + t - 3, \ldots, i_0, i_0 - 1)$, i.e., one candidate is eliminated in each stage. There is a rational number $\frac{n_1}{n_2}$ such that $\frac{p_{i_0}}{p_1} > \frac{n_1}{n_2} >$

1 and $n_2 > 2$. Consider the set of voters V_1 with approval ballots described below:

$$n_1 \times \{b, d_1, \ldots, d_{i_0-2}\} \quad n_1 \times \{c, d_1, \ldots, d_{i_0-2}\}$$
$$n_2 \times \{a, b, d_1, \ldots, d_{i_0-2}\}$$
$$n_2 \times \{a, c, d_1, \ldots, d_{i_0-2}\}$$
$$1 \times \{e_1\} \quad 1 \times \{e_2\} \quad \cdots \quad 1 \times \{e_{t-2}\}.$$

After candidates $e_1, e_2, \ldots, e_{t-2}$ are eliminated in the first $(t-2)$ stages, $\mathcal{R}$ chooses $\{a, b, d_1, \ldots, d_{i_0-2}\}$ or $\{a, c, d_1, \ldots, d_{i_0-2}\}$ in the $(t-1)$-th stage. Then a will be eliminated in the last stage as $n_2 < n_1$. Hence the final winning committee is $\{b, d_1, \ldots, d_{i_0-2}\}$ or $\{c, d_1, \ldots, d_{i_0-2}\}$.
We construct a set of voters V_2 as below:

$$n_1 \times \{a, d_1, \ldots, d_{i_0-2}\} \quad n_1 \times \{c, d_1, \ldots, d_{i_0-2}\}$$
$$n_2 \times \{a, b, d_1, \ldots, d_{i_0-2}\}$$
$$n_2 \times \{b, c, d_1, \ldots, d_{i_0-2}\}$$
$$1 \times \{e_1\} \quad 1 \times \{e_2\} \quad \cdots \quad 1 \times \{e_{t-2}\}.$$

$\mathcal{R}$ chooses $\{a, b, d_1, \ldots, d_{i_0-2}\}$ or $\{b, c, d_1, \ldots, c_{i_0-2}\}$ in the $(t-1)$-th stage. In the last stage, b will be eliminated, and thus the final winning committee is $\{a, d_1, \ldots, d_{i_0-2}\}$ or $\{c, d_1, \ldots, d_{i_0-2}\}$. Hence $\mathcal{R}((C, V_1), \boldsymbol{v}) \cap \mathcal{R}((C, V_2), \boldsymbol{v}) = \{\{c, d_1, \ldots, d_{i_0-2}\}\}$.
However, if we combine V_1 and V_2, $\mathcal{R}$ will choose $\{a, b, d_1, \ldots, d_{i_0-2}\}$ as winning committee in the $(t-1)$-th stage since $p_{i_0} n_2 > p_1 n_1$. In the last stage, committee $\{c, d_1, \ldots, d_{i_0-2}\}$ must not be in $\mathcal{R}((C, V_1 + V_2), \boldsymbol{v})$, and thus $\mathcal{R}$ does not satisfy Consistency. $\square$

Theorem 9 (Pareto Efficiency does not preserve in a multi-stage approval-based rule). *Let $t \geq 2$ be an integer and $\mathcal{R}$ be a t-stage ω-Thiele. If*

$$\omega(i_0 - 1) - \omega(i_0 - 2) > \omega(i_0) - \omega(i_0 - 1) \tag{1}$$

holds for some i_0 in the function ω, then $\mathcal{R}$ does not satisfy Pareto Efficiency.

Note that if we consider multi-stage voting with vector $(m-1, m-2, m-3, \ldots, k)$, we indeed reverse sequential Thiele rules. It is well-known that reverse sequential proportional approval voting does not satisfy Pareto Efficiency [17]. Theorem 9 generalizes the negative result to any ω-Thiele rule which has some i_0 satisfying in Eq. 1.

Proof (Proof of Theorem 9). For the function ω parameterizing $\mathcal{R}$, let p_i denote the difference between $\omega(i)$ and $\omega(i-1)$ for each $i \geq 1$. Our plan is to construct an election $E = (C, V)$ and a vector $\boldsymbol{v} = (k_1, k_2, \ldots, k_t)$ to show that $\mathcal{R}$ does not satisfy Pareto Efficiency. Consider an election that eliminates one candidate in each stage, i.e., $\boldsymbol{v} = (i_0 + t, i_0 + t - 1, \ldots, i_0 + 2, i_0 + 1)$. There exist integers $n_1, n_2, n_3 > 1$ satisfying $\frac{n_1}{n_2} \leq \frac{p_{i_0-1} - p_{i_0}}{p_{i_0}}$ and $\frac{n_3}{n_2} \geq \frac{p_{i_0-1}}{p_{i_0+1}}$. Let us consider the set of voters V with approval ballots as below:

$$n_1 \times \{a, f_1, ..., f_{i_0-2}\} \quad n_2 \times \{a, b, f_1, ..., f_{i_0-2}\} \quad n_2 \times \{a, c, f_1, ..., f_{i_0-2}\}$$
$$n_3 \times \{b, d, e, f_1, ..., f_{i_0-2}\} \quad n_3 \times \{c, d, e, f_1, ..., f_{i_0-2}\}$$
$$1 \times \{g_1\} \quad 1 \times \{g_2\} \quad \cdots \quad 1 \times \{g_{t-2}\}.$$

It is straightforward that candidate $g_1, ..., g_{t-3}$ and g_{t-2} are eliminated in the first $(t-2)$ stages. It can be verified that in the $(t-1)$-th stage, one of the winning committees (of size $i_0 + 2$) is $\{b, c, d, e, f_1, ..., f_{i_0-2}\}$. In the last stage, we will choose either $\{b, c, d, f_1, ..., f_{i_0-2}\}$ or $\{b, c, e, f_1, ..., f_{i_0-2}\}$ as these have maximal score among all $(i_0 + 1)$-subsets of $\{b, c, d, e, f_1, ..., f_{i_0-2}\}$. However, these two subsets are both dominated by $\{a, d, e, f_1, ..., f_{i_0-2}\}$, so Pareto Efficiency fails. □

B   Empirical results

We present a comparative experiment on single-stage and two-stage voting in this section, aiming to gain insight into the fairness of election outcomes. We conduct simulations on synthetic data and employ a variety of commonly used score-based voting rules. The empirical findings indicate that two-stage voting may result in a fairer selection of winning committees when compared to the single-stage method.

B.1   Setup

Sampling Candidates and Voters. We generate 200 random points on the plane $\mathbb{R}^2$ to represent candidates, where 80 of them come from a Gaussian distribution centered at $(1, 0)$ with a standard deviation of 0.5, and the other 120 ones are distributed uniformly in the square $[-2, 1] \times [-2, 1]$. In addition, we uniformly sample 400 points on a disc centered at $(0, 0)$ with a radius of 2 to represent voters. A voter's preference order is determined by the Euclidean distance between the voter and each candidate. That is, given a pair of candidates $c_i, c_j \in \mathbb{R}^2$ and a voter $v \in \mathbb{R}^2$, v prefer c_i to c_j if $d(c_i, v) < d(c_j, v)$, where $d(\cdot, \cdot)$ stands for the Euclidean distance.

Target Committee Sizes. For an election with two stages, we set the target committee size k_2 in the second stage as 20. Then the size k_1 of the winning committee for the first stage ranges between 20 and 200. We run a separate election for each value of k_1. In particular, two-stage voting degenerates into single-stage voting when $k_1 = 20$ or $k_1 = 200$.

Voting Rules. We apply the following score-based voting rules in our experiment: SNTV, Borda, Bloc, and CC. The definitions of these rules can be found in Sect. 2. We implement the CC rule through integer linear programming (ILP) solving.

The Measure of Fairness. We use the Gini index to evaluate the fairness of voting. If n_i represents the number of final winners in the i-th quadrant, then the Gini index G is given by $G = \left(\sum_{i=1}^{4} \sum_{j=1}^{4} |n_i - n_j| \right) / \left(2 \sum_{i=1}^{4} \sum_{j=1}^{4} n_i \right)$. The lower Gini index means a fairer election.

B.2 Results

Figure 1 illustrates the mean and standard deviation of the scores of the winning committees, as determined through 500 random trial elections applying SNTV in each stage. Similarly, Fig. 2 presents the mean and standard deviation of the Gini indices obtained from the same 500 random trials using the SNTV rule. The results of the remaining rules can be found in the appendix.

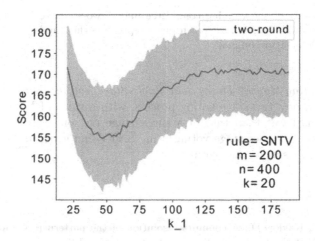

Fig. 1. Score of the winning committee under two-stage voting. The blue line represents the score, and the green shade represents the standard deviation of the score. (Color figure online)

The Winners Produced by Single-Stage and Two-Stage Voting Can Be Distinct. As shown in Fig. 1, the score curve displays a trend of initial decrease followed by an increase. It is intuitive to infer that the score of the winning committee obtained through the two-stage voting process will be approximately equivalent to that obtained by a single-stage election as the value of k_1 approaches k_2 or m. This is because, in such cases, the two-stage election degenerates into a single-stage one that chooses 20 winners from 200 candidates. In particular, it can be observed that the score of the winning committee obtained through a two-stage voting process with a k_1 astage 50 is lower than that obtained by a single-stage election on average. This discrepancy in scores illustrates that the winners produced by single-stage and two-stage voting are different.

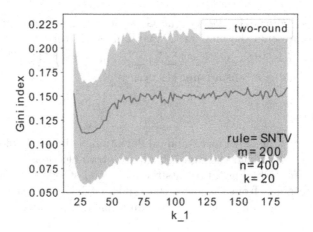

Fig. 2. Gini index of the winning committee under two-stage voting. The blue line represents the Gini index, and the green shade represents the standard deviation of the Gini index. (Color figure online)

Two-Stage Voting May Be Fairer than Single-Stage Voting. Figure 2 shows that the Gini index of a two-stage voting process, with a value of k_1 approximately equal to 30, is lower than that of a single-stage one on average. This observation suggests that two-stage voting using SNTV can generate fairer winning committees than single-stage voting.

References

1. Ablow, C., Kaylor, D.: A committee solution of the pattern recognition problem (Corresp.). IEEE Trans. Inf. Theory **11**(3), 453–455 (1965)
2. Arora, S., Hazan, E., Kale, S.: The multiplicative weights update method: a meta-algorithm and applications. Theory Comput. **8**(1), 121–164 (2012)
3. Aziz, H., Brill, M., Conitzer, V., Elkind, E., Freeman, R., Walsh, T.: Justified representation in approval-based committee voting. Soc. Choice Welfare **48**(2), 461–485 (2017)
4. Barberà, S., Coelho, D.: How to choose a non-controversial list with k names. Soc. Choice Welfare **31**(1), 79–96 (2008)
5. Bartholdi, J.J., Orlin, J.B.: Single transferable vote resists strategic voting. Soc. Choice Welfare **8**, 341–354 (1991)
6. Borodin, A., Lev, O., Shah, N., Strangway, T.: Primarily about primaries. Proc. AAAI Conf. Artif. Intell. **33**, 1804–1811 (2019)
7. Brill, M., Laslier, J.F., Skowron, P.: Multiwinner approval rules as apportionment methods. J. Theor. Polit. **30**(3), 358–382 (2018)
8. Chamberlin, J.R., Courant, P.N.: Representative deliberations and representative decisions: proportional representation and the Borda rule. Am. Polit. Sci. Rev. **77**(3), 718–733 (1983)
9. Davies, J., Narodytska, N., Walsh, T.: Eliminating the weakest link: making manipulation intractable? Proc. AAAI Conf. Artif. Intell. **26**, 1333–1339 (2012)

10. Dummett, M.: Voting Procedures. Clarendon Press (1984)
11. Elkind, E., Faliszewski, P., Skowron, P., Slinko, A.: Properties of multiwinner voting rules. Soc. Choice Welf. **48**(3), 599–632 (2017)
12. Faliszewski, P., Skowron, P., Slinko, A., Talmon, N.: Multiwinner voting: a new challenge for social choice theory. Trends Comput. Soc. Choice **74**(2017), 27–47 (2017)
13. Fishburn, P.C., Pekec, A.: Approval voting for committees: threshold approaches. Technical report (2004)
14. Freund, Y.: Boosting a weak learning algorithm by majority. Inf. Comput. **121**(2), 256–285 (1995)
15. Janson, S.: Phragmén's and Thiele's election methods. Technical report (2016)
16. Khachay, M.: Committee polyhedral separability: complexity and polynomial approximation. Mach. Learn. **101**(1), 231–251 (2015)
17. Lackner, M., Skowron, P.: Multi-winner voting with approval preferences. arXiv preprint arXiv:2007.01795 (2020)
18. Lu, T., Boutilier, C.: Budgeted social choice: from consensus to personalized decision making. In: Twenty-Second International Joint Conference on Artificial Intelligence (2011)
19. Narodytska, N., Walsh, T.: Manipulating two stage voting rules. In: Proceedings of the 2013 International Conference on Autonomous Agents and Multi-Agent Systems, pp. 423–430 (2013)
20. Niou, E.M.S.: A note on Nanson's rule. Public Choice **54**(2), 191–193 (1987)
21. Pareto, V.: Manuale di economia politica: con una introduzione alla scienza sociale, vol. 13. Società editrice libraria (1919)
22. Procaccia, A.D., Rosenschein, J.S., Zohar, A.: On the complexity of achieving proportional representation. Soc. Choice Welfare **30**(3), 353–362 (2008)
23. Sánchez-Fernández, L., Fisteus, J.A.: Monotonicity axioms in approval-based multi-winner voting rules. arXiv preprint arXiv:1710.04246 (2017)
24. Skowron, P., Faliszewski, P., Lang, J.: Finding a collective set of items: from proportional multirepresentation to group recommendation. Artif. Intell. **241**, 191–216 (2016)
25. Skowron, P., Faliszewski, P., Slinko, A.: Achieving fully proportional representation: approximability results. Artif. Intell. **222**, 67–103 (2015)
26. Skowron, P., Faliszewski, P., Slinko, A.: Axiomatic characterization of committee scoring rules. J. Econ. Theory **180**, 244–273 (2019)
27. Smith, J.H.: Aggregation of preferences with variable electorate. Econometrica J. Econometric Soc., 1027–1041 (1973)
28. Sornat, K., Williams, V.V., Xu, Y.: Near-tight algorithms for the Chamberlin-Courant and Thiele voting rules. arXiv preprint arXiv:2212.14173 (2022)
29. Thiele, T.N.: Om flerfoldsvalg. Oversigt over det Kongelige danske Videnskabernes Selskabs Forhandlinger, pp. 415–441 (1895)
30. Woodall, D.: Properties of preferential election rules. Voting Matters **3**, 8–15 (1994)

Production and Infrastructure Construction in a Resource Region: A Comparative Analysis of Mechanisms for Forming a Consortium of Subsoil Users

Sergey Lavlinskii[(✉)], Artem Panin, Alexander Plyasunov, and Alexander Zyryanov

Sobolev Institute of Mathematics, Novosibirsk, Russia
{lavlin,apljas}@math.nsc.ru

Abstract. The paper considers the model of formation of a program for the development of the mineral resource base of a resource region based on public-private partnership. The key element of the program is the mechanism of creating a consortium of private investors, jointly implementing projects for the construction of the necessary production infrastructure. Two variants of consortium formation are considered. In the first case, the consortium is formed "top-down" by the government, which creates a management company. Its goal is to find a compromise between the interests of the government and subsoil users, providing the maximum possible budgetary flow. The second variant substantially assumes the formation of the consortium "from the bottom up" on the initiative of subsoil users. In this case, private investors themselves form a management company. It ensures rational sharing of infrastructure costs and, in dialogue with the government, seeks to maximize the amount of natural resource rent they receive. Both mechanisms formally fit into the Stackelberg game scheme. The corresponding bilevel mathematical programming problems are solved using metaheuristics based on coordinate descent. The numerical experiment carried out on real data of the Zabaikalsky Krai allows us to compare the ways of consortium formation and formulate practical recommendations for the economically unobvious choice of partnership architecture.

Keywords: Stackelberg game · bilevel mathematical programming problems · strategic planning · public-private partnership · a consortium of subsoil users, metaheuristics, coordinate descent

1 Introduction

The mechanism of public-private partnership (PPP) is actively used in the world and allows to achieve a compromise of interests in various spheres of the economy.

In the Russian mineral sector, the main area of PPP application is transport and energy infrastructure projects. To remove the barriers to efficient subsoil use, the government in practice has not been very successful in using forms of partnership based on budget financing of production infrastructure construction projects and large-scale assistance to the private investor. In the context of a budget deficit, this approach is ineffective, and the government is attempting to transform the PPP institution towards its classical forms.

In the classic public-private partnership model, an investor agrees with the government on a list of infrastructure projects that "open up" the field development projects of interest to it and implements these infrastructure projects at its own expense. The government compensates its costs with a certain lag. Such a scheme requires a delicate mechanism of coordination of interests that determines, at a minimum, the algorithm of compensation payments. It also presupposes the existence of a toolkit that allows for the formation of a comprehensive subsoil development program on this basis, which defines the concrete obligations of the actors involved in the partnership.

In the previous works of the authors [1,2] the classical PPP model was supplemented by the clustering of fields and consortium mechanism, which activates horizontal relations of private investors and the effects of resource consolidation. Under these conditions, the partnership mechanism and the subsoil development program become as concrete and targeted as possible and can be used directly in management practice.

This PPP model assumed that the consortium is formed "top-down" by the government, creating management companies (MC), the main leitmotif of whose behavior is the search for a compromise between the interests of the government and subsoil users, ensuring the maximum possible budgetary flow. An alternative approach has been left out of the field of view of the works, meaningfully suggesting the formation of consortia "from the bottom up", at the initiative of subsoil users, who receive compensation for non-core infrastructure costs within the framework of the classical scheme of public-private partnership. In this case, private investors themselves form management companies, which provide a rational division of infrastructure costs in the cluster and, in dialogue with the government, seek to maximize the amount of natural resource rent received by them.

Both statements formally fit into the Stackelberg game scheme and describe the natural resource rent sharing. In the corresponding bilevel mathematical programming problem at the upper level, the leader (government) maximizes budgetary effects. The follower (the system of consortia) acts either in cooperation with the leader, providing him with the largest possible part of the rent (formation of MC "top-down"), or pulls the rent in the direction of subsoil users (initiative formation of consortia by private investors).

From the economic point of view, the choice of the way to form consortia is not obvious. It depends on a lot of factors, such as the level of favorable investment climate in the region, liberal investment policy, the mechanism of reimbursement of infrastructure costs to the subsoil user in the partnership, etc.

For each of the formulations, a model analysis of their effectiveness in these or those conditions is required. This will make it possible to formulate practical recommendations for the choice of partnership architecture.

The present paper is devoted to this very issue. The analysis of the results of numerical experiments on real data serves as a basis for answering the question of which model of consortium formation ("top" or "bottom") is preferable, and under what conditions. Such meaningful conclusions largely determine the overall strategy for the development of raw material territories and will be useful in stimulating the entry of private investors in underdeveloped Russian regions with a rich resource base.

2   Mathematical Model

We assume that fields within the region form a system of non-overlapping clusters. The transport and energy infrastructure required for development is built by a system of consortia localized in these clusters, formed either by the government or the initiative of private investors. Each cluster has its own management company, which organizes and coordinates the co-financing of infrastructure construction by private investors in the consortium and compensates the subsoil users for their expenses using budget funds.

It is assumed that the MC in the consortium formed "top-down" will take on all the functions of the government in the procedures for coordinating the interests of the investor and the government in the part of mining projects (pre-project analysis, environmental control, monitoring, etc.). The MC of the initiative consortium works directly with the government, representing the interests of investors in the cluster. The output of the models is a targeted plan for the formation of a program to develop the mineral resource base. It defines for each consortium (cluster) a list of infrastructure projects to be implemented, and for each investor (field) a schedule of costs for the creation of the necessary production infrastructure and their compensation from the budget.

The method of consortia formation and the regulations of its work are reflected, first of all, in the way of accounting for transaction costs (TC) accompanying the relations of economic agents [3]. In the general list of transaction costs, we will distinguish transaction costs "before" (TC ex ante) and transaction costs "after" (TC ex post) beginning of the project implementation. On the part of the investor participating in the consortium, these are the development of environmental impact assessment, maintenance of services responsible for interaction with the MC and relevant supervisory authorities, etc. On the part of the management company of the consortium, created "top-down" and representing mainly the interests of the government, the TC includes, in particular, control and monitoring (including technical and environmental), as well as the costs of improving the supporting public institutions.

The list of transaction costs of the management company of an initiative-formed consortium is considerably smaller and contains, mainly, the costs of coordinating the processes of distribution of infrastructure costs and their compensation between the consortium members. In this case, most of the TC arising in the process of subsoil development and construction of production infrastructure fall directly on the government's shoulders.

We use the following notation:

$T = \{-T_1, \ldots, 0, 1, \ldots, T_2\}$ is the time horizon; T_0 is the time lag of reimbursement of infrastructure costs by the private investor; I is a set of production projects; J is a set of infrastructure development projects; K is a set of consortia. Each production project has its own private investor.

Mining project i in year t:

CFP_i^t is the operating cashflow; DBP_i^t are the tax revenues of the budget from the project; $ITCP_i^t$ and $MTCP_i^t$ are, respectively, the TC of investor i and the TC incurred by the MC of the investor's "host" consortium, which arise during the preparation(ex $ante$, $t = -T_1, \ldots, 0$) and implementation (ex $post$, $t = 1, \ldots, T_2$) of the mining project.

Infrastructure project j in year t:

ZI_j^t is the schedule of investment costs necessary for implementing project; VDI_j^t are the off-project revenues of the budget from implementing project j, which are associated with the overall development of the local economy; $MTCI_j^t$ is the schedule of the TC incurred by the MC of the consortium implementing project j; $ITCI_j^t$ is the schedule of the total TC of the investors partaking in the implementation of project j.

Outside the planning horizon ($t = -T_1, \ldots, 0$), the model parameters CFP_i^t, DBP_i^t, ZI_j^t and VDI_j^t are assumed to be zero.

Interproject connection:

μ_{ij} is the indicator of technological cohesion between production and infrastructure projects; $i \in I$, $j \in J$:

$$\mu_{ij} = \begin{cases} 1, \text{ if the implementation of production project } i \\ \quad \text{necessarily requires the implementation of infrastructure project } j, \\ 0 \text{ otherwise.} \end{cases}$$

The sets of production and infrastructure projects are divided in a non-overlapping (mutually disjoint) manner into NC consortia, based on the location of the field clusters.

Project ownership by consortia is determined by the following parameters:

α_{kj} is the parameter indicating whether infrastructure project j is attributed to consortium k; this parameter equals to one if the attribution is valid and zero otherwise.

β_{ki} is the parameter indicating whether the investor of production project i is attributed to consortium k; this parameter equals to one if the attribution is valid and zero otherwise.

The discounts of the government and the investor: DG and DI respectively. $BudG^t$, $BudI_i^t$ are the budget constraints of, respectively, the government and investors.

We introduce the following integer variables:

$$z_i = \begin{cases} 1, \text{ if the investor } i \text{ launches production project,} \\ 0 \text{ otherwise;} \end{cases}$$

$$c_j = \begin{cases} 1, \text{ if infrastructure project } j \text{ is implemented by one of the consortia,} \\ 0 \text{ otherwise;} \end{cases}$$

Real-valued variables of the model:

$\bar{W}_k^t$ and W_k^t are the compensation schedules, respectively, offered by the leader (government) and actualized in reality; the compensations reimburse the infrastructure costs incurred by the private investors partaking in consortium k and the expenses associated with the functioning of the MC;

R_i^t is the compensation schedule (determined by the consortium's MC) for the costs of private investor i;

D_{ij} is the share of investor i in the costs of implementing infrastructure project j.

The PPP formation model can be represented as the following problem of bilevel mathematical programming.

Model A (consortium formation "top-down")

The upper-level problem $\widetilde{\mathcal{PG}}$ can be formulated as follows:

$$\sum_{t \in T}\left(\sum_{i \in I} DBP_i^t z_i + \sum_{j \in J} VDI_j^t c_j - \sum_{k \in K} W_k^t\right)/(1+DG)^t \to \max_{\bar{W},z,c,W,R,D} \quad (1)$$

subject to:

$$\sum_{k \in K} \bar{W}_k^t \le BudG^t; t \in T; \quad (2)$$

$$\bar{W}_k^t \ge 0; t \in T; k \in K; \quad (3)$$

$$(z, c, W, D, R) \in \mathcal{F}^*(\bar{W}). \quad (4)$$

The set $\mathcal{F}^*$ is a set of optimal solutions of the following low-level parametric consortia problem $\widetilde{\mathcal{PC}}(\bar{W})$:

$$\gamma \sum_{t \in T}\left(\sum_{i \in I}((CFP_i^t - ITCP_i^t)\, z_i - \sum_{j \in J}(ZI_j^t + ITCI_j^t)D_{ij} + R_i^t)\right)/(1+DI)^t \quad (5)$$

$$+ (1-\gamma)\sum_{t \in T}\left(\sum_{i \in I} DBP_i^t z_i + \sum_{j \in J} VDI_j^t c_j - \sum_{k \in K} W_k^t\right)/(1+DG)^t \to \max_{z,c,W,R,D}$$

subject to:

$$W_k^t \le \bar{W}_k^t; t \in T; k \in K; \quad (6)$$

$$\sum_{t \in T}\left(\sum_{i \in I} DBP_i^t z_i \beta_{ki} + \sum_{j \in J} VDI_j^t c_j \alpha_{kj} - W_k^t\right)/(1+DG)^t \ge 0; k \in K; \quad (7)$$

$$R_i^t = 0; -T_1 \leq t \leq T_0; i \in I; \tag{8}$$

$$R_i^t \geq 0; T_0 + 1 \leq t \leq T_2; i \in I; \tag{9}$$

$$\sum_{i \in I} R_i^t \beta_{ki} + \sum_{j \in J} MTCI_j^t c_j \alpha_{kj} + \sum_{i \in I} MTCP_i^t z_i \beta_{ki} \leq W_k^t; k \in K; t \in T; \tag{10}$$

$$0 \leq D_{ij} \leq \mu_{ij}; i \in I; j \in J; \tag{11}$$

$$\sum_{i \in I} D_{ij} \beta_{ki} = \alpha_{kj} c_j; k \in K; j \in J; \tag{12}$$

$$\sum_{t \in T} \Big(\sum_{i \in I} (CFP_i^t - ITCP_i^t) z_i - \sum_{j \in J} (ZI_j^t + ITCI_j^t) D_{ij} \tag{13}$$

$$+ R_i^t \Big) / (1 + DI)^t \geq 0; i \in I;$$

$$c_j \geq \mu_{ij} z_i; i \in I; j \in J; \tag{14}$$

$$-(CFP_i^t - ITCP_i^t) z_i + \sum_{j \in J} (ZI_j^t + ITCI_j^t) D_{ij} \tag{15}$$

$$- R_i^t \leq BudI_i^t; t \in T; i \in I.$$

Management companies of the consortia formed "top-down" seek to distribute the costs of infrastructure construction and compensation from the budget in such a way as to realize a compromise of the partners' interests (5). The degree to which the interests of the private investor are taken into account is determined by the parameter γ.

The consortium formed "top-down" is meaningful for the government only when its MCs and investors can provide an increase in the objective function of the government (1). This circumstance is fixed in (7) and explains the inefficiency of forming a single consortium for the full set of fields – in some cases, preference should be given to a system of several consortia covering the entire territory and taking into account the spatial reference of planning objects.

Starting from the year $(T_0 + 1)$, management companies begin compensation payments (8), (9). Transaction costs incurred by the government in the process of harmonizing interests in production and infrastructure construction are borne by the MC (12). The compensation schedule should provide the government with a balance of budget revenues and transfers to the consortia MCs (7), as well as compensate each investor for infrastructure costs at a discount (16).

The procedure for sharing infrastructure costs (including TC) among consortium members is described in (13)–(15). A mining project cannot be started unless the necessary production infrastructure is in place (17). Budget constraints (2) and (18) limit the choice of projects.

The solution to the problem determines the program for the development of the mineral resource base $\{c_j, z_i, R_i^t, D_{ij}\}$, which fixes for each consortium the list of infrastructure and production projects to be launched, the mechanism for the implementation of equity construction and the main parameters of the compensation policy.

Model B (consortium formation "from the bottom up")

The upper-level problem $\widetilde{\mathcal{PG}}$:

$$\sum_{t \in T} \Big(\sum_{i \in I} (DBP_i^t - MTCP_i^t) z_i + \sum_{j \in J} (VDI_j^t - MTCI_j^t) c_j \tag{16}$$

$$- \sum_{k \in K} W_k^t \Big) / (1 + DG)^t \to \max_{\bar{W}, z, c, W, R, D}$$

subject to:

$$(2) - (4) \tag{17}$$

Low-level parametric consortia problem $\widetilde{\mathcal{PC}}(\bar{W})$:

$$\sum_{t \in T} \Big(\sum_{i \in I} ((CFP_i^t - ITCP_i^t) z_i - \sum_{j \in J} (ZI_j^t + ITCI_j^t) D_{ij} \tag{18}$$

$$+ R_i^t) \Big) / (1 + DI)^t \to \max_{z, c, W, R, D}$$

subject to:

$$\sum_{i \in I} R_i^t \beta_{ki} \le W_k^t; k \in K; t \in T; \tag{19}$$

$$(6), (8), (9), (11) - (15) \tag{20}$$

In this statement, the main TC on the part of the government is carried directly by the budget. Transaction costs of initiative MCs (expenditures of coordinating the processes of distribution of infrastructure costs and their compensation between consortium participants) are attributed to subsoil users. There is no assessment of consortium feasibility for the government (constraint (7) in model A) (consortia are formed proactively in case of economic reasonability for subsoil users), and the objective functions of the upper and lower-level problems are oriented in different directions and share the natural resource rent. The output of the model is the same, as in model A, address development program $\{c_j, z_i, R_i^t, D_{ij}\}$, which defines the list of infrastructure and mining projects to be launched, and a balanced plan for cost allocation and compensation.

How do these programs correlate in a numerical experiment on real data?

3   Numerical Experiment

Model Polygon. Five non-overlapping clusters of polymetallic fields, 20 mining projects and 10 infrastructure projects are chosen as a model polygon in the experiments. The infrastructural part of the test site is based on the "2030 Program" of the Zabaikalsky Krai – power lines, roads and the Naryn-Lugokan railway (including the second segment to the Solonechenskoye and Lugokanskoye fields). Forecast period – 2015–2034, data from feasibility studies of the projects were used as a source of baseline information on mining projects. The model polygon is built in such a way that if a full infrastructure program is implemented, all field development projects can be launched.

Based on the model polygon, a database of bilevel planning models A and B was formed. For this purpose, forecast simulation models of various types of project implementation are used – development of field, construction of transport and energy infrastructure, etc. [4].

Solution Method. In the work of the authors [1] it was shown that for the formulated models A and B the upper-level problem belongs to the class of Σ_2^P-hard problems. In this case, the search for optimal solutions is unlikely to succeed even on relatively small dimensions, and the search for approximate solutions with a guaranteed estimate of the accuracy of the relative deviation from the optimal solution in polynomial time simply does not make sense [1,5–7]. The fact that the structure of the model has the most general form, and the dimensionality of the investor's problem is rather large, caused the authors to choose a solution method based on metaheuristics using a stochastic approximate hybrid algorithm with coordinate descent.

The basic construct of the metaheuristics algorithm can be represented as follows.

1. Choose a starting solution for the government W_k^t. Put $\bar{W}_k^t$ - the current best solution equal to W_k^t, and take the government objective function value at $\bar{W}_k^t$ as a record, $k \in K; t \in T$.
2. Put $i := 0$.
   2.1. If $i < I$, then $i := i + 1, k' := 1, t' := 1$, otherwise go to step 3.
   2.2. With probability p go to step 2.2.1, otherwise go to step 2.3.
      2.2.1. Put $N := 1$.
      2.2.2. From a random uniform distribution from 0 to $BudG$, choose the value of the variable $\hat{W}_{k'}^{t'}$.
      2.2.3. If the government objective function value at $\hat{W}_k^t$ is above the record, then update the record and put $\bar{W}_k^t := \hat{W}_k^t, i := 0$.
      2.2.4. If $N < N_{samples}$, then $N := N + 1$ and go to Sect. 2.2.2, otherwise go to Sect. 2.3
   2.3. If $t' < T$, then $t' := t' + 1$ and go to step 2.2, otherwise if $k' < NC$, then $k' := k' + 1, t' := 1$ and go to step 2.2, otherwise go to step 2.1.
3. Return the best solution $\bar{W}_k^t$ found.

Thus, the algorithm starts with some solution per government and iteratively tries to improve the starting solution. At each iteration, the variables W_k^t are enumerated, and with probability p the algorithm considers $N_{sapmles}$ values of the variable taken randomly from a uniform distribution from 0 to $BudG$. If the considered value of the variable government objective function is higher than the record, we update the record and the best solution found. The algorithm stops after I consecutive iterations without improving the record.

Comparative Analysis. The numerical experiment is based on the process of studying the Stackelberg equilibrium properties in the course of analyzing the sensitivity of solutions of the corresponding bilevel mathematical programming problem to changes in the main parameters of the model. The key issue is the dependence of the efficiency of the development program formed in models A and B on the level of favorable investment climate, the degree of liberality investment policy pursued by the government and the level of TC.

In real life, the level of favorable investment climate is determined by a whole set of factors – compliance with contracts and property rights, the presence of transparent and stable institutions, free and fair competition, the stability of the tax system, the price of credit resources, etc. All this affects the investor's propensity to invest his own resources in the project, expecting a cash flow and profitability level corresponding to the integral risks. That is why we assume that the investor adequately assesses the whole set of conditions and reflects the assessment of the level of favorable investment climate in its own discount (the level of expected profitability of the project). The value of $DI = 0.07$ in the calculations corresponds to favorable investment conditions, today's invest-ment climate in the Zabaikalsky Krai can be expertly assessed at the level of $DI = 0.13$–0.15.

The degree of liberality of the government's investment policy reflects its dis-count DG. Liberal investment policy operates with a small discount at the level of 0.01. This corresponds to the understanding that the return from investing budget funds in infrastructure (in the form of compensation) is realized on a long-term horizon, primarily due to multiplicative effects. The alternative is a high discount for the subsoil owner with a short-term horizon of assessment of positive effects on the plane of net present value.

For a comparative analysis of the performance indicators of the programs generated by model A and model B, we will consider four scenarios for the implementation of the government's investment policy depending on the emerg-ing investment climate for private investors in the territory:

LF – government's liberal investment policy, favorable investment climate;

LU – government's liberal investment policy, unfavorable investment climate;

CF – government's conservative investment policy, favorable investment cli-mate;

CU – government's conservative investment policy, unfavorable investment climate.

Formally, the set of LF, LU, CF and CU corresponds to the set of pairs of government and investor discounts (0.01; 0.07), (0.01; 0.15), (0.07; 0.07) and

(0.07; 0.15). For each scenario (pair of discounts) we will compare the performance of the solutions of models A and B at different levels of TC, expressed as a percentage of the volume of investment in the projects.

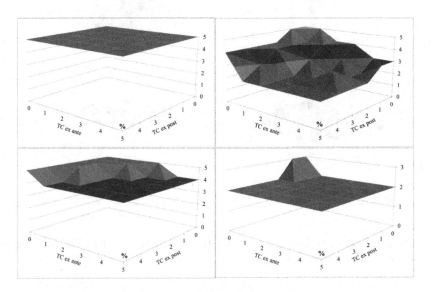

Fig. 1. Number of consortia initiated, model A: Scenario LF (top left), LU ((top right), CF (bottom left), CU (bottom right)

The process of consortia formation proceeds differently in models A and B. Figures 1 and 2 show the dependence of the number of formed consortia on the scenario and TC.

Under a favorable investment climate, both models generate approximately the same number of initiated consortia. All five potential consortia are formed at small TC even under the conditions of the conservative investment policy of the government. In conditions unfavorable for the investor, model A generates a significantly higher number of consortia, especially when the government discount is small.

Such differences between models A and B in the intensity of the consortium formation process give rise to differences in the number of infrastructure and production projects to be launched in the respective subsoil development programs. For the sake of clarity, the following figures show the results of a comparison of the same-type program parameters in the form of the difference between the corresponding solution indicators of models A and B. The analysis of such figures allows us to identify the areas of TC values for which, within a fixed scenario, model A gives better (worse) results than model B (the difference is positive (negative)).

In favorable conditions, the intensity of infrastructure construction is approximately the same, except for the highest TC (Fig. 3). In conditions of an

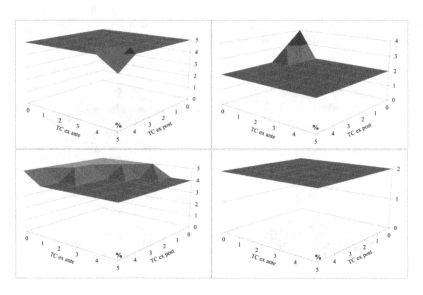

Fig. 2. Number of consortia initiated, model B: Scenario LF (top left), LU ((top right), CF (bottom left), CU (bottom right)

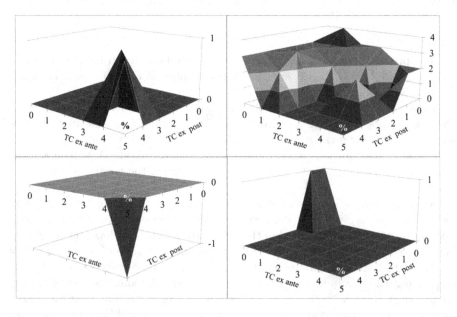

Fig. 3. Number of infrastructure projects implemented, difference between models A and B: Scenario LF (top left), LU ((top right), CF (bottom left), CU (bottom right)

unfavorable investment climate, the government's choice of a liberal policy, coupled with the organization of consortia "top-down," leads to a greater number of implemented infrastructure projects than in model B. The difference appears to be the greater the lower the TC level. The transition to a conservative policy (CU scenario) eliminates the difference in infrastructure programs for almost all levels of TC, except for the smallest.

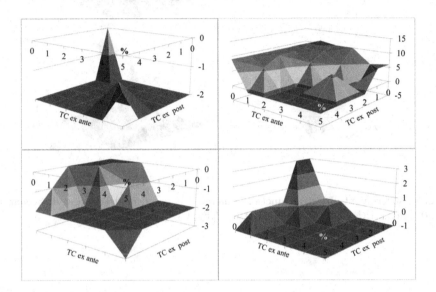

Fig. 4. Number of mining projects implemented, difference between models A and B: Scenario LF (top left), LU ((top right), CF (bottom left), CU (bottom right)

For the production part of the development program, the differentiation of model A and B indicators is determined to the greatest extent by the investor's discount. Under conditions of a favorable investment climate, the initiative of private investors in forming consortia provides a steadily larger front of production projects (Fig. 4). However, when conditions for the investor deteriorate (an increase of its discount), model A in the range of small TC, determined by the liberal investment policy, leads to a significantly more extensive program of subsoil use.

Differences in the parameters of development programs in models A and B naturally give rise to differences in the economic measures of partnership effectiveness. Figure 5 captures the fact that model A is almost always preferable in terms of government function. It provides the greatest benefit at small TC in an unfavorable investment climate with a government pursuing a liberal investment policy. Model B turns out to be preferable only for high TC, small investor discount and conservative government.

How do the required volumes of compensation payments compare in different scenarios? Cost minimization is a competing criterion for partnership efficiency

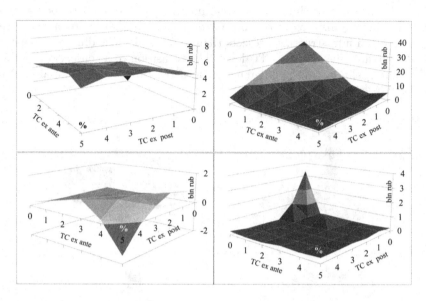

Fig. 5. Government objective function, difference between models A and B: Scenario LF (top left), LU ((top right), CF (bottom left), CU (bottom right)

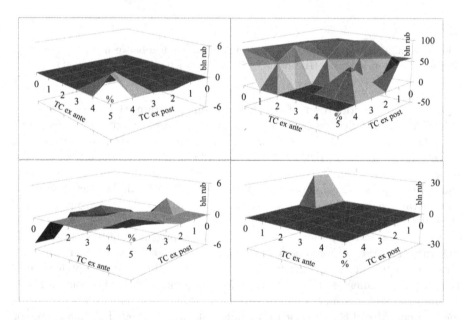

Fig. 6. Government compensation costs, difference between models A and B: Scenario LF (top left), LU ((top right), CF (bottom left), CU (bottom right)

for today's deficit budget. We see that the consortium formed "top-down" consumes substantially more resources than model B in the LU scenario (Fig. 6). The transition to a conservative CU policy stabilizes the situation - except for the area of small TC, rarely seen in an unfavorable investment climate, model A requires less budget tension. In the LF and CF scenarios, the high investor discount leads to the fact that model A requires less compensation for low TC and, favoring a liberal policy, the government should prefer a consortium formed "top-down".

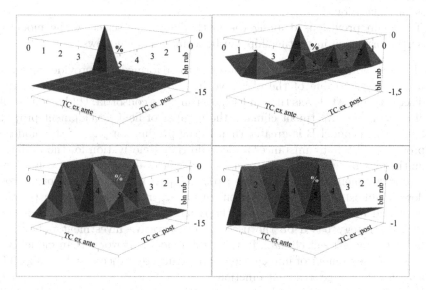

Fig. 7. Investors objective function, difference between models A and B: Scenario LF (top left), LU ((top right), CF (bottom left), CU (bottom right)

Figure 7 shows that for private investors the concept of an initiative-forming consortium is almost always reasonable - only within the framework of a conservative investment policy of the government at a low level of TC the part of rent received by investors is the same as in the model A. In all other cases, investors, having managed to organize themselves, achieve significantly higher profitability.

4   Results and Discussion

The numerical experiment carried out on real data of the Zabaikalsky Krai allows us to answer the question of which model of consortia formation ("top-down" or "from the bottom up") is preferable, and under what conditions. From the economic point of view, the choice of consortium formation method is not obvious and represents a problem for the departments of natural resources of regional administrations. This choice is the key to launching a program of subsoil development based on public-private partnerships.

It is PPP that makes it possible to break the vicious circle in which an investor cannot develop fields because there is no infrastructure, and the government cannot undertake infrastructure projects because it is not sure that the created capacities will be fully used. For the government, the implemented partnership creates jobs, tax revenues to budgets at all levels and prospects for sustainable development of the regional economy. Investors within the consortium gain access to subsoil resources and natural resource rent. The only thing left to do is to optimize the partnership architecture and achieve a compromise of interests within the current realities of the mineral resource complex of the Russian Federation.

The main provisions defining the methodology for selecting the model of consortia formation in partnership can be formulated as follows.

1. Model B reduces the number of consortia to be formed and provides investors with a higher value of the objective function on a wide range of partner discounts and TC levels than in the government's consortia. Under conditions of a favorable investment climate, the number of field development projects launched in model B is greater than in the production part of the model A program. These circumstances are an additional motivation for the initiative on the part of subsoil users, especially under favorable investment conditions.
2. The government, which forms the partnership as a system of consortia "top-down", almost always provides itself with a larger infrastructure construction front and budgetary flow than in the B model. The exception is a not very logical strategy based on the use of a conservative investment policy in a favorable investment climate. In this case, model A is worse than model B in terms of the number of implemented infrastructure projects and, at high TC, also in terms of government function.
3. From the point of view of budget savings in conditions of a favorable investment climate and low TC, model A together with a liberal policy is preferable for the government. When investment conditions for the subsoil user deteriorate and TC is high, the conservative option of model A requires fewer compensation payments than B and therefore will be preferable for the government.
4. For the Zabaikalsky Krai in its current state - unfavorable investment climate and high TC - model A is preferable. Compared to a partnership in which the initiative to form consortia is given to private investors, the A design provides a higher budgetary flow with lower compensation costs.

Even though the conclusions of the paper are based on the information base of a specific resource region, they can be extended to most of the raw material territories of Siberia and the Far East. Their problems and characteristics of fiscal capacity are similar to those of the Zabaikalsky Krai. The substantial conclusions formulated above largely determine the overall strategy for the development of raw material territories and will be useful in stimulating the entry of private investors into underdeveloped Russian regions with a rich resource base.

Acknowledgements. The study was funded by a grant Russian Science Foundation No. 23-28-00849, https://rscf.ru/project/23-28-00849.

References

1. Lavlinskii, S., Panin, A., Pliasunov, A.: Public-private partnership model with a consortiump. In: Khachay, M., Kochetov, Y., Eremeev, A., Khamisov, O., Mazalov, V., Pardalos, P. (eds.) MOTOR 2023. CCIS, vol. 1881, pp. 231–242. Springer, Cham (2023). https://doi.org/10.1007/978-3-031-43257-6-18
2. Lavlinskii, S., Zyryanov, A.: Model for long-term partnerships between the government and subsoil users in production and infrastructure construction. In: 19th International Asian School-Seminar on Optimization Problems of Complex Systems (OPCS) (2023). https://doi.org/10.1109/opcs59592.2023.10275768
3. Marshall, G.R.: Transaction costs, collective action and adaptation in managing socio-economic system. Ecol. Econ. **88**, 185–194 (2013). https://doi.org/10.1016/j.ecolecon.2012.12.030
4. Lavlinskii, S.M.: Public-private partnership in a natural resource region: ecological problems, models, and prospects. Stud. Russ. Econ. Dev. **21**(1), 71–79 (2010). https://doi.org/10.1134/S1075700710010089
5. Panin, A.A., Pashchenko, M.G., Plyasunov, A.V.: Bilevel competitive facility location and pricing problems. Autom. Remote. Control. **75**(4), 715–727 (2014)
6. Dempe, S., Zemkoho, A. (eds.): Bilevel Optimization. SOIA, vol. 161. Springer, Cham (2020). https://doi.org/10.1007/978-3-030-52119-6
7. Talbi, E-G. (ed.): Metaheuristics for Bi-Level Optimization. Studies in Computational Intelligence, vol. 482 (2013). https://doi.org/10.1007/978-3-642-37838-6

Automated and Automatic Systems of Management of an Optimization Programs Package for Decisions Making

Aidazade Kamil[1,3] and Samir Quliyev[1,2(✉)]

[1] Institute of Control Systems, Baku 1141, Azerbaijan
azcopal@gmail.com
[2] Azerbaijan State Oil and Industry University, Baku 1010, Azerbaijan
[3] Azerbaijan University of Architecture and Construction, Baku 1073, Azerbaijan

Abstract. It is known that, in spite of a large number of methods for numerical solutions to various classes of problems, the choice of the most efficient method for solving a particular problem under specific values of its parameters requires a large number of comparative experiments. As a rule, end users tend to have difficulty both in carrying out such experiments, which require knowledge of the domain of applicability of various numerical methods, and in properly conducting the comparative analysis of the results, which is quite time-consuming. We carry out the analysis of techniques and algorithms for managing a computational process to solve difficult or complex optimization problems with the use of computer systems. In this work, for the class of multivariate optimization problems, we propose two approaches for facilitating the use of available applied software packages using modern multi-processor and/or multicore computer systems. One of the approaches involves the active work of the user with the optimization program package in dialogue mode. The other approach involves packet control by means of a specially developed control program in automatic mode. The work contains the protocols and results of computer-based experiments for the class of unconstrained optimization problems on the basis of the developed software package.

Keywords: Decision Making · Optimization Methods · Optimal Control Methods · Parallel Computing · Multiprocessor/Multicore Systems · Dialog Systems

1 Introduction

Modeling and optimization are widely recognized as the fundamental tools for the scientific analysis of real-life systems. Applications of modeling and optimization theories span across various disciplines, including finance, marketing, information and communication technology, data mining, environment, and life sciences, among others. Optimization problems encountered in practice often involves objective functions/functionals which are rather complex and characterized by a number of traits like multi-modality, non-differentiability, non-continuity, ill-conditioning, etc. Practical optimization problems exhibit a diverse range, encompassing both smooth and non-smooth problems, varying in dimension from small to high, featuring ill-conditioned type objective functions/functionals, unimodal and multimodal characteristics, among other things. Numerous optimization methods and algorithms have been developed to deal with these peculiarities, leading to the development of various optimization software packages. Despite the abundance of numerical solution methods for various optimization problem classes, determining the most efficient method for solving a specific problem with given parameter values necessitates numerous comparative experiments. Typically, end users face challenges in conducting these experiments, as it necessitates knowledge about the domain of applicability of different numerical methods. Additionally, conducting a thorough comparative analysis of the results is time-consuming. Managing the problem-solving process, involving the creation of an efficient procedure by combining optimization methods and adjusting parameter values, constitutes its own optimization task, the resolution of which is usually delegated to the end user. The paper describes two approaches to enhance the utilization of existing applied software packages for multivariate (un)constrained optimization problems, specifically designed for modern multi-processor (multicore) computer systems. One approach entails active user interaction with the optimization program package in a dialogue mode, while the alternative approach involves automatic control of the package through a specially developed control program. We suggest a strategy for solving optimization problems, incorporating an automatic selection of an effective optimization algorithm at each stage. The implemented solution strategy takes the form of a software dialogue system designed for the automated control of a package of programs for (un)conditional finite-dimensional optimization, utilizing parallel computations. In the problem-solving process, dialogue optimization systems enable the alteration of not only optimization methods but also the substitution of parameter values based on the information gathered during the search. The systems developed are primarily intended for professionals in the field of optimization. Key attributes of the developed systems include: 1) The user's representation of the optimization problem solved, namely, the systems enable end users to employ any programming language, provided it supports the creation of a library program module (file with the.dll extension). 2) Diverse modes of "user-computer" dialogue.

It is worth mentioning that since the 1970s, various academic institutions have been engaged in developing intelligent algorithms for controlling software packages. For instance, the school led by Yu.G. Evtushenko created dialogue control systems for unconstrained optimization ("DISO") and optimal control ("DISOPT") [1, 2]. In their studies [3, 4], they designed control systems for solving vector optimization problems ("DIVO") and global optimization problems ("GLOPT"). This kind of systems undergo continuous enhancements and modifications in response to the advancements in information, computer, and software technologies. The outcomes presented in this paper are a direct extension of the concepts utilized in the previously mentioned systems, considering the new capabilities offered by contemporary computing technologies. Leveraging prior experiences in developing systems of this nature, we have outlined fundamental requirements for the development of the dialog optimization system under consideration:

- Flexibility in specifying initial values and various configurations, with the ability to dynamically adjust parameters such as steps and criteria;
- Manual configuration management and the dynamic modification of the problem dimension during optimization;
- Utilization of visual aids to depict results through optimization trajectory mapping;
- Incremental recording of results, displaying optimization outcomes with the flexibility to alter configuration parameters at any step, along with storing a configuration tree for rollback capabilities;
- Capacity for ongoing development, allowing easy implementation of new algorithms, rules, criteria, and types of tasks;
- Adherence to modern software implementation standards, including the potential creation of graphical modules for visualization of results and integration with office applications to streamline data entry and reporting. Object-oriented programming stands as the primary approach for implementing the dialog system.

2 Problem Statement

Consider $P = \{p_i(x) : i \in N\}$ as the category of optimization problems (tasks), where N is a specified set delineating individual problems within the class. The variable $x \in D_i \subset \mathbb{R}^n$ represents the arguments for each individual problem, with values drawn from a predefined feasible set D_i tailored to each specific optimization problem. The assumption is made that for each problem $p_i(x)$, there is a subset of extrema $D_i^* \subset D_i$ where D_i^* is non-empty. The problem $p_i(x)$ involves determining at least one point $x^* \in D_i^*$. The set D_i^* is referred to as the set of solutions for the problem $p_i(x)$. To address all the problems within the class P, there typically exists a corresponding set of methods $M = \{M_j : j \in J\}$. Each method within this family is designed to solve the problems $p_i(x)$ belonging to the specified class, meaning they identify a point $x^* \in D_i^*$. Furthermore, each method M_j exhibits varying levels of efficiency, considering factors such as time usage, solution accuracy, when addressing the problem $p_i(x)$. For our systems, we employ direct search methods (zero-order methods), gradient-based methods (first-order methods), and Newton-type methods (second-order methods) [5–8]. These methods come with numerous configurable settings, offering the flexibility to swiftly adapt the system to any process. Additionally, the integration of direct search, gradient-based, and Newton-type methods enables us to achieve an optimal solution with a reduced number of steps and/or objective function calculations – a crucial aspect in terms of the optimization process cost. The paper outlines potential principles for managing an optimization software package while addressing a specific problem $p_i(x) \in P$. This approach aims to enhance the overall efficiency of problem-solving by mixing or alternating optimization methods during the solution process and leveraging a multiprocessor (multicore) computer system.

2.1 Sequential Implementation

Implementation for a sequential (single-core) architecture has an important independent meaning and can be considered as the basic block of implementations for multi-processor or multi-core and multi-node architectures. Consider $M_1, M_2, \ldots, M_k$ as a compilation of optimization methods, encompassing algorithms within the (un)constrained optimization software package. Inclusion of various methods in the list is reasonable, particularly when the structure of the objective function is not conclusively known. The problem-solving process unfolds in stages, each comprising training and working steps. The initial step aims to identify the locally efficient algorithm from a provided list. Subsequently, the working step involves solving the problem solely with the algorithm proven to be most efficient in the initial step. Both training and working steps occur within a designated timespan. There are two options available for the training step:

1. To assess the local efficiency of methods, the optimization process initiates from the same point x^0. Although this approach consumes machine time somewhat wastefully, the training step serves the sole purpose of identifying a locally efficient algorithm;
2. The training step serves the dual purpose of not only finding an efficient algorithm but also progressing to an extreme point. This is achieved by using the current point, instead of the original point, to train each subsequent algorithm.

During the training step, all algorithms from the initial list $M_1, M_2, ..., M_k$ have the chance to operate within the designated initial timespan τ, except for those methods identified as the least efficient for two consecutive training steps. These methods are not allocated any timespan and are temporarily excluded from the list. The initial magnitude of the timespan is contingent on the nature of the minimizable function, specifically, on the computing system's time expenditure for a single computation of the objective function and the number of its variables. In other words, $\tau = \tau(n, \chi)$, where χ represents the time spent on one computation of the function (measured in milliseconds). For instance, the formula $\tau = (5n(n + 1) + 20)\chi$ can be used to determine the timespan. In this context, the coefficient χ is determined based on the assumption that the number of function computations per iteration for Powell's conjugate directions method serves as a foundation. Powell's method necessitates $n(n + 1)/2$ one-dimensional minimizations (minimizations in specified directions), and each of these minimizations involves an average of 10 function computations. Consequently, the learning stage determines the most efficient algorithm, which is subsequently employed in the working step. The duration of the working step, denoted as $T_i = T_i(n, \chi, \tau)$, is chosen according to the formula

$$T_0 = \alpha\tau, T_i = \delta_i T_{i-1} + T_0, \alpha > 1. \tag{1}$$

The value of the parameter α is defined by the complexity and dimension of the objective function and can be chosen arbitrarily. Conducting tests at various values of α is both desirable and advisable. The value of δ_i is set to 1 when the same method proves to be the most efficient in two consecutive stages. Alternatively, if no method emerges as the most efficient in several consecutive stages, we set δ_i to 0. In other words, the duration of the working step may extend under such circumstances. This implies that, for the specific objective function, we have identified the method that attains an optimal point in the least amount of time. Therefore, in this particular (and generally, ideal) scenario, further training becomes unnecessary. To compute the local efficiencies of the methods, we utilize the following formula:

$$E_i = \frac{\left|f\left(x^{k+1}\right) - f\left(x^k\right)\right|}{\left|f\left(x^k\right)\right| + \varepsilon_1} + \frac{\left\|x^{k+1} - x^k\right\|}{\left\|x^k\right\| + \varepsilon_2}, \tag{2}$$

where E_i represents the local efficiency of the i th algorithm, x^{k+1} and x^k denote the final and initial points obtained through the i th algorithm, and $f\left(x^{k+1}\right)$ and $f\left(x^k\right)$ are the respective objective function values at these points; the symbol "$\|\ \|$" denotes the Euclidean norm; ε_1 and ε_2 are small positive numbers.

If the function value does not decrease within the timespan, the local efficiency of that algorithm is deemed as 0. If, during the training stage, all methods in the list exhibit zero efficiency, the search concludes, and the procedure terminates. This scenario may occur when the list of methods is not tailored to address the specific problem at hand, such as in cases where the function has a large condition number, and the list consists of only gradient and coordinate-wise descent methods. Inclusion of a diverse set of methods in the list can potentially prevent such situations. It is important to observe that local zero efficiency of a method may occur when an arithmetic interrupt takes place during its operation. In the software system, when implementing exception handling for such cases, it is essential to ensure a transition from the current method to the subsequent one in the list. As a result, the current method may not complete its operation within the allocated timespan, leading to a zero local efficiency value.

The presented procedure terminates when exit criteria are satisfied for all methods. Ultimately, the user obtains comprehensive information about the search process, encompassing the optimal sequence of methods employed during working steps, the total search time for solutions, the values of the objective function and of the decision variables, and local efficiencies of the methods obtained during the training step. The schematic block diagram illustrating the proposed procedure for solving the optimization problem on a single-core machine is given in Fig. 1.

2.2 Parallel Implementation on a Multicore Architecture

A straightforward implementation of a multi-threaded version for solving the given optimization problem appears to involve multiple threads independently executing the operations outlined in the sequential algorithm described earlier. The process of solving an unconstrained optimization problem unfolds in stages, with each stage involving the implementation of the following steps:

- In the initial step, a subset of algorithms is chosen randomly from the list of available unconstrained optimization algorithms $(M_1, M_2, \ldots, M_k)$. These selected algorithms are denoted as $M_{s_1}, M_{s_2}, \ldots, M_{s_m}$, and the number m of these algorithms is set equal to the number of CPU cores in the computer system (alternatively, m can be chosen as an integer multiple of the number of CPU cores);
- During the working step, we determine the most effective algorithms. The duration T_i of the working step may extend if a method consistently proves to be the most efficient across multiple consecutive stages;

- The present values of the local efficiencies E_i for the methods are computed. We eliminate half of the working algorithms from the list, specifically those that have demonstrated the lowest efficiency;
- Next, we append to the list of working algorithms an equal number of algorithms that were previously excluded, and proceed to repeat steps 2 through 4 once more.

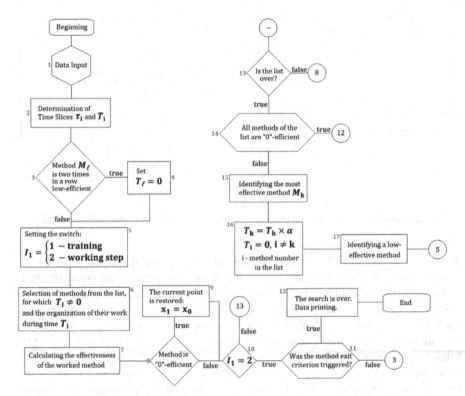

Fig. 1. Flowchart of the sequential process of finding a minimum on a single-core machine.

In the outlined procedure, all methods from the list are forcefully halted after the designated timespan, even though for some method(s), the ongoing iteration may not have finished by then. A potential modification to this procedure involves allowing all methods to conclude the iterations that have commenced or complete a whole number of iterations within the vicinity of the specified time interval. In the latter scenario, a slight adjustment to the formula for calculating the local efficiencies of methods becomes necessary. The schematic block diagram illustrating the proposed procedure for solving the optimization problem on a multicore machine is given in Fig. 2.

2.3 Parallel Implementation in a Distributed Environment

In contrast to parallel systems, distributed systems exhibit a hierarchical structure, comprising diverse nodes, each of which can, in turn, function as a parallel system following

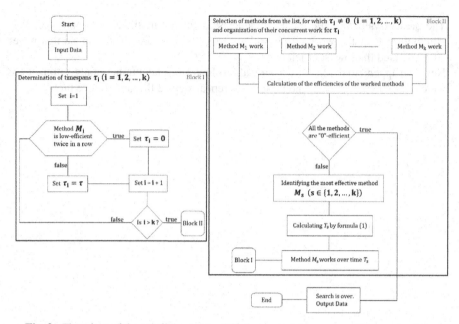

Fig. 2. Flowchart of the parallel process of finding a minimum on a multi-core machine.

the models discussed earlier. It is reasonable to presume that the highest efficiency is attained when the computational process aligns with this hierarchical organization. In this approach, computations at each node follow the most suitable scheme for that specific node. Essentially, each node in the distributed system executes an independent application, referred to as a "solver," to carry out the operations of the chosen optimization algorithm. Interaction among multiple applications (nodes) is coordinated at the subsequent level of the hierarchy through a dedicated central control process, known as a "supervisor." The initial step in solving an optimization problem within a distributed computing environment involves synthesizing the computational space formed by instances of solvers. Managing applications manually can be time-consuming, particularly with a large number of nodes. Therefore, it is essential to enable the automated launch of solvers by the supervisor. Remote task launching tools, such as ssh, grid services, etc., provided by a specific system, can be utilized in this scenario.

The established computing space becomes instrumental in problem-solving. During the solution process, it is essential to distribute computations among solvers to optimize application performance in a distributed environment. Exchange of data between the solvers and the supervisor can be executed using the tools designed for interaction with a specific node. When a direct network connection is feasible, exchange methods relying on TCP/IP protocols, such as the socket interface, are employed. Alternatively, in certain situations, data exchange with applications is only achievable by transferring files using the middleware "grid." Load balancing transpires at two tiers: at the higher level, the supervisor allocates the computational load among the solvers, while at the lower level (within a computing node), the solver distributes the workload among methods designed for a specific type of computing node.

Both approaches of the aforementioned search algorithm facilitate the automatic selection of an efficient, expedient optimization method from the available list to address a specific problem by virtue of the self-learning capability inherent in the utilized methods.

When utilizing the dialogue and automatic systems, the user adheres to standard requirements by expressing an optimization problem in any programming language as a module (dynamic link library). Subsequently, the user incorporates it into the system by indicating the full path to the created library file. By utilizing directives (instructions), the user initiates the most suitable algorithms from the library of modules, adjusting various settings according to their preference. The control program will orchestrate the interaction among the modules in the package, oversee the input of initial and current information, interpret user directives (instructions), dynamically load optimization modules into the computer memory, and present the computation results on the display (alternatively, results can also be obtained on a printer) in a specified format.

Upon analyzing the computation results, the user decides on subsequent calculations, thereby gaining the ability to monitor the progression of problem-solving, promptly intervene in the computation process, select the working methods, and adjust their parameters as needed. The user specifies the frequency and format for displaying results on the screen, and subsequently conducts calculations using a predefined set of directives.

The optimization systems developed were constructed modularly, considering the potential expansion of their functionalities. These systems come equipped with a comprehensive library of optimization programs. The interactive service offered within the dialog system empowers the user to oversee the problem-solving process on a computer system. Based on the current calculation results, the user can choose the most optimal sequence of methods, adjusting method parameters if needed. Additionally, the user has the capability to make modifications in the formulation of the problems being solved.

Throughout the system's operation, the user and system activity modes alternate. When computations are interrupted, the user engages in any of the following actions:

- Analyzing and correcting the current state;
- Computing the values of the objective function, constraints, and their gradients at the current point; Selecting solution method(s);
- Adjusting control parameters of the method(s), specifying stopping criteria, and determining information delivery forms;
- Initiating method(s);
- Recording the current state in text (or binary) files.

3   Concise Description of the Software

The software for the optimization systems was coded entirely in the C# programming language within the Microsoft Visual Studio IDE. The design of the developed software follows a modular approach and incorporates the principles of object-oriented programming. The software includes 2 forms and a multitude of modules targeted for solving one-dimensional, multi-dimensional unconstrained and constrained optimization problems, which all adhere to a standard notation and possess a unified structure. The library of the systems includes various optimization algorithms, a set of standard programs for calculating gradients and Hessian matrices with different degrees of accuracy, auxiliary programs for displaying information. The user of the systems can either describe his special differentiation procedures (based on analytical formulas or finite-difference approximations), or use standard procedures stored in the software system library. Each module has the following procedures which act as interface methods between the coded methods and the main part of the program:

- "Objective Function": provides the value of the objective function;
- " Objective Gradient": provides the gradient of the objective function;
- " Objective Hessian": provides the Hessian matrix of the objective function.

If either of the last two mentioned procedures is not present in the dynamic link library (file with the.dll extension), finite-difference approximation formulas are employed in calculations.

When the application starts, the main form (window) of the application is loaded (see Fig. 3). The user works with this form by providing the following information:

- In the "Epsilon" field, the accuracy of the solution to the minimization problem is entered; this enforces convergence criteria $\|x^k - x^*\|_{\mathbb{R}^n} \leq \max(\epsilon, \epsilon \|x^k\|_{\mathbb{R}^n})$ and $\|\nabla f(x^k)\|_{\mathbb{R}^n} \leq \epsilon$.
- In the "Delta" field, the accuracy of the solution to the one-dimensional minimization problem is entered; this enforces convergence criteria similar to the ones given above.
- In the "Step" field, he enters the initial value of the step used for heuristic selection of the initial interval of uncertainty.
- In the "Increment" field, he enters the initial value of the increment of the arguments used for finite-difference approximation of both the gradient and the Hessian matrix of the objective function.
- In the "Time (milliseconds)" field, he enters the initial value of the timespan (in milliseconds) during which each selected multivariate minimization method works.

- In the "Update" field, he enters the number of iterations, after which such first-order methods as conjugate gradient methods, quasi-Newton methods, etc. are restarted to ensure their convergence when minimizing non-quadratic functions.
- In the "Parameter Initial Values" field, the initial guess (vector) for the optimizable variables of the objective function is specified either manually or by loading the contents of a text file.
- From the "One-dimensional Minimization Procedure" list, he selects one of the exact or inexact line search procedures.
- One or more multivariate minimization methods are selected from the "Multi-dimensional Minimization Procedures" list.
- In the "Max Iterations" field, he enters the maximum admissible number of iterations of multivariate minimization methods. (If this field is left empty, the number of iterations is unlimited.)
- The "Exit Criterion" group is intended to indicate the termination criterion for the multivariate minimization procedure. "Vector" is the termination criterion based on the norm of the difference between the vector of optimizable variables at two consecutive iterations; "Functional" is the termination criterion based on the absolute value of the difference between the objective function's values at two consecutive iterations; "Gradient" is the termination criterion based on the norm of the objective function's gradient;
- The "Evaluation" group contains three buttons for calculating the objective function's value, gradient, and Hessian matrix at the current point.
- Double-clicking on the "Functional File" field initiates a dialog box for selecting a file with the.dll extension, which contains code for calculating the value of the objective function, as well as its gradient and Hessian matrix. A file with the.dll extension can be written in any programming language and compiled in any programming system. For clarity, we present the structure (contents) of this file in the Object Pascal programming language:

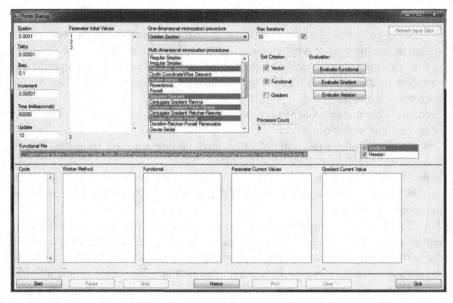

Fig. 3. Main window of the dialog system application.

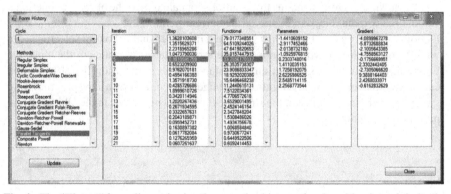

Fig. 4. The "History" form allows viewing the operation history of each multivariate minimization method at each cycle of the minimization process.

```
library ProjectFunctional;
uses
  System.SysUtils,
  System.Classes;
{$R *.res}
type
  Vector = array [0..10] of System.Double;
  Matrix = array [0..10,0..10] of System.Double;
const
  Dimension : System.Integer = 10;
function ObjectiveFunction(x : Vector) : System.Double;
                                         export; cdecl;
var
  i : System.Integer; value : System.Double;
begin
  value := Sqr(x[1] - 1);
  for i := 1 to Dimension do
    value := value + i * Sqr(x[i+1] - x[i]);
  result := value;
end;

procedure ObjectiveGradient(x : Vector;
                       var grad : Vector); export; cdecl;
var
  i : System.Integer;
begin
  for i := 1 to Dimension do
  begin
  if (i = 1) then
    grad[i] := 2.0*(x[i] - 1) - 2.0*i*(x[i+1] - x[i]);
  if (i > 1) and (i < Dimension) then
    grad[i] := 2.0*(i - 1)*(x[i] - x[i-1]) -
                          2.0*i*(x[i+1] - x[i]);
  if (i = Dimension) then
    grad[i] := 2.0*(i - 1)*(x[i] - x[i-1]);
  end;
end;

exports ObjectiveFunction, ObjectiveGradient;

Begin
//empty initialization section
End.
```

- The "Cycle" list displays consecutive cycle numbers when minimizing the objective function.

- The "Worker Method" list displays the names of multivariate minimization methods that turned out to be the best in the next minimization cycle.
- The "Functional" list displays the minimum values of the objective function obtained in the next minimization cycle.
- The components of the optimizable vector of the current minimization cycle are displayed in the "Parameter Current Values" list.
- The components of the objective function's gradient of the current minimization cycle are displayed in the "Gradient Current Value" list.
- Pressing the "Start" button initiates the minimization process, which runs in the "background" mode, independent of the UI application thread.
- At any time, the user can pause the minimization process by clicking on the "Pause" button, then changing, if necessary, some minimization parameters (the upper fields of the form), and starting (more precisely, continuing) the minimization process again.
- At any time, the user can finish the minimization process by clicking on the "Stop" button.
- When the user clicks on the "History" button, the "Form History" window of the application opens (see Fig. 4). Here the user can see the progression of each method at each cycle of the search process.

During the search process and upon completion, the program generates several output files, which the user can use to watch some statistics and to initiate (continue) the search process at another time.

4   Results of Numerical Experiments

The outcomes of computations are heavily influenced by the control parameter values; by meticulously selecting them, one can significantly influence the progression of computations and achieve noteworthy results. Presented below are the results of computation records conducted in dialog mode. The computations primarily used the parameters of the methods specified for the "default" operation mode; only in a few instances were two or three additional computations performed with different parameters, and the best result is showcased in the tables. It's important to note that the data provided in the tables are far from being "record." To simplify the description of numerical experiments, basic test problems were employed. Therefore, the computation results offer limited insights into the effectiveness of the utilized methods. The methods need to undergo a comparative analysis based on the solution of diverse and more difficult problems, like the ones described in [9–14]. To bring the computations somewhat closer to reality, numerical differentiation of the functions to be optimized is applied in all conducted numerical experiments. The computational complexity is primarily determined by the number n, which corresponds to the number of calls to the function (including those needed for first and second derivatives computation).

4.1   Example 1

Consider the minimization of the function.

	Cycle	Method	Function Value	Function Count
Experiment #1	1	Conjugate Gradients Polak-Ribiere variant	0.605243	3901
	2	Newton's method (modification)	1.6658×10^{-18}	80027
	3	Newton's method (original)	2.5174×10^{-27}	10006
Final State of the Optimizable Vector: $x^* = (1.000000)$, i = 1,2, ... ,50				
Experiment #2	1	Conjugate Gradients Polak-Ribiere variant	44.542616	4633
	2	Quasi-Newton Davidon-Fletcher-Powell	32.056547	6411
	3	Parallel Tangents	26.305256	6468
	4	Powell's method Conjugate Directions	17.588626	66370
	5	Powell's method Conjugate Directions	7.611967	8500
	6	Newton's method (modification)	0.008171	220064
	7	Newton's method (original)	1.7558×10^{-15}	40015
Final State of the Optimizable Vector: $x^* = (1.000000)$, i = 1,2, ... ,50				

Fig. 5. Results of numerical experiments for the test function (3).

$$f(\mathrm{x}) = (1 - x_1)^2 + \sum_{i=1}^{49}\left[100\left(x_{i+1} - x_i^2\right)^2\right] \tag{3}$$

from different initial points. During the minimization of this objective function, the optimization parameters were set as follows: the accuracy of solving the multidimensional minimization problem is $\epsilon = 10^{-4}$; the accuracy of solving the one-dimensional minimization problem is $\delta = 10^{-5}$; each algorithm in the list was allocated 0.5 s for execution; for calculating the gradient and Hessian of the objective functional, a finite-difference formula with an approximation step of $\Delta = 10^{-6}$ was employed. Minimizing the function (3) using all the available unconstrained optimization algorithms in the package from the initial point $x_i^0 = 10$, $i = 1, 2, \ldots, 50$ (with an initial function value $f(x^0) \approx 3.97 \times 10^8$), resulted in the protocol of the minimization process given in the top half of Fig. 5. Minimizing the function (3) using all the available unconstrained optimization algorithms in the package from the initial point $x_i^0 = -26+i$, $i = 1, 2, \ldots, 50$ (with an initial function value $f(x^0) \approx 3.97 \times 10^8$), resulted in the protocol of the minimization process given in the bottom half of Fig. 5.

4.2 Example 2

Consider the minimization of the function.

$$f(x) = x_1^2 + \left(x_2 - x_1^2 - 1\right)^2 + \sum_{i=2}^{30}\left\{\sum_{j=2}^{9}(j-1)x_j\left(\frac{i-1}{29}\right)^{j-2}\right.$$

$$\left. - \left[\sum_{j=1}^{9}x_j\left(\frac{i-1}{29}\right)^{j-1}\right]^2 - 1\right\}^2 \tag{4}$$

from different initial points. For minimization of this objective function, the optimization parameters were set as follows: the accuracy of solving the multidimensional minimization problem is $\epsilon = 10^{-6}$; the accuracy of solving the one-dimensional minimization problem is $\delta = 10^{-7}$; each algorithm in the list was allocated 0.5 s for execution; for calculating the gradient and Hessian of the objective functional, a finite-difference formula with an approximation step of $\Delta = 10^{-8}$ was employed. Minimizing the function (4) using all the available unconstrained optimization algorithms in the package from the initial point $x_i^0 = 10$, $i = 1, 3, 5, 7, 9$ and $x_i^0 = -10$, $i = 2, 4, 6, 8$, (with an initial function value $f(x^0) \approx 1.18 \times 10^5$), resulted in the protocol of the minimization process given in the upper half of Fig. 6. Minimizing the function (4) using all the available unconstrained optimization algorithms in the package from the initial point $x_i^0 = 20 - i$, $i = 1, 2, \ldots, 9$ (with an initial function value $f(x^0) \approx 1.11 \times 10^9$), resulted in the protocol of the minimization process given in the bottom half of Fig. 6.

	Cycle	Method	Function Value	Function Count
Experiment #1	1	Powell's method Conjugate Directions	0.017328	14704
	2	Hooke-Jeeves	0.009339	4528
	3	Quasi-Newton Davidon-Fletcher-Powell	3.68×10^{-5}	2422
	4	Levenberg–Marquardt	1.39×10^{-6}	15998
	5	Newton's method (modification)	1.17×10^{-6}	17690
Final State of the Optimizable Vector: $x^* =$ (0.0000; 0.9997; 0.0147; 0.1463; 1.0010; −2.6181; 4.1048; −3.1438; 1.0526)				
Experiment #2	1	Hooke-Jeeves	0.107453	7469
	2	Rosenbrock's method	0.026808	6736
	3	Quasi Newton BFGS	4.06×10^{-5}	17192
	4	Newton's method (modification)	1.17×10^{-6}	17070
Final State of the Optimizable Vector: $x^* =$ (0.0000; 0.9997; 0.0147; 0.1463; 1.0008; −2.6177; 4.1044; −3.1436; 1.0526)				

Fig. 6. Results of numerical experiments for the test function (4).

4.3 Example 3

Consider the minimization of the function.

$$f(x) = \sum_{i=1}^{29}\left[\mathrm{e}^i\left(x_{i+1} - x_i^2\right)^2 + (1 - x_i)^2\right] \tag{5}$$

from different initial points. During the minimization of this objective function, the optimization parameters were set as follows: the accuracy of solving the multidimensional minimization problem is $\epsilon = 10^{-8}$; the accuracy of solving the one-dimensional minimization problem is $\delta = 10^{-9}$; each algorithm in the list was allocated 6.0 s for execution; for calculating the gradient and Hessian of the objective functional, a finite-difference formula with an approximation step of $\Delta = 10^{-10}$ was employed. Minimizing the function (5) using all the available unconstrained optimization algorithms in the

	Cycle	Method	Function Value	Function Count
	1	Hooke-Jeeves	10.401169	17181
	2	Newton's method (original)	0.203044	4914490
	3	Newton's method (original)	0.026706	5343247
Experiment #1	4	Newton's method (original)	0.006686	6510619
	5	Newton's method (second modification)	0.000951	6023236
	6	Newton's method (second modification)	2.0175×10^{-5}	5957248
	7	Newton's method (second modification)	4.2264×10^{-10}	10004512
	**Final State of the Optimizable Vector: $x^* = (1.000000)$, $i = 1,2,...,30$**			
	1	Newton's method (modification)	0.386688	6625915
	2	Newton's method (modification)	0.092663	6017008
Experiment #2	3	Newton's method (modification)	0.002265	7411369
	4	Newton's method (modification)	0.000282	7984246
	5	Newton's method (second modification)	7.3566×10^{-6}	4413862
	6	Newton's method (second modification)	4.4589×10^{-6}	4952764
	7	Newton's method (second modification)	1.9541×10^{-10}	5429344
	**Final State of the Optimizable Vector: $x^* = (1.000000)$, $i = 1,2,...,30$**			

Fig. 7. Results of numerical experiments for the test function (5).

package from the initial point $x_i^0 = 10$, $i = 1, 2, \ldots, 30$ (with an initial function value $f(x^0) \approx 5.03 \times 10^{16}$), resulted in the protocol of the minimization process given in the top half of Fig. 7. Minimizing the function (5) using all the available unconstrained optimization algorithms in the package from the initial point $x_i^0 = -16 + i$, $i = 1, 2, \ldots, 30$ (with an initial function value $f(x^0) \approx 1.27 \times 10^{17}$), resulted in the protocol of the minimization process given in the bottom half of Fig. 7.

5 Conclusion

In this paper, we have described an approach to conduct the computational process involved in solving difficult practical problems, illustrated through the context of multivariate optimization problems, by using available numerical algorithms, tailored for modern multi-processor and multicore computing systems. The developed system aims to optimize the problem-solving process by dynamically selecting the most effective optimization algorithm at each stage, thereby enhancing overall efficiency. The proposed approach significantly simplifies the end users' tasks in utilizing standard software packages, catering to varying levels of user expertise in the algorithms embedded within the software packages. Through sequential implementation on single-core architectures, we outline a systematic approach for identifying locally efficient algorithms from a provided list and subsequently solving the problem using the most efficient algorithm within a designated time frame. Parallel implementation on multicore and/or multinode architectures further enhances computational efficiency by distributing computations among multiple threads, with each thread executing operations independently. Numerical experiments conducted in dialogue mode showcase the effectiveness of the proposed approach, although we must acknowledge the need for further comparative analysis on more complex problems to fully assess their performance. Overall, the paper contributes valuable insights into the development of automated and automatic systems for optimization management, offering practical solutions to enhance decision-making processes across diverse domains.

References

1. Evtushenko, Y.: Methods of solving extreme problems and their application in optimization systems. Nauka, Moscow (1982)
2. Evtushenko, Y., Mazurik, V.P.: Optimization Systems Software. Znanie, Moscow (1989)
3. Aidazade, K.R., Sidorenko, N.S.: An approach to the construction of combined optimization algorithms. Tech. Cybern. **6**, 87–93 (1982)
4. Aidazade, K.R., Novruzbekov, I.G.: Dialog system of multicriteria optimization. Proceedings of the Academy of Sciences of Azerbaijan SSR, series of Physical-Technical and Mathematical Sciences (2) (1987)
5. Vasiliev, F.P.: Optimization methods. MTsNMO, Moscow (2011)
6. Nocedal, J., Wright, S.: Numerical Optimization. Springer, New-York (2003)
7. Polyak, B.T.: Introduction to Optimization. Lenand, Moscow (2014)
8. Larichev, O.I., Horvitz, G.G.: Methods of Searching for a Local Extremum of Ravine Functions. Nauka, Moscow (1990)

9. Aida-Zade, K.R., Kuliev, S.Z.: Hydraulic resistance coefficient identification in pipelines. Autom. Remote. Control. **77**(7), 1225–1239 (2016). https://doi.org/10.1134/S00051179160 70092
10. Aidazade, K.R., Guliyev, S.Z.: Numerical solution of nonlinear inverse coefficient problems for ordinary differential equations. Comput. Math. Math. Phys. **51**(5), 803–815 (2011)
11. Aidazade, K.R., Guliyev, S.Z.: On numerical solution of one class of inverse problems for discontinuous dynamic systems. Autom. Remote. Control. **73**(5), 786–796 (2012)
12. Aidazade, K.R., Guliyev, S.Z.: Zonal control synthesis for nonlinear systems under nonlinear output feedback. Autom. Inf. Sci. **47**(1), 51–66 (2015)
13. Guliyev, S.Z.: Synthesis of control in nonlinear systems with different types of feedback and strategies of control. Autom. Inf. Sci. **45**(7), 74–86 (2013)
14. Guliyev, S.Z.: Numerical solution of a zonal feedback control problem for the heating process. IFAC-Papers Online **51**(30), 251–256 (2018)

Filtering Correction for Robotic Arms Multipurpose Regulators

Ruslan Sevostyanov$^{(\boxtimes)}$

Saint-Petersburg State University, Universitetskaya 7-9, 199034 Saint-Petersburg, Russia
sevostyanov.ruslan@gmail.com

Abstract. The paper is devoted to the problem of compensating the external disturbances while stabilizing the robotic arm in the specified position by controlling motor torques directly. Disturbance in form of the polyharmonic function is of particular interest here. For example, it can represent the hull vibrations or the motion oscillations in case when arm is mounted on the moving platform. The goal is to minimize the control reaction to such disturbance while maintaining the same stabilization properties in order to save the resource of the robot motors.

Stabilization of the robot is achieved through the usage of the special multipurpose control structure. It is most useful in case when there is a set of requirements specified for the motion of the plant in different operating modes. The main advantage of such structure is that it can be synthesized step by step moving from one mode to another, so the initial complicated problem is divided into several simpler tasks.

Robotic arms have nonlinear mathematical models so in order to use multipurpose regulator here feedback linearization is applied first. The main result of the paper is the synthesis method of the dynamic corrector which is important part of the multipurpose regulator. The efficiency and the corresponding problems of the proposed method are demonstrated by computer model experiments.

Keywords: Feedback Linearization · Multipurpose Regulator · External Disturbances · Robotic Arms

1 Introduction

Robotic arms usage is extremely important currently in many industrial fields such as automated factories, hazardous production, automated storages, even in some medical applications. Production rates imply more and more strict requirements for the precise motion of robotic arms in different operating modes. Especially in case of presence of the external disturbances of any kind such as unaccounted tool weight or hull vibrations.

There are quite a lot of papers on stabilization of the different kinds of plants, including the problems of satisfying a set of dynamics requirements, for example [1–5]. But usually such papers consider only one specific operating mode and not the whole regulator.

In order to take into account a specified set of requirements to the controlled motion dynamics it might be useful to apply a special control structure called multipurpose

A. Eremeev et al. (Eds.): MOTOR 2024, LNCS 14766, pp. 408–420, 2024.
https://doi.org/10.1007/978-3-031-62792-7_27

regulator [6, 7]. It allows considering each operating mode "one by one" starting from some simplest case and then adding next requirement on top of it. This makes the whole synthesis process a lot simpler, dividing initial complex problem into sequential simpler subtasks. One of the components of the multipurpose regulator is the dynamic corrector. Its main goal is to generate the reaction to the external disturbances. In case of the polyharmonic disturbance the dynamic corrector can either minimize the effect of the disturbance to the controlled variables dynamics or it can minimize the reaction of the control itself [8] – for example, in order to prolongate the lifecycle of the actuator motors. The latter one is of the main interest in this paper.

Most of the papers on the multipurpose regulators consider plants with linear mathematical models or the models with particular known nonlinearities such as rotation matrices in the problem of dynamic positioning. Robot arms have fully nonlinear complicated models. In order to apply multipurpose approach in this case we can use feedback linearization method [9] to compensate the nonlinearities and use classical multipurpose regulator then. However it can be the source of some other problems which we will address later in this paper.

This paper focuses mainly on the periodic disturbance mode and proposing the method of dynamic corrector synthesis to minimize the control action. The results of the computational experiments with the two-link robotic arm computer model are provided to demonstrate the efficiency and the problems with the proposed approach.

2 Problem Statement

Consider the following mathematical model of the plant's dynamics [10]:

$$\mathbf{M}(\boldsymbol{\theta})\ddot{\theta} + \mathbf{C}(\boldsymbol{\theta}, \dot{\theta}) + \mathbf{g}(\boldsymbol{\theta}) = \tau + \tau_e, \tag{1}$$

where $\mathbf{M} = \mathbf{M}^{\mathrm{T}} \in E^{n \times n}$ is a positive definite inertia matrix, $\boldsymbol{\theta} \in E^n$ is a vector of generalized coordinates, $\tau \in E^n$ is control input, $\tau_e(t) \in E^n$ is external disturbance, $\mathbf{C} \in E^n$ is a vector of combined centripetal and Coriolis forces, $\mathbf{g} \in E^n$ is a vector of gravitational forces.

In general the control goal is ensuring the following condition:

$$\lim_{t \to +\infty} \boldsymbol{\theta}(t) = \boldsymbol{\theta}_d, \tag{2}$$

i.e. stabilizing the current plant position $\boldsymbol{\theta}$ in a given position $\boldsymbol{\theta}_d$. Along with this goal let us consider three operating modes with special requirements for the system dynamics.

1) With $\tau_e(t) = \mathbf{0}$ the system in the *unperturbed motion mode* and besides condition (2) there might be some requirements to the character of the transient process such as given constraints to the overshoot or the settling time.
2) With $\tau_e(t) = \tau_{e0}$ where τ_{e0} is a constant value, the system is in the *motion under constant disturbances mode*. In this mode it is assumed that the system is already in the desired position $\boldsymbol{\theta}_d$ and the goal is to provide astatism of the system and to constrain the maximum deviation of the current state from the $\boldsymbol{\theta}_d$ under the disturbance influence.

3) With $\tau_e(t)$ in the form of polyharmonic function the system is in the *motion under periodic disturbance mode*. Main requirement here is the minimization of the control reaction to such disturbance with known frequencies. Formally it means that we need to provide zero value of the transfer matrix from the disturbance to the control output on these frequencies. It is also assumed that $\theta_d = \theta_0$ i.e. the system initial state corresponds to the desired position.

Thus let us formulate the problem of the synthesis of such feedback that can handle all the goals and constrains formulated previously for the presented operation modes.

3 Multipurpose Feedback Synthesis

First of all let us consider system (1) without the external disturbance. In this case we can achieve linearization by using control of the form

$$\tau = C(\theta, \omega) + g(\theta) + M(\theta)u, \tag{3}$$

where $u = u(t) \in E^n$ is some vector function. Substituting (3) to (1) we get linear system

$$\ddot{\theta} = u,$$

which in turn can be transformed to a first-order system

$$\dot{\theta} = \omega,$$
$$\dot{\omega} = u. \tag{4}$$

Now in order to build such function u that provides satisfaction of all the requirements in all of the operating modes including ones with disturbance, we can directly apply multipurpose regulator of the form

$$\dot{z}_\theta = z_\omega + H_\theta(\theta - z_\theta),$$
$$\dot{z}_\omega = u + H_\omega(\theta - z_\theta),$$
$$\dot{p} = \alpha p + \beta(\theta - z_\theta),$$
$$\xi = \gamma p + \mu(\theta - z_\theta),$$
$$u = -K_\theta(z_\theta - \theta_d) - K_\omega z_\omega + \xi, \tag{5}$$

where $z_\omega \in E^n$ and $z_\theta \in E^n$ are asymptotic observer state vectors; $p \in E^{n_p}$ is dynamic corrector state vector; $\xi \in E^n$ is an output vector of a dynamical corrector. Constant matrices H_θ, H_ω, K_θ, K_ω, α, β, γ, μ are subjects to search during the synthesis of the regulator. Asymptotic observer outputs serve as input to the dynamic corrector which is responsible for the control reaction to the external disturbance, i.e. it can provide astatism and filtration of the periodic disturbance or any other desired behavior. One notable thing is that corrector can be turned off in the modes without the disturbance.

Specific values of the above matrices depend on certain requirements in different operating modes, but we can formulate necessary conditions to provide asymptotic stability of the closed-loop system. Due to the fact that system (4) is linear, it can be easily shown that matrices $\mathbf{H}_\theta$, $\mathbf{H}_\omega$, $\mathbf{K}_\theta$ and $\mathbf{K}_\omega$ must provide Hurwitz property of the matrices

$$\mathbf{K}_b = \begin{pmatrix} \mathbf{0} & \mathbf{E} \\ -\mathbf{K}_\theta & -\mathbf{K}_\omega \end{pmatrix}, \mathbf{H}_o = \begin{pmatrix} -\mathbf{H}_\theta & \mathbf{E} \\ -\mathbf{H}_\omega & \mathbf{0} \end{pmatrix},$$

and the matrix $\boldsymbol{\alpha}$ must be Hurwitz.

In operating modes with disturbance it is essential to provide astatism of the system. Using reasoning similar to one in [11], we can show that for that purpose transfer matrix of the dynamic corrector must satisfy the condition

$$\mathbf{F}(0) = -\mathbf{K}_\theta - \mathbf{K}_\omega \mathbf{H}_\theta - \mathbf{H}_\omega, \tag{6}$$

where $\mathbf{F}(s) = \boldsymbol{\gamma}(\mathbf{E}s - \boldsymbol{\alpha})^{-1}\boldsymbol{\beta} + \boldsymbol{\mu}$ with s being Laplace variable.

With the above conditions multipurpose regulator (5) provides the desired stabilization of the system (1) and astatism property in case of the constant disturbance acting on it.

Now let us consider the problem of filtering periodic external disturbance in the polyharmonic function form

$$w(t) = \sum_{i=1}^{N_\omega} A_i \sin(\omega_i t),$$

$$\mathbf{d}(t) = \mathbf{a}w(t),$$

where N_ω is a harmonics number, A_i are constant amplitudes, ω_i are harmonic frequencies, $\mathbf{a}$ – constant coefficients vector.

Now let us introduce variable $\boldsymbol{\varepsilon}_\theta = \boldsymbol{\theta} - \mathbf{z}_\theta$ for observer estimation error and use dynamic corrector in the operator form. Then we can rewrite regulator (5) as

$$\dot{\mathbf{z}}_\theta = \mathbf{z}_\omega + \mathbf{H}_\theta \boldsymbol{\varepsilon}_\theta,$$
$$\dot{\mathbf{z}}_\omega = \mathbf{u} + \mathbf{H}_\omega \boldsymbol{\varepsilon}_\theta,$$
$$\boldsymbol{\xi} = \mathbf{F}(p)\boldsymbol{\varepsilon}_\theta,$$
$$\mathbf{u} = -\mathbf{K}_\theta \mathbf{z}_\theta - \mathbf{K}_\omega \mathbf{z}_\omega + \boldsymbol{\xi}, \tag{7}$$

where $\mathbf{F}(p)$ is corrector transfer matrix, $p = d/dt$. From the first two equations we have

$$\dot{\mathbf{z}}_\theta = \mathbf{z}_\omega + \mathbf{H}_\theta \boldsymbol{\varepsilon}_\theta,$$
$$\dot{\mathbf{z}}_\omega = -\mathbf{K}_\theta \mathbf{z}_\theta - \mathbf{K}_\omega \mathbf{z}_\omega + \boldsymbol{\xi} + \mathbf{H}_\omega \boldsymbol{\varepsilon}_\theta,$$

which can be represented in matrix form

$$\begin{pmatrix} \dot{\mathbf{z}}_\theta \\ \dot{\mathbf{z}}_\omega \end{pmatrix} = \mathbf{A}_s \begin{pmatrix} \mathbf{z}_\theta \\ \mathbf{z}_\omega \end{pmatrix} + \mathbf{B}_s \begin{pmatrix} \boldsymbol{\varepsilon}_\theta \\ \boldsymbol{\xi} \end{pmatrix}, \tag{8}$$

where

$$\mathbf{A}_s = \begin{pmatrix} \mathbf{0} & \mathbf{E} \\ -\mathbf{K}_\theta & -\mathbf{K}_\omega \end{pmatrix}, \mathbf{B}_s = \begin{pmatrix} \mathbf{H}_\theta & \mathbf{0} \\ \mathbf{H}_\omega & \mathbf{E} \end{pmatrix}.$$

We can represent (8) in the tf-form as

$$\begin{pmatrix} \mathbf{z}_\theta \\ \mathbf{z}_\omega \end{pmatrix} = \mathbf{P}(s) \begin{pmatrix} \boldsymbol{\varepsilon}_\theta \\ \boldsymbol{\xi} \end{pmatrix}, \tag{9}$$

with

$$\mathbf{P}(s) = (\mathbf{E}s - \mathbf{A}_s)^{-1} \mathbf{B}_s = \begin{pmatrix} \mathbf{P}_{11}(s) & \mathbf{P}_{12}(s) \\ \mathbf{P}_{21}(s) & \mathbf{P}_{22}(s) \end{pmatrix} \tag{10}$$

From (9) and (10) we can see that

$$\begin{aligned} \mathbf{z}_\theta &= \mathbf{P}_{11}\boldsymbol{\varepsilon}_\theta + \mathbf{P}_{12}\boldsymbol{\xi}, \\ \mathbf{z}_\omega &= \mathbf{P}_{21}\boldsymbol{\varepsilon}_\theta + \mathbf{P}_{22}\boldsymbol{\xi}. \end{aligned} \tag{11}$$

If we substitute (11) into control signal equation in (7), we get

$$\begin{aligned} \mathbf{u} &= -\mathbf{K}_\theta(\mathbf{P}_{11}\boldsymbol{\varepsilon}_\theta + \mathbf{P}_{12}\boldsymbol{\xi}) - \mathbf{K}_\omega(\mathbf{P}_{21}\boldsymbol{\varepsilon}_\theta + \mathbf{P}_{22}\boldsymbol{\xi}) + \boldsymbol{\xi} = \\ &= -(\mathbf{K}_\theta\mathbf{P}_{11} + \mathbf{K}_\omega\mathbf{P}_{21})\boldsymbol{\varepsilon}_\theta - (\mathbf{K}_\theta\mathbf{P}_{12} + \mathbf{K}_\omega\mathbf{P}_{22} - \mathbf{E})\boldsymbol{\xi}. \end{aligned} \tag{12}$$

Introducing notation $\mathbf{T}(s) = \mathbf{K}_\theta\mathbf{P}_{12}(s) + \mathbf{K}_\omega\mathbf{P}_{22}(s) - \mathbf{E}$ let us rewrite (12) as

$$\begin{aligned} \mathbf{u} &= -(\mathbf{K}_\theta\mathbf{P}_{11} + \mathbf{K}_\omega\mathbf{P}_{21})\boldsymbol{\varepsilon}_\theta - \mathbf{T}\boldsymbol{\xi} \\ &= -(\mathbf{K}_\theta\mathbf{P}_{11} + \mathbf{K}_\omega\mathbf{P}_{21})\boldsymbol{\varepsilon}_\theta - \mathbf{T}\mathbf{F}\boldsymbol{\varepsilon}_\theta \\ &= -(\mathbf{K}_\theta\mathbf{P}_{11} + \mathbf{K}_\omega\mathbf{P}_{21} + \mathbf{T}\mathbf{F})\boldsymbol{\varepsilon}_\theta. \end{aligned}$$

The expression in front of the $\boldsymbol{\varepsilon}_\theta$ is in fact the transfer matrix $\mathbf{G}(s) = -[\mathbf{K}_\theta\mathbf{P}_{11}(s) + \mathbf{K}_\omega\mathbf{P}_{21}(s) + \mathbf{T}(s)\mathbf{F}(s)]$ from the input $\boldsymbol{\varepsilon}_\theta$ to the output $\mathbf{u}$. Therefore if our goal to minimize the control signal response to the external polyharmonic disturbance with known frequencies $\omega_i, i = \overline{1, N_\omega}$ it is necessary to ensure the condition $\mathbf{G}(j\omega_i) = 0, i = \overline{1, N_\omega}$. This can be achieved by means of choosing the right transfer matrix $\mathbf{F}(s)$ of the dynamic corrector, i.e.:

$$\mathbf{F}(j\omega_i) = -\mathbf{T}^{-1}(j\omega_i)[\mathbf{K}_\theta\mathbf{P}_{11}(j\omega_i) + \mathbf{K}_\omega\mathbf{P}_{21}(j\omega_i)] \tag{13}$$

Now we need to consider the problem of synthesizing the dynamic corrector with the transfer matrix $\mathbf{F}(s) = \begin{pmatrix} \mathbf{F}_1(s) & \mathbf{F}_2(s) & \dots & \mathbf{F}_n(s) \end{pmatrix}^{\mathrm{T}}$ that satisfies the conditions (6) and (13). Let us introduce constant matrices

$$\begin{aligned} \mathbf{F}^0 &= \begin{pmatrix} \mathbf{F}_1^0 & \mathbf{F}_2^0 & \dots & \mathbf{F}_n^0 \end{pmatrix}^{\mathrm{T}} = \mathbf{F}(0), \\ \mathbf{F}_i^* &= \begin{pmatrix} \mathbf{F}_{i1}^* & \mathbf{F}_{i2}^* & \dots & \mathbf{F}_{in}^* \end{pmatrix}^{\mathrm{T}} = \mathbf{F}(j\omega_i), \\ i &= \overline{1, N_\omega}. \end{aligned}$$

Now we can consider the dynamic corrector as a set of independent correctors, each corresponding to individual generalized coordinate θ_k:

$$\xi_k = \mathbf{F}_k(s)\varepsilon_\theta,$$
$$k = \overline{1, n}, \tag{14}$$

which can be represented as state space system

$$\dot{\mathbf{p}}_k = \alpha_k \mathbf{p}_k + \beta_k \varepsilon_\theta,$$
$$\xi_k = \gamma_k \mathbf{p}_k + \mu_k \varepsilon_\theta, \tag{15}$$

where $\mathbf{p}_k \in E^{2N_\omega}$ is a state vector of the corresponding individual corrector. From the (14) and (15) it is obvious that $\mathbf{F}_k(s) = \gamma_k(\mathbf{E}_{2N_\omega}s - \alpha_k)^{-1}\beta_k + \mu_k$, so we can rewrite conditions (6) and (13) in the form

$$-\gamma_k\alpha_k^{-1}\beta_k + \mu_k = \mathbf{F}_k^0,$$
$$\gamma_k(\mathbf{E}_{2N_\omega}j\omega_i - \alpha_k)^{-1}\beta_k + \mu_k = \mathbf{F}_{ik}^*, \tag{16}$$
$$i = \overline{1, N_\omega}, k = \overline{1, n}.$$

Next, let us pick arbitrary non-zero vectors γ_k with dimensions $1 \times 2n$ and arbitrary Hurwitz matrices α_k with dimensions $2n \times 2n$. Such choice of α_k ensures that matrices $\mathbf{E}_{2N_\omega}j\omega_i - \alpha_k$ from (16) are non-singular. Now let us introduce values $\mathbf{R}_{ik} = \mathrm{Re}\mathbf{F}_{ik}^*$ and $\mathbf{I}_{ik} = \mathrm{Im}\mathbf{F}_{ik}^*$ for real and imaginary parts of $\mathbf{F}_{ik}^*$ correspondingly, subtract the first equation of (16) from the second one and write separate equations for real and imaginary parts, achieving system

$$\gamma_k(\alpha_{ik}^R + \alpha_k^{-1})\beta_k = \mathbf{R}_{ik} - \mathbf{F}_k^0,$$
$$\gamma_k\alpha_{ik}^I\beta_k = \mathbf{I}_{ik}, \tag{17}$$
$$i = \overline{1, N_\omega}, k = \overline{1, n},$$

where

$$\alpha_{ik}^R = \mathrm{Re}(\mathbf{E}_{2N_\omega}j\omega_i - \alpha_k)^{-1},$$
$$\alpha_{ik}^I = \mathrm{Im}(\mathbf{E}_{2N_\omega}j\omega_i - \alpha_k)^{-1}.$$

Note that (17) is in fact linear system

$$\mathbf{A}_k\beta_k = \mathbf{B}_k \tag{18}$$

with

$$\mathbf{A}_k = \begin{pmatrix} \gamma_k(\alpha_{1k}^R + \alpha_k^{-1}) \\ \gamma_k\alpha_{1k}^I \\ \cdots \\ \gamma_k(\alpha_{N_\omega k}^R + \alpha_k^{-1}) \\ \gamma_k\alpha_{N_\omega k}^I \end{pmatrix}, \mathbf{B}_k = \begin{pmatrix} \mathbf{R}_{1k} - \mathbf{F}_k^0 \\ \mathbf{I}_{1k} \\ \cdots \\ \mathbf{R}_{N_\omega k} - \mathbf{F}_k^0 \\ \mathbf{I}_{N_\omega k} \end{pmatrix}, k = \overline{1, n}$$

System (18) has $2N_\omega \times n$ equations and $2N_\omega \times n$ variables for each fixed k. Recall that $\mathbf{A}_k$ are non-singular matrices, therefore (18) has unique solution for matrix $\boldsymbol{\beta}_k$. Finally, $\boldsymbol{\mu}_k$ can be calculated using the first equation of the (16):

$$\boldsymbol{\mu}_k = \mathbf{F}_k^0 + \boldsymbol{\gamma}_k \boldsymbol{\alpha}_k^{-1} \boldsymbol{\beta}_k, \quad k = \overline{1, n}. \tag{19}$$

Thus, if we take particular values for matrices $\boldsymbol{\alpha}_k$ and vectors $\boldsymbol{\gamma}_k$ as it was stated above, then we can get $\boldsymbol{\beta}_k$ matrices as solutions of system (18) and $\boldsymbol{\mu}_k$ vectors from (19). These four elements define dynamic corrector transfer matrix $\mathbf{F}$ satisfying conditions (6) and (13) i.e. providing astatism and polyharmonic disturbance filtering for known frequencies.

4 Numerical Example

To demonstrate the efficiency of the proposed methods let us consider two-link manipulator as a plant (Fig. 1).

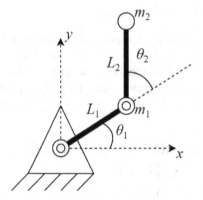

Fig. 1. Two-link manipulator.

In this case the matrices of system (1) take the form [10]

$$\mathbf{M}(\boldsymbol{\theta}) = \begin{bmatrix} m_1 L_1^2 + m_2(L_1^2 + 2L_1 L_2 \cos \theta_2 + L_2^2) & m_2(L_1 L_2 \cos \theta_2 + L_2^2) \\ m_2(L_1 L_2 \cos \theta_2 + L_2^2) & m_2 L_2^2 \end{bmatrix},$$

$$\mathbf{C}(\boldsymbol{\theta}, \dot{\theta}) = \begin{bmatrix} -m_2 L_1 L_2 \cos \theta_2 (2\dot{\theta}_1 \dot{\theta}_2 + \dot{\theta}_2^2) \\ m_2 L_1 L_2 \dot{\theta}_1^2 \sin \theta_2 \end{bmatrix},$$

$$\mathbf{g}(\boldsymbol{\theta}) = \begin{bmatrix} (m_1 + m_2)L_1 g \cos \theta_1 + m_2 g L_2 \cos(\theta_1 + \theta_2) \\ m_2 g L_2 \cos(\theta_1 + \theta_2) \end{bmatrix},$$

where m_1 and m_2 are point masses at the link ends, L_1 and L_2 are link lengths, g is a gravitational acceleration, $\boldsymbol{\theta} = \begin{pmatrix} \theta_1 & \theta_2 \end{pmatrix}^{\mathrm{T}}$ is a vector of joint angles. We will take $m_1 = m_2 = 1$ kg, $L_1 = L_2 = 1$ m, $g = 9.8$ m/s^2.

Multipurpose regulator (5) was synthesized with the following matrices:

$$\mathbf{K}_\theta = \begin{pmatrix} 8 & 0 \\ 0 & 8 \end{pmatrix}, \mathbf{K}_\omega = \begin{pmatrix} 8 & 0 \\ 0 & 8 \end{pmatrix},$$

$$\mathbf{H}_\theta = \begin{pmatrix} 16 & 0 \\ 0 & 16 \end{pmatrix}, \mathbf{H}_\omega = \begin{pmatrix} 32 & 0 \\ 0 & 32 \end{pmatrix}.$$

The matrices α_k of the dynamic corrector were taken as Frobenius matrices with eigenvalue $\lambda = -3$ of multiplicity 6. The values γ_k were chosen as vectors $\gamma_k = (0\ 0\ 0\ 0\ 0\ 1)$. All of the provided values were chosen manually to provide better visibility of the regulator effects in different modes. In real problems constant matrices of the control signal and of the asymptotic observers are obtained during some optimization process depending on particular requirements to the dynamics. For example, it can be done via LQR synthesis.

A constant disturbance in the corresponding mode was set in the form.
$\tau_e(t) = (0.1\ 0.1)^T$.

Finally, the vector τ_e in the periodic disturbance mode was chosen as

$$\tau_e(t) = \left(w(t)\ w(t) \right)^T,$$
$$w(t) = \sin(\omega_1 t) + \sin(\omega_2 t) + \sin(\omega_3 t), \tag{20}$$

with $\omega_1 = 31.416 \text{s}^{-1}$, $\omega_2 = 37.699 \text{s}^{-1}$ and $\omega_3 = 43.982 \text{s}^{-1}$.

The computer model was developed in the Octave environment in order to test the regulator in different modes. All the numerical experiments were carried out on the desktop computer with AMD Ryzen 7 3700X processor and 16 Gb RAM. Now let us demonstrate theses experiments consequently.

Let us start with the operating mode without external disturbance. In this mode we don't need the dynamic corrector. Assume that initial position is $\theta_0 = (0\ 0)^T$, $\omega_0 = (0\ 0)^T$. Let us take the desired position as $\theta_d = (45\ 60)^T$ degrees. Figure 2 demonstrates the system dynamics in this case – the desired position has been achieved.

Let us set the initial position corresponding to the desired one, i.e. $\theta_0 = \theta_d = (45\ 60)^T$ degrees and consider the constant disturbance mode. Let us apply constant disturbance (31) still without the dynamic corrector. As it can be seen from Fig. 3, now the position deviates from the desired one and fails to return back. Therefore astatism is not provided without the dynamic corrector.

Let us now turn the dynamic corrector on within the same conditions. Figure 4 demonstrates that in this case the corrector actually ensures astatism and despite the initial deviation, the robot state returns back to the desired position.

Finally, let us consider the influence of polyharmonic disturbance (20). By the means of better effect visibility, the first 10s of system motion will be without dynamic corrector. From Fig. 5 it can be seen that the periodic disturbance also moves robot state from the initial position, but turning corrector on brings it back, while the intensity of the θ output oscillations stays the same. But if we take a look at Fig. 6 with the dynamics of the control signal τ we can notice only a slight decrease in the intensity of control action due to the corrector being turned on.

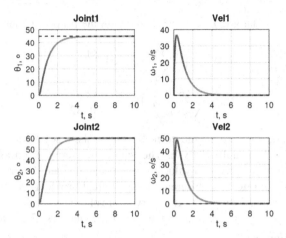

Fig. 2. Dynamics of the system without external disturbance.

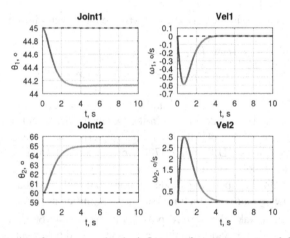

Fig. 3. Dynamics of a system under the influence of a constant external disturbance.

Such small effect can be explained by the following reasoning. Due to the feedback linearization the multipurpose regulator is applied actually to linearized system (4) and it filtrates the disturbance just for the linear part **u** of the control signal τ.

The dynamics of the linear part **u** of the control signal τ is shown in Fig. 7. It can be seen that the influence of external disturbance is completely compensated by turning the dynamic corrector on, just as expected.

We can reduce the control signal τ reaction if we use outputs of asymptotic observers z_θ and z_ω instead of actual perturbed measurements θ and ω in the calculation of the values of matrices **M**, **C** and **g** which are used in feedback linearization. System dynamics obtained with this approach is shown in Figs. 8 and 9. It can be seen that the joint angles θ oscillations amplitude becomes slightly smaller compared to Fig. 5, while a more significant decrease in the intensity of the control signal τ is noticeable after turning on

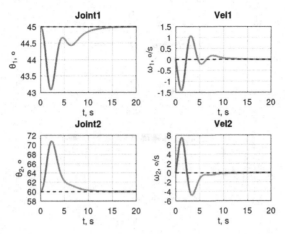

Fig. 4. Dynamics of a system with constant external disturbance and dynamic corrector.

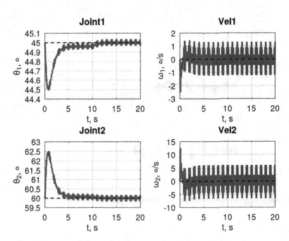

Fig. 5. Dynamics of a system with a polyharmonic external disturbance and with a dynamic corrector enabled at the moment $t = 10$ s.

the corrector compared to Fig. 5. The influence of the disturbance on the linear part **u** in this case is also completely compensated, as shown in Fig. 10. We can also notice form Fig. 8 that astatism is still provided as joint values goes straight to the desired position after corrector turning on. Therefore we can conclude that usage of the asymptotic observer outputs actually can increase filtration quality.

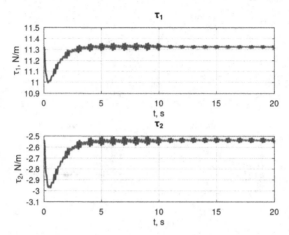

Fig. 6. Dynamics of the control signal τ with a polyharmonic external disturbance.

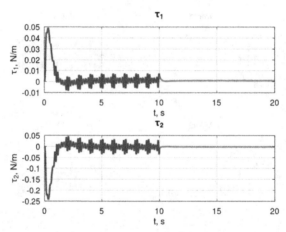

Fig. 7. Dynamics of the linear part **u** of the control signal τ with polyharmonic external disturbance.

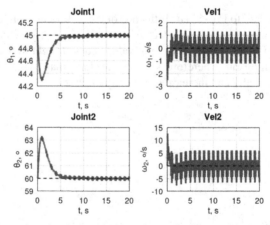

Fig. 8. System dynamics when using observer outputs in control signal matrices calculation.

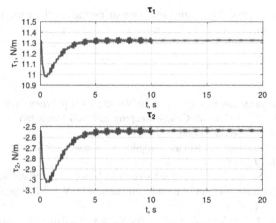

Fig. 9. Control signal τ dynamics when using observer outputs in matrices calculation.

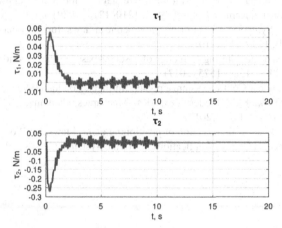

Fig. 10. Dynamics of the linear part **u** of the control signal τ when using observer outputs to calculate matrices in the control signal.

5 Conclusions

This paper presented the method of minimizing the control signal reaction to the external periodic disturbance while providing astatism in the problem of robotic arm stabilization in the desired position. Multipurpose regulator with feedback linearization and special dynamic corrector was considered as a feedback and synthesis process of such regulator was discussed. Experiments with computer model of the two-link robotic manipulator in different modes were presented. Analysis of the obtained results shows that using multipurpose regulator directly actually provides quite poor minimization of the control action. Though this result can be enhanced by usage the output of the asymptotic observers in calculation of the system matrices to provide better feedback linearization. But probably even better solution might be usage of nonlinear asymptotic observers in the multipurpose regulator. This topic and comparing the results with the ones obtained in this work is the goal of our nearest research.

References

1. Fossen, T.I.: Guidance and Control of Ocean Vehicles, 494 p. Wiley, New York (1994)
2. Perez, T.: Ship Motion Control: Course Keeping and Roll Stabilization Using Rudder and Fins, p. 300. Springer-Verlag, London (2005)
3. Xian-zhou, W., Han-zhen, X.: Robust autopilot with wave filter for ship steering. J. Marine Sci. Appl. **5**, 24–29 (2006)
4. Han, Q., Liu, Z., Su, H., Liu, X.: Filter-based disturbance observer and adaptive control for euler-lagrange systems with application to a quadrotor UAV. IEEE Trans. Industr. Electron. **70**(8), 8437–8445 (2023)
5. Bilal, H., Yin, B., Aslam, M.S., et al.: A practical study of active disturbance rejection control for rotary flexible joint robot manipulator. Soft Comput. **27**, 4987–5001 (2023)
6. Veremey, E.I., Korchanov, V.M.: Multiobjective stabilization of a certain class of dynamic systems. Automation Remote Control **49**(9), 1210–1219 (1989)
7. Veremey, E.I.: Synthesis of multi-objective control laws for ship motion. Gyroscopy Navigation **1**(2), 119–125 (2010)
8. Veremey, E.I.: Separate filtering correction of observer-based marine positioning control laws. Int. J. Control **90**(8), 1561–1575 (2017)
9. Isidori, A.: Nonlinear Control Systems, p. 549. Springer, London (1995)
10. Lynch, K.M., Park, F.C.: Modern Robotics: Mechanics, Planning, and Control, 624 p. Cambridge University Press (2017)
11. Veremey, E.I.: Dynamical correction of positioning control laws. In: Proceedings of the 9th IFAC Conference on Control Applications in Marine Systems (CAMS-2013), Japan, pp. 31–36 (2013)

On the Optimal Management of Energy Storage

Alexander Vasin[ID] and Olesya Grigoryeva[✉][ID]

Lomonosov Moscow State University, Moscow 119991, Russia
olesyagrigez@gmail.com

Abstract. In the present work, we consider a problem of optimizing storage management for a storage device, whose actions do not affect the market prices of electricity. An owner of the device can buy energy at fixed tariffs or at wholesale market prices and sell it back to the network at these prices. He aims to maximize his profit from the energy resale in a given planning interval. We obtain an efficient algorithm for calculating the optimal storage strategy. This algorithm permits to search for a solution to the problem in an "almost analytical" form. The solution is of interest for consumers using storage devices, as well as for a centralized regulation of electricity networks.

Keywords: Energy storages · Optimal control · Efficient algorithms

1 Introduction

The development of the electrical power industry is an important task in terms of accelerating economic growth rates. This involves the utilization of new economic and technological tools to optimize electricity generation and consumption. Among such tools, there are renewable energy sources and electricity storage. To reduce costs, consumers may utilize renewable energy sources, such as solar panels. However, the amount of energy produced from these sources is a random variable that depends on weather conditions. Therefore, an energy storage system can play an instrumental role in minimizing costs. By utilizing an energy storage device, consumers can accumulate energy at low prices when there is excess supply and utilize it at higher prices during periods of peak demand. This approach can help to manage fluctuations in energy demand and supply and optimize the overall efficiency of the electricity system.

Optimal algorithms for energy storage systems in microgrid applications are widely discussed in literature. The paper [1] explores the benefits of flexible demand sources in microgrids for optimal trading of energy in the day-ahead and real-time markets. Electric vehicles and batteries are identified as sources with flexible demand, and the paper does not specify the corresponding stochastic optimization problem but suggests a two-step heuristic optimization method. In work [2], a multi-energy microgrid with wind, solar, and combined electricity,

© The Author(s), under exclusive license to Springer Nature Switzerland AG 2024
A. Eremeev et al. (Eds.): MOTOR 2024, LNCS 14766, pp. 421–431, 2024.
https://doi.org/10.1007/978-3-031-62792-7_28

heat, and gas loads is considered, and the article [3] examines various types of existing energy storage systems and discusses their characteristics and development trends. In the paper [4], microgrid network models are studied in which drivers are treated as independent market participants. An optimization problem has been formulated for optimal energy trading on oligopolistic energy markets, both for the day ahead and in real-time, for an individual producer. The use of the Karush–Kuhn–Tucker theorem and conditions of strong duality allow us to reduce this optimization problem to that of mixed-integer programming. In the article [5], the problem of optimizing energy consumption by a production complex within a microgrid is investigated. Two cases are considered, where the production complex can utilize a storage device to store purchased energy or use solar panels as an additional self-generation source. The main results of this work pertain to optimizing the electricity purchase schedule to minimize total costs associated with purchasing energy from the grid, as well as storing energy in the case where a storage device is utilized, or producing solar power. However, the case where both a storage device and solar panels may be used simultaneously has not been analyzed. As in [4], the model does not account for random factors related to future energy demands and its supply through solar panels and other renewable sources. These factors are assumed to be known in advance.

In [6], we present our results on finding the optimal consumption plan, taking into consideration shiftable loads and the potential use of an energy storage device. We discuss a case where the storage management system is based on a reliable forecast of random variables for the planning period, and also address the corresponding stochastic optimization problem, where future values of the random variables are characterized by their probability distribution. The optimal market strategy is obtained for the case, where the customer can sell excess energy at the current market price. We prove the possibility of reducing the optimization problem for a customer using a storage device to solving separate optimization problems for consumption and the storage control.

In our work [7], we found solutions to some special cases of the problem of optimizing the storage control strategy from the point of view of energy resale and showed that this problem is a convex programming problem. Using Lagrange's theorem, relationships for calculating optimal parameters of the storage were obtained.

In the present work, we obtain a finite algorithm for calculating the optimal storage strategy for quasi-continuous price dynamics, that is, for any time, the difference of prices in subsequent period is sufficiently small. This algorithm allows us to search for a solution to the problem in an "almost analytical" form. Below, Sect. 2 establishes necessary conditions for an optimal strategy of the storage management. Proceeding from these conditions, Sect. 3 describes a method for calculating the optimal strategy.

2   Optimal Strategy of Energy Storage

Let's formulate for the storage device the optimal control problem with discrete time t during the planning interval $t \in \{\overline{1, T}\}$, corresponding to one day. Denote by E — the storage volume, V — the maximum rate of charging and discharging, η_{ch} and η_{dis} — charging and discharging efficiency coefficients v^t_{Bat} — the amount of energy the storage charging (if > 0) or discharging (if < 0) during period t. Below, the main focus of the article is on a case where the storage device has the ability to sell its stored energy back to the grid at the current market price. In this case, the management strategy is determined by vector $\vec{v}_{Bat} = (v^t_{Bat}, t = \overline{1,T})$, satisfying the following constraints: $\forall t$

$$|v^t_{Bat}| \leq V; \tag{1}$$

$$0 \leq \sum_{k=1}^{t} v^k_{Bat} \leq E; \tag{2}$$

$$\sum_{t=1}^{T} v^t_{Bat} = 0. \tag{3}$$

Let prices $p^t, t \in \{\overline{1,T}\}$, be known in the planning period. The optimal storage strategy aims to maximize the profit from energy resale under these prices and restrictions (1–3):

$$\vec{v}^*_{Bat} = \arg\max_{\vec{v}_{Bat}}(-\sum_{t=1}^{T} p^t v^t_{Bat} \eta(v^t_{Bat})), \tag{4}$$

where $\eta^t(v^t_{Bat}) = \begin{cases} \eta_{ch}, & \text{if } v^t_{Bat} > 0 \text{ (battery is charging);} \\ \frac{1}{\eta_{dis}}, & \text{if } v^t_{Bat} < 0 \text{ (battery is discharging).} \end{cases}$

The problem (4) is also of interests from a perspective of optimizing the operations of the power grid as a whole. According to a well-known theorem, (Welfare Theorem, see [8]), its solution forms a part of an optimal plan that ensures the maximum social welfare. Denote $\hat{\eta} = \eta_{ch} \cdot \eta_{dis}$. The solution to the problem (4) has been obtained in certain special cases [3]. Let the limit on storage volume specified in (2) never be reached, $\tau_1, \tau_2, \ldots, \tau_T$ denote the sequence of time periods arranged in ascending order by price (i.e.$\{p^{\tau_1} \leq p^{\tau_2} \ldots \leq p^{\tau_T}\}$), while $\overline{\tau}_1, \overline{\tau}_2, \ldots, \overline{\tau}_T$ represent the sequence of time periods organized in descending order by price within the planning horizon $\overline{1,T}$, m denotes the maximum number of indices j such that $\hat{\eta} p^{\tau_j} < p^{\overline{\tau}_j}$.

Proposition 1. (see. [7, proposition 1]). *The optimal strategy of the problem (4) is:*

$$v^{t*}_{Bat} = V, \text{ under } t = \tau_1, \ldots, \tau_m; \ v^{t*}_{Bat} = -V, \text{ under } t = \overline{\tau}_1, \ldots, \overline{\tau}_m, \ v^{t*}_{Bat} = 0 \text{ under other } t. \tag{5}$$

A more typical case is when both constraints, on the charging rate (1) and on the battery capacity (2), may be active, that is, at some time t, condition (1) is satisfied as equality and, at some other time t', one of the conditions of constraint (2) is satisfied as equality too. We define significant extremes $t_1 < \bar{t}_1 < t_2 < \bar{t}_2 < \ldots < t_k < \bar{t}_k$, such that $p^{t_1}, \ldots, p^{t_k}$ — are local minima prices and, $p^{\bar{t}_1}, \ldots, p^{\bar{t}_k}$ — are local maxima prices meeting the following conditions:

$$\forall t \in (t_j, \bar{t}_j) \ \ p^{\bar{t}_j} \geq p^t \geq p^{t_j}, \ \ \forall t' \in (t, \bar{t}_j) \ \hat{\eta} p^{t'} > p^t,$$

$$\forall t \in (\bar{t}_j, t_{j+1}) \ \ p^{\bar{t}_j} \geq p^t \geq p^{t_{j+1}}, \ \ \forall t' \in (t, t_{j+1}) \ \ p^{t'} \leq \hat{\eta} p^t,$$

$$\forall j \ \ p^{\bar{t}_j} > \hat{\eta} p^{t_j}, \ \ p^{\bar{t}_j} > \hat{\eta} p^{t_{j+1}}, \ \ \text{where} \ \ k+1 := 1.$$

Let $\bar{t}_{j,0} = \bar{t}_j, \bar{t}_{j,1}, \bar{t}_{j,2}, \ldots$ — the ordering of time periods from local minimum t_j to the next local minimum t_{j+1} in descending order of prices; $t_{j,0} = t_j, t_{j,1}, t_{j,2}, \ldots$ — the ordering in ascending order from local maximum $\bar{t}_{j-1}$ to the next local maximum $\bar{t}_j$. Denote $e = [E/V] + 1$ — as the required number of periods for fully charging or discharging the storage (where $[x]$ represents x rounded down), $\forall j = 1, \ldots, k \ l(j) = \max\{l | \hat{\eta} p^{t_{j,l}} < p^{\bar{t}_{j,l}}\}$, $\bar{l}(j) = \max\{l | \hat{\eta} p^{t_{j+1,l}} < p^{\bar{t}_{j,l}}\}$. The first condition shows that buying at period $t_{j,l}$ and selling at period $\bar{t}_{j,l}$ respectively, is profitable. Similarly, the second condition states that selling at period $\bar{t}_{j,l}$ and followed buying at period $t_{j+1,l}$ is also profitable.

Note 1. In the case of equality in these conditions, we consider that the transaction would be unprofitable, due to the existence of small fixed costs associated with the purchases and sales involved.

We will first determine the optimal strategy for the case where the storage volume is low at a given price level and charging rate, that is, when the following condition holds:

$$e \leq e^* = \min_j \min\{l(j), \bar{l}(j)\}. \tag{6}$$

Proposition 2. (see. [7, proposition 3]). *Under the given conditions, the optimal solution to the problem (4) takes the following form:* $\forall j = 1, \ldots, k \ v_{Bat}^{t_{j,r}}{}^* = V, v_{Bat}^{\bar{t}_{j,r}}{}^* = -V$ under $r = 1, \ldots, e-1, v_{Bat}^{t_{j,e}}{}^* = -v_{Bat}^{\bar{t}_{j,e}}{}^* = E - (e-1)V, v_{Bat}^{t*} = 0$ under other t.

Therefore, the battery is fully charged in the vicinity of each significant local minimum and discharged in the vicinity of the next significant local maximum. Now, we consider the case where the condition (6) does not hold. Next, let us assume that $e = E/V$ — is an integer, (the case of a non-integer will be considered later in the article). We will describe a method for calculating an optimal strategy $\vec{v}^*_{Bat}(e) = (v^{t*}_{Bat}(e), t = \overline{1, T})$ under $e > e^*$. For given e, this strategy is characterized by a set of connected intervals, where a connected interval is a maximum time interval (t', t''), during which the constraints (2) on the battery energy level are not active under the strategy $\overrightarrow{v}^*_{Bat}(e) : \forall t \in (t', t'')$ $0 < e(t) < e$, where $e(t)V$ — is the battery energy level after period t under the given strategy. There are four types of connected intervals characterized by the energy levels at the beginning and at the end of each interval: 1) $e(t') = 0$ and $e(t'') = e$; 2) $e(t') = e$ and $e(t'') = 0$; 3) $e(t') = e(t'') = 0$; 4) $e(t') = e(t'') = e$. To calculate the optimal control strategy, it would be convenient to determine each connected interval by its first significant local extremum (θ') and its last significant local extremum (θ'').

Lemma 1. *For any connected interval of type 1, $\theta', \theta'' \in \{t_1, t_2, .., t_k\}$, that is θ', θ'' are significant local minimums. For any connected interval of type 2 $\theta', \theta'' \in \{\bar{t}_1, \bar{t}_2, \ldots, \bar{t}_k\}$, that is θ', θ'' are significant local maximums. For any connected interval of type 3 $\theta' \in \{t_1, t_2, \ldots, t_k\}$, $\theta'' \in \{\bar{t}_1, \bar{t}_2, \ldots, \bar{t}_k\}$. For any connected interval of type 4 $\theta' \in \{\bar{t}_1, \bar{t}_2, \ldots, \bar{t}_k\}$, $\theta'' \in \{t_1, t_2, \ldots, t_k\}$.*

Proof. Consider the first statement of the lemma. It means that $\theta' \in [\bar{t}_{j-1}, t_j)$. Assume form the contrary that the first charge happens in $[t_j, \bar{t}_j)$. If the previous discharge happens in period $t \in [t_j, t']$ then, according to the definition of a significant local maximum, it is profitable to cancel this sale and the purchase at time t'. If $t < t_j$ then it is profitable to shift this sale from t' to t_j. The rest statements are justified in a similar way.

For optimal strategy $\overrightarrow{v}^*_{Bat}(e)$, each connected interval $I = (t', t'')$ is characterized by the periods τ_1 and τ_2 which are the least profitable periods for the purchase and sale of energy, respectively. In this interval, the maximum purchase price is achieved at p^{τ_1}, and the minimum sale price is achieved at p^{τ_2}. Then, $\forall t \in I$ $(p^t \in (p^{\tau_1}, p^{\tau_2})) \Rightarrow v^{t*}_{Bat}(e) = 0$, which means that the state of the battery does not change in this interval. Note that there may be no energy sales (i.e., unloading of the battery) in a connected interval of type 1, and there may be no energy purchases (i.e., loading of the battery) in a connected interval of type 2. We will refer to these intervals as simple, and we will assume that $p^{\tau_2} = \infty$ for intervals of type 1 and $p^{\tau_1} = -\infty$ for intervals of type 2 (Fig. 1).

Lemma 2. *Under optimal strategy $\overrightarrow{v}^*_{Bat}(e)$, for any connected interval (t', t'') $\hat{\eta} p^{\tau_1} < p^{\tau_2}$ and $\forall t \in (t', t'')$ $p^t < p^{\tau_1} \Rightarrow v^{t*}_{Bat}(e) = V$, buying is being made; $p^t > p^{\tau_1} \Rightarrow v^t_{Bat}(e) = -V$, selling is being made.*

Proof. Assume from the contrary that inequality $\hat{\eta}^{\tau_1} > p^{\tau_2}$ holds. According to the strategy, we made a purchase in period τ_1, made a sale in τ_2, but due to the fixed losses, this operation is unprofitable. If we cancel the purchase and the sale, it would be beneficial to us. No restrictions would be violated, since we are in the range $(t', t")$, where constraint (2) is not binding.

Based on the definition of intervals of various types, the optimal strategy meets the following conditions for the storage charging balance:

for type 1 intervals,

$$|\{t \in (t', t'')|v^{t*}_{Bat}(e) = V\}| - |\{t \in (t', t'')|v^{t*}_{Bat}(e) = -V\}| = e;$$

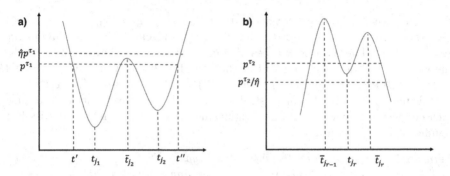

Fig. 1. a) Simple interval type 1; b) Simple interval type 2.

for type 2 intervals

$$|\{t \in (t', t'')|v^{t*}_{Bat}(e) = V\}| - |\{t \in (t', t'')|v^{t*}_{Bat}(e) = -V\}| = -e;$$

for intervals of type 3 and 4

$$|\{t \in (t', t'')|v^{t*}_{Bat}(e) = V\}| = |\{t \in (t', t'')|v^{t*}_{Bat}(e) = -V\}|.$$

Now, we specify the conditions that must hold for prices p^{τ_1} and p^{τ_2} for adjacent connected intervals. Note that an interval of Type 1 or Type 4 may be followed by an interval of Type 2 or Type 4, while an interval of Type 2 or Type 3 may be followed by an interval of Type 1 or Type 3 (see Fig. 2 for the examples). Let $S(e) = \{I_1(e), \dots, I_{k(e)}(e)\}$ be a sequence of connected intervals corresponding to optimal strategy $\overrightarrow{v}^*_{Bat}(e)$.

Lemma 3. *Under optimal strategy $\overrightarrow{v}^*_{Bat}(e)$ any adjacent connected intervals I_l and I_{l+1} meet the following conditions:*

1) *For interval I_l of type 1 or 4, and the following interval I_{l+1} of type 2 or 4, the inequality $\hat{\eta} \cdot p^{\tau_1}(e, I_l) < p^{\tau_2}(e, I_{l+1})$ holds.*
2) *For interval I_l, of type 2 or 3, and the following interval I_{l+1}, of type 1 or 3, the inequality $p^{\tau_2}(e, I_l) > \hat{\eta} \cdot p^{\tau_1}(e, I_{l+1})$ holds.*

Proof. Is similar to the proof of Lemma 2.

3   The Algorithm for Computation of the Optimal Strategy of the Storage

Let us describe a method for calculating optimal strategy $\overrightarrow{v}_{Bat}$: consider how the strategy changes when the available storage capacity increases from e to $e + 1$ under $e \geq e^*$. Optimal strategy $\overrightarrow{v}_{Bat}(e)$ is determined by set $S(e)$ of consecutive connected intervals $(I_1(e), \dots, I_{k(e)}(e))$, their boundary local extremes $\theta'(I_l(e))$, $\theta''(I_l(e))$, maximum purchase price $p^{\tau_1}(e, I_l(e))$ and minimum sale price $p^{\tau_2}(e, I_l(e))$.

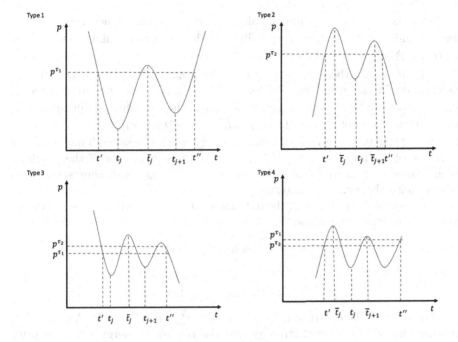

Fig. 2. Intervals of types 1–4.

For each interval $I_l(e)$, given by his boundary local extremes, we determine the maximum purchase price $\tilde{p}^{\tau_1}(e+1, I_l(e))$ and sale price $\tilde{p}^{\tau_2}(e+1, I_l(e))$, based on the assumption that the interval $I_l(e)$ will continue to be part of the structure of connected intervals under volume $e+1$.

Lemma 4. *If interval $I_l(e)$ of type 3 or 4 is preserved in structure $S(e+1)$, then it follows that $\tilde{p}^{\tau_i}(e+1, I_l(e)) = \tilde{p}^{\tau_i}(e, I_l(e))$ for $i = 1, 2$. At the same time, the boundary points of the interval $I_l(e+1)$ are the same as those of the interval $I_l(e)$: $t'(e, I_l(e)) = t'(e+1, I_l(e))$, $t''(e, I_l(e)) = t''(e+1, I_l(e))$.*

Note 2. As shown below, for connected intervals of type 1 and 2, when they are stored in the structure $S(e+1)$, boundaries $t'(e+1, I_l(e))$ and $t''(e+1, I_l(e))$ may differ from $t'(e, I_l(e))$ and $t''(e, I_l(e))$.

Consider an interval of the first type corresponding to local extremum $\theta' = t_{j'}$, $\theta'' = t_{j''}$, with prices $p^{\tau_i}(e, I_l(e))$, $i = 1, 2$. If this interval is simple, then the optimal strategy $\vec{v}_{Bat}(e+1)$ for volume $e+1$ is determined as follows: find

$$\bar{t} = \underset{t \in (t_{j'-1}, t_{j''}): v^t_{Bat}(e)=0}{\arg\min} (p^t), \tag{7}$$

and set $v^{\bar{t}}_{Bat}(e+1) = V$, i.e. the storage is charged at time $\bar{t}$. For any other $t \in I_l(e)$ the strategy does not change, $v^t_{Bat}(e+1) = v^t_{Bat}(e)$. The new price is then $\tilde{p}^{\tau_1}(e+1, I_l(e)) = p^{\bar{t}}$. Obviously, $\tilde{p}^{\tau_1}(e+1, I_l(e)) \geq p^{\tau_1}(e, I_l(e))$.

For a generic type 1 interval, along with calculating $\bar{t}$ according to (7), we determine $\bar{\bar{t}} \in I_l(e)$, corresponding to the minimum selling price: $p^{\bar{\bar{t}}} = p^{\tau_2}(e, I_l(e))$; $v_{Bat}^{\bar{\bar{t}}}(e) = -V$.

If $\hat{\eta} \cdot p^{\bar{t}} < p^{\bar{\bar{t}}}$, then strategy $\overrightarrow{v}_{Bat}(e+1)$ and price $\tilde{p}^{\tau_1}(e+1, I_l(e))$ are determined in the same way as for a simple connected interval of type 1: $\tilde{p}_{\tau_1}(e+1, I_l(e)) = p^{\bar{t}}$ and $v_{Bat}^{\bar{t}}(e+1) = V$. The minimum selling price for this interval does not change: $\tilde{p}^{\tau_2}(e+1, I_l(e)) = p^{\tau_2}(e, I_l(e))$. If $\hat{\eta} \cdot p^{\bar{t}} \geq p^{\bar{\bar{t}}}$, then $v_{Bat}^{\bar{\bar{t}}}(e+1) = 0$, for any other t the strategy stays the same. That is, if the purchase of additional volume V is profitable, it is made, and if the purchase is unprofitable, then the sale at the lowest price is canceled, thus saving this volume until the end of the interval.

For interval $I_l(e)$ of type 2, the optimal strategy under volume $e+1$ is determined symmetrically: we find

$$\bar{t} = \underset{t \in (\bar{t}_{j'}, \bar{t}_{j''+1}): v_{Bat}^t(e)=0}{\arg\max} (p^t);$$

$$\bar{\bar{t}}: \ p^{\bar{\bar{t}}} = p^{\tau_1}(e, I_l(e)), \ v_{Bat}^{\bar{\bar{t}}}(e) = V.$$

If $\hat{\eta} \cdot p^{\bar{\bar{t}}} < p^{\bar{t}}$, then $\tilde{p}^{\tau_2}(e+1, I_l(e)) = p^{\bar{t}}$ and $v_{Bat}^{\bar{\bar{t}}}(e+1) = -V$. All the other components of the optimal strategy and the minimum energy purchase price $\tilde{p}^{\tau_1}(e+1, I_l(e)) = p^{\tau_1}(e, I_l(e))$ in this interval remain unchanged. If $\hat{\eta} \cdot p^{\bar{\bar{t}}} \geq p^{\bar{t}}$, then $v_{Bat}^{\bar{t}}(e+1) = 0$, $\tilde{p}^{\tau_1}(e+1, I_l(e))$ — the maximum purchase price among the remaining t such that: $v_{Bat}^t(e) = V$, $\tilde{p}^{\tau_2}(e+1, I_l(e)) = p^{\tau_2}(e, I_l(e))$. For all other t, the strategy does not change. That is, if the sale of additional volume is profitable, it is made, and if it is unprofitable, then the purchase at the highest price is canceled.

After the recalculations of the threshold prices $\tilde{p}^{\tau_i}(e+1, I_l)$ are completed for every interval $I_l \in S(e)$, we check the boundary conditions given in Lemma 3 for these new prices $\tilde{p}$.

Based on Lemma 1, let us denote an interval of the first type as $I(e, t_j, t_{j+r})$, an interval of the second type as $I(e, \bar{t}_j, \bar{t}_{j+r})$, an interval of the third type as $I(e, t_j, \bar{t}_{j+r})$; and an interval of the fourth type as $I(e, \bar{t}_j, t_{j+r})$. Consider a situation where the boundary conditions are not met for the interval $I_l(e, t_{j'}, t_{j''})$ of the first type, followed by the interval $I_{l+1}(e, \bar{t}_{j''}, \bar{t}_{j''+r})$ of the second type, i.e., $\hat{\eta} \cdot \tilde{p}^{\tau_1}(e+1, I_l(e)) \geq \tilde{p}^{\tau_2}(e+1, I_l(e))$. In this case, the intervals are united in the structure $S(e+1)$ and the resulting interval is of the third type: $I_l(e+1, t_{j'}, \bar{t}_{j''+r})$. In this new interval, the optimal control does not change: $v_{Bat}^t(e+1) = v_{Bat}^t(e)$ for any t, and volume restrictions do not apply. The threshold prices are calculated as follows:

$$p^{\tau_1}(e+1, t_{j'}, \bar{t}_{j''}) = \max\{p^{\tau_1}(e, t_{j'}, t_{j''}), p^{\tau_1}(e, \bar{t}_{j'}, \bar{t}_{j''})\},$$

$$p^{\tau_2}(e+1, t_{j'}, \bar{t}_{j''}) = \min\{p^{\tau_2}(e, t_{j'}, t_{j''}), p^{\tau_2}(e, \bar{t}_{j'}, \bar{t}_{j''})\}$$

Lemma 5. *Strategy* $\overrightarrow{v}_{Bat}(e+1)$ *defined in this way, meets the necessary optimality conditions of Lemma 2 within the limits of the new interval.*

When an interval of Type 1 is followed by an interval of Type 4, and the specified boundary conditions are violated, the intervals are united into a single interval of Type 1. For volume $e + 1$, the optimal strategy should be modified. The strategy should involve either increasing the purchase or reducing the sale, depending on the specific circumstances. Proceeding from the necessary conditions for the optimal purchase strategy $v_{Bat}(e)$, it is unprofitable to increase the purchase within the interval I_l. Therefore, either an additional purchase should be made, or the least profitable sales within the interval I_{l+1} should be cancelled. Similarly, if other types of intervals are involved, the strategy should be recalculated in case of boundary condition violations.

Lemma 6. *When constructing the structure of $S(e + 1)$, the following relationships hold. For subsequent intervals of type 1 and type 4, type 3 and type 1, type 2 and type 3, and type 4 and type 2, if boundary conditions are not met, optimal strategies are re-calculated and these intervals are united according to the aforementioned rules. When union intervals of types 1 and 4 as well as type 3 and type 1, a type 1 interval is formed. Similarly, when union intervals of types 2 and 3 as well as types 4 and 2, a type 2 interval is generated. Successive intervals of types 1 and 2, as well as those of types 2 and 1, are united when boundary conditions are disobeyed, without altering strategies within these intervals, and they become intervals of types 3 and 4 respectively. For neighboring intervals of types 3 and 4 as well as those of type 4 and 3, there is no union and the strategy remains unchanged in them unless they are united with their other neighboring intervals.*

Based on this lemma, the re-calculation of the strategy begins by allocating intervals of types 1 and 2, and determining the need for union them with neighboring intervals. After the first round of union, the same algorithm is applied to the newly formed structure, and so on until union of intervals of type 1 and type 2 with neighboring intervals no longer occurs.

Step by step description of the algorithm for computation of the optimal strategy $\overrightarrow{v}_{Bat}(e)$ for $e \geq e^*$ is as follows. For a given available storage capacity e, the optimal strategy $\overrightarrow{v}_{Bat}(e)$ is determined by set $S(e)$ of consecutive connected intervals $(I_1(e), ..., I_k(e)(e))$, their boundary local extremes $\theta'(I_l(e))$, $\theta''(I_l(e))$, maximum purchase prices $p^{\tau_1}(e, I_l(e))$, and minimum sale prices $p^{\tau_2}(e, I_l(e))$.

Step 1. For storage capacity e^* specified in (6), find the optimal strategy according to Proposition 2. The following steps form the cycle for recalculating the optimal strategy as the available storage capacity increases from e to $e + 1$.

Step 2. For each interval $I_l(e) \in S(e)$ of type 1 or 2, compute purchase price $\tilde{p}^{\tau_1}(e + 1, I_l(e))$, sale price $\tilde{p}^{\tau_2}(e + 1, I_l(e))$ and the optimal strategy for volume $e + 1$ according to relations after Lemma 4. For each interval $I_l(e) \in S(e)$ of type 3 or 4, the prices and the strategy remain the same.

Step 3. For each interval $I_l(e) \in S(e)$ of type 1 or 2, check the boundary conditions from Lemma 3 for the new prices.

Step 4. Using Lemma 6, merge intervals in the structure $S(e)$ according to the boundary conditions. If no intervals merge, go to the next step. Otherwise, return to step 2 with the new structure.

Step 5. The resulting strategy is optimal for volume $e + 1$. If $e + 1 = e$ then stop. Otherwise, set $e := e + 1$ and return to step 2.

Note 3. In general, the resulting strategy meets necessary, but not sufficient optimality conditions. For instance, for successive intervals I_l and I_{l+1} of types 1 and 2 in the structure meeting Lemmas 2 and 3, it might be profitable to cancel the purchase at price $p^{\tau_1}(e, I_l)$ and shift it to time $\bar{t}$ in the interval I_{l+1}, determined according to (7). Lemmas 2 and 3 provide sufficient conditions for the storage strategy to be optimal in the case where the price dynamics is quasi-continuous, that is, for any t, the difference of $|p^{t+1} - p^t|$ is sufficiently small. Then any changes of the strategy are unprofitable. Under this condition, we obtain the following result.

Theorem 1. *For any quasi-continuous price dynamics, the specified rules for recalculating strategies and joining neighboring connected intervals completely determine the method of recalculating the optimal strategy when increasing the storage volume from e to $e + 1$.*

In general, the following estimate is true: if, for any t, the difference $|p^{t+1} - p^t| < d$ for some $d > 0$ then the strategy determined by the given algorithm is $2dk(e)$ - optimal, where $k(e)$ is the number of connected intervals in the final structure. That is, the profit under this strategy differs from the maximal profit less than $2dk(e)$.

Note 4. In the case of a non-integer value of $e = E/V$, the optimal strategy can be determined in a similar way, with the following modification. We consider $[E/V] + 1$ as the value of e, and define all components of the strategy in the same way for all cases, except for the final operation when the vehicle is fully charging or empty. For the final purchase or sale, the quantity of the transaction should be equal to the remainder of the division of E/V.

4   Conclusion

The paper considers the problem of optimizing the management of a storage device, the actions of which do not affect market prices for electricity. The owner of the device can buy energy at fixed rates or at wholesale market prices and sell it back to the grid at these prices. He aims to maximize his profit from energy resale in a given planning interval. We obtain a method for calculating the optimal storage control strategy for quasi-continuous price dynamics. In the general case, the method determines a strategy that meets the necessary optimality conditions specified in Lemmas 2 and 3. We demonstrate the practical application of the method in an "almost analytical" form, highlighting its relevance not only for individual consumers with storage devices but also for the broader goal of centralized electricity network regulation. Our study contributes to the existing literature by providing a comprehensive framework for understanding and optimizing energy storage management. It underscores the importance of strategic energy storage and resale, especially in the context of fluctuating market prices and the increasing integration of renewable energy sources. The important tasks for the future are to complete the method for calculating the optimal storage control strategy in the general case and to take into account energy losses in the storage. These losses are significant for usual batteries.

References

1. Eseye, A.T., Lehtonen, M., Tukia, T., Uimonen, S., Millar, R. J.: Optimal energy trading for renewable energy integrated building microgrids containing electric vehicles and energy storage batteries. IEEE Access **7**, 106092–106101 (2019). https://doi.org/10.1109/ACCESS.2019.2932461
2. Shen, Y., Hu, W., Liu, M., Yang, F., Kong, X.: Energy storage optimization method for microgrid considering multi-energy coupling demand response. J. Energy Storage **45** (2022). https://doi.org/10.1016/j.est.2021.103521
3. Nadeem, F., Suhail Hussain, S.M., Kumar Tiwari, P., Kumar Goswami, A., Selim Ustun, T.: Comparative review of energy storage systems, their roles, and impacts on future power systems. IEEE Access **7**, 4555–4585 (2018). https://doi.org/10.1109/ACCESS.2018.2888497
4. Gazijahani, F.S., Esmaeilzadeh, R.: Increasing the penetration of renewables by releasing merchant energy storage flexibility. Intern. J. Modern Power Syst. **1** (2022)
5. Aleksandrov, A.A., Zakharov, M.N., Kuts, M.S.: Optimization of industrial complex power supply by means of renewable energy sources. Herald of the Bauman Moscow State Technical University. Series Mech. Eng. **1**, 85–102 (2021)
6. Vasin, A.A., Grigoryeva, O.M.: Optimal strategies of consumers with energy storages in electricity market. In: Kochetov, Y., Eremeev, A., Khamisov, O., Rettieva, A. (eds.) MOTOR 2022. Communications in Computer and Information Science, vol. 1661. Springer, Cham (2022). https://doi.org/10.1007/978-3-031-16224-4_21
7. Vasin, A.A., Grigoryeva, O.M., Seregina, I.Y.: Optimization of storage parameters for consumers at the electricity market. Mosc. Univ. Comput. Math. Cybern. **15**(1), 21–27 (2023)
8. Debreu, G.: Valuation equilibrium and pareto optimum. Proc. Natl. Acad. Sci. U.S.A. **40**(7), 588–592 (1954)

On the Application of Saddle-Point Methods for Combined Equilibrium Transportation Models

Demyan Yarmoshik[1,2]([✉]) [iD] and Michael Persiianov[1,2,3] [iD]

[1] Moscow Institute of Physics and Technology, Dolgoprudny, Russia
{yarmoshik.dv,persiianov.mi}@phystech.edu
[2] Institute for Information Transmission Problems, Moscow, Russia
[3] The Skolkovo Institute of Science and Technology, Moscow, Russia

Abstract. Travel demand modeling is an essential tool in urban planning and transportation system management. Existing practically efficient algorithms for solving multistage travel demand problems are variations of the heuristic sequential procedure. We propose a novel approach that applies saddle-point methods to a combined convex optimization formulation of the problem. Unlike all previous methods, our algorithm does not require costly shortest-paths calculations, and can be seamlessly scaled on GPUs. We show that in some cases it drastically outperforms the sequential procedures.

Keywords: Convex optimization · Trip distribution · Traffic assignment · Combined transportation model · First-order methods · Affine constraints

1 Introduction

In today's rapidly evolving urban landscapes, understanding and forecasting travel demand play pivotal roles in shaping efficient transportation systems and sustainable urban planning. Essentially, the purpose of a travel demand model is to predict the equilibrium traffic matrix and distribution of passenger and cargo flows through the transportation network, based on the network's characteristics. This information enables authorities to forecast the impact of changes in the transportation network, thus facilitating more informed decision-making.

The conventional four-stage approach [13,17]: trip generation (travel choice), trip distribution (destination choice), modal split (travel mode choice), and traffic assignment (route choice) stages, a widely used method in travel-demand modeling, has long been the cornerstone of transportation planning and policy-making. However this approach, despite its prevalence, has a number of limitations, e.g. there is no convergence guarantee [7,12,51] due to tendency to overlook consistent consideration of travel times and congestion effects across its various stages [24,45]. An alternative strategy to tackle this challenge involves

© The Author(s), under exclusive license to Springer Nature Switzerland AG 2024
A. Eremeev et al. (Eds.): MOTOR 2024, LNCS 14766, pp. 432–448, 2024.
https://doi.org/10.1007/978-3-031-62792-7_29

adopting a mathematical model formulated as an optimization problem that can subsequently be solved via numerical optimization methods. Such models are commonly referred to as *combined* or *integrated* models.

1.1   Combined Models

For decades, there has been extensive research into models that combine various stages. The pioneering work [6] introduced the first constrained convex optimization formulation for the user-equilibrium assignment problem with elastic demand. This formulation treated travelers between every origin-destination (OD) pair as a function of the travel service for that specific OD pair. Building upon this foundation, [19,21] expanded the convex optimization formulation to include destination choice through a combined distribution and assignment (CDA) model. In this model, trip distribution follows a gravity model with a negative exponential deterrence function, while traffic assignment adheres to a user-equilibrium model. To account for congestion effects at destinations, [50] introduced variable destination costs into the CDA model.

Subsequent research extended the CDA model in various directions. [9,37], and [29] refined the CDA model to accommodate multiple user classes, while [20] incorporated modal split within the CDA framework. [23] proposed an equivalent optimization problem for combined multiclass distribution, assignment, and modal split, eliminating symmetry restrictions. [60] put forth an optimization model that combines distribution, hierarchical mode choice, and assignment networks with multiple user and mode classes. [2] developed a comprehensive combined model by integrating all four steps of sequential demand forecasting based on random utility theory. [51] further extended this by simultaneously considering travel-destination-mode-route choice, utilizing the multinomial logit model within a hierarchical structure where each traveler is perceived as a customer of urban trips, with choices influenced by utility and budget constraints. Recent study [34], employing the entropy/utility maximization framework, proposed a convex programming approach equivalent to the hierarchical extended logit model which leads for the simultaneous estimation of all travel levels, thereby overcoming the input-output discrepancies in each level unlike traditional models with "feedback" loops [53].

1.2   CDA Model Applications

Among the discussed combined models, the CDA model holds particular significance in numerous transportation applications. It aids transportation planners in understanding the intricate interactions between land use and transportation, informing future urban development and transportation system improvement plans. For instance, researchers have utilized the CDA model within a bilevel programming framework to address various land use and transportation concerns. [62] applied the CDA model to analyze the capacity and level of service of urban transportation networks, while [56] used it to determine the maximum number of cars given road network capacity and available parking spaces. This model has been utilized to estimate the capacity of urban transportation networks incorporating rapid transit [16] and to identify critical links within road networks [18]. Besides, [15] introduced network-based accessibility measures to assess the vulnerability of degradable transportation networks, leveraging the CDA model. Additionally, [42] employed the CDA model in a land use-network design problem to analyze integrated layouts of land uses, public facilities, transport networks, and travel demands. [40] investigated equity issues associated with land use development in terms of changes in equilibrium OD travel costs, while [63] used the CDA model to optimize network reliability concerning residential and employment allocations and network enhancements. Furthermore, [28] embedded the CDA model of housing locations and traffic equilibrium to determine optimal housing provision patterns in a continuum transportation system.

1.3 Solution Algorithms

When Beckmann et al. [6] initially proposed in 1956 the formulation of a model of elastic demand and route choice, no solution algorithm was provided. Twenty years later, Ph.D. student Suzanne Evans [19] introduced a partial linearization method for CDA problem so that only the objective function was linearized, and the OD choice functions was unchanged. While [20,21] presented another algorithm based on the Frank and Wolfe [22] method. However, according to [11,39], Evans' algorithm exhibits superior convergence characteristics.

Founded on two unproven conjectures, [30] introduced a modification to the Evans' algorithm aimed at reducing computational times and memory requirements. However, as highlighted by [31], Horowitz's adapted algorithm doesn't consistently converge to the optimal solution. In response, [31] proposed further refinements to Horowitz's approach, providing rigorous proof of convergence.

On the other hand, [44] proposed a hybrid solution method, merging the Evans' algorithm with the disaggregate simplicial decomposition (DSD) method developed by [38]. Expected to offer enhanced computational efficiency and greater solution accuracy, this method operates within the path-flow domain. However, combining the Evans' algorithm with the second-order DSD algorithm may lead to non-convergence.

More recently, [4,5] extended the origin-based (OB) algorithm [3] for traffic assignment for tackling the CDA model as a fixed-point problem. This algorithm replaced the link-based assignment Evans' algorithm with an origin-based procedure to update the solution of the assignment problem, which demonstrated superior effectiveness in achieving highly accurate solutions. Notably, during the implementation of the OB algorithm, it was observed that the determination of step-size in the OD flow update step significantly impacts the algorithm's performance. To enhance the efficiency of the OB algorithm, [61] adopted the modified line search strategy proposed by [37] for the Evans' algorithm. In contrast, the work of [35] instead of using OB algorithms for convergence speed up utilized more prominent primal-dual methods [48] for traffic assignment sub-problem and [26] for trip distribution.

All previously mentioned algorithms can be divided into two groups: link-based (e.g., [19,22,35]) and origin-based (e.g., [4,5]). *Link-based* algorithms prioritize aggregate link flows over individual path flows, making them memory-efficient and suitable for early computing environments. While they are easier to parallelize and implement, they can be slow, especially when high precision is required, due to convergence challenges. *Origin-based* algorithms focus on path flow vectors, offering faster convergence by retaining more information. Despite their higher memory demand and coding complexity, they excel in precision-critical scenarios by exploiting the vast number of paths in a network.

In this paper, we apply a convex optimization algorithm from [54] to solve the CDA problem. This approach offers several advantages over existing methods:

- The algorithm does not require solving the costly shortest-paths subproblem
- A single algorithm is used to solve the entire CDA problem, simplifying implementation and reducing the hustle of parameter tweaking
- The computation-heavy operations are element-wise function calculations and matrix-vector multiplication, which allows to write high-performance code in high-level programming languages like Python and seamlessly employ GPU acceleration via modern powerful frameworks such as [14].

The rest of the paper is organized as follows: Sect. 2 provides notation and mathematical model background. The algorithm for the doubly constrained CDA problem is presented in Sect. 3. Experiments are demonstrated in Sect. 4. Concluding remarks are discussed in Sect. 5.

2 Preliminaries

2.1 Notation

By $\sigma_{\max}(A), \sigma_{\min}^{+}(A)$ we denote maximal and minimal positive singular values of a linear operator A respectively. $\lambda_{\max}(A), \lambda_{\min}(A)$ denote maximal and minimal eigenvalues of a linear transformation A respectively. We use $\langle x, y \rangle := \sum_{i=1}^{d} x_i y_i$ to define the inner product of $x, y \in \mathbb{R}^d$. $\|x\| := \sqrt{\langle x, x \rangle}$ is the Euclidean norm of $x \in \mathbb{R}^d$, and $\|X\| := \sqrt{\langle X, X \rangle} = \sqrt{\operatorname{Tr}(X^\top X)}$ is the Frobenius norm of $X \in \mathbb{R}^{m \times n}$, where $\operatorname{Tr}(X)$ denotes the trace of X. By $\operatorname{Proj}_X(x_0) := \arg\min_{x \in X} \|x - x_0\|$ we mean Euclidean projection of x_0 on a set X. $1_{m \times n} \in \mathbb{R}^{m \times n}$, $1_n \in \mathbb{R}^n$, $0_{m \times n} \in \mathbb{R}^{m \times n}$, $0_n \in \mathbb{R}^n$ stand for matrices and vectors of all ones and zeros respectively. Indenity matrix is denoted by I_n. For $X \in \mathbb{R}^{n \times n}$ we define $\operatorname{diag}(X) \in \mathbb{R}^n$ to be the vector of diagonal elements of a matrix X; and if $x \in \mathbb{R}^n$, then $\operatorname{diag}(x) \in \mathbb{R}^{n \times n}$ is defined to be the diagonal matrix, constructed from a vector x.

2.2 Traffic Assignment. Beckmann Model

The traffic assignment problem is to find the equilibrium flow and travel time for each road given traffic matrix and traffic network. Following the outline of [35], let the traffic network be represented by a directed graph $\mathcal{G} = (\mathcal{V}, \mathcal{E})$, $|\mathcal{V}| = n$, $|\mathcal{E}| = m$, where vertices $\mathcal{V}$ correspond to intersections or centroids [55] and edges $\mathcal{E}$ correspond to roads. Suppose we are given the traffic matrix d of travel demands: namely, let $d_{ij}(\text{veh/h})$ be a trip rate from source $i \in \mathcal{V}$ to destination $j \in \mathcal{V}$. Here, to simplify the notation, we assume without loss of generality that all the vertices are sources and targets, and some of them can have zero travel demand or supply. We denote by P_{ij} the set of all simple paths from source $i \in V$ to target $j \in V$, and $P = \bigcup_{i,j \in V \times V} P_{ij}$ the set of all paths between sources and targets.

Agents traveling from node i to node j are distributed among paths from P_{ij}, i.e. for any $p \in P_{ij}$ there is a flow $x_p \in \mathbb{R}_+$ along the path p, and $\sum_{p \in P_{ij}} x_p = d_{ij}$. Flows from origin nodes to destination nodes create the traffic in the entire network $\mathcal{G}$, which can be represented by an element of

$$X = X(d) = \left\{ x \in \mathbb{R}_+^{|P|} : \sum_{p \in P_{ij}} x_p = d_{ij}; \ i,j \in V \right\}. \tag{1}$$

Then, the flow on edge e

$$f_e(x) = \sum_{p \in P} \delta_{ep} x_p \text{ for } e \in \mathcal{E},$$

where δ_{ep} equals 1 if $e \in p$ and 0 otherwise.

The problem of finding equilibrium load of a transportation network, known as traffic assignment problem, traditionally formulates [6] as

$$\Psi(x) = \sum_{e \in \mathcal{E}} \underbrace{\int_0^{f_e} \tau_e(z)dz}_{\sigma_e(f_e)} = \sum_{e \in \mathcal{E}} \sigma_e \left(\sum_{p \in P} \delta_{ep} x_p \right) \longrightarrow \min_{\substack{\sum_{p \in P_{ij}} x_p = d_{ij}, \\ x_p \geq 0}}, \tag{2}$$

where $\tau_e(f_e)$ is a cost function, usually it is BPR function [58]

$$\tau_e(f_e) = \bar{t}_e \left(1 + \rho(f_e/\bar{f}_e)^{1/\mu} \right), \tag{3}$$

where $\bar{t}_e$ are travel times at zero flow, and $\bar{f}_e$ are road capacities of a given network's link e.

Since the number of all possible paths between two nodes in a typical transportation network is enormous, direct manipulations with vector of path flows x is unapproachable. Therefore, numeric solvers for traffic assignment problem typically use edge flows f_e as variable or apply column generation techniques [41]. These approaches rely on calculating shortest paths to make progress in minimizing Ψ while maintaining feasible edge flows w.r.t. traffic matrix d.

However, it is also possible to write down the constraint imposed by traffic matrix via Kirchhoff law. This formulation, which is commonly used for multicommodity flow problems [1], is our key technique to apply new saddle-point methods for CDA. Let us fix source $i \in V$ and consider some node $j \in V$. Define f to be the matrix of edge flows f from given source to all targets: the element f_{ei} of matrix f is the amount of flow originating at source i and traversing (oriented) edge e. Then, consider the flow originating from i that settles down in j. It writes as

$$\Delta_{ij} = \sum_{e \in \mathcal{E}_{in}(j)} f_{ei} - \sum_{e \in \mathcal{E}_{out}(j)} f_{ei}. \tag{4}$$

If $i \neq j$, this amount equals the flow running from i to j, i.e. $\Delta_{ij} = d_{ij}$. If $i = j$, this amount is the sum of flows originating from i with inverse sign, i.e. $\Delta_{ii} = -\sum_{k \neq i} d_{ik}$.

Further, let N denotes the *incidence matrix* of $\mathcal{G}$, i.e.

$$N_{ke} = \begin{cases} -1, & k = i, \\ 1, & k = j, \\ 0, & \text{otherwise,} \end{cases} \tag{5}$$

for each edge $e \in \mathcal{E}$ with source $i \in V$ and destination $j \in V$. Note, that $(Nf)_{ji} = \sum_e N_{je} f_{ei} = \Delta_{ij}$. On the other hand, $\Delta_{ij} = -(\text{diag}(d 1_n) - d)_{ij}$.

At last, the Kirchhoff law can be written as $Af + Bd = 0$, where A denotes a linear operator $A : \mathbb{R}^{m \times n} \to \mathbb{R}^{n \times n}$ such that

$$Af = (Nf)^\top, \tag{6}$$

and B denotes a linear operator such that

$$Bd = \text{diag}(d 1_n) - d. \tag{7}$$

Bd is also known as a *traffic Laplacian*.

Our final formulation of the *traffic assignment problem* is

$$\Psi(f) = \sum_{e \in \mathcal{E}} \sigma_e \left(\sum_i f_{ei} \right) \to \min_{\substack{Af+Bd=0 \\ f \geq 0}}. \tag{8}$$

According to Theorem 4 from [49], problem (8) has following dual problem

$$Q(t) = \sum_{ij \in OD} d_{ij} T_{ij}(t) - \sum_{e \in \mathcal{E}} \sigma_e^*(t_e) \to \max_{t \geq \bar{t}}, \tag{9}$$

where $T_{ij}(t) = T_{p_{ij}}(t) = \min_{p \in P_{ij}} T_p(t)$ is the travel cost from source i to destination j and

$$\sigma_e^*(t_e) = \sup_{f_e \geq 0} \{ t_e f_e - \sigma_e(f_e) \} = \bar{f}_e \left(\frac{t_e - \bar{t}_e}{\bar{t}_e \rho} \right)^\mu \frac{(t_e - \bar{t}_e)}{1 + \mu}$$

is the convex conjugate function of $\sigma_e(f_e)$, $e \in \mathcal{E}$.

Subsequently, we can employ the duality gap

$$\Delta(f, t) = \Psi(f) - Q(t) \tag{10}$$

to serve as a convergence measure for traffic assignment problem solvers. It is invariably nonnegative and vanishes only at the solution (f^*, t^*). This quantity is also known as simply *gap* in the traffic assignment literature [4, 10, 46].

2.3   Trip Distribution. Entropy Model

Trip distribution problem is to find the traffic matrix if total input and output flow is known for each vertex along with travel times between each pair of vertices. Now, we derive the entropy-based single mode trip distribution model of [59]. The basic assumption is that the probability of the distribution d_{ij} is proportional to the number of states of the system with particular distribution, and which satisfy the constraints

$$Q(l, w) = \left\{ d \geq 0 \mid \sum_j d_{ij} = l_i, \sum_i d_{ij} = w_j \right\}. \tag{11}$$

By defining $h = \begin{pmatrix} l \\ w \end{pmatrix}$ we can rewrite $Q(l, w)$ via an affine constraint as

$$Q(l, w) = \left\{ d \geq 0 \mid Kd = h \right\}, \tag{12}$$

where linear operator K acts as

$$Kd = \begin{pmatrix} d1_n \\ d^\top 1_n \end{pmatrix}. \tag{13}$$

Suppose

$$D = \sum_i l_i = \sum_j w_j \tag{14}$$

is the total number of trips. The number of distinct arrangements of agents $W(d_{ij})$, which produce distribution d_{ij}

$$W(d_{ij}) = \frac{D!}{d_{11}!(D - d_{11})!} \cdot \frac{(D - d_{11})!}{d_{12}!(D - d_{11} - d_{12})!} \cdots = \frac{D!}{\prod_{ij} d_{ij}!}. \tag{15}$$

To find the most probable distribution of trips d_{ij} we solve

$$\log W(d_{ij}) \to \max_{d \in Q(l,w)}, \tag{16}$$

sing Stirling's approximation (i.e., $\log D! = D \log D - D$), the objective function can be written as

$$\log W(d_{ij}) = \log D! - \sum_{ij} \log d_{ij}! \approx \log D! - \sum_{ij} d_{ij} \log d_{ij} + \sum_{ij} d_{ij}. \tag{17}$$

Taking into account that $\log D!$ and $\sum_{ij} d_{ij} = D$ are constants, the problem can be rewritten as

$$\sum_{ij} d_{ij} \log d_{ij} \to \min_{d \in Q(l,w)} \tag{18}$$

It is also common to assume that larger travel time between two nodes corresponds to lower trip rate between them. Accounting for the travel time factor we obtain

$$\sum_{ij} d_{ij} \log d_{ij} - \gamma \sum_{ij} T_{ij} d_{ij} \to \min_{d \in Q(l,w)}, \tag{19}$$

where $\gamma > 0$ is a scalar parameter, and T_{ij} is the generalized cost of single trip from $i \in V$ to $j \in V$, in the simplest case defined as the shortest path travel time between i and j. [59] justifies the addition of the travel time term by interpreting it as the Lagrangian relaxation of the "budget constraint" $\sum_{ij} T_{ij} = T$. Another argument for incorporating the budget constraint is to maintain correspondence with the well-known gravity model: entropy model and gravity model are equivalent when the exponential gravity function is utilized.

Note, that to obtain T_{ij} one needs to solve the traffic assignment problem, and the traffic assignment problem clearly requires the traffic matrix d as its parameter. This is the reason why CDA is usually solved in practice by sequentially solving the two subproblems.

2.4   CDA

CDA is to find such pair (f, d), that d is the solution of entropy model subproblem, where travel times are calculated w.r.t. f, and f is the solution of equilibrium traffic assignment subproblem with traffic matrix being d [8]. This fixed point problem can also be formulated as a convex optimization problem.

Since trip distribution problem (18) and dual traffic assignment problem (9) share the only term $d_{ij}T_{ij}$ which couples travel times t and trip rates d, solution of the problem

$$S(t,d) = \frac{1}{\gamma}\sum_{ij} d_{ij}\log d_{ij} - \sum_{ij} T_{ij}d_{ij} + \sum_{e\in\mathcal{E}}\sigma_e^*(t_e) \to \min_{\substack{t\geq\bar{t}\\ d\in Q(l,w)}},$$

is the solution to both (18) and (9). By the Lagrangian duality we obtain that problem

$$P(f,d) = \sum_e \sigma_e(f_e) + \frac{1}{\gamma}\sum_{ij} d_{ij}\ln d_{ij} \to \min_{f,d} \qquad (20)$$

$$\text{s.t. } Af + Bd = 0, \quad Kd = h,$$
$$f \geq 0, \; d \geq 0.$$

combines both trip distribution problem (18) and (primal) traffic assignment problem (8) [25]. Problem (20) is our final formulation of CDA.

3   Application of Saddle-Point Algorithms

Our formulation (20) of CDA is an affine-constrained convex optimization problem with matrix variables of reasonable dimension (typically, $n \sim 10^3$, $m \sim 10^4$–10^5 in detailed transportation models of large cities). This formulation allows to employ any black-box algorithm for convex optimization with affine constraints. We propose to solve CDA via an optimal first-order algorithm for strongly convex optimization under affine constraints [54], which we will call here OFAC for short. This is an accelerated variant of Proximal Alternating Predictor-Corrector

(PAPC) algorithm which employs Nesterov's and Chebyshev's acceleration mechanisms to achieve convergence rates which match lower bounds for this class of problems. Nesterov's acceleration is used to improve the dependency on the objective's condition number, while Chebyshev's acceleration allows to separate gradient calculation and matrix multiplication complexities and improve the dependency on the affine constraints' condition number. Algorithm 1 and Algorithm 2 provide listings of the OFAC algorithm and its Chebyshev's acceleration subroutine respectively.

Algorithm 1 OFAC

1: **Input data:** $x^0 \in X, \mathbf{M}, b \in \operatorname{Im}\mathbf{M}, \operatorname{Proj}_X, \nabla F$.
2: **Parameters:** $\lambda_1 \geq \sigma_{\max}^2(\mathbf{M}), \lambda_2 \leq \sigma_{\min}^{+}{}^2(\mathbf{M}), L \geq \max_{x \in X} \lambda_{\max}\left(\nabla^2 F(x)\right),$
   $\quad \mu \leq \min_{x \in X} \lambda_{\min}\left(\nabla^2 F(x)\right)$
3: $\tau := \sqrt{\frac{19\mu}{44L}}, \eta := \frac{1}{4\tau L}, \theta := \frac{15}{19\eta}, \alpha := \mu,$
4: $x_f^0 := x^0, u^0 := 0_X$
5: **for** $k = 0, 1, \ldots$ **do**
6: $\quad x_g^k := \tau x^k + (1 - \tau) x_f^k$
7: $\quad x^{k+\frac{1}{2}} := \operatorname{Proj}_X\left((1 + \eta\alpha)^{-1}\left(x^k - \eta(\nabla F(x_g^k) - \alpha x_g^k + u^k)\right)\right)$
8: $\quad r^k := \theta\left(x^{k+\frac{1}{2}} - \operatorname{Chebyshev}(x^{k+\frac{1}{2}}, \mathbf{M}, b, \lambda_1, \lambda_2)\right)$
9: $\quad u^{k+1} := u^k + r^k$
10: $\quad x^{k+1} := \operatorname{Proj}_X\left(x^{k+\frac{1}{2}} - \eta(1 + \eta\alpha)^{-1} r^k\right)$
11: $\quad x_f^{k+1} := x_g^k + \frac{2\tau}{2-\tau}(x^{k+1} - x^k)$
12: **end for**

Algorithm 2 Chebyshev iteration

1: **Input data:** $z^0 \in X, \mathbf{M}, b \in \operatorname{Im}\mathbf{M}$
2: **Parameters:** $\lambda_1 \geq \sigma_{\max}^2(\mathbf{M}), \lambda_2 \leq \sigma_{\min}^{+}{}^2(\mathbf{M}).$
3: $N := \left\lceil \sqrt{\frac{\lambda_1}{\lambda_2}} \right\rceil, \rho := \left(\lambda_1 - \lambda_2\right)^2/16, \nu := (\lambda_1 + \lambda_2)/2$
4: $\gamma^0 := -\nu/2$
5: $p^0 := -\mathbf{M}^\top(\mathbf{M}z^0 - b)/\nu$
6: $z^1 := z^0 + p^0$
7: **for** $i = 1, \ldots, N - 1$ **do**
8: $\quad \beta^{i-1} := \rho/\gamma^{i-1}$
9: $\quad \gamma^i := -(\nu + \beta^{i-1})$
10: $\quad p^i := \left(\mathbf{M}^\top(\mathbf{M}z^i - b) + \beta^{i-1}p^{i-1}\right)/\gamma^i$
11: $\quad z^{i+1} := z^i + p^i$
12: **end for**
13: **Output:** z^N

Clearly, we can apply OFAC to the problem (20) by denoting $x = \begin{pmatrix} f \\ d \end{pmatrix}$,

$$F(x) = P(f, d), \quad X = \mathbb{R}_+^{(m+n) \times n}, \quad \mathbf{M} = \begin{pmatrix} A & B \\ 0_{n \times m} & K \end{pmatrix} \text{ and } b = \begin{pmatrix} 0_n \\ h \end{pmatrix}.$$

For actual implementation of OFAC we need to derive formulas for multiplication by $\mathbf{M}^\top$ and specify the values of parameters $\lambda_1, \lambda_2, L, \mu$. The formulas for other scalar parameters in Algorithm 1 and the formula for N in Algorithm 2 are taken from Theorem 2 of [54] and Sect. 6.3.1 of [54] respectively. Note, that the reference has a typo in the formula $\tau = \min \left\{ 1, \frac{1}{2} \sqrt{\frac{19}{15\kappa}} \right\}$. 19/15 should be 19/11, as stated in Sect. 6.3.1, and because $\kappa = L/\mu \geq 1$, we get $\tau = \sqrt{\frac{19\mu}{44L}}$.

Since $Af = (Nf)^\top$, and by the definition of adjoint operator $A^\top$ it must holds $\langle y, Af \rangle = \langle y, (Nf)^\top \rangle = \langle yN, f^\top \rangle = \langle (yN)^\top, f \rangle = \langle A^\top y, f \rangle$, we obtain

$$A^\top y = (yN)^\top. \tag{21}$$

Similarly, $\langle y, Bd \rangle = \langle y, \text{diag}(d1_n) - d \rangle = \langle \text{diag}(y)1_n^\top - y, d \rangle = \langle B^\top y, d \rangle$ and

$$B^\top y = \text{diag}(y)1_n^\top - y.$$

Finally, denoting $y = (y_l^\top \ y_w^\top)^\top$ we get $\langle y, Kd \rangle = \langle y_l, d1_n \rangle + \langle y_w, d^\top 1_n \rangle = \langle y_l 1_n^\top, d \rangle + \langle y_w 1_n^\top, d^\top \rangle = \langle y_l 1_n^\top + 1_n y_w^\top, d \rangle = \langle K^\top y, d \rangle$ and

$$K^\top y = y_l 1_n^\top + 1_n y_w^\top.$$

To choose the values of parameters λ_1, λ_2 we need to estimate the maximal and the minimal positive singular values of $\mathbf{M}$.

Lemma 1. *For K defined by (13) maximal and minimal positive singular values are given by $\sigma_{\max}^2(K) = 2n$, $\sigma_{\min}^{+}{}^2(K) = n$.*

Proof.

$$\sigma_{\max}^2(K) := \max_{\|d\| \leq 1} \|Kd\|^2. \tag{22}$$

This is a convex optimization problem, and by Karush-Kuhn-Tucker's theorem [33, 36] d^* is its solution iff (d^*, λ^*) is a stationary point of the Lagrangian

$$\mathcal{L}(d, \lambda) = \|d1_n\|^2 + \|d^\top 1_n\|^2 + \lambda (\|d\|^2 - 1).$$

For $(d^*, \lambda^*) = (1_n 1_n^\top / n, -2)$ we have

$$\nabla_d \mathcal{L}(d^*, \lambda^*) = 2(d^* 1_n 1_n^\top + 1_n 1_n^\top d^* + \lambda^* d^*) = 2(2d^* + \lambda^* d^*) = 0_{n \times n},$$
$$\nabla_\lambda \mathcal{L}(d^*, \lambda^*) = \|d^*\|^2 - 1 = 0.$$

Thus (d^*, λ^*) is a stationary point of $\mathcal{L}$ and maximum in (22) is attained at d^*:

$$\sigma_{\max}^2(K) = \|Kd^*\|^2 = \|d^* 1_n\|^2 + \|d^{*\top} 1_n\|^2 = 2n. \tag{23}$$

The minimal positive singular value is defined as

$$\sigma_{\min}^{+}{}^{2}(K) := \min_{d \in \ker^{\perp} K} \frac{\|Kd\|^2}{\|d\|^2}. \tag{24}$$

It is well known that for a real-valued matrix K the orthogonal complement to its kernel is equal to the image of its transpose: $\ker^{\perp} K = \operatorname{Im} K^{\top}$. Thus, by (3)

$$\sigma_{\min}^{+}{}^{2}(K) = \min_{y} \frac{\|KK^{\top}y\|^2}{\|K^{\top}y\|^2} = \min_{y} \frac{\|K(y_l 1_n^{\top} + 1_n y_w^{\top})\|^2}{\|(y_l 1_n^{\top} + 1_n y_w^{\top})\|^2}. \tag{25}$$

Define $s_w = 1_n^{\top} y_w$, $s_l = 1_n^{\top} y_l$, then

$$K(y_l 1_n^{\top} + 1_n y_w^{\top}) = \begin{pmatrix} y_l 1_n^{\top} 1_n + 1_n y_w^{\top} 1_n \\ 1_n y_l^{\top} 1_n + y_w 1_n^{\top} 1_n \end{pmatrix} = \begin{pmatrix} n y_l + 1_n s_w \\ n y_w + 1_n s_l \end{pmatrix},$$

$$\|K(y_l 1_n^{\top} + 1_n y_w^{\top})\|^2 = n^2 \|y\|^2 + 4n s_w s_l + n(s_w^2 + s_l^2),$$

$$\|y_l 1_n^{\top} + 1_n y_w^{\top}\|^2 = \|y_l 1_n^{\top}\|^2 + 2\langle y_l 1^{\top}, 1 y_w^{\top} \rangle + \|1_n y_w^{\top}\|^2 = n\|y\|^2 + 2 s_l s_w,$$

Substituting this into the (25) we obtain

$$\sigma_{\min}^{+}{}^{2}(K) = n \frac{n\|y\|^2 + 4 s_l s_w + (s_w^2 + s_l^2)}{n\|y\|^2 + 2 s_l s_w} \geq n,$$

and the inequality becomes equality e.g. when $s_w = s_l = 0$.   □

The flattened representation of $Af + Bd$ can be obtained by multiplying flattened representations of f and d by a block-diagonal matrix:

$$\operatorname{Flatten}(Af + Bd) = \begin{pmatrix} N f_{\bullet 1} + \left[(1_n \, 0_n \ldots 0_n)^{\top} - I_n \right] d_{1 \bullet} \\ \vdots \\ N f_{\bullet n} + \left[(0_n \ldots 0_n 1_n)^{\top} - I_n \right] d_{n \bullet} \end{pmatrix}$$

where $X_{\bullet i}, X_{i \bullet}$ are column vectors made from i-th column and row of X respectively. Therefore

$$\sigma_{\min}^{+}(A \; B) = \min_{i=1,\ldots,n} \sigma_{\min}^{+} \left(N \left[e_i^{\top} \otimes 1_n - I_n \right] \right),$$

$$\sigma_{\max}(A \; B) = \max_{i=1,\ldots,n} \sigma_{\max} \left(N \left[e_i^{\top} \otimes 1_n - I_n \right] \right), \tag{26}$$

what can be computed numerically via SVD.

To finish the derivation of λ_1, λ_2 we will use the fact that for a vertical block matrix $M = \begin{pmatrix} M_1 \\ \vdots \\ M_k \end{pmatrix}$ it holds that $\sigma_{\max}(M) \leq \sqrt{\sum_{i=1}^{k} \sigma_{\max}^2(M_i)}$. We can also expect $\sigma_{\min}^{+}(M)$ to be of the same order as $\min_{i=1,\ldots,k}\{\sigma_{\min}^{+}(M_i)\}$. Thus

$\sigma_{\max}(\mathbf{M}) \leq \sqrt{4n^2 + \sigma_{\max}^2(A\ B)}\}$ and $\sigma_{\min}^+(\mathbf{M}) \sim \min\{n, \sigma_{\min}^+(A\ B)\}$, where $\sigma_{\min}^+(A\ B)$ and $\sigma_{\max}(A\ B)$ can be computed via SVD using (26).

Unfortunately, we cannot a priori estimate strong convexity and Lipschitz smoothness constants μ and L since the objective $P(f, d)$ is not Lipschitz smooth nor strongly convex on its domain. Therefore, convergence properties of a first-order method are defined by local convexity and smoothness properties of the objective in a region around the solution. In this situation we resort to some manual tweaking of the parameters L, μ or a grid search.

4 Experiments

To evaluate the performance of OFAC we plot its convergence curve against the most popular algorithm to date [53] — the sequential procedure, which is realized by sequentially solving the trip distribution problem (18) and the traffic assignment problem (8) in a loop. The sequential procedure requires an inertia-like feedback mechanism for convergence [47,53]. For that, we average travel times T_{ij} with their values from the previous iteration before feeding them into the trip distribution subproblem (18).

The traffic assignment subproblem in the sequential procedure is solved by a 3-conjugate Frank-Wolf (3CFW) [32] algorithm, which is a variant of the conjugate Frank-Wolf algorithm [46], that showed the best performance on the given network, comparing with other Frank-Wolf algorithm's modifications. 3CFW is terminated then the duality gap (10) meets the threshold. We plot convergence curves for various values of the threshold, see "Gap" values at the legend of Fig. 1. The trip distribution subproblem is solved via Sinkhorn algorithm, which converges to very high precision and its runtime is negligible comparing to that of 3CFW.

The convergence plots are shown in Fig. 1. We used a well-known [43,46,47] Sioux-Falls transportation network [57] with $n = 24$ nodes and $m = 76$ edges. To estimate the distance to the solution, OFAC was run with an increased runtime to obtain an approximate solution which is several orders more accurate according to the primal-dual optimality conditions. We choose the distance to the solution as the metric because OFAC and Sinkhorn are primal-dual algorithms that violate the constraints during the optimization process and satisfy them precisely only asymptotically. Consequently, the function value at some iteration can be strictly less than the optimal function value. In such cases, we believe that distance to the solution is the most sensible and comprehensible metric to use.

From Fig. 1 we see that OFAC converges much faster and does not suffer from stucking on a limited precision determined by the method parameter values.

Since we plot the convergence metric against the runtime, it is important to discuss the implementation details. In both approaches, the Python language is used to implement the high-level algorithm logic and data processing, while all computationally expensive calculations are performed by efficient open-source libraries implemented in C/C++. A high-performance implementation (in C++) of Dijkstra algorithm from the graph-tool library [52] is used for shortest path

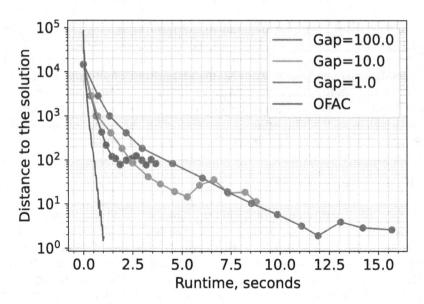

Fig. 1. Comparison of OFAC and the sequential procedure on the Sioux-Falls [57] network

finding in 3CFW, and all other operations with vectors and matrices are done via the NumPy framework [27], whose backend is written in C. For large networks, GPU acceleration of OFAC can be done by a drop-in replacement of NumPy with the jax.numpy module [14]. Unfortunately, on larger networks, the OFAC algorithm, designed for strongly convex functions, exhibits poor convergence due to the high variation of edge flow density in the solution, which means very poor conditioning of the objective function. Further research is required to design a first order primal-dual algorithm with adaptivity and/or preconditioning properties to enhance the efficiency of our approach on large networks. The code is available on GitHub[1].

5   Conclusion

We propose a new algorithmic approach for solving the combined transportation model, offering multiple implementation and computational advantages. Promising convergence properties have been demonstrated through numerical experiments on a real network. However, further research into the application of adaptive preconditioning and the exploitation of other problem-specific features is necessary for practical use and scalability to larger networks.

---

[1] https://github.com/niquepolice/mmo_tm/tree/sp-combined.

Acknowledgments. The work of Yarmoshik D. is supported by the Ministry of Science and Higher Education of the Russian Federation (Goszadaniye) No. 075-03-2024-117, project No. FSMG-2024-0025. The work of Persiianov M. is supported by the annual income of the Endowment Fund of Moscow Institute of Physics and Technology (target capital no. 5 for the development of artificial intelligence and machine learning, https://fund.mipt.ru/capitals/ck5/).

References

1. Ahuja, R.K., Magnanti, T.L., Orlin, J.B.: Network flows (1988)
2. Ali Safwat, K.N., Magnanti, T.L.: A combined trip generation, trip distribution, modal split, and trip assignment model. Transp. Sci. **22**(1), 14–30 (1988)
3. Bar-Gera, H.: Origin-based algorithm for the traffic assignment problem. Transp. Sci. **36**(4), 398–417 (2002)
4. Bar-Gera, H., Boyce, D.: Origin-based algorithms for combined travel forecasting models. Trans. Res. Part B: Methodol. **37**(5), 405–422 (2003)
5. Bar-Gera, H., Boyce, D.: Solving a non-convex combined travel forecasting model by the method of successive averages with constant step sizes. Trans. Res. Part B: Methodol. **40**(5), 351–367 (2006)
6. Beckmann, M., McGuire, C.B., Winsten, C.B.: Studies in the economics of transportation. Tech. rep. (1956)
7. Boyce, D.: Is the sequential travel forecasting paradigm counterproductive? J. Urban Planning Developm. **128**(4), 169–183 (2002)
8. Boyce, D.: Network equilibrium models for urban transport. Handbook of regional science, pp. 247–275 (2021)
9. Boyce, D., Bar-Gera, H.: Multiclass combined models for urban travel forecasting. Netw. Spat. Econ. **4**, 115–124 (2004)
10. Boyce, D., Ralevic-Dekic, B., Bar-Gera, H.: Convergence of traffic assignments: how much is enough? J. Transp. Eng. **130**(1), 49–55 (2004)
11. Boyce, D.E.: Urban transportation network-equilibrium and design models: recent achievements and future prospects. Environ Plan A **16**(11), 1445–1474 (1984)
12. Boyce, D.E., Zhang, Y.F., Lupa, M.R.: Introducing "feedback" into four-step travel forecasting procedure versus equilibrium solution of combined model. Transp. Res. Rec. **1443**, 65 (1994)
13. Boyles, S.D., Lownes, N.E., Unnikrishnan, A.: Transportation Network Analysis, vol. 1 (2023), edition 0.91
14. Bradbury, J., et al.: JAX: composable transformations of Python+NumPy programs (2018). http://github.com/google/jax
15. Chen, A., Yang, C., Kongsomsaksakul, S., Lee, M.: Network-based accessibility measures for vulnerability analysis of degradable transportation networks. Netw. Spat. Econ. **7**, 241–256 (2007)
16. Cheng, L., Du, M., Jiang, X., Rakha, H.: Modeling and estimating the capacity of urban transportation network with rapid transit. Transport **29**(2), 165–174 (2014)
17. de Dios Ortúzar, J., Willumsen, L.G.: Modelling transport. John wiley & sons (2011)
18. Du, M., Jiang, X., Cheng, L.: Alternative network robustness measure using system-wide transportation capacity for identifying critical links in road networks. Adv. Mech. Eng. **9**(4), 1687814017696652 (2017)

19. Evans, S.P.: Derivation and analysis of some models for combining trip distribution and assignment. Transp. Res. **10**(1), 37–57 (1976)
20. Florian, M., Nguyen, S.: A combined trip distribution modal split and trip assignment model. Transp. Res. **12**(4), 241–246 (1978)
21. Florian, M., Nguyen, S., Ferland, J.: On the combined distribution-assignment of traffic. Transp. Sci. **9**(1), 43–53 (1975)
22. Frank, M., Wolfe, P., et al.: An algorithm for quadratic programming. Naval Res. Logist. Q. **3**(1–2), 95–110 (1956)
23. Friesz, T.L.: An equivalent optimization problem for combined multiclass distribution, assignment and modal split which obviates symmetry restrictions. Trans. Res. Part B: Methodol. **15**(5), 361–369 (1981)
24. Garrett, M., Wachs, M.: Transportation planning on trial: The Clean Air Act and travel forecasting. Sage Publications (1996)
25. Gasnikova, E., et al.: Sufficient conditions for multi-stages traffic assignment model to be the convex optimization problem. arXiv preprint arXiv:2305.09069 (2023)
26. Guminov, S., Dvurechensky, P., Tupitsa, N., Gasnikov, A.: On a combination of alternating minimization and nesterov's momentum. In: International Conference on Machine Learning, pp. 3886–3898. PMLR (2021)
27. Harris, C.R., et al.: Array programming with NumPy. Nature **585**(7825), 357–362 (2020). https://doi.org/10.1038/s41586-020-2649-2
28. Ho, H., Wong, S.: Housing allocation problem in a continuum transportation system. Transportmetrica **3**(1), 21–39 (2007)
29. Ho, H., Wong, S., Loo, B.P.: Combined distribution and assignment model for a continuum traffic equilibrium problem with multiple user classes. Trans. Res. Part B: Methodol. **40**(8), 633–650 (2006)
30. Horowitz, A.J.: Tests of an ad hoc algorithm of elastic-demand equilibrium traffic assignment. Trans. Res. Part B: Methodol. **23**(4), 309–313 (1989)
31. Huang, H.J., Lam, W.H.: Modified evans' algorithms for solving the combined trip distribution and assignment problem. Trans. Res. Part B: Methodol. **26**(4), 325–337 (1992)
32. Ignashin, I., Yaramoshik, D.: Modifications of the frank-wolfe algorithm in the problem of finding the equilibrium distribution of traffic flows. Math, Model, Numerical Simulat. **10**(1), 10–25 (2024)
33. Karush, W.: Minima of functions of several variables with inequalities as side constraints. M. Sc. Dissertation. Dept. of Mathematics, Univ. of Chicago (1939)
34. Kim, Y., Samaranayake, S., Wischik, D.: A combined convex model for travel demand forecasting with hierarchical extended logit model. arXiv preprint arXiv:2308.01817 (2023)
35. Kubentayeva, M., et al.: Primal-dual gradient methods for searching network equilibria in combined models with nested choice structure and capacity constraints. CMS **21**(1), 15 (2024)
36. Kuhn, H.W., Tucker, A.W.: Nonlinear programming. In: Neyman, J. (ed.) Proceedings of the Second Berkeley Symposium on Mathematical Statistics and Probability (1951)
37. Lam, W.H., Huang, H.J.: A combined trip distribution and assignment model for multiple user classes. Trans. Res. Part B: Methodol. **26**(4), 275–287 (1992)
38. Larsson, T., Patriksson, M.: Simplicial decomposition with disaggregated representation for the traffic assignment problem. Transp. Sci. **26**(1), 4–17 (1992)
39. LeBlanc, L.J., Farhangian, K.: Efficient algorithms for solving elastic demand traffic assignment problems and mode split-assignment problems. Transp. Sci. **15**(4), 306–317 (1981)

40. Lee, D.H., Wu, L., Meng, Q.: Equity based land-use and transportation problem. J. Adv. Transp. **40**(1), 75–93 (2006)
41. Leventhal, T., Nemhauser, G., Trotter, L., Jr.: A column generation algorithm for optimal traffic assignment. Transp. Sci. **7**(2), 168–176 (1973)
42. Lin, J.J., Feng, C.M.: A bi-level programming model for the land use-network design problem. Ann. Reg. Sci. **37**, 93–105 (2003)
43. Liu, Z., Yin, Y., Bai, F., Grimm, D.K.: End-to-end learning of user equilibrium with implicit neural networks. Trans. Res. Part C: Emerging Technol. **150**, 104085 (2023)
44. Lundgren, J.T., Patriksson, M.: An algorithm for the combined distribution and assignment model. In: Transportation Networks: Recent Methodological Advances. Selected Proceedings of the 4th EURO Transportation Meeting Association of European Operational Research Societies (1999)
45. McNally, M.G.: The activity-based approach (2000)
46. Mitradjieva, M., Lindberg, P.O.: The stiff is moving-conjugate direction frank-wolfe methods with applications to traffic assignment. Transp. Sci. **47**(2), 280–293 (2013)
47. Najmi, A.: Interaction of demand and supply in transport planning model systems: A comprehensive revisit. Ph.D. thesis, UNSW Sydney (2020)
48. Nesterov, Y.: Universal gradient methods for convex optimization problems. Math. Program. **152**(1), 381–404 (2015). https://doi.org/10.1007/s10107-014-0790-0
49. Nesterov, Y., De Palma, A.: Stationary dynamic solutions in congested transportation networks: summary and perspectives. Netw. Spat. Econ. **3**, 371–395 (2003)
50. Oppenheim, N.: Equilibrium trip distribution/assignment with variable destination costs. Trans. Res. Part B: Methodol. **27**(3), 207–217 (1993)
51. Oppenheim, N., et al.: Urban travel demand modeling: from individual choices to general equilibrium. John Wiley and Sons (1995)
52. Peixoto, T.P.: The graph-tool python library. figshare (2014). https://doi.org/10.6084/m9.figshare.1164194
53. Reeder, P., Bhat, C., Lorenzini, K., Hall, K., et al.: Positive feedback: exploring current approaches in iterative travel demand model implementation (2012)
54. Salim, A., Condat, L., Kovalev, D., Richtárik, P.: An optimal algorithm for strongly convex minimization under affine constraints. In: International Conference on Artificial Intelligence and Statistics, pp. 4482–4498. PMLR (2022)
55. Sheffi, Y.: Urban transportation networks, vol. 6. Prentice-Hall, Englewood Cliffs, NJ (1985)
56. Tam, M., Lam, W.H.: Maximum car ownership under constraints of road capacity and parking space. Trans. Res. Part A: Policy Pract. **34**(3), 145–170 (2000)
57. Transportation Networks for Research Core Team: Transportation networks for research (2024). https://github.com/bstabler/TransportationNetworks (Accessed 29 Feb 2024)
58. US Bureau of Public Roads: Traffic Assignment Manual. Washington D.C, Department of Commerce, Urban Planning Division (1964)
59. Wilson, A.G.: The use of entropy maximising models, in the theory of trip distribution, mode split and route split. J. Trans. Econ. Policy, 108–126 (1969)
60. Wong, K.I., Wong, S., Wu, J., Yang, H., Lam, W.H.: A combined distribution, hierarchical mode choice, and assignment network model with multiple user and mode classes. Urban and regional transportation modeling, pp. 25–42 (2004)
61. Xu, M., Chen, A., Gao, Z.: An improved origin-based algorithm for solving the combined distribution and assignment problem. Eur. J. Oper. Res. **188**(2), 354–369 (2008)

62. Yang, H., Bell, M.G., Meng, Q.: Modeling the capacity and level of service of urban transportation networks. Trans. Res. Part B: Methodol. **34**(4), 255–275 (2000)
63. Yim, K.K., Wong, S., Chen, A., Wong, C.K., Lam, W.H.: A reliability-based land use and transportation optimization model. Trans. Res. Part C: Emerging Technol. **19**(2), 351–362 (2011)

Stadium Antennas Deployment Optimization

Alexander Yuskov[1]([✉]) [iD], Igor Kulachenko[2][iD], Andrey Melnikov[2][iD],
and Yury Kochetov[2][iD]

[1] Novosibirsk State University, Novosibirsk, Russia
a.yuskov@g.nsu.ru
[2] Sobolev Institute of Mathematics of Siberian Branch of Russian Academy of
Sciences, Novosibirsk, Russia
{ink,melnikov,jkochet}@math.nsc.ru

Abstract. The stadium is divided into sectors. Each sector is split into
cells. Users in the cells must be provided with a certain quality of signal
from antennas assigned to their sector. Our goal is to select antenna
types, their location, assignment to sectors, and orientation to optimize the signal distribution, measured by three different metrics under
some technical constraints. The quality metrics are signal quality, average signal-to-interference ratio (SIR), and consistency. Each variant of
antenna deployment is evaluated by a simulator. Thus, we deal with a
constrained black-box optimization problem with three objectives. To
tackle the problem, we design a three-stage algorithmic approach. In
the first stage, we apply a fast constructive heuristic. Later on, a local
improvement procedure is called. Finally, a VNS metaheuristic is used
to get high-quality solutions. The approach demonstrates strong performance and ability to improve signal quality by 7% and SINR by at least
14% without worsening the given consistency threshold for test instances
with up to 7 antenna types, 19 sectors, and 4426 cells.

Keywords: black box optimization · simulation · local search ·
wireless network · quality of signal

1 Introduction

Wireless network service in crowded locations such as stadiums is delivered via a
specially organized system of antennas. As a rule, the area is divided into sectors.
Each sector is split into small pieces called cells. Users located in the cells must be
provided with a certain quality of signal from antennas assigned to service their
sector. The goal is to select antenna types, their location, assignment to sectors,
and downtilt and azimuth orientation to optimize the signal distribution, measured by three different metrics under some technical constraints. The quality
metrics are called signal quality, average signal-to-interference-plus-noise ratio
(SINR), and consistency. The signal quality shows the number of cells, where
SINR exceeds a certain threshold. The average SINR is computed straightforwardly by averaging SINR across all the cells. Finally, the consistency indicates

A. Eremeev et al. (Eds.): MOTOR 2024, LNCS 14766, pp. 449–461, 2024.
https://doi.org/10.1007/978-3-031-62792-7_30

how many cells get the strongest signal from antennas assigned to a sector the cell belongs to. Each variant of antenna deployment is evaluated within a simulator, computing metrics of signal quality provided. Together with technical constraints on antenna placement, the Stadium Antennas Deployment problem (SAD) falls into the class of mixed integer non-linear constrained black-box optimization problems with multiple objectives [1,8,11,12].

To tackle the problem, we use a composition of three separate procedures applied sequentially. At the first stage, a constructive heuristic builds a first-stage solution. Later on, a local improvement procedure derives a second-stage solution from the first-stage one. Further, a general Variable Neighborhood Search (VNS) with Variable Neighborhood Descent (VND) [2,5,6,10] are running to get high-quality solutions, which are polished by an additional run of the local improvement procedure. The procedures from each of the stages use a weighted sum of criteria. So, different solutions could be obtained from multiple runs of the procedure sequence, using different weight values. The multiple-run scheme is called CPVNS. The scheme demonstrates high performance and efficiency. Two real test instances were used to evaluate the performance of the algorithmic approach. The instances are supplemented with baseline solutions. The first instance has 4426 cells, 17 sectors, 7 antenna types and 27 antennas initially. The second instance has 3551 cells, 19 sectors, 4 antenna types and 47 antennas initially. The developed algorithms appeared to improve signal quality by 7% and SINR by at least 14% without worsening the consistency.

The rest part of the paper is organized as follows. In Sect. 2, we introduce the notation, problem statement, and the Mixed Integer Non-Linear Programming (MINLP) model. In Sect. 3, we present a fast constructive heuristic to create a feasible solution. We decompose the problem by sectors and rely on geometrical intuition about the signal distribution properties for antenna location and orientation selection. Local improvement procedures based on antenna type selection, its locations and modifications of space orientation are described in Sect. 4. The VNS and VND algorithms with eight neighborhood structures are presented. The final algorithmic scheme with constraint repairing is included here too. In Sect. 5, we describe our test instances and show the performance of the algorithms. The last Sect. 6 concludes the paper.

2   Problem Statement

Let F be a set of levels of the stadium, S be a set of sectors in all levels, C be a set of cells in all sectors, and T be a set of available antenna types. We call *antenna placement* a quadruple $x = \langle l, t, o, s \rangle$, where l is a slot for antenna location, $t \in T$ is its type, $o \in [0, 360] \times [0, 90]$ is space orientation, and $s \in S$ is a sector assigned to it. We assume that the sets of locations and orientations are finite.

For a set of indices J, we define a set of antenna placements $X = \{x_j\}$, $x_j = \langle l_j, t_j, o_j, s_j \rangle$, $j \in J$, and $C' \subseteq C$ as a subset of cells on the stadium. We will use a black-box function $\boldsymbol{R}(C', X)$ computing values $R_{cs}(C', X)$ as the

Reference Signal Received Power (RSRP) in cell $c \in C'$ from antennas assigned to sector $s \in S$. Now we can define the SINR [3] value $Q_c(C', X)$ for each cell $c \in C'$ by formula

$$Q_c(C', X) = 10 \cdot \log \frac{\max_{s \in S} R_{cs}(C', X)}{\sum_{s \in S} R_{cs}(C', X) - \max_{s \in S} R_{cs}(C', X)}. \tag{1}$$

We do not consider external noise in the formula. So, it also can be called the Signal-to-Interference Ratio (SIR). The key procedure $Q(C', X)$, computing the quality of antenna deployment, calls the black-box function $R(C', X)$ and computes the SINR values by formula (1). The quality is measured according to metrics, which would be denoted and referred further as Q—*signal quality*, S—*average SINR*, and C—*consistency*.

We define the signal quality Q as the number of cells, where SINR is at least the threshold Q_0

$$Q(C', X) = |\{c \in C' \mid Q_c(C', X) \geq Q_0\}|. \tag{2}$$

The value S of average SINR is computed straightforwardly

$$S(C', X) = \frac{1}{|C'|} \sum_{c \in C'} Q_c(C', X). \tag{3}$$

Finally, the consistency C is the number of cells that receive the highest signal from antennas assigned to its sector. Let C_s be a set of all cells in sector s. Given a cell $c \in C'$, its sector would be denoted by s_c. Thus, we can compute the consistency C as follows

$$C(C', X) = |\{c \in C' \mid R_{cs_c}(C', X) \geq R_{cs}(C', X) \; \forall s \in S\}|. \tag{4}$$

In the SAD problem, we want to maximize the signal quality

$$Q(C, X) = (Q(C, X), S(C, X), C(C, X)) \to \max_X \tag{5}$$

subject to the following three sets of constraints.

A cell is covered if its RSRP exceeds a threshold R_0. We need at least U_0 covered cells

$$|\{c \in C \mid R_{cs}(C, X) \geq R_0\}| \geq U_0. \tag{6}$$

We can use at most N_0 antennas in total

$$|X| \leq N_0. \tag{7}$$

At each level, we can apply at most two types of antennas. Let $S_f \subseteq S$ be the subset of sectors on level $f \in F$. Hence, we have the following

$$|\{t \in T \mid \exists j \in J \text{ such that } t_j = t \text{ and } s_j \in S_f\}| \leq 2, \quad f \in F. \tag{8}$$

A. Yuskov et al.

Let us introduce a finite set P of antenna configurations and assume that for each configuration $p \in P$ we know the space orientation of the antenna. Discretization of down tilt and azimuth orientation can be different for antenna types. Thus, we consider subset $P_t \subseteq P$ of possible orientations for antennas with type $t \in T$. Denote by r_{citp} the value of the RSRP received by cell c if antenna of type t with configuration p is placed in slot i. We will use these output parameters of simulation to calculate the total value R_{cs} of the RSRP received by cell c from all antennas assigned to sector s.

We use the following decision variables.

x_{itp} equals one, if slot i is occupied by antenna of type t with configuration p, and zero otherwise;

y_{ft} equals one, if antennas of type t are used to service sectors on level f, and zero otherwise;

z_{is} equals one, if slot i is assigned to service sector s, and zero otherwise;

u_c equals one, if cell c is covered, and zero otherwise;

v_c equals one, if cell c has a good signal quality, and zero otherwise;

w_c equals one, if cell c gets the highest RSRP value from antennas servicing sector s_c, and zero otherwise.

If constant M is large enough, we can formulate the Stadium Antenna Deployment (SAD) model as follows.

$$\max \mathcal{Q} = \frac{1}{|C|} \sum_{c \in C} v_c \tag{9}$$

$$\max \mathcal{S} = \frac{1}{|C|} \sum_{c \in C} Q_c \tag{10}$$

$$\max \mathcal{C} = \frac{1}{|C|} \sum_{c \in C} w_c \tag{11}$$

$$\sum_{i \in I} \sum_{t \in T} \sum_{p \in P_t} x_{itp} \leq N_0 \tag{12}$$

$$\sum_{t \in T} y_{ft} \leq 2, \quad f \in F \tag{13}$$

$$\sum_{t \in T} \sum_{p \in P_t} x_{itp} \leq 1, \quad i \in I \tag{14}$$

$$x_{itp} \leq 1 - z_{is} + y_{ft}, \quad t \in T, i \in I, f \in F, s \in S_f \tag{15}$$

$$\sum_{s \in S} z_{is} = 1, \quad i \in I \tag{16}$$

$$R_{cs} = \sum_{i \in I} \sum_{t \in T} \sum_{p \in P_t} r_{citp} x_{itp} z_{is}, \quad c \in C, s \in S \tag{17}$$

$$u_c R_0 \leq \max_{s \in S} R_{cs}, \quad c \in C \tag{18}$$

$$Q_c = 10 \log \frac{\max_{s \in S} R_{cs}}{\sum_{s \in S} R_{cs} - \max_{s \in S} R_{cs}}, \quad c \in C \tag{19}$$

$$v_c Q_0 \leq Q_c, \quad c \in C \tag{20}$$

$$R_{cs} + M(1 - w_c) \geq R_{cs'}, \quad s, s' \in S, s \neq s', c \in C_s \tag{21}$$

$$\sum_{c \in C} u_c \geq U_0 \tag{22}$$

$$x_{itp}, y_{ft}, z_{is}, u_c, v_c, w_c \in \{0, 1\}. \tag{23}$$

Objective functions (9)–(11) are the corresponding objectives of the SAD problem. Constraint (12) states that we can place at most N_0 antennas. Constraints (13) forbid choosing more than two types of antennas per level. Constraints (14) enforce to place at most one antenna in each slot. Constraints (15) say that the antenna placements must be coordinated with both the assignment of slots to sectors and the choice of antenna types for each level. Constraints (16) states that we must assign each antenna to exactly one sector. Equations (17) define the values of total RSRP based on the placements. Equations and inequalities (18)–(21) define the components of the objective functions. Finally, constraint (22) says that at least U_0 cells must get the necessary RSRP level.

3 Constructive Heuristic

We implement a constructive heuristic to build a solution by decomposing the problem into sectors and relying on geometrical intuition about the signal distribution properties to select antenna location and orientation. We present the pseudo-code of the heuristic in Algorithm 1.

The construction is performed level by level. To reduce the search space, we use only one type of antenna per level (line 4). Each sector is optimized independently. Thus, we can do that in a sequential or parallel manner. Since this heuristic is an enumeration scheme, we evaluate at most 4 antennas per sector. The maximal number of antennas $maxAnt$ is defined at line 6. The value $|C_s|$ corresponds to the total number of cells in the current sector s, the value $\texttt{diameter}(s)$ corresponds to the geometric diameter of cell coordinates in sector s, whereas $cellsPerAnt$ and $diameterPerAnt$ are its expected values. We use $cellsPerAnt = 120$ and $diameterPerAnt = 16$ in our experiments.

For each sector, antennas are placed regularly in slots from one border to another (see Fig. 1). But there may be different placement patterns depending on the number of antennas n (line 8). For example, if $n = 3$, then we can place three antennas uniformly, or all three of them in the centre. Only central symmetrical placement patterns are considered. Some antennas (or none) are located in the centre, while others are distributed uniformly.

To reduce running time, we assume that each antenna may be oriented at one of nine possible points of the sector (line 10). These points are positioned at 14,

---

Algorithm 1: Constructive heuristic

Parameters: *cellsPerAnt, diameterPerAnt*

1 **Function** constructiveHeuristic()
2      $best \leftarrow \emptyset$;
3   **for** *level* $f \in F$ **do**
4   **for** *antenna type* $t \in T$ **do**
5   **for** *sector* $s \in S_f$ **do**
6                  $maxAnt \leftarrow \min \left\{ 4, \max \left\{ \lceil \frac{|C_s|}{\text{cellsPerAnt}} \rceil, \lceil \frac{\text{diameter}(s)}{\text{diameterPerAnt}} \rceil \right\} \right\}$;
7   **for** $n \in 1 \ldots maxAnt$ **do**
8   **for** *placement pattern* p **do**
9   Place n antennas of type t using pattern p;
10   Generate configurations o_j for each antenna $j \in 1 \ldots n$;
11   Find best antenna configurations among $o_1 \times \cdots \times o_n$;
12   **if** *resulting solution is better than best* **then**
13   Update *best*;

14   **return** *best*;

50, and 86 percentiles of the convex hull for the sector (Fig. 1). All the considered antenna configurations for sector s are enumerated and evaluated with respect to $\frac{\sum_{c \in C_s} R_{cs}}{\sum_{c \in C \backslash C_s} R_{cs}}$ measure. Note that we ignore other sectors here and use this new measure instead of the objective functions (2)–(4). According to this measure, we try to concentrate the signal power inside of the current sector and reduce it outside.

The resulting solution of the constructive heuristic may include a different number of antennas in some sectors. As a result, we can violate the constraint (7) for the total number of antennas. Thus, we apply an iterative repair procedure. At each step, we remove the antenna with minimal deterioration of the combined aggregated measure. We present the exact definition of the measure in the next section. We remove one antenna at a step until fulfill the constraint.

4 Local Improvement Procedures

Minor modifications of antenna orientation could significantly improve the quality of a solution. This motivated the implementation of a relatively fast local descent procedure called hereinafter postoptimization. Further exploration of local search scheme capabilities has led to the development of a variable neighborhood search, that needs a larger computational budget, but can avoid being trapped in local optima and provides better results.

In local improvement procedures, we use the following combined improvement measure, computed with respect to some reference solution X_{ref}:

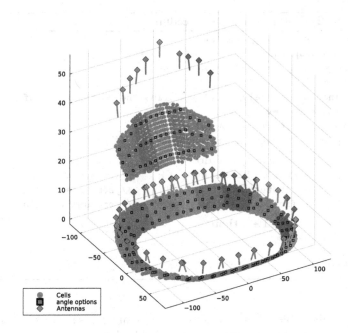

Fig. 1. Antenna placements of the constructive heuristic

$$f(X, X_{\text{ref}}) = \left(\alpha \frac{\mathcal{Q}(X)}{\mathcal{Q}(X_{\text{ref}})} + \beta \frac{\mathcal{S}(X)}{\mathcal{S}(X_{\text{ref}})} + \gamma \frac{\mathcal{C}(X)}{\mathcal{C}(X_{\text{ref}})} \right) / (\alpha + \beta + \gamma) \qquad (24)$$

We say X is better than X_{ref} if $f(X, X_{\text{ref}}) > 1$.

After building an initial solution with the constructive heuristic, we can easily achieve noticeable improvements by applying just slight modifications to it. This is due to the fact that the constructive heuristic explores only a small set of antenna configurations. The pseudo-code for this procedure is given in Algorithm 2 and the basic components are described below.

At each iteration, we perform the following operations:

- Optimize the antenna types (line 4). Now, we consider that two types of antennas can be used at each level. We divide the antennas of each level into two groups: those, that are closer to the assigned sector's center, and the rest. The best pair of types is chosen for each level by a full enumeration provided that the antennas in each group get the same type.
- For each antenna in the solution, we first try to duplicate it and check if this improves the aggregated objective function. If no improvement is achieved, we try to remove the antenna and check again if it is beneficial (line 6). The duplication is performed only if the number of antennas is less than the

---

Algorithm 2: Postoptimization procedure

Parameters: ε, stopping condition

1 **Function** postoptimize(X)

2   **while** *stopping condition is not met* **do**

3         $X^* \leftarrow X$;

4   findBestAntennaTypes(X) ;   // find the best 2 types of antennas for each level

5   **for** *each antenna* $a_j \in X$ **do**

6             $x_j \leftarrow$ checkDuplicationOrRemoval(x_j, X);

7   **for** *each antenna* $a_j \in X$ **do**

8             $l_j \leftarrow$ findBestPlacement(l_j, X);   // some closest slots

9             $o_j \leftarrow$ findBestDowntilt(o_j, X);   // some small angle offsets

10            $o_j \leftarrow$ findBestAzimuth(o_j, X);   // some small angle offsets

11   **if** $f(X, X^*) < 1 + \varepsilon$ **then**

12   break;

13   **return** X^*;

---

maximum allowable. Moreover, if we have two antennas in the same slot, then move the new antenna to the closest available slot.

- After that, we optimize individual antenna positions and orientations (line 7–10). Only six closest slots are considered for changing position (generally, three in one direction and three in another). For downtilt and azimuth, we also consider only small changes: $\pm 4°$, $\pm 2°$, $\pm 1°$ and $\pm 16°$, $\pm 8°$, $\pm 4°$, $\pm 2°$, $\pm 1°$, respectively. We repeatedly change each placement parameter while it improves the solution.

The stopping conditions for the postoptimization procedure are the following:

- Aggregated objective improvement achieved during the iteration is less than predefined threshold (line 11). We use $\varepsilon = 0.001$ in our experiments.
- The time or iteration limit is exceeded (line 2).

Local search schemes can be enabled to find high-quality solutions [7,9]. We implement the scheme of variable neighborhood search for these purposes. Starting from an initial solution, the VNS consists of the following phases: shaking procedure, improvement procedure and neighborhood change step. In shaking and improvement procedures, we use two different sets of neighborhoods $(\mathcal{N}_k^s)$, $k = 1, \ldots, k_{max}$ and $(\mathcal{N}_l^i)$, $l = 1, \ldots, l_{max}$, respectively. The pseudo-code of the VNS algorithm is presented by Algorithm 3. We use maximum running time as the stopping criterion (step 2) and the VND as an improvement procedure. The neighborhood change step makes a decision on which neighborhood will be explored next. Sequential neighborhood change is applied, steps of which are given in Algorithm 5.

Neighborhoods used in the VND algorithm are presented below.

- $\mathcal{N}_1^i(X)$ (Placement): move an antenna to one of the closest slots.

Algorithm 3: Variable Neighborhood Search

1 **Function** VNS(X, $k_{\max}$, $l_{\max}$, $(\mathcal{N}_k^s)$, $(\mathcal{N}_l^i)$)
2   **while** *stopping criterion is not reached* **do**
3       $k \leftarrow 1$;
4   **while** $k \le k_{\max}$;   // We use $k_{\max} = 8$
5   **do**
6          $X' \leftarrow$ shake(X, k, $\mathcal{N}_k^s$);
7          $X'' \leftarrow$ VND(X', $l_{\max}$, $(\mathcal{N}_l^i)$));
8   neighborhoodChangeSequential(X, X'', k);
9   **return** best-found solution X;

Algorithm 4: Variable Neighborhood Descent

1 **Function** VND(X, $l_{\max}$, $(\mathcal{N}_l^i)$)
2    $X' \leftarrow X$;
3   **repeat**
4   *stop* $\leftarrow$ true;
5       $l \leftarrow 1$;
6   **repeat**
7          $X'' \leftarrow \arg\min_{x \in \mathcal{N}_l^i(X')} f(x, X')$;
8   neighborhoodChangeSequential(X', X'', l);
9   **until** $l = l_{\max}$;
10   **if** $f(X', X) > 1$ **then**
11   *stop* $\leftarrow$ false;
12          $X \leftarrow X'$;
13   **until** *stop* = *true*;
14   **return** X;

- $\mathcal{N}_2^i(X)$ (Downtilt): change downtilt of an antenna by a small offset.
- $\mathcal{N}_3^i(X)$ (Azimuth): change azimuth of an antenna by a small offset.
- $\mathcal{N}_4^i(X)$ (Angle): change downtilt and azimuth of an antenna simultaneously.
- $\mathcal{N}_5^i(X)$ (Duplicate): duplicate an antenna, if it is possible.
- $\mathcal{N}_6^i(X)$ (Remove): remove an antenna
- $\mathcal{N}_7^i(X)$ (Remove and Duplicate): remove an antenna and duplicate another.
- $\mathcal{N}_8^i(X)$ (Antenna types): change the type of antenna.

The first improvement scheme is used for random 10% of each neighborhood.

Shaking procedure shake (step 6) diversifies the search and tries to prevent getting stuck in a local optimum. It moves to a neighbouring solution no matter if it is better or worse. Neighborhood $\mathcal{N}_k^s(X)$ used in the shaking procedure is achieved by repeating k times the following steps.

- Move an antenna to a new slot at the distance $3k$ from the current one.
- Change the downtilt of an antenna by $3k$ degrees and the azimuth of the same antenna by $6k$ degrees.

Algorithm 5: Sequential neighborhood change step

---

1 **Procedure** neighborhoodChangeSequential(X, X', k)
2   **if** $f(X', X) > 1$ **then**
3         $X \leftarrow X'$;
4         $k \leftarrow 1$;
5   **else**
6         $k \leftarrow k + 1$;

- If $k \geq 3$, remove one antenna and duplicate another.

Finally, we change the type of k antennas on a level to the other one of the same level. If $k \geq 5$, then we change one type to another one for all antennas of the same type on a level.

The final algorithmic framework uses the following sequence: Constructive heuristic, Postoptimization, and Variable Neighborhood Search (CPVNS). It guarantees that the solutions satisfy conditions (7) and (8) but can violate (6). We try to satisfy it implicitly by using as many antennas as possible and distributing them uniformly in each sector. In fact, we get feasible solutions but cannot guarantee this property. We guess that it is an interesting line for further research.

To provide multiple non-dominated solutions from the Pareto front, the optimization is run multiple times with aggregated objective functions representing a weighted sum of quality measures. The weights (α, β, γ) are changed from one run to another.

5 Numerical Experiments

All experiments in this section were performed on a computer equipped with Intel Core i7-8700 CPU 3.20 GHz and 32 GiB of RAM, running Microsoft Windows 10 Pro operating system. The algorithms utilized all 12 threads available.

For the computational experiments, we use two test instances for the SAD problem. We call them 0830 and 0918. The first instance has 4426 cells and 27 antennas in the baseline solution. The second instance has 3551 cells and 47 antennas initially. Figure 2 shows the structure of the instances. Each sector here has its own colour. In the case 0918, the first level is like a circle. The second level is small in cells and located on the left side. It has five sectors in blue colors. In case 0830, the first level has seven sectors in deep and light blue. The second level has ten sectors with a large number of cells.

Table 1 presents the main characteristics of the baseline solutions for the instances. Note that the solution for instance 0830 is infeasible according to constraints (6) or (22). The threshold U_0 is defined in such a way to guarantee 95 % of cells have the desired level of RSRP. The baseline solution has 92% only. Nevertheless, we use it for comparison as a solution with strong objective function values.

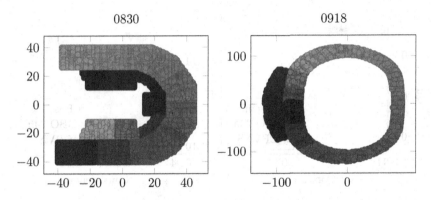

Fig. 2. Structure of the stadiums

Table 1. Values of objective functions for baseline solutions

Instance	SINR	Quality	Consistency
0830	10.16	0.486	0.946
0918	14.28	0.735	0.961

We ran the proposed CPVNS scheme for two hours with four sets of weights: $W = \{(1, 1, 1.5), (1, 1, 2.5), (2, 1, 1), (3, 1, 1)\}$. All non-dominated solutions are collected and returned. We compare the results with baseline solutions and package BlackBoxOptim[1] running for 24 h. BlackBoxOptim is an open-source global optimization package. It supports both multi- and single-objective optimization problems and is focused on (meta-)heuristic/stochastic algorithms (DE, NES etc.). It also supports parallel evaluation to speed up optimization for black-box functions that are slow to evaluate. We used the Borg MOEA algorithm [4] from the package. If the BlackBoxOptim is given less time then the quality of resulting solutions is worse. Figure 3 shows the values of obtained results for instance 0830 in different axes. Figure 4 shows results for instance 0918. We can see that the algorithmic approach can find strong solutions for all three objectives. The improvements achieved by the CPVNS approach over the baseline results are presented in Table 2. This table illustrates the relative improvements in terms of SINR, Quality, and Consistency for 10 diverse points from the approximation of the Pareto front. Specifically, instance 0830 shows notable improvements across all measured metrics, reflecting the efficacy of the method. Similarly, instance 0918 demonstrates consistent gains, thereby underscoring the robustness of the approach across different scenarios.

---

[1] https://github.com/robertfeldt/BlackBoxOptim.jl.

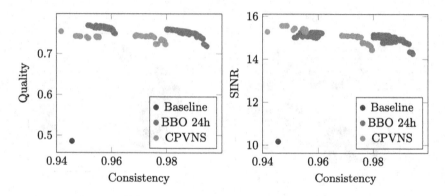

Fig. 3. Algorithm results for instance 0830

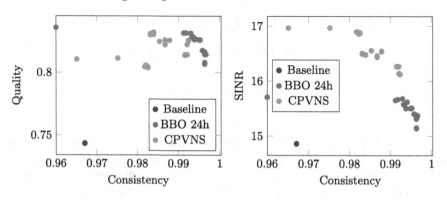

Fig. 4. Algorithm results for instance 0918

Table 2. Relative improvements of the CPVNS approach compared to baseline

	Instance 0830			Instance 0918		
#	SINR, %	Quality, %	Consist., %	SINR, %	Quality, %	Consist., %
1	42	24	3	8	8	3
2	45	25	3	9	7	2
3	48	26	3	11	8	2
4	48	26	3	11	9	2
5	48	26	3	13	6	2
6	48	26	2	14	7	0
7	50	27	0	15	5	−1
8	52	26	1	17	8	−1
9	52	25	1	17	9	−1
10	53	26	0	17	8	−2

6 Conclusion

We have considered the Stadium Antenna Deployment problem and presented polynomial-time constructive and local search algorithms. The resulting scheme CPVNS creates an initial solution using a special decomposition idea for antenna locations and orientations. It uses an aggregated weighted objective function and several weight vectors to find the best solution for each weight vector. The procedure of optimizing each weight vector consists of running postoptimization heuristic, local search, and postoptimization again. The best obtained points simultaneously improve baseline solutions by 7% in terms of quality and by 14% in terms of SINR without worsening the consistency value.

Acknowledgement. The work is carried out within the framework of the state contract of the Sobolev Institute of Mathematics (project no. FWNF-2022-0019).

References

1. Audet, C., Hare, W.: Derivative-Free and Blackbox Optimization. Springer, Cham (2017). https://doi.org/10.1007/978-3-319-68913-5
2. Davydov, I., Kochetov, Y., Tolstykh, D., Xialiang, T., Jiawen, L.: Hybrid variable neighborhood search for automated warehouse scheduling. Optim. Lett. **17**(9), 2185–2199 (2023). https://doi.org/10.1007/s11590-022-01921-6
3. Goldsmith, A.: Wireless Communications. Cambridge University Press, Cambridge (2005)
4. Hadka, D., Reed, P.: Borg: an auto-adaptive many-objective evolutionary computing framework. Evol. Comput. **21**(2), 231–259 (2013). https://doi.org/10.1162/EVCO_a_00075
5. Hansen, P., Mladenović, N.: Variable Neighborhood Search, pp. 1–29. Springer, Cham (2016). https://doi.org/10.1007/978-3-319-07153-4_19-1
6. Hansen, P., Mladenović, N.: Variable neighborhood search: principles and applications. Eur. J. Oper. Res. **130**(3), 449–467 (2001). https://doi.org/10.1016/S0377-2217(00)00100-4
7. Mirjalili, S., Gandomi, A.H. (eds.): Comprehensive Metaheuristics. Algorithms and Applications. Academic Press (2023). https://doi.org/10.1016/C2021-0-01466-8
8. Pardalos, P.M., Rasskazova, V., Vrahatis, M.N. (eds.): Black Box Optimization, Machine Learning, and No-Free Lunch Theorems. Springer Cham (2021). https://doi.org/10.1007/978-3-030-66515-9
9. Salhi, S.: Heuristic Search. The Emerging Science of Problem Solving, 1st edn. Palgrave Macmillan, Cham (2017). https://doi.org/10.1007/978-3-319-49355-8
10. Yuskov, A., Kochetov, Y.: Local search heuristics for the identical parallel machine scheduling with transport robots. Int. J. Artif. Intell. **21**(2), 130–149 (2023)
11. Yuskov, A., Kulachenko, I., Melnikov, A., Kochetov, Y.: Decomposition approach for simulation-based optimization of inventory management. In: Khachay, M., Kochetov, Y., Eremeev, A., Khamisov, O., Mazalov, V., Pardalos, P. (eds.) Mathematical Optimization Theory and Operations Research: Recent Trends, pp. 259–273. Springer, Cham (2023). https://doi.org/10.1007/978-3-031-43257-6_20
12. Yuskov, A., Kulachenko, I., Melnikov, A., Kochetov, Y.: Two-stage algorithm for bi-objective black-box traffic engineering. In: Olenev, N., Evtushenko, Y., Jaćimović, M., Khachay, M., Malkova, V. (eds.) Optimization and Applications, pp. 110–125. Springer, Cham (2023). https://doi.org/10.1007/978-3-031-47859-8_9

Author Index

A. Eremeev et al. (Eds.): MOTOR 2024, LNCS 14766, pp. 463–464, 2024.
https://doi.org/10.1007/978-3-031-62792-7

Printed in the United States
by Baker & Taylor Publisher Services